International Conference on Strongly Correlated Electron Systems (SCES 2016)

Journal of Physics: Conference Series Volume 807

Hangzhou, China
9 – 13 May 2016

ISBN: 978-1-5108-3901-4
ISSN: 1742-6588

Printed from e-media with permission by:

Curran Associates, Inc.
57 Morehouse Lane
Red Hook, NY 12571

Some format issues inherent in the e-media version may also appear in this print version.

Printed by Curran Associates, Inc. (2017)

For permission requests, please contact the Institute of Physics
at the address below.

Institute of Physics
Dirac House, Temple Back
Bristol BS1 6BE UK

Phone: 44 1 17 929 7481
Fax: 44 1 17 920 0979

techtracking@iop.org

Additional copies of this publication are available from:

Curran Associates, Inc.
57 Morehouse Lane
Red Hook, NY 12571 USA
Phone: 845-758-0400
Fax: 845-758-2633
Email: curran@proceedings.com
Web: www.proceedings.com

TABLE OF CONTENTS

HEAVY FERMION PHYSICS

KONDO EFFECT AND INTERMEDIATE VALENCE

QUANTUM CRITICALITY AND NON-FERMI LIQUID PHENOMENA

QUANTUM MAGNETISM AND FRUSTRATION

UNCONVENTIONAL SUPERCONDUCTIVITY

METAL-INSULATOR TRANSITIONS

FERROICS AND MULTIFERROICS

INTERFACES, SUPERLATTICES AND ARTIFICIAL STRUCTURES

International Conference on Strongly Correlated Electron Systems
Hangzhou, China May 8th-13th, 2016

The 18th International Conference on Strongly Correlated Electron Systems, SCES 2016, took place in Hangzhou, China from May 8th - 13th, 2016. The meeting was held under the auspices of the Department of Physics and the Center for Correlated Matter of Zhejiang University. SCES is an international conference series that has developed into the primary meeting for those interested in all aspects of strongly correlated matter. Previous meetings were held in Sendai (1992), San Diego (1993), Amerstam (1994), Goa (1995), Zürich (1996), Paris (1998), Nagano (1999), Ann Arbor(2001), Krakow (2002), Karlsruhe (2004), Vienna (2005), Houston (2007), Buzios (2008), Santa Fe (2010), Cambridge (2011), Tokyo (2013), and Grenoble (2014). Every three years, SCES is part of the triennial international conference on magnetism (ICM). The SCES 2017 will be held in Prague.
SCES 2016 was the first SCES meeting in China. The total number of participants was 542 from 19 different countries. 249 out of the 542 participants came from the mainland of China. Together with the participants from Japan (103), Germany (59), the US (36), and South Korea (23) this accounts for more than 85% of SCES 2016 participation.
More information on the international conference on strongly correlated electron systems in Hanghzou, SCES 2016, can be found at the conference website

http://www.sces2016.org

The present volume of IOP Conference Series contains the contributed papers presented at the Conference. The sections S1 to S10 reflect the categories covered by SCES 2016 and mirror the current research focus within the vast field of strongly correlated fermion systems. These were

S01 Heavy fermion physics
S02 Kondo effect and intermediate valence
S03 Quantum criticality and non-Fermi liquid phenomena
S04 Quantum magnetism and frustration
S05 Unconventional superconductivity
S06 Metal-insulator transitions
S07 Ferroics and multiferroics
S08 Interfaces, superlattices, and artificial structures
S09 Strongly correlated systems away from equilibrium
S10 Topological aspects of strongly correlated systems

The proceedings of the plenary and invited talks of SCES 2016, including the conference summaries by J.T. Thompson (Los Alamos) and P. Coleman (Rutgers), can be found in the special issue "Correlated Electrons 2016" of the Philosophical Magazine.
The work of the guest editors of the present volume of IOP Conference Series has been aided by the members of the publication committee of SCES 2016, i.e.,

G.-H. Cao (Hangzhou, co-chair)
X. Dai (Beijing)
K. Ishida (Kyoto)
D. Kaczorowski (Wrocław)
S. Kirchner (Hangzhou, chair)
C. Krellner (Frankfurt)

IOP Conf. Series: Journal of Physics: Conf. Series **807** (2017) 011001 doi:10.1088/1742-6596/807/0/011001

S. Y. Li (Shanghai)
T. Park (Suwon, co-chair)
D. Qian (Shanghai)
F. Ronning (Los Alamos)
A. M. Strydom (Johannesburg)
S. Wirth (Dresden)
Y. F. Yang (Beijing)
G. M. Zhang (Beijing)

We thank all colleagues who supported SCES 2016 by serving in various committees and functions and all participants of SCES 2016 for helping to make the meeting a success. We are in particular grateful for those who decided to contribute to this volume.
We also would like to especially express our gratitude for the staff at IOP Conference Series, notably Anete Ashton and Sarah Toms, for their competent help and continuous support. Finally, we thank Yanting Lai from the School of International Studies at Zhejiang University for helping with issues related to manuscript formatting.

Stefan Kirchner, Guang-Han Cao, and Tuson Park

IOP Conf. Series: Journal of Physics: Conf. Series **807** (2017) 011002 doi:10.1088/1742-6596/807/0/011002

Peer review statement

All papers published in this volume of *Journal of Physics: Conference Series* have been peer reviewed through processes administered by the proceedings Editors. Reviews were conducted by expert referees to the professional and scientific standards expected of a proceedings journal published by IOP Publishing.

IOP Conf. Series: Journal of Physics: Conf. Series **807** (2017) 012001 doi:10.1088/1742-6596/807/1/012001

Fermi Surface and Magnetic Properties in Ferromagnet CoS_2 and Paramagnet $CoSe_2$ with the Pyrite-type Cubic Structure

A Teruya[1], F Suzuki[1], D Aoki[2], F Honda[2], A Nakamura[2], M Nakashima[3], Y Amako[3], H Harima[4], K Uchima[5], M Hedo[6], T Nakama[6], and Y Ōnuki[6]

[1]Graduate School of Engineering and Science, University of the Ryukyus, Nishihara, Okinawa 903-0213, Japan
[2]Institute for Materials Research, Tohoku University, Oarai, Ibaraki 311-1313, Japan
[3]Faculty of Science, Shinshu University, Matsumoto, Nagano 390-8621, Japan
[4]Graduate School of Science, Kobe University, Kobe 657-8501, Japan
[5]General Education, Okinawa Christian Junior College, Nishihara, Okinawa 903-0207, Japan
[6]Faculty of Science, University of the Ryukyus, Nishihara, Okinawa 903-0213, Japan

E-mail: `uchima@ocjc.ac.jp`

Abstract. We succeeded in growing high-quality single crystals of pyrite-type cubic compounds $CoSe_2$ and CoS_2 using a transport agent of $CoBr_2$ and measured the electrical resistivity, specific heat, magnetic susceptibility, magnetization, and the de Haas–van Alphen (dHvA) effect. We confirmed that $CoSe_2$ is an exchange-enhanced paramagnet revealing a broad maximum at around 50 K in the temperature dependence of the magnetic susceptibility. The electronic specific heat coefficient is moderately large, $\gamma = 18$ mJ/(K^2·mol). On the other hand, CoS_2 is a ferromagnet with a Curie temperature $T_C = 122$ K and an ordered moment $\mu_s = 0.93\ \mu_B$/Co. The γ value of 21 mJ/(K^2·mol) in CoS_2 is slightly larger than that of $CoSe_2$. A large ordered moment, together with a large γ value, is a characteristic feature in CoS_2 because CoS_2 is a half-metallic spin state in the ferromagnetic state. Correspondingly, we detected a main dHvA branch with a large cyclotron effective mass of $13m_0$ in the dHvA experiments. The detected dHvA branches in CoS_2 and $CoSe_2$ are discussed on the basis of the results of energy band calculations, revealing a broken four-fold-symmetry in the angular dependence of the dHvA frequency.

1. Introduction

CoS_2 and $CoSe_2$ belong to the pyrite-type compounds, which were once studied extensively and are still interestingly studied from a viewpoint of a magnetic transition under pressure [1] and the application of spintronics [2]. CoS_2 is a ferromagnet with a Curie temperature $T_C = 124$ K. In CoS_2, the antibonding Co-e_g band is occupied by one $3d$ electron, revealing a quarter filling. An ordered moment of $\mu_s = 0.85\ \mu_B$/Co with an easy-axis along the [111] direction in CoS_2 is close to 1 μ_B/Co in a low-spin state of Co^{2+} ($S = 1/2$ and $g = 2$). In fact, the magnetic susceptibility follows the Curie–Weiss law above 400 K, with an effective magnetic moment of $\mu_{eff} = 1.76\ \mu_B$/Co, which corresponds to 1.73 μ_B of Co^{2+} [3, 4]. The de Haas–van Alphen (dHvA) experiment was carried out, detecting the mainly five kinds of dHvA branches with the largest cyclotron mass of about $5m_0$ [5]. On the other hand, $CoSe_2$ is an exchange-enhanced paramagnet. The magnetic susceptibility follows the Curie–Weiss law above 300 K, with $\mu_{eff} = 1.72$ μ_B/Co [3], which also corresponds to 1.73 μ_B/Co of Co^{2+}. These characteristic features including the temperature-induced local moments at high temperatures and itinerant 3*d*-electrons at low temperatures are explained by the spin fluctuation theory [6].

In the present dHvA experiments, we detected a large cyclotron effective mass $m_c^* = 13m_0$ for a main Fermi surface of CoS_2. This is unexpectedly large because CoS_2 possesses a large ordered moment. The cyclotron mass of CoS_2 is compared with that of $CoSe_2$. The Fermi surfaces obtained from the angular dependences of dHvA frequencies in CoS_2 and $CoSe_2$ are also compared with the results of energy band calculations.

2. Experimental Procedure

Single crystals of CoS_2 were previously grown by the vapor transport method using the chlorine (Cl_2) gas [4, 5]. Recently, $CoBr_2$ was employed as a transport agent [7], revealing the residual resistivity $\rho_0 = 0.86$ $\mu\Omega$·cm and the residual resistivity ratio RRR[$= \rho_{RT}/\rho(5\ K)$] = 160. The CoS_2 powder was set in a hot zone at 700 °C and single crystals were grown in a cold zone at 640 °C. Following this method, we grew single crystals of CoS_2 with 3–4 mm on an edge. The best sample was obtained with $\rho_0 = 0.42$ $\mu\Omega$·cm and RRR = 410. Note that RRR was estimated to be 60 in the vapor transport method using Cl_2 in the previous report [4, 5]. Under the same experimental condition, we also grew single crystals of $CoSe_2$. As far as we know, the previous measurements of $CoSe_2$ were carried out using polycrystalline samples. The present results shown below were the first ones using single crystal samples. The direction of the sample was determined by the X-ray Laue method.

The electrical resistivity was measured by the conventional four-terminal DC method. The magnetic susceptibility and the magnetization were measured using a commercial superconducting quantum interference device (SQUID) magnetometer. The specific heat was measured by the quasi-adiabatic heat-pulse method. The dHvA experiment was carried out by the standard field-modulation method with a modulation frequency of 60 Hz and a modulation field of 80 Oe in strong magnetic fields up to 150 kOe.

3. Experimental Results

First, we confirmed the fundamental properties of a paramagnet $CoSe_2$ and a ferromagnet CoS_2 measuring the electrical resistivity, specific heat, magnetic susceptibility, and magnetization. Figure 1(a) shows the temperature dependence of the electrical resistivity for the current along the [100] direction. The resistivity in $CoSe_2$ decreases almost linearly with decreasing temperature. The ρ_0 and RRR are $\rho_0 = 2.1$ $\mu\Omega$·cm and RRR = 49, revealing a relatively good single crystal sample.

Next, we measured the specific heat C below 15 K. Figure 1(b) shows the T^2-dependence of the specific heat in the form of C/T. The specific heat in $CoSe_2$ consists of the electronic specific heat γT and phonon contribution $C_{ph} = \beta T^3$ because the magnetic susceptibility becomes approximately constant below 15 K, shown below. The electronic specific heat coefficient γ and the Debye temperature Θ_D are obtained as $\gamma = 18$ mJ/(K^2·mol) and $\Theta_D = 340$ K. These values are compared with $\gamma = 11$ mJ/(K^2·mol) and $\Theta_D = 345$ K for the polycrystalline sample [8].

Next, we show in Fig. 1(c) the temperature dependence of the magnetic susceptibility χ for three principal directions, [100], [110], and [111] in $CoSe_2$. The magnetic susceptibility increases with decreasing temperature, possesses a broad peak at around 50 K and becomes approximately constant below 15 K. Anisotropy of the susceptibility was not observed among the three directions. Note that the Curie–Weiss law with the effective magnetic moment $\mu_{eff} = 1.72$ μ_B/Co is observed above 300 K [3], as mentioned in Introduction.

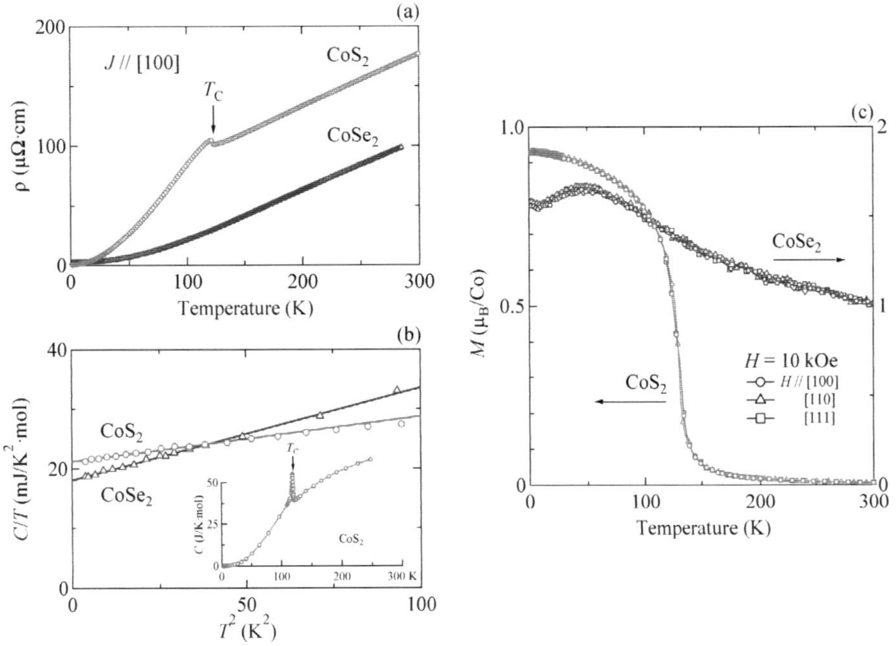

Figure 1. (Color online) Temperature dependences of (a) the electrical resistivity ρ, (b) specific heat C in the form of C/T, and (c) magnetic susceptibility χ and magnetization M in $CoSe_2$ and CoS_2.

Similarly, we measured the electrical resistivity, specific heat, and magnetization M in the ferromagnet CoS_2, as shown in Fig. 1. The ferromagnetic ordering at $T_C = 122$ K is the first-order-like phase transition, revealing sharp peaks in the resistivity and specific heat, as shown in Figs. 1(a) and 1(b). The ρ_0 and RRR are $\rho_0 = 0.42$ $\mu\Omega$·cm and RRR = 410, revealing a high-quality sample. The $\gamma = 21$ mJ/(K^2·mol) is slightly larger than 18 mJ/(K^2·mol) in $CoSe_2$. Note that the previous γ values in CoS_2 are 25.8 mJ/(K^2·mol) [4] and 21.6 mJ/(K^2·mol) [9] for the single crystal samples. The ordered moment μ_s is $\mu_s = 0.93$ μ_B/Co at 2 K, as shown in Fig. 1(c). Note that the saturated moments at 2 K are 0.933 μ_B/Co for $H \parallel [110]$ and 0.928 μ_B/Co for $H \parallel [100]$ and [111], revealing a small difference of 0.005 μ_B/Co between $H \parallel [110]$ and $H \parallel [100]$ and [111]. These values are slightly larger than the previous values of 0.89 μ_B/Co for $H \parallel [100]$ and [111] in the magnetization curve at 4.2 K [4], but are close to 0.91 μ_B/Co in the magnetization curve at 5 K [10] and 0.92(4) μ_B/Co obtained from the polarized neutron scattering experiment at 1.8 K [11]. These cited values were obtained for single crystal samples, which are larger than the value of 0.85 μ_B/Co for the polycrystalline sample mentioned in Introduction [3].

Next, we carried out the dHvA experiment for $CoSe_2$ and CoS_2. Figures 2(a) and 2(b) show the typical dHvA oscillations for $H \parallel [111]$ and the corresponding fast Fourier transformation (FFT) spectrum, respectively, for $CoSe_2$. Four dHvA branches named β, γ, $\varepsilon_{1,2}$, and δ branches are observed. The dHvA frequency F is in the range from 0.548×10^7 to 2.517×10^7 Oe. Here, the dHvA frequency $F(= c\hbar S_F/2\pi e)$ is proportional to the maximum or minimum cross-sectional area of the Fermi surface S_F, which is shown as a unit of magnetic field. We determined the cyclotron masses m_c^* for these branches from the temperature dependences of the dHvA amplitudes. The cyclotron mass, which is described in Fig. 2(b), is in the range from 0.9 to

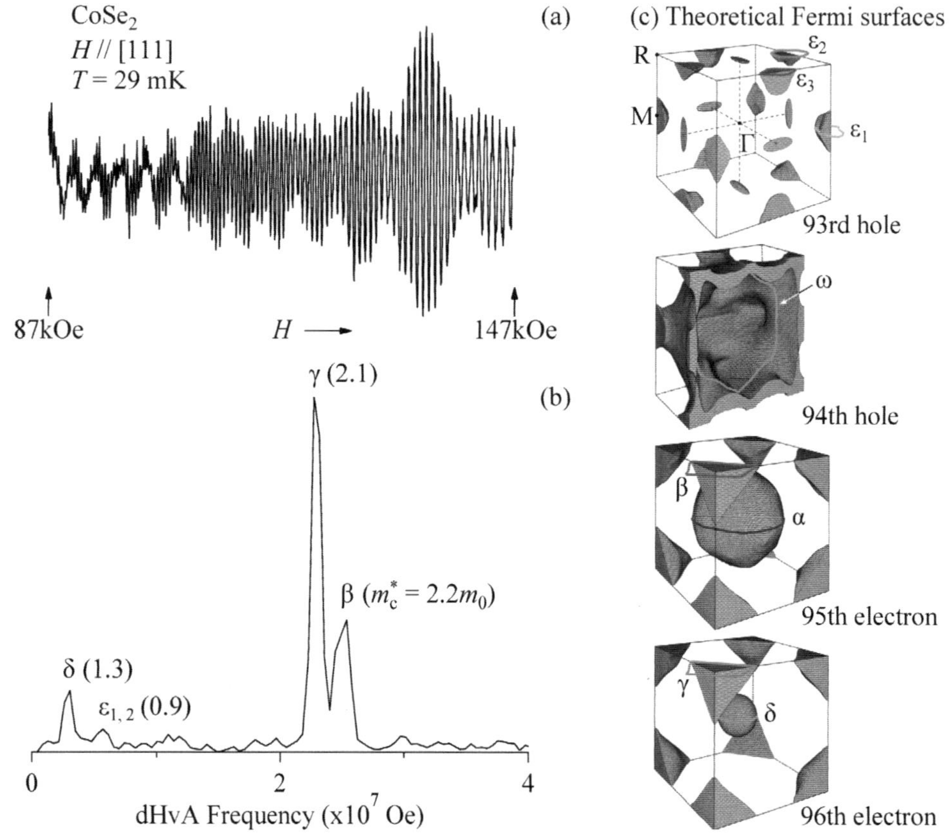

Figure 2. (Color online) (a) Typical dHvA oscillations and (b) the corresponding FFT spectrum for $H \parallel [111]$, and (c) theoretical Fermi surfaces in $CoSe_2$. Figures in parentheses are cyclotron effective masses for dHvA branches.

$2.2m_0$, revealing moderately large masses because the corresponding dHvA frequencies are small in magnitude, namely corresponding to small Fermi surfaces.

To clarify the present experimental results, we carried out the full-potential linearized augmented plane wave (FLAPW) energy band calculations with local density approximation (LDA) for the exchange correlation potential for $CoSe_2$ using the lattice constant $a = 5.8593$ Å and the x parameter of 0.3795 for Se in ref. [12]. In the band calculations, the scalar relativistic effect was taken into account for all the electrons and the spin–orbit interaction was included self-consistently for all the valence electrons as in a second variational procedure. $3p^6 3d^7 4s^2$ electrons for Co and $3d^{10} 4s^2 4p^4$ electrons for Se are treated as valence electrons. Thus, the obtained theoretical Fermi surfaces are shown in Fig. 2(c). The Fermi surfaces consist of band 93rd and 94th hole Fermi surfaces and the compensated band 95th and 96th electron Fermi surfaces. Note that the unit cell contains four molecules of $CoSe_2$ and then electrons are even in number, meaning equal volumes of the hole and electron Fermi surfaces. The detected dHvA branches are identified from the results of band calculations, as follows: (1) branch β is due to a nearly spherical band 95th electron Fermi surface centered at the R point, (2) branch γ is due to a nearly spherical band 96th electron Fermi surface centered at the R point, (3) branches ε_i (i = 1, 2, 3) are due to nearly spherical band 93rd hole Fermi surfaces centered at the M point,

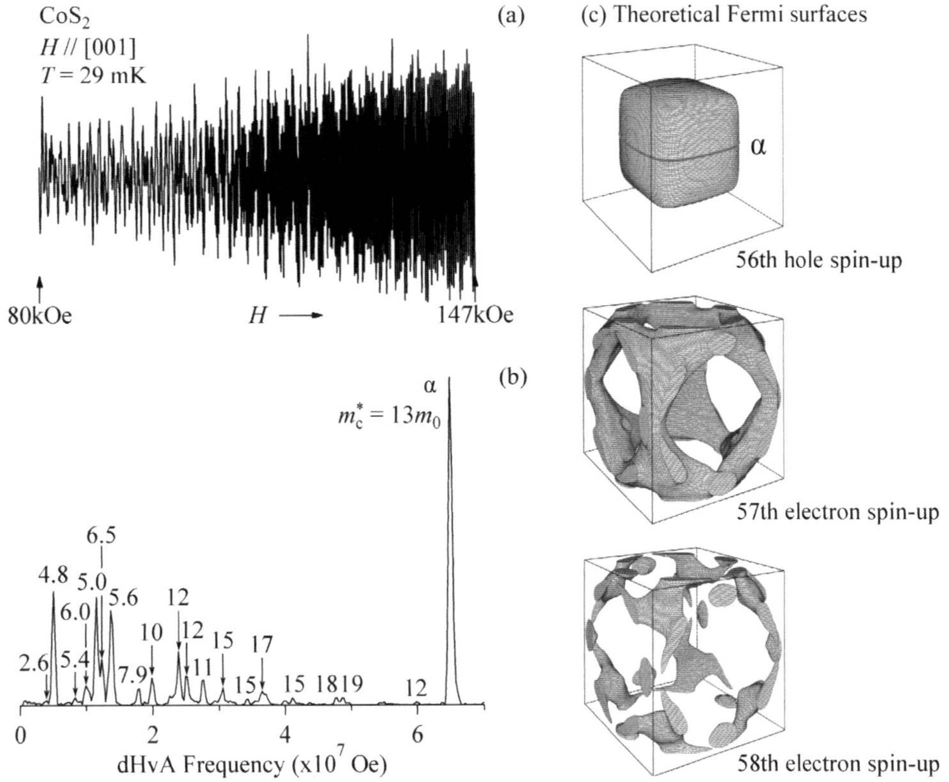

Figure 3. (Color online) (a) Typical dHvA oscillations and (b) the corresponding FFT spectrum, and (c) theoretical Fermi surfaces in CoS_2. Figures in (b) correspond to the cyclotron effective masses, for example $13m_0$ for branch α.

(4) branch δ is due to a spherical band 96th electron Fermi surface centered at the Γ point, although the theoretical Fermi surface is larger than the experimental one in volume, and (5) theoretically, much larger Fermi surfaces exist, which are due to a nearly spherical band 95th electron Fermi surface named α and a multiply-connected band 94th hole Fermi surfaces with vacant space in center. These branches were not observed experimentally mainly because of large cyclotron masses of (10–$20m_0$) and a quality of the sample (RRR = 49).

We show later in Fig. 4(a) the total and partial densities of states in $CoSe_2$. As discussed in Introduction, the bonding e_g band spreads widely in the energy range from 0 to 0.35 Ry, the t_{2g} band forms a very narrow band around 0.40–0.45 Ry, and the antibonding e_g band spreads from 0.5 to 0.7 Ry in energy. All these bands are formed mainly from contributions of Co-3d and Se-4p electrons. A dominant contribution is due to Co-3d electrons. A quarter-filled e_g band is characteristic in the density of states, revealing a large density of states at Fermi energy in a narrow band. The theoretical γ_b value is $\gamma_b = 6.93$ mJ/(K^2·mol). The experimental γ value of 18 mJ/(K^2·mol) is by 2.6 times larger than the theoretical one.

Similarly, we carried out the dHvA experiments for the ferromagnet CoS_2. Figures 3(a) and 3(b) show the typical dHvA oscillations in the ferromagnetic state of CoS_2 and the corresponding FFT spectrum, respectively. A main dHvA branch named α possesses a dHvA frequency of 6.506×10^7 Oe. The cyclotron effective mass is unexpectedly large, $13m_0$. The other branches are also large, ranging from 2.6 to $19m_0$.

Figure 3(c) shows the theoretical Fermi surfaces. Here, the present band calculations were carried out using the FLAPW method within the local spin density approximation where the ferromagnetic state is assumed, but the spin-orbit interaction is neglected for simplicity. The lattice parameters are adopted from ref. [13], namely a = 5.5385 Å and x = 0.38988 for S. $3p^6 3d^7 4s^2$ electrons for Co and $3s^2 3p^4$ electrons for S are treated as valence electrons.

It is characteristic that Fermi surfaces are constructed only from the spin-up band. The Fermi energy is located just below the antibonding e_g band with the spin-down state, revealing a half-metallic spin state. The Fermi surfaces in Fig. 3(c) is thus based on the 56th hole spin-up band and the compensated 57th electron spin-up and 58th electron spin-up bands. Note that the theoretical dHvA frequency of branch α for $H \parallel [001]$ is $F_b = 6.352 \times 10^7$ Oe, which is close to the experimental value of $F = 6.506 \times 10^7$ Oe, while the corresponding band mass of $m_b = 0.68 m_0$ is compared with the experimental value of $m_c^* = 13 m_0$. Theoretically, Co-$3d$ electrons produce a magnetic moment of 0.88 μ_B/Co.

4. Concluding Remark

Finally, we discuss the ferromagnetic ordering and the Fermi surface properties in CoS_2. Figures 4(a), 4(b), 4(c), and 4(d) show the total and partial densities of states in $CoSe_2$, the Co-$3d$ partial density of states in the paramagnetic state of CoS_2, and the Co-$3d$ partial densities of states in the ferromagnetic spin-up and -down states of CoS_2, respectively. Note that the Co-$3d$ partial density of states is roughly the same between CoS_2 and $CoSe_2$ in the paramagnetic state. Correspondingly, the theoretical γ_b value of 6.62 mJ/(K^2·mol) in the paramagnetic state of CoS_2 is approximately the same as γ_b = 6.93 mJ/(K^2·mol) in $CoSe_2$. Important is that the band width below the Fermi energy in the paramagnetic state of CoS_2 is much narrower than that in $CoSe_2$. Moreover, there exists a much larger Co-$3d$ partial density of states below the Fermi energy in the paramagnetic state of CoS_2, which produces a ferromagnetic ordering at low temperatures in CoS_2. Note that the theoretical γ_b value in the ferromagnetic state is 3.03 mJ/(K^2·mol), which is smaller than γ_b = 6.62 mJ/(K^2·mol) in the paramagnetic state. The present small theoretical value of γ_b = 3.03 mJ/(K^2·mol) in the ferromagnetic state is compared with the experimental value of γ = 21 mJ/(K^2·mol), revealing a large enhancement of the γ value, $\gamma/\gamma_b = 7$. From these experimental results, an electronic state of CoS_2 is close to a quantum critical point, which is defined as an electronic state of $T_C \to 0$. In fact, the Curie temperature becomes zero in $Co(S_{1-x}Se_x)_2$ with $x = 0.11$ [14].

Moreover, as shown in Fig. 4(c), only the spin-up state in the e_g band is occupied by the Co-$3d$ electrons, while the spin-down state is empty, revealing the half-metallic spin state. This is a main reason why the large ordered moment of μ_s = 0.93 μ_B/Co is realized in CoS_2. It is thus concluded that a large ordered moment, together with a large γ value, is a characteristic feature in CoS_2.

Acknowledgments

We are very grateful to Prof. K. Ueda for helpful discussions. This work was supported by JSPS KAKENHI Grant Numbers 15H05884, 25247055, and 15H05886.

References

[1] Sidorov V A, Krasnorussky V N, Petrova A E, Utyuzh A N, Yuhasz W M, Lograsso T A, Thompson J D and Stishov S M 2011 *Phys. Rev.* B **83** 060412
[2] Leighton C, Manno M, Cady A, Freeland J W, Wang L, Umemoto K, Wentzcovitch R M, Chen T Y, Chien C L, Kuhns P L, Hoch M J R, Reyes A P, Moulton W G, Dahlberg E D, Checkelsky J and Eckert J 2007 *J. Phys.: Condens. Matter* **19** 315219
[3] Adachi K, Sato K and Takeda M 1969 *J. Phys. Soc. Jpn.* **26** 631
[4] Hiraka H and Endoh Y 1994 *J. Phys. Soc. Jpn.* **63** 4573

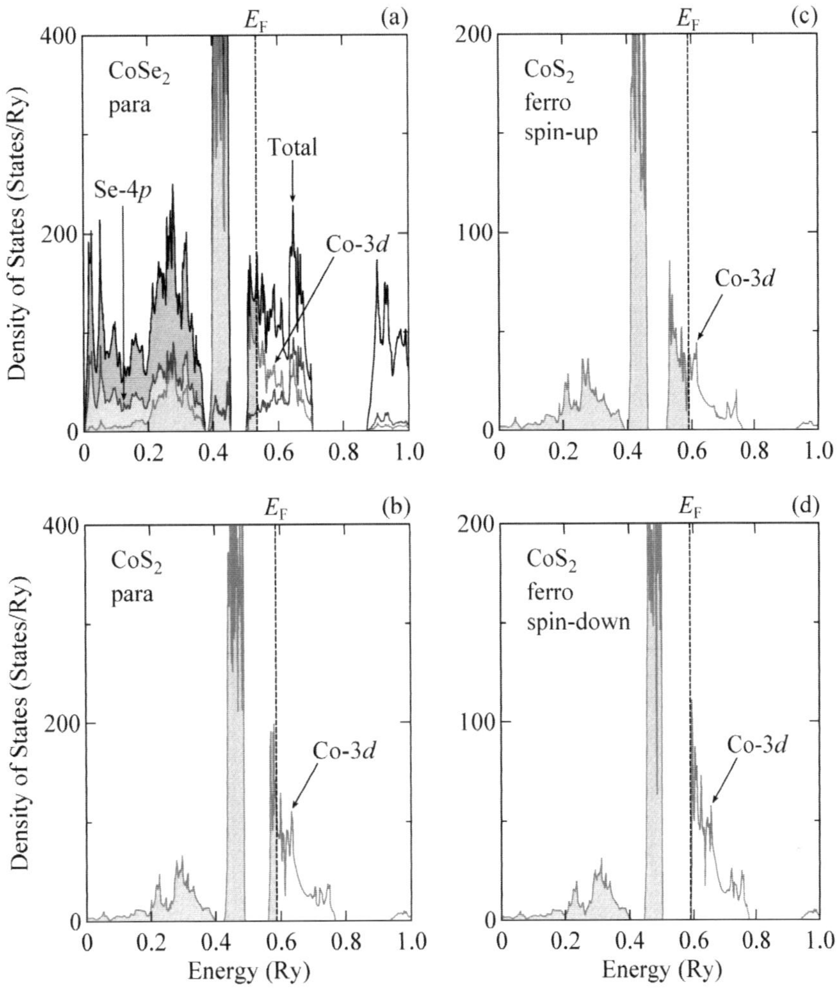

Figure 4. (Color online) Total and partial densities of states in (a) the paramagnetic $CoSe_2$, (b) Co-3d partial densities of states in the paramagnetic CoS_2, and (c) and (d) the ferromagnetic CoS_2.

[5] Ishiguro A, Sawada A, Suzuki M, Hiraka H, Yamada K, Endoh Y and Komatsubara T 1998 *J. Magn. Magn. Mater.* **177–181** 1355
[6] Moriya T 1985 *Spin Fluctuations in Itinerant Electron Magnetism* (Berlin Heidelberg: Springer-Verlag) p 152
[7] Wang L, Chen T Y, Chien C L and Leighton C 2006 *Appl. Phys. Lett.* **88** 232509
[8] Ogawa S and Nishihara Y 1977 *J. Phys. Soc. Jpn.* **42** 343
[9] Otero-Leal M, Rivadulla F, García-Hernández M, Piñeiro A, Pardo V, Baldomir D and Rivas J 2008 *Phys. Rev.* B **78** 180415
[10] Yamamoto R, Machida A, Moritomo Y and Nakamura A 1999 *Phys. Rev.* B **59** R7793
[11] Brown P J, Neumann K-U, Simon A, Ueno F and Ziebeck K R A 2005 *J. Phys.: Condens. Matter* **17** 1583
[12] Furuseth S, Kjekshus A and Andresen A F 1969 *Acta Chem. Scand.* **23** 2325
[13] Nowack E, Schwarzenbach D and Hahn T 1991 *Acta Crystallogr.* B **47** 650
[14] Goto T, Shindo Y, Takahashi H and Ogawa S 1997 *Phys. Rev.* B **56** 14019

IOP Conf. Series: Journal of Physics: Conf. Series **807** (2017) 012002 doi:10.1088/1742-6596/807/1/012002

Ultrasonic Study on the Hexagonal Antiferromagnet $Dy_3Ru_4Al_{12}$

I Ishii[1†], K Takezawa[1], H Goto[1], S Kamikawa[1], A V Andreev[2], D I Gorbunov[2,3], M S Henriques[2] and T Suzuki[1,4,5‡]

[1]Department of Quantum Matter, ADSM, Hiroshima University, Higashi-Hiroshima 739-8530, Japan
[2]Institute of Physics, Academy of Sciences, Na Slovance 2, 182 21 Prague, Czech Republic
[3]Dresden High Magnetic Field Laboratory, Helmholtz-Zentrum Dresden-Rossendorf, D-01314 Dresden, Germany
[4]Institute for Advanced Materials Research, Hiroshima University, Higashi-Hiroshima 739-8530, Japan
[5]Cryogenics and Instrumental Analysis Division, N-BARD, Hiroshima University, Higashi-Hiroshima 739-8526, Japan

E-mail: †`ish@hiroshima-u.ac.jp`, ‡`tsuzuki@hiroshima-u.ac.jp`

Abstract. In the distorted kagome lattice antiferromagnet $Dy_3Ru_4Al_{12}$ with $T_N = 7$ K, a crystal electric field (CEF) effect is expected at high temperatures. To investigate the CEF effect and the phase transition at T_N, we performed ultrasonic measurements on a single-crystalline sample. At high temperatures, both the longitudinal elastic modulus C_{11} and the transverse modulus C_{44} increase monotonically with decreasing temperature. Below 60 K a characteristic elastic softening is observed in C_{44} in contrast to C_{11} with monotonic hardening down to T_N. We analyzed C_{44} using the Curie-Weiss-type equation and obtained a negative parameter: Θ which is proportional to a quadrupole-quadrupole coupling constant under the hexagonal CEF. With further decreasing temperature, both moduli exhibit abrupt elastic hardening at T_N due to a magnetostriction.

1. Introduction

Many compounds of which magnetic ions form a kagome lattice have attracted considerable interest for their fascinating magnetic properties originating from geometrical frustration of spins [1, 2, 3, 4]. The ternary rare-earth compounds $R_3Ru_4Al_{12}$ (R: rare-earth) crystallize in the hexagonal $Gd_3Ru_4Al_{12}$-type structure (space group $P6_3/mmc$) [5, 6, 7, 8, 9]. In this hexagonal structure, constituent atoms are stacked in layers perpendicular to the c-axis with R-Al and Ru-Al layers alternately, and R atoms form a distorted kagome net.

In the Dy-based compound $Dy_3Ru_4Al_{12}$, the electrical resistivity shows a metallic behavior [8]. The specific heat exhibits a clear peak at $T_N = 7$ K indicating the phase transition. The transition is of first-order and is not accompanied by a structural transformation [9]. A cusp-type anomaly is observed at T_N in the magnetic susceptibility along the magnetically easy c-axis. Magnetization experiments manifested that there are one and two meta-magnetic phase transitions in the magnetic field along the a- and c-axes, respectively, suggesting rather complicated magnetic structure. The phase transition at T_N is reported as an antiferromagnetic ordering with two possible noncollinear magnetic structures by neutron diffraction experiments [8].

At high temperatures, the magnetic susceptibility obeys the Curie-Weiss law above 100 K along the a-axis and above 170 K along the c-axis, suggesting an almost localized character of $4f$-electrons. The estimated effective magnetic moments (10.1 μ_B for both axes) are close to the value of the free Dy^{3+} ion (10.6 μ_B) [8]. A crystal electric field (CEF) effect is expected in $Dy_3Ru_4Al_{12}$ and magnetic anisotropy can be attributed to the CEF effect. Here, the sixteen-fold multiplet of the Dy^{3+} ion splits into eight Kramers doublets in the hexagonal CEF, where the total angular momentum J is equal to 15/2. In the preset work, we carried out ultrasonic measurements on $Dy_3Ru_4Al_{12}$ in order to study the CEF effect on the elastic moduli and the phase transition at T_N.

2. Experimental

A single crystal of $Dy_3Ru_4Al_{12}$ was grown by a modified Czochralski method [8]. The elastic moduli C_{11} and C_{44} were measured as a function of the temperature T from 4.2 to 150 K using the phase comparison-type pulse echo method [10]. The modulus C_{11} is the longitudinal mode propagating along the a-axis, and C_{44} is the transverse mode propagating along the a-axis with the polarization direction along the c-axis. We used $LiNbO_3$ transducers with the fundamental resonance frequency of about 30 MHz. The modulus C_{ii} was calculated using the relation $C_{ii} = \rho v^2$ with a room-temperature mass density $\rho = 6.30$ g/cm^3, where v is the sound velocity in a sample. The absolute value of v is estimated at 4.2 K by using the sample length and a time interval between pulse echoes.

3. Results and discussion

Figure 1 shows the T dependence of the longitudinal elastic modulus C_{11} in $Dy_3Ru_4Al_{12}$. The modulus C_{11} increases monotonically with decreasing T down to T_N. An abrupt elastic hardening is detected at T_N, as shown in the inset of Fig. 1.

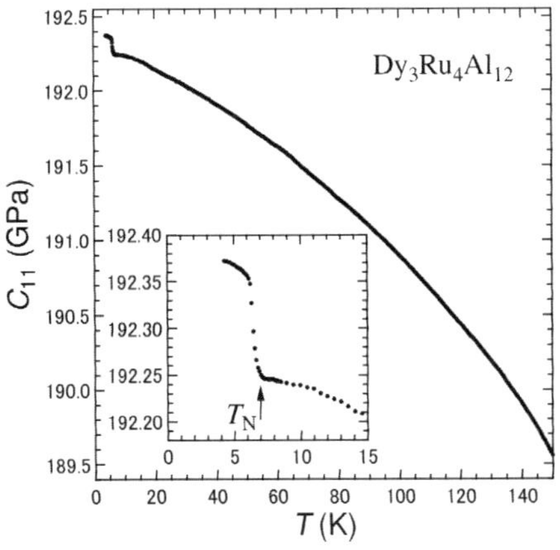

Figure 1. Temperature dependence of the longitudinal elastic modulus C_{11}. The inset represents the same data in an expanded scale below 15 K.

The T dependence of the transverse modulus C_{44} is shown in Fig. 2. At high temperatures, monotonic hardening is observed above 60 K. The modulus C_{44} turns into soften below 60 K. The elastic softening stops at T_N, and then C_{44} exhibits abrupt hardening at T_N, as shown in the inset of Fig. 2. Further elastic softening appears below 6 K.

The softening between T_N and 60 K in C_{44} is a characteristic behavior originating from a quadrupole interaction under the CEF. To simplify the analysis, we performed the theoretical

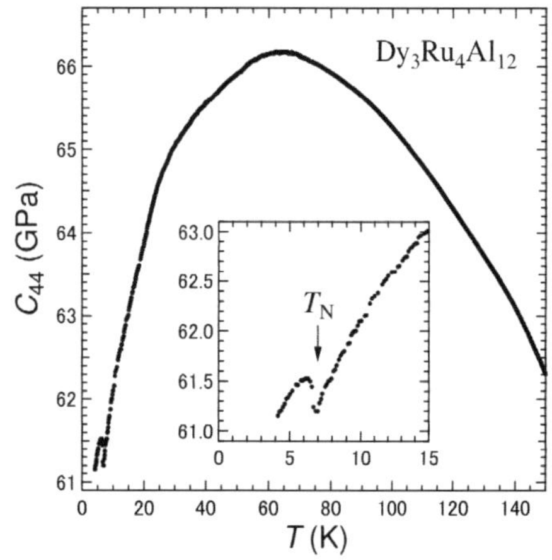

Figure 2. Temperature dependence of the transverse elastic modulus C_{44}. The inset represents the same data in an expanded scale below 15 K.

fitting based on the Curie-Weiss-type formula for the co-operative Jahn-Teller effect [11].

$$C(T) = C_0 \frac{T - T_c}{T - \Theta}, \tag{1}$$

where Θ is proportional to a quadrupole-quadrupole coupling constant and $T_c - \Theta$ is a measure of Jahn-Teller energy including a strain-quadrupole coupling constant. We assumed the back ground stiffness C_0 as

$$C_0 = a + bT^2 + cT^4, \tag{2}$$

where a, b, and c are constants.

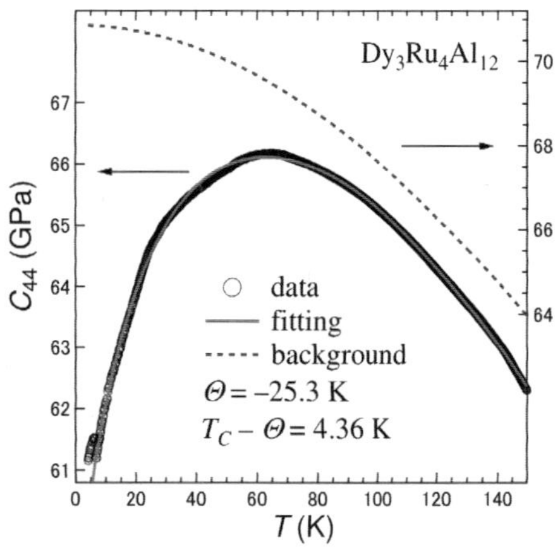

Figure 3. Temperature dependence of C_{44}. The red solid and blue broken curves demonstrate the fitting result and the background stiffness, respectively.

The softening of C_{44} is well reproduced above T_N with fitting parameters listed in Table 1, as shown in Fig. 3. This analysis reveals that the softening of C_{44} arises from the quadrupole interaction. Furthermore, the negative value of Θ suggests that the interaction is of antiferroquadrupolar-type. Under the hexagonal CEF with the Dy^{3+} ion ($J = 15/2$), no elastic

Table 1. Fitting parameters of C_{44}: Θ (K), $T_c - \Theta$ (K), a (GPa), b ($\times 10^{-4}$ GPa/K^2), and c ($\times 10^{-9}$ GPa/K^4).

Θ	$T_c - \Theta$	a	b	c
-25.3	4.36	70.9	-3.37	1.32

softening is expected by only the ground doublet, since the Kramers doublet has no quadrupole degeneracy. The softening is caused by an indirect quadrupole interaction between the ground doublet and the excited doublets, such as our earlier results of YbIrGe and YbPtGe [12, 13]. Consequently, we obtained information of the quadrupole interaction due to $4f$-electronic states of the Dy^{3+} ion under the hexagonal CEF from the T dependence of C_{44}. In the future works, we are planning to measure the elastic moduli of other modes (C_{33} and C_{66}) and perform the theoretical fitting using the hexagonal CEF to reproduce the elastic moduli and the magnetic susceptibility.

As for the phase transition at T_N, antiferromagnetic ordering is reported by neutron diffraction experiments [8]. In thermal expansion measurements, anisotropic step-wise lattice contraction due to a magneto-elastic coupling is detected at T_N [9]. Both moduli C_{11} and C_{44} show the abrupt elastic hardening at T_N. Our results also indicate that a strain strongly couples to a magnetic order parameter. These hardening might originate from the magnetostriction and can be explained by the thermodynamic theory of elastic modulus with the Landau theory [14].

4. Conclusion

The elastic moduli C_{11} and C_{44} were measured in the distorted kagome lattice antiferromagnet $Dy_3Ru_4Al_{12}$. We found characteristic elastic softening due to the quadrupole interaction below 60 K in the transverse modulus C_{44} in contrast to the longitudinal modulus C_{11} without the softening. The negative Θ obtained by the Curie-Weiss-type fitting suggests that there is the antiferroquadrupolar interaction. The indirect quadrupole interaction between the ground doublet and the excited doublets under the hexagonal CEF plays a central role for the softening. The abrupt elastic hardening, owing to the strong coupling between the strain and the magnetic order parameter, is observed at T_N in both modes.

5. Acknowledgments

This work was supported by JSPS KAKENHI Grant Numbers 26800189, 262870830A, and 2624706001. This work was also supported by JSPS Core-to-Core Program, A. Advanced Research Networks. We acknowledge the support of the Czech Science Foundation (Projects 14-03276S and 16-03593S) and High Magnetic Field Laboratory (HLD) at Helmholtz-Zentrum Dresden-Rossendorf (HZDR), a member of the European Magnetic Field Laboratory (EMFL).

References

[1] Ramirez A P 1994 *Annu. Rev. Mater. Sci.* **24** 453
[2] Moessner R and Chalker J T 1998 *Phys. Rev. B* **58** 12049
[3] Okamoto Y, Yoshida H and Hiroi Z 2009 *J. Phys. Soc. Jpn.* **78** 033701
[4] Sengupta K, Forthaus M K, Kubo H, Katoh K, Umeo K, Takabatake T and Abd-Elmeguid M M 2010 *Phys. Rev. B* **81** 125129
[5] Niermann J and Jeitschko W 2002 *Z. Anorg. Allg. Chem.* **628** 2549
[6] Ge W, Ohta H, Michioka C and Yoshimura K 2012 *J. Phys.: Conf. Ser.* **344** 012023
[7] Nakamura S, Toyoshima S, Kabeya N, Katoh K, Nojima T and Ochiai A 2014 *J. Phys. Soc. Conf. Proc.* **3** 014004
[8] Gorbunov D I, Henriques M S, Andreev A V, Gukasov A, Petříček V, Baranov N V, Skourski Y, Eigner V, Paukov M, Prokleška J and Gonçalves A P 2014 *Phys. Rev. B* **90** 094405

[9] Henriques M S, Gorbunov D I, Kriegner D, Vališka M, Andreev A V and Matěj Z 2016 *J. Magn. Magn. Mater.* **400** 125

[10] Lüthi B, Bruls G, Thalmeier P, Wolf B, Finsterbusch D and Kouroudis I 1994 *J. Low Temp. Phys.* **95** 257

[11] Lüthi B 2005 *Physical Acoustics in the Solid State* (Verlag Berlin Heidelberg: Springer) p 121

[12] Ishii I, Noguchi Y, Kamikawa S, Goto H, Fujita T K, Katoh K and Suzuki T 2014 *J. Phys. Soc. Jpn.* **83** 043601

[13] Xi X, Ishii I, Noguchi Y, Goto H, Kamikawa S, Araki K, Katoh K and Suzuki T 2015 *J. Phys. Soc. Jpn.* **84** 124602

[14] Rehwald W 1973 *Adv. Phys.* **22** 721

Temperature dependence of Yb valence in the sub-surface of YbB_{12}(001)

Kenta Hagiwara,[1] Yusuke Takeno,[1] Yoshiyuki Ohtsubo,[1,2] Ryu Yukawa,[3] Masaki Kobayashi,[3] Koji Horiba,[3] Hiroshi Kumigashira,[3] Julien Rault,[4] Patrick Le Fèvre,[4] François Bertran,[4] Amina Taleb-Ibrahimi,[4,5] Fumitoshi Iga,[6] Shin-ichi Kimura[1,2]

[1]*Department of Physics, Graduate School of Science, Osaka Unviersity, Toyonaka 560-0043, Japan*

[2]*Graduate School of Frontier Biosciences, Osaka University, Suita 565-0871, Japan*

[3]*Photon Factory, Institute of Materials Structure Science, High Energy Accelerator Research Organization (KEK), 1-1 Oho, Tsukuba 305-0801, Japan*

[4]*Synchrotron SOLEIL, Saint-Aubin-BP 48, F-91192 Gif sur Yvette, France*

[5]*UR1/CNRS-Synchrotron SOLEIL, Saint-Aubin, F-91192 Gif sur Yvette, France*

[6]*College of Science, Ibaraki University, Mito 310-8512, Japan*

E-mail: kimura@fbs.osaka-u.ac.jp

Abstract. The typical Kondo insulator, YbB_{12}, is a candidate of topological Kondo insulators. To clarify the surface electronic structure of not only bulk but also surface by photoelectron measurements, we developed a method to obtain a clean surface of YbB_{12}(001) by annealing at 1650 K in an ultra-high vacuum. By surface-sensitive photoelectron measurements, it was known that the surface consists of only B atoms without Yb. In the sub-surface region, Yb atoms exist and the temperature dependence of the valence agrees with those of previous HAXPES results. Here we report the cleaning method to obtain well-defined YbB_{12}(001) surfaces and the temperature-dependent Yb valence obtained from Yb 3*d* core-level and valence photoelectron spectra.

1. Introduction

YbB_{12} is a typical Kondo insulator with the Kondo temperature (T_K) of about 230 K [1, 2]. The bulk of YbB_{12} possesses a clear energy gap with the gap size of about 40 meV detected by an optical conductivity measurement [3]. However, it was reported that a finite density of states (DOS) has been observed at lower temperatures than that of the full gap opening [4]. The observed finite DOS at E_F is considered to originate from a metallic surface state. In fact, recently, the metallic surface state has been confirmed by an electrical transport measurement [5]. A theoretical study in which the metallic surface conductivity originates from a topological surface state is one of hot topics in condensed-matter physics recently [6]. So far, the bulk electric structure of YbB_{12} has been studied by hard-X-ray angle-"integrated" photoelectron spectroscopy (HAXPES) and soft-X-ray angle-"integrated" photoelectron spectroscopy (SXPES) measurements [7, 8]. However, angle-"resolved" photoelectron spectroscopy (ARPES) on a well-defined YbB_{12} surface has not been performed because the well-defined clean surface can not be obtained by conventional methods such as cleaving.

Then, to clarify the surface metallic state as well as to obtain the band structure of surface and bulk, we performed ARPES of a well-defined YbB_{12}(001) surface prepared by using a newly developed high-temperature annealing method in an another paper [9]. In this paper, we report the method to obtain the well-defined surface and the character of the obtained surface and sub-surface state by photoelectron spectroscopies (PESs) of Yb 3*d* core level and valence band. We show that the obtained YbB_{12}(001) surface consists of only B_{12} molecules without Yb atoms. Yb atoms exist at the sub-surface and the temperature dependence of the valence is consistent with those of previous HAXPES data [7]. This result suggests that the electronic structure and physical properties of the obtained sub-surface state is the same as those of the bulk.

2. Experimental

The single crystalline YbB_{12} was grown by the floating-zone method by using an image furnace with four xenon lamps [1]. The crystal was cut by a diamond saw along (001) plane based on an *in-situ* Laue pattern and then polished in the air by diamond lapping films until a mirror-like shining plane was obtained. The polished YbB_{12}(001) crystal was heated up to 1650 K in ultra-high vacuum chambers ($P < 2 \cdot 10^{-8}$ Pa) for ~10 sec. For the heating, we adopted a SiC wafer laid under YbB_{12}(001) as depicted in Fig. 1. The sample temperature was monitored by a radiation thermometer.

PES measurements were performed with synchrotron radiation at the BL-2A Musashi of Photon Factory, KEK, Japan, and CASSIOPÉE beamline of synchrotron SOLEIL, France. The photon energies of 80 eV and 1600 eV were used for a wide valence and Yb $3d_{5/2}$ core level (the binding energy is ~1540 eV) measurements, respectively, because of the high surface and sub-surface sensitivity of the photoelectron with the kinetic energy of 50–100 eV [10]. The incident-photons were linearly polarized and the electric field of photons lies in the incident plane (so-called *p* polarization). The photoelectron kinetic energy at the Fermi level (E_F) and the overall energy resolution of the PES setup (~100 meV) were calibrated by the Fermi edge of the photoelectron spectra from Ta foils attached to the sample. The base pressure of the PES main chamber was about $5 \cdot 10^{-9}$ Pa.

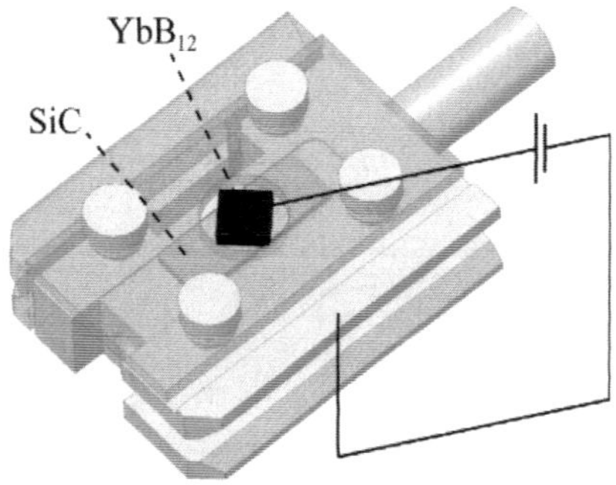

Fig. 1. A schematic drawing of the sample mounting on a sample folder for the PES measurement at KEK-PF. An SiC wafer laid below the YbB_{12} sample is the heater for annealing. Direct current for heating up flows through the sample and the SiC wafer.

3. Results and discussion

Figure 2 shows the annealing-temperature dependence of wide valence PES spectra of YbB_{12} with the photon energy of 80 eV at 20 K. At each spectrum, 4*f* peaks of Yb^{2+} and Yb^{3+} are observed at the similar energy positions to those in previous papers [7, 8] measured with scraped or cleaved YbB_{12} single crystals. The most evident difference between these spectra are the absence of Yb^{2+} 4*f* peaks at around 0.9 and 2.2 eV, which are assigned to be from the surface Yb atoms [11]. In addition, the intensity ratio of the Yb^{3+} 4*f* multiplet peaks to the Yb^{2+} spin-orbit peaks increases with increasing annealing

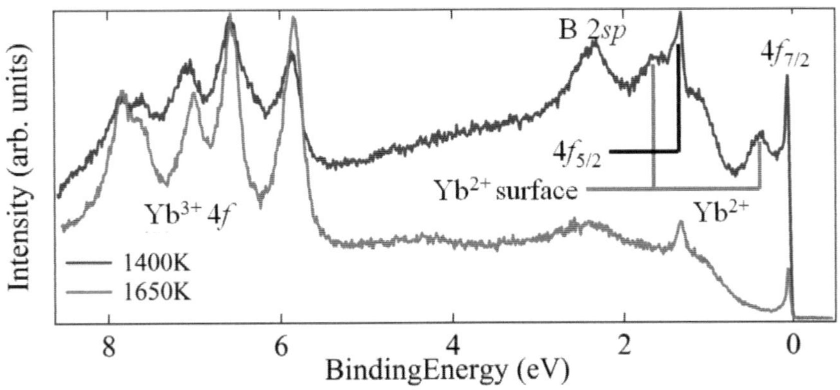

Fig. 2. Annealing-temperature-dependent PES spectra in a wide valence region taken by 80 eV photons at 20 K. Each spectrum is normalized to its maximum value.

temperature. The Yb^{3+} features should be from the bulk states, because most of valence-fluctuating materials, such as Yb and Eu compounds, tend to have smaller valence values at surface, where the coordination number is usually smaller than that in the bulk and the lattice constant increases [11]. Both of these results suggest that the Yb atoms are absent at the topmost surface, possibly because they were desorbed from the surface during the heating process up to 1650 K.

We measured the temperature dependence of the Yb $3d_{5/2}$ core level spectra of YbB_{12} with the photon energy of 1600 eV to obtain the temperature-dependent mean valence of the sub-surface. Figure 3 (a) shows the spectra after the subtraction of parabolic backgrounds. Since the binding energy of the Yb $3d_{5/2}$ core level is about 1540 eV, the kinetic energy of the Yb $3d_{5/2}$ electrons becomes about 60 eV, which is sensitive to surface and sub-surface state because of the short electron mean free path at the kinetic energy. In Fig. 3 (a), a single peak at the kinetic energy of 65 eV and peaks at 50 – 60 eV originate from the Yb^{2+} 3*d* state ($3d^9 4f^{14}$ final state) and the Yb^{3+} 3*d* state ($3d^9 4f^{13}$ final state), respectively. The latter has a multiplet structure.

From the integrated intensities of the Yb^{2+} state [$I^{(2+)}$] and the Yb^{3+} state [$I^{(3+)}$], the mean valence can be calculated according to the function $3 - I^{(2+)}/(I^{(2+)} + I^{(3+)})$ as shown in Fig. 3 (b). The mean valences at 20 and 300 K were evaluated as 2.92±0.01 and 2.90±0.01, respectively. In the figure, the Yb valence evaluated from the Yb 3*d* HAXPES spectra at $h\nu$ = 8000 eV (the most bulk sensitive because the kinetic energy is about 6500 eV) and Yb 4*f* SXPES spectra at $h\nu$ = 700 eV (it contains both of bulk and surface characters) in previous papers [7, 8] are also plotted. These Yb 4*f* data were taken by using a cleaved surface. In all cases, Yb valence clearly shows a decrease on cooling from room temperature to 20 K. The origin of such decrease can be explained to be owing to the appearance of the Kondo effect because the hybridization between conduction bands and Yb 4*f* states, namely *c*-*f* hybridization, appears below the Kondo temperature of ~220 K, which band dispersion has been observed at lower temperatures by our ARPES measurement directly [9]. The Yb valence evaluated from the Yb 3*d* core level is almost the same as those from previous HAXPES results [7] but is much different from the lower mean valence of SXPES results [8]. The HAXPES spectra contain only the bulk property but the SXPES spectra contain properties not only from the bulk but also from the surface because the surface Yb ions are divalent. Therefore, it is confirmed that the Yb^{2+} ions at the topmost surface prepared in this study are desorbed from the surface with annealing up to 1650 K.

4. Conclusions

To summarize, we established the method to obtain a well-defined YbB_{12}(001) surface by using a high-temperature annealing method to elucidate the origin of the metallic surface electronic structure of

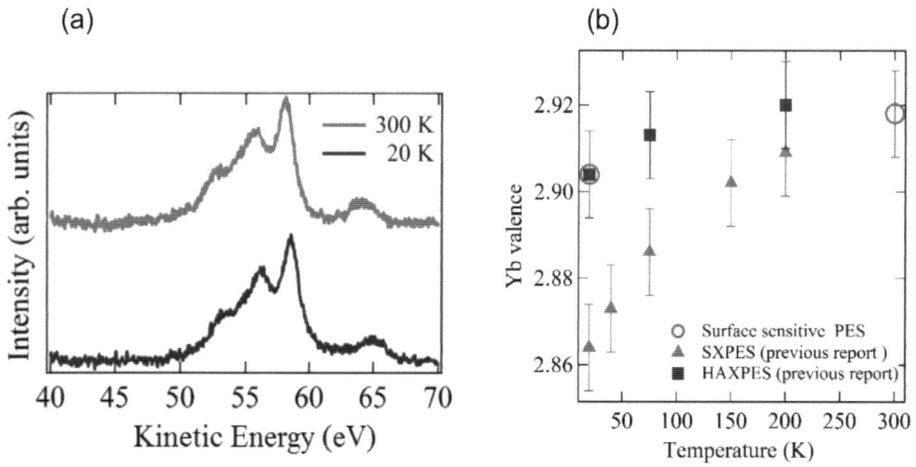

Fig. 3 (a) Yb 3*d* PES spectra taken with 1600 eV photons. Parabolic backgrounds were subtracted. (b) Yb valence evaluated in this work (PES) compared with the previous data taken by HAXPES and SXPES measurements [7, 8].

the Kondo insulator YbB_{12}. We observed that the topmost surface state contains only B atoms except for Yb atoms and Yb atoms exists at the sub-surface state by the cleaning method. The Yb valence at the sub-surface is 2.92±0.01 and 2.90±0.01 at 300 and 20 K, respectively, which are consistent with those of bulk observed by a HAXPES measurement.

Acknowledgments
We also acknowledge D. Ragonnet and F. Deschamps for their support during the experiments on the CASSIOPÉE beamline at the SOLEIL synchrotron. This work was performed by Photon Factory Proposal No. 2015G540. The JSPS Grant-in-Aid also supported this work for Scientific Research (B) (Grant No. 15H03676).

References

[1] F. Iga, N. Shimizu, and T. Takabatake, *J. Magn. Magn. Mater.* **177-181**, 337 (1998).
[2] T. Takabatake, F. Iga, T. Yoshino, Y. Echizen, K. Katoh, K. Kobayashi, M. Higa, N. Shimizu, Y. Bando, G. Nakamoto, H. Fujii, K. Izawa, T. Suzuki, T. Fujita, M. Sera, M. Hiroi, K. Maezawa, S. Mock, H.v. Löhneysen, A. Brückl, K. Neumaier, and K. Andres, *J. Magn. Magn. Mater.* **177-181**, 277 (1998).
[3] H. Okamura, T. Michizawa, T. Nanba, S. Kimura, F. Iga, and T. Takabatake, *J. Phys. Soc. Jpn.* **74**,1954 (2005).
[4] M. Okawa, Y. Ishida, M. Takahashi, T. Shimada, F. Iga, T. Takabatake, T. Saitoh, and S. Shin, *Phys. Rev. B* **92**, 161108(R) (2015).
[5] F. Iga, private communication.
[6] H. Weng, J. Zhao, Z. Wang, Z. Fang, and X. Dai, *Phys. Rev. Lett.* **112**, 016403 (2014).
[7] J. Yamaguchi, A. Sekiyama, S. Imada, H. Fujiwara, M. Yano, T. Miyamachi, G. Funabashi, M. Obara, A. Higashiya, K. Tamasaku,M. Yabashi, T. Ishikawa, F. Iga, T. Takabatake, and S. Suga, *Phys. Rev. B* **79**, 125121 (2009).
[8] A. Shigemoto, J. Yamaguchi, A. Sekiyama, S. Imada, A. Yamasaki, A. Irizawa, T. Ukawa, T. Muro, F. Iga, T. Takabatake, and S. Suga, *J. Electron Spectrosc. Relat. Phenom.* **156-158**,

472 (2007).

[9] K. Hagiwara, Y. Ohtsubo, M. Matsunami, S. Ideta, K. Tanaka, H. Miyazaki, J. Rault, P. Févre, F. Bertran, A. Ibrahimi, R. Yukawa, M. Kobayashi, K. Horiba, H. Kumigashira, F. Iga, and S. Kimura, arXiv:1602.08173.

[10] C.J. Powell, A. Jablonski, *Nucl. Instr. and Meth. Phys. Res. A* **601**, 54 (2009).

[11] K. Kummer, Yu. Kucherenko, S. Danzenbächer, C. Krellner, C. Geibel, M. G. Holder, L. V. Bekenov, T. Muro, Y. Kato, T. Kinoshita, S. Huotari, L. Simonelli, S. L. Molodtsov, C. Laubschat, and D. V. Vyalikh, *Phys. Rev. B* **84**, 245114 (2011).

IOP Conf. Series: Journal of Physics: Conf. Series **807** (2017) 012004 doi:10.1088/1742-6596/807/1/012004

NMR studies on heavy fermion and conventional metal superlattices $CeCoIn_5/YbCoIn_5$

T. Yamanaka[1], M. Shimozawa[2], R. Endo[1], Y. Mizukami[3], H. Shishido[4]. T. Terashima[5], T. Shibauchi[1,3], Y. Matsuda[1], K. Ishida[1]

[1]Department of Physics, Kyoto University, Kyoto 606-8502, Japan
[2]Institute for Solid State Physics, University of Tokyo, Kashiwa, Chiba 277-8581, Japan
[3]Department of Advanced Materials Science, University of Tokyo, Kashiwa, Chiba 277-8561, Japan
[4]Department of Physics and Electronics, Osaka Prefecture University, Sakai, Osaka 599-8531, Japan
[5]Research Center for Low Temperature and Materials Sciences, Kyoto University, Kyoto 606-8501, Japan

E-mail: t-yamanaka@scphys.kyoto-u.ac.jp

Abstract.
We performed nuclear magnetic resonance (NMR) experiments on heavy fermion and conventional metal superlattices $CeCoIn_5/YbCoIn_5$. We succeeded in identifying the signals arising from $CeCoIn_5$ and $YbCoIn_5$ block layers (BLs) by comparing the spectra of the $CeCoIn_5/YbCoIn_5$ superlattices, $CeCoIn_5$ thin film, and $YbCoIn_5$ thin film. Furthermore, we found that the signals of Ce-BLs could be divided into signals arising from interfacial layers and inner layers even in one Ce-BL by comparing the spectra of two $CeCoIn_5/YbCoIn_5$ superlattices with different thickness of the Ce-BLs. A comparison of the spectra of the superlattices with different Ce-BL thickness and field dependence of nuclear-spin lattice relaxation rate $1/T_1$ indicate that antiferromagnetic spin fluctuations at the interfacial site are more suppressed than those at the inner layer site even in the Ce-BLs.

1. Introduction

The physics of strongly correlated electron systems is remarkably rich, and in such materials the entanglement of charge, spin, and orbital degrees of freedom often causes the emergence of exotic quantum phases. It has been shown that in the presence of exotic spin-orbit interaction, spatial inversion symmetry breaking can produce notable effects on these systems through a Fermi surface splitting with different spin structures even in the absence of a magnetic field, giving rise to a plethora of novel phenomena such as anomalous magnetoelectric effects[1, 2] and topological superconducting states[3, 4, 5]. In addition, it has been shown that the electronic properties are changed at the interface of different compounds at which the inversion symmetry is locally broken, even when the global inversion symmetry is preserved in the whole crystal[6]. Among others, the metallic state with the strongest electron correlation is realized in the heavy-fermion systems. Recently, the technique of fabricating epitaxial superlattices consisting of heavy-fermion compounds and conventional metals has been developed and provides us a unique research field to study the interplay between locally inversion symmetry breaking and the heavy-fermion characters[7, 8, 9].

$CeCoIn_5$ [Fig.1(a)], one of the constituents of the $CeCoIn_5/YbCoIn_5$ superlattices studied here, has the highest superconducting transition temperature $T_c = 2.3$ K [10] among Ce-based heavy-fermion superconductors and the superconducting gap symmetry of $CeCoIn_5$ is considered as $d_{x^2-y^2}$[11, 12, 13, 14]. The Cd- and Hg-substituted $CeCoIn_5$ show antiferromagnetic (AFM) ordering[15, 16] and therefore $CeCoIn_5$ is also considered to be located at the vicinity of AFM quantum critical point. These facts suggest that the superconductivity in $CeCoIn_5$ is mediated by AFM fluctuations. $YbCoIn_5$, the other constituent of the $CeCoIn_5/YbCoIn_5$ superlattices, exhibits Fermi liquid behavior in electrical resistivity and magnetic susceptibility[17], indicating that the Yb ions in $YbCoIn_5$ are divalent ions with the closed-shell $4f^{14}$ configuration[18].

In $CeCoIn_5/YbCoIn_5$ superlattices, highly unusual superconducting properties such as the enhanced anisotropy of the upper critical field H_{c2}[8] and anomalous enhancement of H_{c2} regardless of T_c decrease have been observed[19, 20], in particular when the thickness of the Ce-block layers (BLs) is only a few unit cells. Although the importance of the interfaces between Ce- and Yb-BLs has been reported, it is difficult to investigate their physical properties at the microscopic level. Recently, we reported nuclear magnetic resonance (NMR) studies on the $CeCoIn_5/YbCoIn_5$ superlattices, in which we found that the $1/T_1T$ of Ce-BLs was suppressed with decreasing Ce-BL thickness, whereas that of Yb-BLs essentially the same as $YbCoIn_5$ thin film[21]. In this paper, we discuss the NMR-spectrum difference in different Ce-BL thickness and show that the NMR spectra of Ce-BLs in $CeCoIn_5/YbCoIn_5$ superlattices can be divided into the signals arising from interfacial and inner layers.

2. Experimental

We have performed NMR measurements on two thin films of $CeCoIn_5$ and $YbCoIn_5$, and two superlattices $CeCoIn_5(n)/YbCoIn_5(5)$, where $n(= 5$ and $9)$ layers of $CeCoIn_5$ and 5 layers of $YbCoIn_5$ were stacked alternately[Fig. 1(b)]. In the NMR measurements, the frequency was fixed at 114.5 MHz and the magnetic field was applied along the c axis. The samples were grown by molecular beam epitaxial technique[8], and the set of $CeCoIn_5(n)$ and $YbCoIn_5(5)$ in the $CeCoIn_5(n)/YbCoIn_5(5)$ samples repeats 40 times for $n = 5$ and 30 times for $n = 9$. The thickness of the thin films of $CeCoIn_5$ and $YbCoIn_5$ are 500 nm ($\approx$ 600 unit cell layers) and 350 nm ($\approx$ 400 unit cell layers), respectively. The $CeCoIn_5$ thin film was grown on MgF_2 substrate directly. The $YbCoIn_5$ thin film has $CeIn_3$ buffer layers on MgF_2 substrate, and the $CeCoIn_5(n)/YbCoIn_5(5)$ superlattices have $CeIn_3$(28 nm) and $YbCoIn_5$(15-unit-cell thick) as buffer layers.

3. Results and discussions

NMR is one of the most suitable microscopic probes, providing BL-selective information on the magnetic properties. Panels of FIG. 1(c) depict the NMR spectra of the $n = 9$ and $n = 5$ superlattices for $\boldsymbol{H} \parallel c$, together with the spectra of $CeCoIn_5$ [top] and $YbCoIn_5$ [bottom] thin films. From detailed analysis of the field swept spectra we are able to obtain the site-selective NMR information, i.e. spectroscopic information resolved for the Ce- and Yb-BLs separately. In the thin films of $CeCoIn_5$ and $YbCoIn_5$, NMR signals arising from two In sites, In(1) located at the center of the Ce/Yb-In layer and In(2) site located on the lateral faces, and Co site can be clearly identified and be reproduced by NMR-line simulation with the parameters shown in Table I. The parameters of bulk $CeCoIn_5$ are approximately the same values reported in some papers[22, 23, 24]. The largest principal axis of the electric field gradient is parallel to the c axis at the In(1) and Co sites, while it is perpendicular to the lateral face at the In(2) site.

Here we focus on the spectra arising from the central transition ($+1/2 \leftrightarrow -1/2$) of the In(1) site, as shown in panels of FIG. 1(d), which are expanded views in the field range between 11.9 T and 12.4 T. It should be noted that the spectra of the $n = 9$ and $n = 5$ superlattices consist of the spectra of the Ce- and Yb-BLs. Indeed, the positions of the spectra pointed by red and

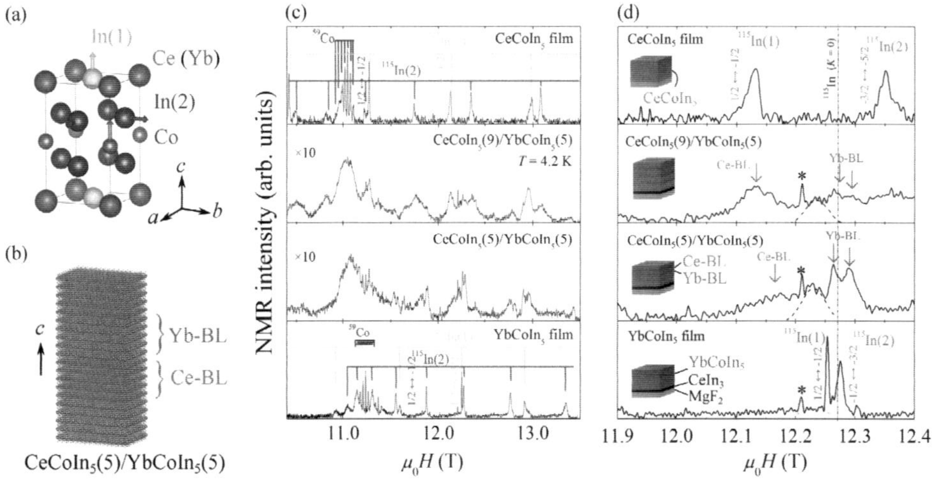

Figure 1. (a) Crystal structure of R(= Ce, Yb)$CoIn_5$.(b) Schematic image of the $CeCoIn_5(5)/YbCoIn_5(5)$ superlattice. (c) Field swept NMR spectra of $CeCoIn_5$ thin film, $CeCoIn_5(9)/YbCoIn_5(5)$ superlattice, $CeCoIn_5(5)/YbCoIn_5(5)$ superlattice, and $YbCoIn_5$ thin films at 4.2 K. (d) Expanded views of NMR spectra from $\mu_0 H = 11.9$ T to 12.4 T of each sample. The signals around 12.22 T pointed by asterisks and dashed lines arise from the buffer layers of $CeIn_3$.

Table 1. NMR parameters of the $CeCoIn_5$ and $YbCoIn_5$ thin films, i.e. Knight shift along the c axis(K_c), electric quadrupole frequency(ν_Q), and asymmetric parameter of the electric quadrupole interaction (η) at Co, In(1), and In(2) sites.

Nucleus	Sample	K_c (%)	ν_Q (MHz)	η
^{59}Co	$CeCoIn_5$ bulk	3.39	0.230	0
	$CeCoIn_5$ film	3.58	0.302	0
	$YbCoIn_5$ film	1.58	0.285	0
^{115}In(1)	$CeCoIn_5$ bulk	1.12	8.15	0
	$CeCoIn_5$ film	1.14	8.04	0
	$YbCoIn_5$ film	0.18	6.22	0
^{115}In(2)	$CeCoIn_5$ bulk	5.16	15.7	0.404
	$CeCoIn_5$ film	5.55	15.5	0.375
	$YbCoIn_5$ film	0.40	14.0	0.490

green arrows almost coincide with the peaks observed in the $CeCoIn_5$ and $YbCoIn_5$ thin films, respectively. The maximum at around 12.22 T observed in the $n = 5$ and 9 superlattices and the $YbCoIn_5$ thin film (asterisk and dashed peaks) is considered to be from the $CeIn_3$ buffer layer because such a peak around 12.22 T was not observed in the $CeCoIn_5$ thin film without the buffer layer. Owing to $\eta = 0$ and ν_{zz} parallel to the c axis at the In(1) site, central-transition peak ($+1/2 \leftrightarrow -1/2$) of the In(1) site is not shifted by the electric quadrupole interaction, but shifted by the hyperfine interaction related to the local spin susceptibility[25]. The Knight shift

$K(T)$ at the fixed frequency ω_0 is defined as

$$K(T) = \left(\frac{H_0 - H_{\rm res}}{H_{\rm res}}\right)_{\omega=\omega_0} = A_{\rm hf}\chi(T)$$

where $H_{\rm res}$ and H_0 are resonant magnetic fields of a sample and a bare nucleus that has the relation of $\omega_0 = \gamma_n H_0$ with the nuclear gyromagnetic ratio γ_n. $A_{\rm hf}$ and $\chi(T)$ are the hyperfine coupling constant and the local spin susceptibility, respectively. The shift from $\mu_0 H_0 = 12.27$ T ($K = 0$) is proportional to the Knight shift arising from the local static susceptibility at the In(1) site.

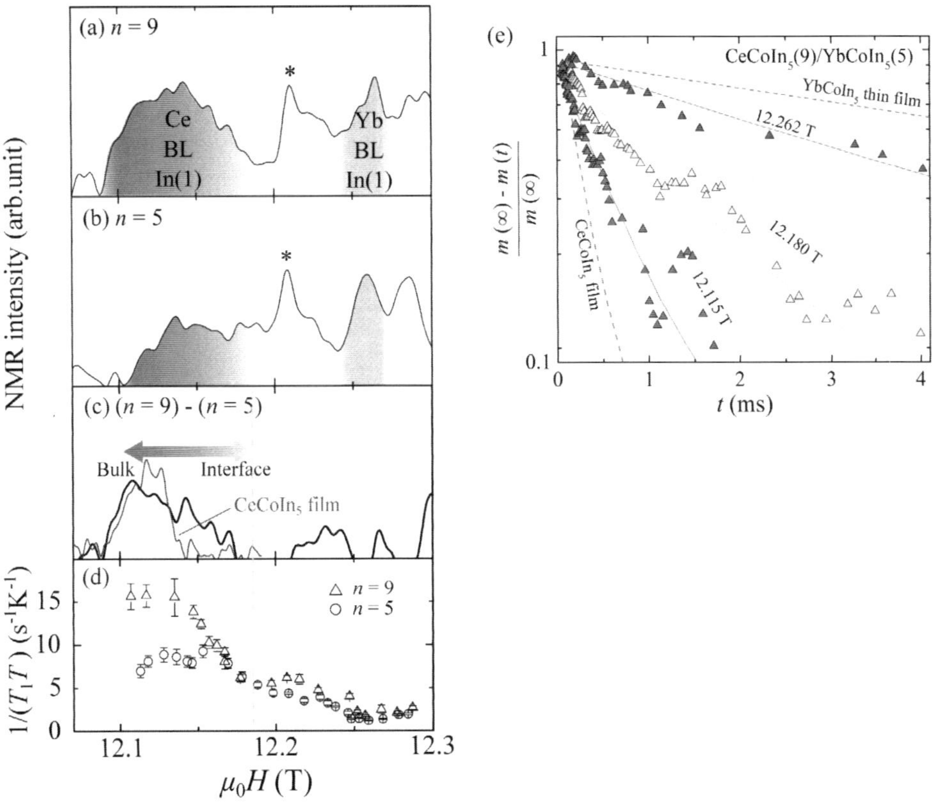

Figure 2. Field swept NMR spectra of the $n = 9$ (a) and $n = 5$ (b) superlattice samples at $T = 3.2$ K. (c) Subtraction of the spectrum of $n = 5$ sample from that of $n = 9$ sample. The NMR spectrum of $CeCoIn_5$ thin film is shown by blue line.(d) Field dependence of $(T_1T)^{-1}$ of the $n = 9$ and $n = 5$ samples. (e) Relaxation curves vs t plot at 12.262, 12.115, and 12.180 T in $CeCoIn_5(9)/YbCoIn_5(5)$ superlattice. The results of fitting are shown by solid line with each color as well as the $CeCoIn_5$ (red-dashed line) and $YbCoIn_5$ (green-dashed line) thin films.

Figures 2(a) and (b) depict the NMR spectra at 3.2 K at around the In(1) central transition of the $n = 9$ and $n = 5$ superlattices. The NMR spectra of these superlattices consist of signals from three different layers, i.e. Ce- and Yb-BLs and $CeIn_3$ buffer layers, each of which has different Knight shift. It should be stressed that the spectra arising from each layer are well separated in this field range. In FIGs. 2(a) and (b), the red to yellow gradient regions represent the spectra of the Ce-BLs, the green regions represent the spectra of the Yb-BLs, and the peak pointed by asterisk is the spectra arising from $CeIn_3$ buffer layers. These assignments are made by the straightforward comparison with the spectra of $CeCoIn_5$ and $YbCoIn_5$ thin films. A salient feature is that the shape of the spectra of Ce-BLs for $n = 9$ is different from that of $n = 5$. The spectrum in the lower-field region for $n = 9$ has a much larger weight than that for $n = 5$. The lower-field signal indeed remains after the subtraction of the spectrum of $n = 5$ from that of $n = 9$, and it looks similar to the peak of the $CeCoIn_5$ thin film as shown in FIG. 2(c). The subtraction is considered to be the process which emphasizes the signal arising from the inner layers. Figure 3 indicates a schematic image of the subtraction. Here we labeled the layers of Ce-BL from "1" to "5" ("3") in $n = 9$ ($n = 5$). The layer "1" is the nearest layer to the Yb-BLs and the layer "2" is the next to the layer "1" as shown in FIG. 3. The number of the interfacial layers in Ce-BL should be the same in both of the $n = 9$ and $n = 5$ superlattices, and the difference between two superlattices is the number of the inner layers. The spectra arising from interfacial layers are canceled out by the subtraction of spectra of $n = 5$ from that of $n = 9$ and the remaining spectra in FIG. 2(c) is mainly arising from inner layer as shown in FIG. 3 (right). Thus, it is naturally concluded that the spectra in higher field regions shaded by yellow [FIG. 2(a) and (b)] arise from the outer $CeCoIn_5$ layers labeled "1" close to the interfaces and that the spectra in the lower field regions shaded by red arise from the inner $CeCoIn_5$ layers labeled "3" to "5".

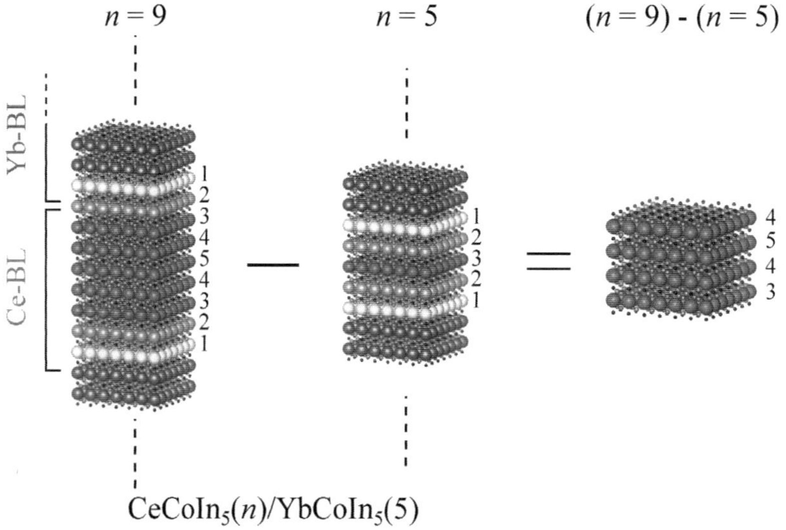

Figure 3. Schematic $CeCoIn_5$- and $YbCoIn_5$- layer arrangements in the superlattices $n = 9$(left) and $n = 5$(middle). The numbers shown at the right side of Ce-BLs represent the distance from the interfaces between Ce- and Yb-BL. In the subtraction $(n = 9)-(n = 5)$ (right), the numbered layers remain.

Next we discuss the field dependence of $(T_1T)^{-1}$ to resolve the layer dependence of the magnetic fluctuations even within a same Ce-BLs. Figure 2(d) depicts the field dependence of $1/T_1T$. The field dependence of $1/T_1T$ is clearly recognized from the recovery of the nuclear magnetization $m(t)$ at a time t after a saturation pulse(FIG.2 (e)). The recovery $(m(\infty) - m(t))/m(\infty)$ at 12.180 T (yellow triangles) is more slowly damped than the recovery at 12.115 T (red triangles). In the data of the $1/T_1T$s of FIG. 2(d), the $(m(\infty) - m(t))/m(\infty)$

curves were fitted by the function for central transition of nuclear spin $I = 9/2$ at the site with $\eta = 0$, given by

$$\begin{aligned}(m(\infty) - m(t))/m(\infty) = C\big[&0.0060606 \exp(-t/T_1) + 0.033566 \exp(-6t/T_1) \\ &+ 0.092308 \exp(-15t/T_1) + 0.215 \exp(-28t/T_1) \\ &+ 0.65306 \exp(-45t/T_1)\big],\end{aligned}$$

where C and T_1 are fitting parameters[26]. The field dependence of $(T_1T)^{-1}$ indicates that AFM fluctuations which are dominant in bulk $CeCoIn_5$[22, 27] have strong spatial dependence even within the same Ce-BL ; fluctuations near at the interfacial layers are more suppressed than those at the inner $CeCoIn_5$ layers. At the inner layer signals around $\mu_0 H = 12.115$ T, the $1/T_1T$ depends on Ce-BL thickness n i.e. the $1/T_1T$ increases with increasing n. On the other hand, at the interfacial signals around $\mu_0 H = 12.180$ T, $1/T_1T$ is strongly suppressed and shows the same behavior in two superlattice samples. This is consistent with the fact that the number of interfacial layers are the same in both of $n = 9$ and $n = 5$ superlattices.

The observed suppression of the AFM fluctuations in the thinner Ce-BLs is opposite to the enhancement of fluctuations expected from two-dimensionalization. Taking into account of the spatial dependence of $1/T_1T$, it should be noted that the suppression of the AFM fluctuations is caused by effects of the interfaces. The suppression can be explained in terms of the effect of the breaking of local inversion symmetry at the interfaces between Ce- and Yb-BLs. In the presence of the local inversion symmetry breaking together with the fact that the Ce atom has a large atomic number, the Rashba-type spin-orbit coupling is expected to be strong in the present superlattices. The Fermi surface splitting due to the Rashba coupling should modify seriously the nesting condition and hence is expected to reduce the commensurate AFM fluctuations. With the reduction of n, the fraction of the noncentrosymmetric interface layers increases rapidly, leading to the suppression of the AFM fluctuations. The importance of the local inversion symmetry breaking at the interface has been emphasized experimentally through the peculiar angular variation of upper critical field, which can be interpreted as a strong suppression of the Pauli pair-breaking effect[19, 20]. The present results suggest that the local inversion symmetry breaking plays a key role for the magnetic properties at the interface, which is also consistent with the suppression of the superconductivity.

4. Conclusion

We have performed NMR measurements of $CeCoIn_5(n)/YbCoIn_5(5)$ superlattices with n = 5 and 9, as well as $CeCoIn_5$ and $YbCoIn_5$ thin films. By comparing the spectra of the superlattices with that of the thin films, we succeeded in identifying the ^{115}In NMR signals arising from the $CeCoIn_5$ and $YbCoIn_5$ layers in the superlattices, separately. Furthermore, we succeeded in distinguishing signals arising from interfacial and inner layers in Ce-BLs by comparing spectra of two superlattices which have different thickness of Ce-BLs. The suppression of $1/T_1T$ was prominent at the interfacial signals, which indicates that the AFM fluctuations at the interface are strongly suppressed than those of the bulk. The suppression of the AFM fluctuations is considered to be an intrinsic properties at the interface.

Acknowledgments

The authors thank S. K. Goh, R. Peters, Y. Tada, Y. Yanase and H. Ikeda for valuable discussions. This work was partially supported by Kyoto Univ. LTM center, Grant-in-Aid from the Ministry of Education, Culture, Sports, Science, Technology(MEXT) of Japan, Grants-in-Aid for Scientific Research (KAKENHI) from Japan Society for the Promotion of Science (JSPS), "J-Physics" (No. 15H05882, No. 15H05884, and No. 15K21732) and "Topological Quantum Phenomena" (No. 25103713) Grant-in-Aid for Scientific Research on Innovative Areas from the MEXT of Japan, and by Grant-in-Aids for Scientific Research (Grants No. 25220710, and No. 15H05745). One of authors would like to acknowledge the support from the Motizuki Fund of Yukawa Memorial Foundation. Figures of crystal structures were shown using the VESTA package.[28]

References

[1] Fujimoto S 2005 *Phys. Rev. B* **72**(2) 024515
[2] Bauer E and MSigrist 2012 *Non-Centrosymmetric Superconductors: Introduction and Overview* (Springer, Heidelberg)
[3] Sato M, Takahashi Y and Fujimoto S 2009 *Phys. Rev. Lett.* **103**(2) 020401
[4] Sau J D, Lutchyn R M, Tewari S and Das Sarma S 2010 *Phys. Rev. Lett.* **104**(4) 040502
[5] Lutchyn R M, Sau J D and Das Sarma S 2010 *Phys. Rev. Lett.* **105**(7) 077001
[6] Hwang H Y, Iwasa Y, Kawasaki M, Keimer B and Nagao N 2012 *Nature materials* **11** 103–113
[7] Shishido H, Shibauchi T, Yasu K, Kato T, Kontani H, Terashima T and Matsuda Y 2010 *Science* **327** 980–983
[8] Mizukami Y, Shishido H, Shibauchi T, Shimozawa M, Yasumoto S, DWatanabe, Yamashita M, Ikeda H, Terashima T, Kontani H and Matsuda Y 2011 *Nature physics* **7** 849–853
[9] Ishii T, Toda R, Hanaoka Y, Tokiwa Y, Shimozawa M, Kasahara Y, Endo R, Terashima T, Nevidomskyy A H, Shibauchi T and Matsuda Y 2016 *Phys. Rev. Lett.* **116**(20) 206401
[10] Petrovic C, Pagliuso P G, Hundley M F, Movshovich R, Sarrao J L, Thompson J D, Fisk Z and Monthoux P 2001 *Journal of Physics: Condensed Matter* **13** L337
[11] Izawa K, Yamaguchi H, Matsuda Y, Shishido H, Settai R and Onuki Y 2001 *Phys. Rev. Lett.* **87**(5) 057002
[12] Stock C, Broholm C, Hudis J, Kang H J and Petrovic C 2008 *Phys. Rev. Lett.* **100**(8) 087001
[13] Zhou B B, Misra S, da Silva Neto E H, Aynajian P, Baumbach R E, Thompson J D, Bauer E D and Yazdani A 2013 *Nature physics* **9** 474–479
[14] Allan M P, Massee F, Morra D K, Dyke J V, Rost A, Mackenzie A P, Petrovic C and Davis J C 2013 *Nature physics* **9** 468–473
[15] Nicklas M, Stockert O, Park T, Habicht K, Kiefer K, Pham L D, Thompson J D, Fisk Z and Steglich F 2007 *Phys. Rev. B* **76**(5) 052401
[16] Booth C H, Bauer E D, Bianchi A D, Ronning F, Thompson J D, Sarrao J L, Cho J Y, Chan J Y, Capan C and Fisk Z 2009 *Phys. Rev. B* **79**(14) 144519
[17] Huy H T, Noguchi S, Hieu N V, Shao X, Sugimoto T and Ishida T 2009 *Journal of Magnetism and Magnetic Materials* **321** 2425 – 2428 ISSN 0304-8853
[18] Zaremba V I, Rodewald U C, Hoffmann R D, Kalychak Y M and Pottgen R 2003 *Zeitschrift fur anorganische und allgemeine Chemie* **629** 1157–1161 ISSN 1521-3749
[19] Goh S K, Mizukami Y, Shishido H, Watanabe D, Yasumoto S, Shimozawa M, Yamashita M, Terashima T, Yanase Y, Shibauchi T, Buzdin A I and Matsuda Y 2012 *Phys. Rev. Lett.* **109**(15) 157006
[20] Shimozawa M, Goh S K, Endo R, Kobayashi R, Watashige T, Mizukami Y, Ikeda H, Shishido H, Yanase Y, Terashima T, Shibauchi T and Matsuda Y 2014 *Phys. Rev. Lett.* **112**(15) 156404
[21] Yamanaka T, Shimozawa M, Endo R, Mizukami Y, Shishido H, Terashima T, Shibauchi T, Matsuda Y and Ishida K 2015 *Phys. Rev. B* **92**(24) 241105
[22] Kohori Y, Yamato Y, Iwamoto Y, Kohara T, Bauer E D, Maple M B and Sarrao J L 2001 *Phys. Rev. B* **64**(13) 134526
[23] Curro N J, Simovic B, Hammel P C, Pagliuso P G, Sarrao J L, Thompson J D and Martins G B 2001 *Phys. Rev. B* **64**(18) 180514
[24] Curro N J 2006 *New Journal of Physics* **8** 173
[25] Carter G C, Bennett L H and Kahan D J 1977 *Metallic Shifts in NMR* (Pergamon, Oxford)
[26] Narath A 1967 *Phys. Rev.* **162**(2) 320–332
[27] Sakai H, Brown S E, Baek S H, Ronning F, Bauer E D and Thompson J D 2011 *Phys. Rev. Lett.* **107**(13) 137001
[28] Momma K and Izumi F 2011 *Journal of Applied Crystallography* **44** 1272–1276 ISSN 1600-5767

IOP Conf. Series: Journal of Physics: Conf. Series **807** (2017) 012005 doi:10.1088/1742-6596/807/1/012005

Growth and NMR evaluation of a single crystal $CePd_2Ga$

Yasuki Kishimoto, Hisashi Kotegawa, Eiichi Matsuoka, Hitoshi Sugawara, and Hideki Tou

Department of Physics, Graduate School of Science, Kobe University, Kobe 657-8501, Japan

E-mail: ykisimoto@stu.kobe-u.ac.jp

Abstract. We have measured the electrical resistivity and the powder x-ray diffraction (XRD) for ascast, 700 °C, and 800 °C annealed samples of the Ce-based antiferromagnetic compound $CePd_2Ga$. From the XRD measurement, the unit-cell volume of 700 °C and 800 °C annealed sample is smaller by 0.52% and 0.87% than ascast sample. In order to check microscopically, we are carried out Ga-Nuclear Magnetic Resonance (NMR) and Nuclear Quadrupole Resonance (NQR).

1. Introduction

Numerous ternary lanthanoids/actinides-based intermetaric compounds have been studied and attracted, because, in Ce, U based heavy-fermion (HF) compounds, unconventional superconductivity or non-Fermi-liquid behaviors emerge frequently near quantum critical point (QCP) by tuning physical (chemical) pressure and/or magnetic field. Such anomalous ground states are caused by a hybridization between conduction electrons and *f*-electrons. It is known that LnT_2X (Ln = lanthanoid, T = transitionmetal) series possesses three kinds of different structures, i.e., $MnCu_2Al$ type (cubic Space group $Fm\bar{3}m$)[1], $ZrPt_2Al$ type (hexagonal, Space group $P6_3/mmc$)[2] and YPd_2Si type (Orthohombic, Space group $Pnma$)[3]. $CePd_2X$ ($X = \mathrm{Al, Ga}$) crystallizes YPd_2Si type structure. Ce ion (local symmetry : .m.) forms zigzag-chain along the a-axis which is similar to ε-TiNiSi type structure, e.g., the ferromagnetic superconductor UCoGe[4], Kondo semiconductor CeNiSn[5].

Porycrystalline samples of $CePd_2Al$ and $Ce_{(1-x)}Y_xPd_2Ga$ have been reported by Das et al.[6, 7]. $CePd_2Al$ exhibits an antiferromagnetic (AFM) order at T_N = 3.3 K. An electronic specific-heat coefficient is as large as γ_e = 110 mJ/mol K^2, and a minimum of the electrical resistivity is observed at $T \sim$ 14 K. These results suggest that $CePd_2Al$ is Kondo-lattice-system. No ordering was observed in electrical resistivity in $Ce_{(1-x)}Y_xPd_2Ga$ for $x > 0.2$. Meanwhile the AFM order clearly appears for $x < 0.2$. Macroscopic measurements, such as electric resistivity, magnetic susceptibility and heat capacity have been performed for the single-crystalline $CePd_2Ga$ [9]. An AFM order was observed at around 3 K from each measurements, which value is close to $CePd_2Al$. The electronic specific-heat coefficient was γ_e = 63 mJ/mol K^2 in the AFM state. The level scheme of the crystal electric field was estimated to the first and the second excited states lie at 66 K and about 280 K, respectively. The estimated Kondo temperature T_K = 2-4 K is comparable with AFM ordering temperature, suggesting the competition between the Kondo effect and RKKY interaction.

2. Experimental

Single-crystalline samples of $CePd_2Ga$ were grown by Czochralski method in a tetra-arc furnace. Starting materials of Ce (3N), Pd (4N) and Ga (6N) were taken in the ratio 1 : 2 : 1. The pulling speed was 10 mm/hr. An ingot was 2-3 mm in diameter and 40 mm in length. In order to investigate annealing effect, single crystals were cut from this ingot and were wrapped in tantalum foil and annealed for two days at different annealing conditions 700, 800 and 900 °C in evacuated quartz tubes. The powder x-ray diffraction (XRD) measurements were carried out using a Cu-$K\alpha$ radiation. The electrical resistivity was measured down to about 1.5 K with a standard four probe method. The ascast sample was crushed for rf field penetration in NMR/NQR measurements. NMR/NQR spectra were obtained by sweeping a magnetic field/Fourier step-sum methods.

3. Experimental result and discussion

3.1. Powder X-ray diffraction

The powder XRD result is shown in Fig.1 with previous report [8]. No impurity phase was detected. We have succeeded in synthesizing single phase of $CePd_2Ga$. Figure 2 shows the powder XRD pattern of ascast (black), 700 °C annealed (red), and 800 °C annealed (blue) sample in the angle range of 2θ from 10° to 90°. The sample annealed at 900 °C was melted. The peak positions of the annealed samples shifted to higher angle, and the peak position of 800 °C annealed sample is higher than that of 700 °C annealed sample slightly. The unit-cell volume is reduced by 0.52% for 700 °C annealing (0.87% for 800 °C annealing) from that of the ascast sample.

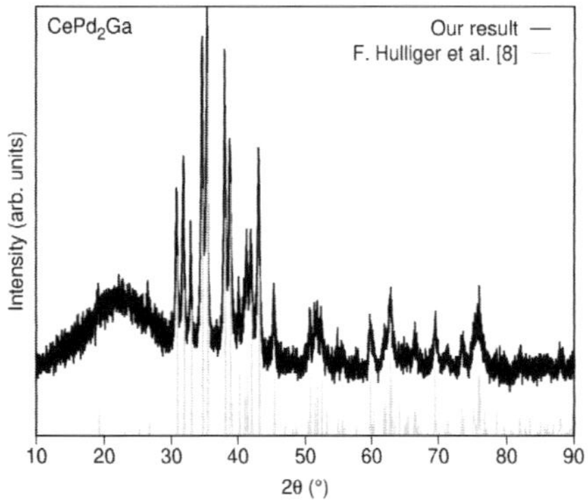

Figure 1. The powder XRD pattern of single crystalline $CePd_2Ga$ in this study (black) with the previous report (green) with lattice constants a = 7.5574 Å, b = 7.0478 Å, and c = 5.7902 Å [8]. Present result is well consistent with the previous report.

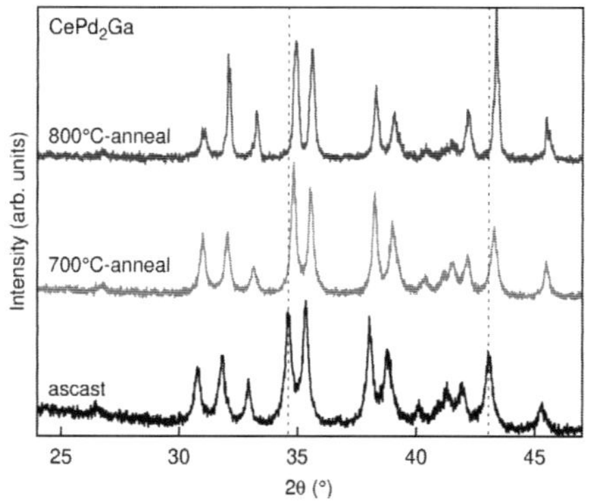

Figure 2. Annealing-temperature dependence of the powder x-ray diffraction patterns for ascast (black), 700 °C(red) and 800 °C(blue) respectively. The peak positions were shifted to higher angle with increasing an annealing temperature (details are shown in the text).

3.2. Electrical resistivity

The effect of annealing for the electrical resistivity ρ is shown in Fig.3, where a current was applied to ab-plane. Although annealing conditions are different, the temperature dependence of ρ exhibits a similar behavior. A broad shoulder was observed at around 50 K, which is roughly corresponding to the crystal electrical field splitting between the ground and the first exited state [9]. A clear kink was observed at around 3 K for all annealing conditions and is ascribed to AFM ordering, which is consistent with the previous report [9]. The present results indicate that Néel temperature (T_N) is independent of an annealing condition, suggesting that the reduction of the unit-cell volume by 0.87% does not effect a magnetic transition.

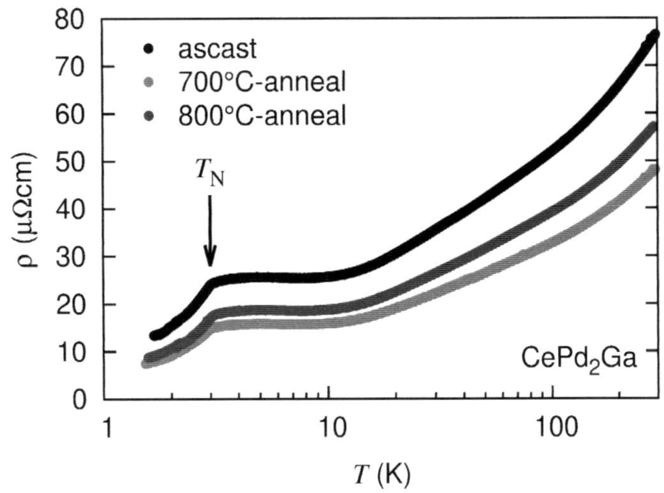

Figure 3. Temperature dependences of the electric resistivity of ascast (black), 700 °C-annealed (red) and 800 °C-annealed (blue) sample. A residual resistivity ratio ρ (300 K)/ρ (0 K) is $\sim$ 6.4, $\sim$ 6.9, and $\sim$ 7.1 for ascast, 700 °C-annealed and 800 °C-annealed sample respectively. The AFM ordering temperature is independent of the annealing condition.

3.3. NMR/NQR measurements

^{69}Ga-NMR (nuclear spin: $I = 3/2$, gyromagnetic ratio: $\gamma_n/2\pi = 10.21885$ MHz/T) measurement of the single-crystalline powder sample was performed at 150 K. The nuclear spin Hamiltonian ($I > 1/2$) under an external magnetic field H is given by

$$\begin{aligned} \mathcal{H} &= \mathcal{H}_z + \mathcal{H}_Q \\ &= -\mu_0\gamma_n\hbar I \cdot H_{\text{eff}} + \frac{h\nu_Q}{6}\left\{3I_Z^2 - I(I+1) + \frac{\eta}{2}\left(I_+^2 + I_-^2\right)\right\} \end{aligned}$$

where H_{eff} is the effective field at the Ga nuclei and is written by $(1+K)\,H$, K is the Knight shift. The nuclear quadrupole resonance frequency ν_Q and the asymmetric parameter η are defined as $\nu_Q \equiv \frac{3e^2qQ}{2I(I-1)h}$ and $\eta \equiv \frac{|V_{YY}|-|V_{XX}|}{|V_{ZZ}|}$ ($|V_{XX}| \leq |V_{YY}| \leq |V_{ZZ}|$). $V_{\alpha\alpha} = \frac{\partial^2 V}{\partial \alpha^2}$ ($\alpha = X, Y, Z$) is the electric field gradient (EFG) for principle axes at the Ga nucleus position. Figure 4 shows the ^{69}Ga-NMR spectrum corresponding to $|-1/2\rangle \leftrightarrow |+1/2\rangle$ transition at a fixed frequency $f_0 = 69.050$ MHz with the simulation result (solid line). The central peak was spread due to the second order perturbation effect of the nuclear quadrupole interaction. The result of the simulation gives a rough estimate of NQR parameters; NQR frequency $^{69}\nu_Q \sim 13.608$ MHz and asymmetric parameter $\eta \sim 0.25$. These parameters allow us to determine the resonance frequency $^{69}\nu_{\text{NQR}} = 13.749$ MHz. Moreover, the nuclear quadrupole moment ratio ($^{69}Q/^{71}Q = 1.598$) is equal to the ratio of the NQR frequency ($^{69}\nu_Q/^{71}\nu_Q$). This relationship gives $^{71}\nu_{\text{NQR}} = 8.652$ MHz. Figure 5 displays the ^{69,71}Ga-NQR spectra (corresponding to $|\pm3/2\rangle \leftrightarrow |\pm1/2\rangle$ transition) at 10 K. ^{69}Ga and ^{71}Ga-NQR signals are observed at$f \sim 13.769$

MHz and 8.668 MHz, respectively. The experimental result is well consistent with those obtained from ^{69}Ga-NMR.

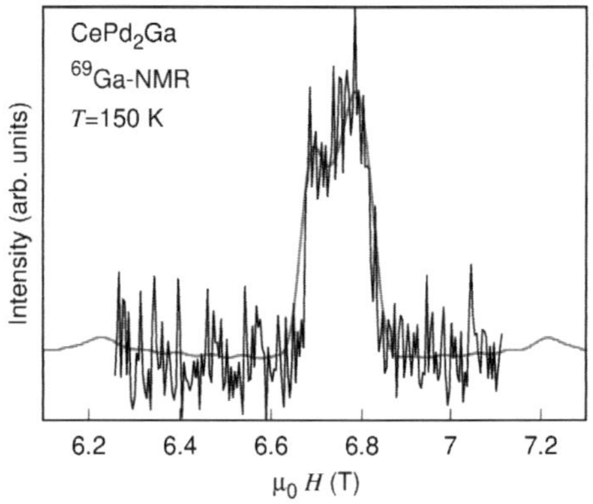

Figure 4. The powder pattern of ^{69}Ga-NMR central line (black) and the simulated spectrum is convoluted with Gaussian function of width of 0.02 T (red). Here, we assumed that a knight shift is 0.5%. The spectrum is spread due to a quadrupole interaction of the second-order perturbation.

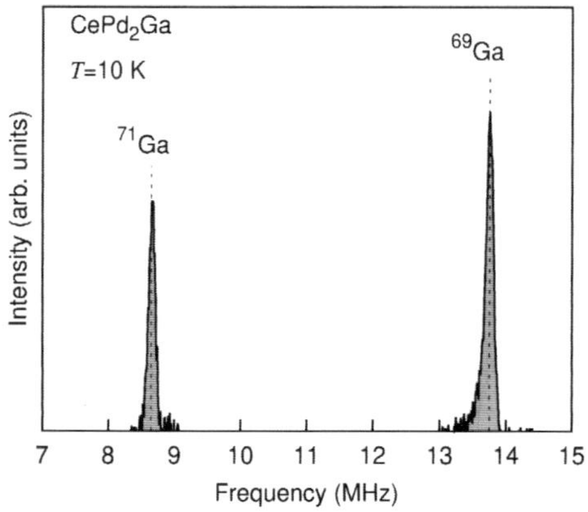

Figure 5. ^{69,71}Ga-NQR signals in $CePd_2Ga$ at 10 K. The dash lines represent the simulations for $^{69}\nu_{NQR}$ and $^{71}\nu_{NQR}$ (details are shown in the text). The narrow full width at half maximum for ^{69,71}Ga ($\sim 200, 150$ kHz) indicates that the sample quality is high.

4. Summary

We have measured the electric resistivity and the powder XRD pattern for ascast, 700 °C annealed and 800 °C annealed $CePd_2Ga$. Furthermore, we have succeeded in observing Ga-NMR/NQR signals for the first time. The unit-cell volume was reduced about 0.52% and 0.87% by annealing of 700 °C and 800 °C, respectively. An absolute value of the electrical resistivity was depended on the anneal temperature, whereas, the behavior does not associate with that. The quadrupole parameters in $CePd_2Ga$ are determined as $^{69}\nu_Q \sim 13.608$ and $\eta \sim 0.25$ from the simulation of the ^{69}Ga-NMR central line. Additionally, we have observed ^{69}Ga and ^{71}Ga-NQR spectra at $f \sim 13.769$ MHz and 8.668 MHz, respectively. In order to investigate microscopic properties of $CePd_2Ga$, the temperature dependence of NMR/NQR spectra and a nuclear spin-lattice relaxation rate should be measured in future studies.

Acknowledgments

This work has been supported by a Grant-in-Aid for Scientific Research on Innovative Areas "J-Physics" (No. 15H05885) and Grants-in-Aid for Scientific Research (Nos. 26400359, 15H05745, and 15H03689) of the Ministry of Education, Culture, Sports, Science and Technology, Japan.

References

[1] Oxley D P, Tebble R S and Williams K C 1963 *J. Appl. Phys.* **34** 1362
[2] Dwight A E 1987 *Matter. Res. Bull.* **22** 201
[3] Moreau J M, Le Roy J and Paccard D 1982 *Acta. Crystallogr. Sect. B* **38** 2446
[4] Huy N T, Gasaparini A, de Nijs D E, Huang Y, Klaasse J C P, Gortrenmulder T, de Visser A, Hamann A, Görlach T and Löhneysen H V 2007 *Phys. Rev. Lett.* **99** 067006
[5] Malik S K and Adroja D T 1991 *Phys. Rev. B* **43** 6277

IOP Conf. Series: Journal of Physics: Conf. Series **807** (2017) 012005 doi:10.1088/1742-6596/807/1/012005

[6] Das I, Sampathkumaran E V and Rajarajan A K 1995 *J. Alloys Compounds* **218** L11
[7] Das I, Sampathkumaran E V, Chari S and Gopalakrishnan K V 1993 *J. Alloys Compounds* **202** L7
[8] Hulliger F, Mattenberger K and Siegrist S 1992 *J. Alloys Compounds* **190** 125
[9] Terayama K, Aoki Y and Sato H 1995 *J. Phys. Cond. Matt.* **7** 6899

Temperature-dependent electronic structure of $EuNi_2P_2$ revealed by angle-resolved photoemission spectroscopy

H. Anzai[1], K. Ichiki[1], E. F. Schwier[2], H. Iwasawa[2], K. Shimada[2], H. Namatame[2], M. Taniguchi[2], A. Mitsuda[3], H. Wada[3], and K. Mimura[1,2]

[1] Graduate School of Engineering, Osaka Prefecture University, Sakai 599-8531, Japan
[2] Hiroshima Synchrotron Radiation Center, Hiroshima University, Higashi-Hiroshima 739-0046, Japan
[3] Graduate School of Science, Kyushu University, Fukuoka, 819-0395 Japan

E-mail: anzai@ms.osakafu-u.ac.jp

Abstract. We studied the electronic structure of $EuNi_2P_2$, which exhibits both heavy-fermion and mixed-valence behaviors, using angle-resolved photoemission spectroscopy. Multiple Ni 3*d* bands were observed near the Fermi energy, and one of them forms a hole-like Fermi surface around the X point of the Brillouin zone. We also found that the spectral weight of the Ni 3*d* states is rapidly enhanced with decreasing temperature, which is consistent with the temperature dependence of the mean valence of Eu ions. Our results thus demonstrate hybridization between the Ni 3*d* and Eu 4*f* electrons in $EuNi_2P_2$.

1. Introduction

In rare-earth compounds, the hybridization effect between conduction and localized-*f* electrons (*c*-*f* hybridization) leads to the formation of a strongly renormalized quasiparticle, and it is the major contributor to the possible mechanisms for the heavy-fermion state. As the *c*-*f* hybridization strength increases, the ground state of rare-earth systems is indeed changed from the magnetic ordered state with localized *f* moments (RKKY interaction) to the nonmagnetic heavy-fermion state with itinerant *f* electrons (Kondo effect) and finally to the mixed-valence state [1]. In this respect, $EuNi_2P_2$ provides a unique opportunity to understand the *c*-*f* hybridization effect. First, the coefficient of the electronic specific heat is large, $\gamma \sim 100$ mJ / (K^2·mol), and $EuNi_2P_2$ is known as the first heavy-fermion compound among Eu-based materials [2]. Second, the mean valence of Eu ions deduced from the Mössbauer isomer shift varies from 2.25 at 300 K to 2.50 at 1 K, which indicates that the $4f^7$ and $4f^6$ electron configurations are mixed in the ground state [3]. Such a large variation in the mean Eu valence is ideal for the study of the temperature dependence of the *c*-*f* hybridization effect.

Angle-resolved photoemission spectroscopy (ARPES) is a direct probe to measure the electronic excitation as functions of energy and momentum. A previous ARPES study of $EuNi_2P_2$ reported that the electronic structure near the Fermi level (E_F) consists of multiple bands and orbitals [4]. Furthermore, an energy gap due to *c*-*f* hybridization between the Ni 3*d* and Eu $4f$ states was observed at $|\omega| \sim 0.6$ eV. These results indicate that multiband electronic structure and the hybridization effect are key to the physics of the heavy-fermion behavior. However, the orbital characters of the bands remain unclear.

In this paper, we report ARPES studies of the electronic structure in $EuNi_2P_2$ by using linearly polarized incident synchrotron radiation. We observe three dispersive bands along the Γ-X direction and identify their possible orbital characters. Quantifying the spectral weight of the Ni $3d$ states, we also show that the spectral evolution with temperature is derived from a hybridization effect between the Ni $3d$ and Eu $4f$ states.

2. Experimental

Single crystals of $EuNi_2P_2$ with $ThCr_2Si_2$ type tetragonal structure were synthesized by the Sn-flux method [5]. The ARPES experiments were performed at BL-1 of the Hiroshima Synchrotron Radiation Center, where two polarization geometries (s and p polarization) are achieved by rotating the ARPES chamber around linearly polarized radiation [6]. All the data were taken with 70 eV photons under s polarization. The total energy resolution was set to 25 meV. The samples were cleaved *in situ* and kept under an ultrahigh vacuum (pressure below 5×10^{-11} Torr) at $T = 15$ K during the measurements.

3. Results and discussion

Figure 1(b) shows a photoemission intensity plot, $I(k, \omega)$, along the Γ-X direction. To better visualize the band dispersions, we also plot the second derivative of the photoemission intensity with respect to the energy, $\partial^2 I(k,\omega)/\partial\omega^2$, in Fig. 1(c). The horizontal stripe seen at $0.1 < |\omega| < 0.8$ eV is assigned to multiplet splitting of the Eu $4f^6$ final state configurations [4]. Around the Γ point, a weak but observable band (labeled as α) crosses E_F, as demonstrated by the momentum distribution curves (MDCs) in Fig. 1(d). We found that this band forms a hole-like Fermi pocket around the X point, as shown in Fig. 1(a). Two other bands, assigned to

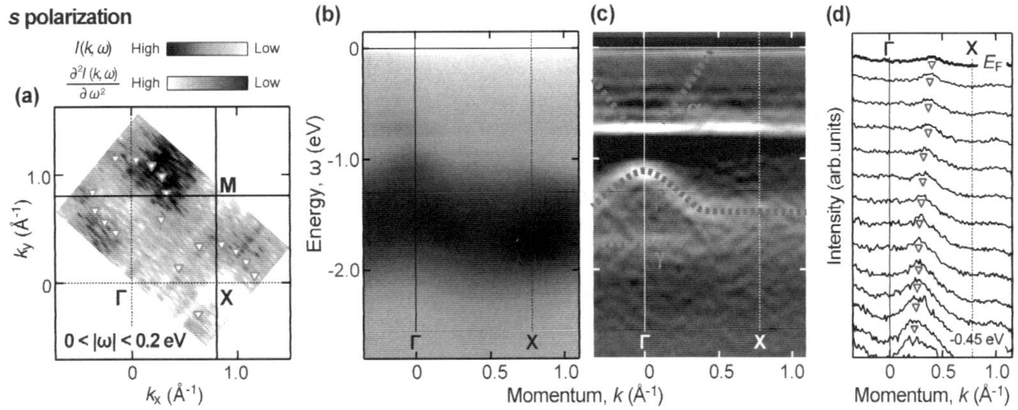

Figure 1. Electronic structure of $EuNi_2P_2$ in the s-polarization geometry. The data were collected at $T = 15$ K with $h\nu = 70$ eV. (a) ARPES intensity map in the Brillouin zone integrated over $0 < |\omega| < 0.2$ eV. Open triangles are Fermi-level crossings determined from the analysis of MDCs. (b) Photoemission intensity plot, $I(k, \omega)$, along Γ-X direction. (c) Second energy derivatives, $\partial^2 I(k,\omega)/\partial\omega^2$, for data in (b). The band dispersions are marked with dashed curves obtained by tracking the peak positions of the MDCs and the local minimum locus of $\partial^2 I(k,\omega)/\partial\omega^2$. (d) MDCs near E_F for data in (b).

β and γ, are determined by tracking the peak positions of the MDCs and the local minimum locus of $\partial^2 I(k,\omega)/\partial\omega^2$. The overall band structure is consistent with previous observations [4].

Polarization-dependent ARPES is a powerful tool for identifying the orbital characters of bands. The photoemission intensity is proportional to the square of the dipole-transition matrix element, $I(k,\omega) \propto |\langle\phi_f^k|\boldsymbol{A}\cdot\boldsymbol{p}|\phi_i^k\rangle|^2$, where ϕ_i^k (ϕ_f^k) is the initial (final) state of the electron wave function, and $\boldsymbol{A}$ and $\boldsymbol{p}$ are the vector potential and momentum operator, respectively. The final state ϕ_f^k can be approximated by a plane-wave state in the mirror plane of the crystal. The overall matrix element and ϕ_f^k should be even. Thus, the initial state ϕ_i^k must have the same symmetry as the dipole operator $\boldsymbol{A}\cdot\boldsymbol{p}$ [7]. In the current experimental geometry (s polarization), $\boldsymbol{A}$ is perpendicular to the mirror plane, so we can selectively observe the odd-symmetry initial state [8].

To determine the orbital characters of the three bands shown in Fig. 1(c), we performed numerical calculations using the density functional theory (DFT) method. The calculated total density of states (DOS) is dominated by the Ni $3d$ states at the energies of $|\omega| > 0.5$ eV, as shown in Fig. 2(a). Considering the spacial symmetry of the $3d$ orbitals, d_{xy} and d_{yz} are odd with respect to the mirror plane, which now lies along the Γ-X direction of the sample. Thus, the β and γ band consist of d_{xy} and/or d_{yz} orbitals. The α band is likely to be hybridized with the Eu $4f$ states, taking into account the large DOS of the Eu $4f$ states near E_F. Further experiments in p-polarization geometry along both the Γ-X and Γ-M directions are necessary to confirm the orbital characters.

The temperature dependence of the energy distribution curves (EDCs) at the Γ point is shown in Fig. 2(b). Two peaks corresponding to the β and γ bands are observed around $|\omega| \sim 1.2$ and 1.7 eV, respectively. The peak intensity of the β band increases strongly with decreasing temperature. On the other hand, the γ band is less sensitive to changes in temperature. The valence or hole-occupation number can be deduced from the photoemission spectral weight [9, 10]. In this study, we quantified the Ni $3d$ spectral weight of the β and γ bands by integrating the intensity of the EDCs over an energy window of $0.9 < |\omega| < 2.5$ eV

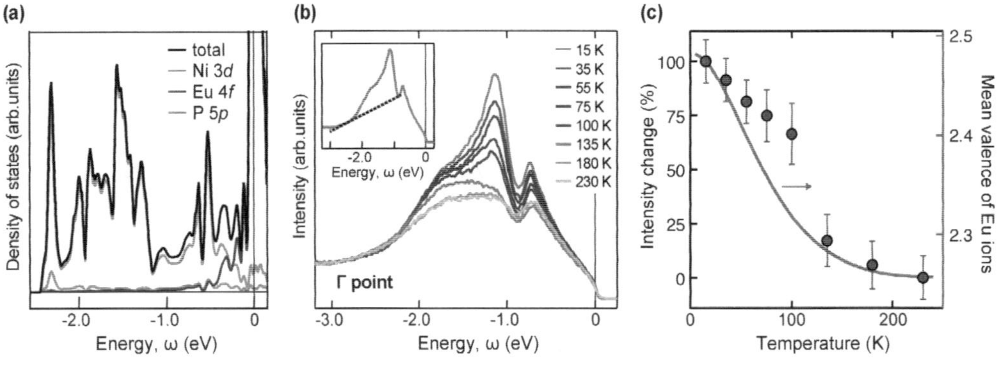

Figure 2. (a) Calculated DOS of $EuNi_2P_2$, where green, blue, and orange curves correspond to the Ni $3d$, Eu $4f$, and P $5p$ partial DOS, respectively. (b) Temperature dependence of EDCs at Γ point taken from the data in Fig. 1(b). Inset shows the EDC at 15 K and a linear background (dashed line). The spectral weight is obtained by integrating the intensities over an energy range of $0.9 < |\omega| < 2.5$ eV after the linear background is subtracted. (c) Evolution of the intensity change (black circles, left axis) and mean valence of Eu ions (red curve, right axis) reported in Ref. 3. The intensity change is normalized to 100% by dividing the difference between the maximum and minimum spectral intensities.

after subtracting the linear background; a typical example is shown in the inset of Fig. 2(b). The difference between the maximum and minimum intensities is normalized to 100% and plotted as a function of temperature in Fig. 2(c) together with the mean Eu valence [3]. Note that the normalized intensity reflects mainly the temperature evolution of the β band, because the background-subtracted intensity of the γ band is approximately constant over a wide temperature range. The intensity change shown in Fig. 2(c) increases with decreasing temperature and is comparable to the temperature variation of the mean Eu valence. This similarity indicates that the Ni $3d$ electron occupation is associated with the mixed-valence state between the $4f^7$ and $4f^6$ configurations of the Eu ions. Therefore, the hybridization between the Ni $3d$ and Eu $4f$ states plays an important role in the heavy-fermion behavior of $EuNi_2P_2$. Similar conclusions were reported from resent studies of ARPES, x-ray magnetic circular dichroism, and optical conductivity [4, 11, 12].

The substantial increase of the Ni $3d$ spectral weight around 120 K is likely related to the thermal evolution of the Kondo effect, as proposed from tunneling spectroscopy measurements [13]. Hiranaka *et al.* has suggested that the heavy fermion state in this system is based on the Kondo effect [14]. The characteristic Kondo temperature is estimated as T_K = 75-80 K, which is below the leading-edge midpoint of the intensity change in Fig. 2(c). This deviation implies a fluctuating precursor of the long-range Kondo effect that develops fully at lower temperature.

4. Conclusions

In conclusion, we have studied the electronic structure of $EuNi_2P_2$ using ARPES. There is one band intersecting the Fermi level, which forms a hole-like pocket surrounding the X point. The spectral weight of the Ni $3d_{xy}$ and/or $3d_{yz}$ band increases significantly with decreasing temperature. On the basis of a comparison with the DFT calculation and the temperature variation of the mean Eu valence, we suggest that this behavior can be attributed to electron transfer between the Ni $3d$ and Eu $4f$ states. Our results indicate that the c-f hybridization effect makes an essential contribution to the heavy-fermion behavior of $EuNi_2P_2$.

5. Acknowledgments

We thank T. Sato and M. Arita for valuable discussions, and A. Hariki and R. Takeshita for assistance. This work was supported by JSPS KAKENHI Grant Number JP26800191. The ARPES experiments were performed under the approval of HSRC (Proposal Nos. 14-A-18 and 15-A-49).

6. References

[1] S. Doniach, Physica B **231**, 91 (1977).
[2] R. A. Fisher, P. Radhakrishna, N. E. Phillips, J. V. Badding, and A. M. Stacy, Phys. Rev. B **52**, 13519 (1995).
[3] R. Nagarajan, G. K. Shenoy, L. C. Gupta, and E. V. Sampathkumaran, Phys. Rev. B **32**, 2846 (1985).
[4] S. Danzenbächer, D. V. Vyalikh, Yu. Kucherenko, A. Kade, C. Laubschat, N. Caroca-Canales, C. Krellner, C. Geibel, A. V. Fedorov, D. S. Dessau, R. Follath, W. Eberhardt, and S. L. Molodtsov, Phys. Rev. Lett. **102**, 026403 (2009).
[5] W. Jeitschko and B. Jaberg, J. Solid State Chem. **35**, 312 (1980).
[6] H. Hayashi, K. Shimada, J. Jiang, H. Iwasawa, Y. Aiura, T. Oguchi, H. Namatame, and M. Taniguchi, Phys. Rev. B **87**, 035140 (2013).
[7] W. Eberhardt and F. J. Himpsel, Phys. Rev. B **21**, 5572 (1980).
[8] A. Damascelli, Z. Hussain, and Z.-X. Shen, Rev. Mod. Phys. **75**, 473 (2003).
[9] L. Tjeng, S.-J. Oh, E.-J. Cho, H.-J. Lin, C. Chen, G.-H. Gweon, J.-H. Park, J. Allen, T. Suzuki, M. Makivić, and D. Cox, Phys. Rev. Lett. **71**, 1419 (1993).
[10] T. Kinoshita, H. P. N. J. Gunasekara, Y. Takata, S. Kimura, M. Okuno, Y. Haruyama, N. Kosugi, K. G.

Nath, H. Wada, A. Mitsuda, M. Shiga, T. Okuda, A. Harasawa, H. Ogasawara, and A. Kotani, J. Phys. Soc. Jpn. **71**, 148 (2002).

[11] Y. H. Matsuda, Z. W. Ouyang, H. Nojiri, T. Inami, K. Ohwada, M. Suzuki, N. Kawamura, A. Mitsuda, and H. Wada, Phys. Rev. Lett. **103**, 046402 (2009).

[12] V. Guritanu, S. Seiro, J. Sichelschmidt, N. Caroca-Canales, T. Iizuka, S. Kimura, C. Geibel, and F. Steglich, Phys. Rev. Lett. **109**, 247207 (2012).

[13] S. Ernst, S. Kirchner, C. Krellner, C. Geibel, G. Zwicknagl, F. Steglich, and S. Wirth, Nature **474**, 362 (2011).

[14] Y. Hiranaka, A. Nakamura, M. Hedo, T. Takeuchi, A. Mori, Y. Hirose, K. Mitamura, K. Sugiyama, M. Hagiwara, T. Nakama, and Y. Onuki, J. Phys. Soc. Jpn. **82**, 083708 (2013).

IOP Conf. Series: Journal of Physics: Conf. Series **807** (2017) 012007 doi:10.1088/1742-6596/807/1/012007

Paramagnetic resonance in $GdRh_2Si_2$

Jörg Sichelschmidt [1], **K. Kliemt** [2], **C. Krellner** [2], **C. Geibel** [1]

[1] Max Planck Institute for Chemical Physics of Solids, Nöthnitzer Straße 40, 01187 Dresden, Germany

[2] Institute of Physics, University of Frankfurt, Frankfurt, Germany

E-mail: joerg.sichelschmidt@cpfs.mpg.de

Abstract. The paramagnetic properties of single-crystalline $GdRh_2Si_2$ are investigated by electron spin resonance (ESR) spectroscopy. The almost isotropic paramagnetic resonance is characterized by a spin dynamics of coupled Gd^{3+} – conduction electron spins. Towards magnetic ordering the spin relaxation is dominated by ferromagnetic in-plane spin fluctuations.

1. Introduction

$GdRh_2Si_2$ belongs to the rare-earth silicides with the tetragonal $ThCr_2Si_2$ structure which show exceptional magnetic properties, e.g. the antiferromagnetic (AFM) Kondo systems $YbRh_2Si_2$ [1] or $CeRh_2Si_2$ [2]. $GdRh_2Si_2$ possesses well localized magnetic moments with a weak magnetic anisotropy and displays AFM order below $T_N = 107$ K [3, 4]. It recently came into focus because of the spin-splitting of itinerant magnetic surface states induced by the ordering of the Gd 4f moments in the bulk [5].

Electron Spin Resonance (ESR) is a bulk sensitive method to investigate the spin dynamics of magnetic moments in metals where Gd^{3+}-spins are commonly used as probes [6]. This allows narrow lines because a direct spin relaxation to the lattice is not possible. The reason is the zero orbital magnetic moment of Gd^{3+} (pure spin ground state) and thus crystalline electric field effects as well as the magnetic anisotropy are weak. In this respect it is worth to note that in the heavy fermion metal $YbRh_2Si_2$, despite a strong magnetic anisotropy, an ESR signal is observed as a consequence of the Kondo interaction [7].

In $GdRh_2Si_2$ well-localized Gd^{3+} magnetic moments serve as ideal probes for a spin resonance. Interestingly, the weak in-plane magnetic anisotropy [8] allows an observation of the in-plane resonance at standard microwave energies (X-band, 9.4 GHz) even in the magnetically ordered region. In this first report we focus on the Gd^{3+} ESR spectra in the paramagnetic region, which is easier to analyze and shall provide the basis for getting an understanding of the ESR signal in the AFM ordered regime.

2. Experiment and Results

For the ESR investigation we used high-quality single-crystalline $GdRh_2Si_2$, the growth and characterization of which is described in Ref.[3]. We used a continuous-wave ESR spectrometer at X-band frequency ($\omega/2\pi = 9.40$ GHz) together with a nitrogen-flow cryostat allowing for temperatures between 100 and 300 K. Additional measurements at Q-band frequency ($\omega/2\pi = 34.07$ GHz) were done for $c \perp H$ (i.e. magnetic field is perpendicularly aligned to the tetragonal c-axis) in the same temperature range by using a He-flow cryostat. In general, an ESR spectrometer allows to measure the absorbed power P of a transversal magnetic microwave field as a function of a static and external magnetic field $\mu_0 H$. A lock-in technique improves the signal-to-noise ratio by a field modulation which then yields the derivative of the resonance signal dP/dH as the measured quantity. The resulting ESR spectra were fitted with a Lorentzian function including the influence of the counter-rotating component of the linearly polarized microwave field [9]. From the fit we obtained the resonance field H_{res} and the linewidth ΔH (half-width at half maximum). In the paramagnetic region the resonance condition simply reads: $\omega/\gamma = H_{res}$ where $\gamma = g\mu_B/\hbar$ is the gyromagnetic ratio and g is the spectroscopic splitting factor (the "ESR g-factor").

In the paramagnetic regime, i.e. for $T > T_{\rm N} = 107$ K, the ESR spectra and their temperature dependence display a behavior as typically expected for well-defined local moments in a metallic environment and for temperatures nearby magnetic ordering [6, 10]. Fig. 1 shows $H_{res}(T)$ and $\Delta H(T)$ for two orientations of the external magnetic field relative to the tetragonal c-axis. An anisotropy within the tetragonal basal plane could not be detected above $T_{\rm N}$. The spectra shapes are well defined Lorentzians (inset of Fig. 1). They are asymmetric due to a microwave penetration depth (≈ 0.003 mm) which is much smaller than the sample thickness (≈ 0.5 mm). The temperature dependences of H_{res} and ΔH result from the interaction between the magnetic moments of Gd^{3+} and conduction electrons. The anisotropy in the paramagnetic susceptibility $\chi \propto (T + \Theta)^{-1}$ leads to the observed weak temperature dependence of $H_{res}^{\|,\perp}$ in terms of the difference in the Weiss temperatures $\Theta_\perp - \Theta_\|$ [11]. The dashed grey lines in the upper frame of Fig. 1 fit the data with

$$g_\|^{\rm eff} = g_\|^0 \left(1 + \frac{\Theta_\perp - \Theta_\|}{T + \Theta_\|}\right) \quad , \quad g_\perp^{\rm eff} = g_\perp^0 \left(1 - \frac{0.5\,(\Theta_\perp - \Theta_\|)}{T + \Theta_\perp}\right) \tag{1}$$

taken from Ref. [11] with $\Theta_\| = -6.5$ K and $\Theta_\perp = -9.2$ K which also consistently describe the paramagnetic susceptibility data (see Fig. 6 in [3]).[12] The g-factors, $g_\|^0 = 1.995$ and $g_\perp^0 = 2.001$, indicated in Fig. 1 are the high-temperature limits of Eqs. (1) and are consistent with the value reported for powdered samples $g = 1.995 \pm 0.01$ [13].

Towards high temperatures, the linewidth linearly depends on temperature and is determined by the relaxation of the coupled Gd^{3+} - conduction electron spin system towards the lattice (so-called "bottlenecked relaxation") [13]. Approaching $T_{\rm N}$, due to the decreasing fluctuations of the Gd^{3+} moments, the exchange-narrowing [14] gets suppressed. This leads to a critical divergence of $\Delta H^{\rm crit}$ which is generally described by a power-law behavior (see p. 354 in [10]). We describe the total temperature dependence of ΔH in the paramagnetic regime as shown with the dashed line in Fig. 1 according to the following equation:

$$\Delta H = \Delta H^{\rm crit} + b \cdot T + \Delta H_0 \ \text{ with } \ \Delta H^{\rm crit} = C_0 \left(\frac{T}{T_{\rm N}} - 1\right)^{-p} . \tag{2}$$

Here, C_0 and ΔH_0 are temperature independent parameters, $C_0^\| = 5.8$ mT, $C_0^\perp = 2.3$ mT and $\Delta H_0 = 5.5$ mT. The critical linewidth divergence is fitted with $p = 1$ and $T_{\rm N} = 107$ K. $p = 1$ corresponds to a model for a 3D-Heisenberg ferromagnet whereas a 3D-Heisenberg antiferromagnet would correspond to $p = 1.7$ (see p. 358 in [10]). Therefore, ferromagnetic fluctuations in the plane seem to dominate the spin dynamics in agreement with in-plane ferromagnetically ordered layers but in contrast to their AFM staggered order [8]. Although $\Delta H^{\rm crit}$ describes the data within experimental error reasonably well it cannot account for the fact that the maximum of the observed linewidth occurs a few degrees above $T_{\rm N}$. This effect is observed also in other ferromagnets and is related to the field-dependence of the linewidth-divergence, see Ref. [15], for instance. The linear parts fit to the slopes $b_\perp = 0.18$ mT/K and

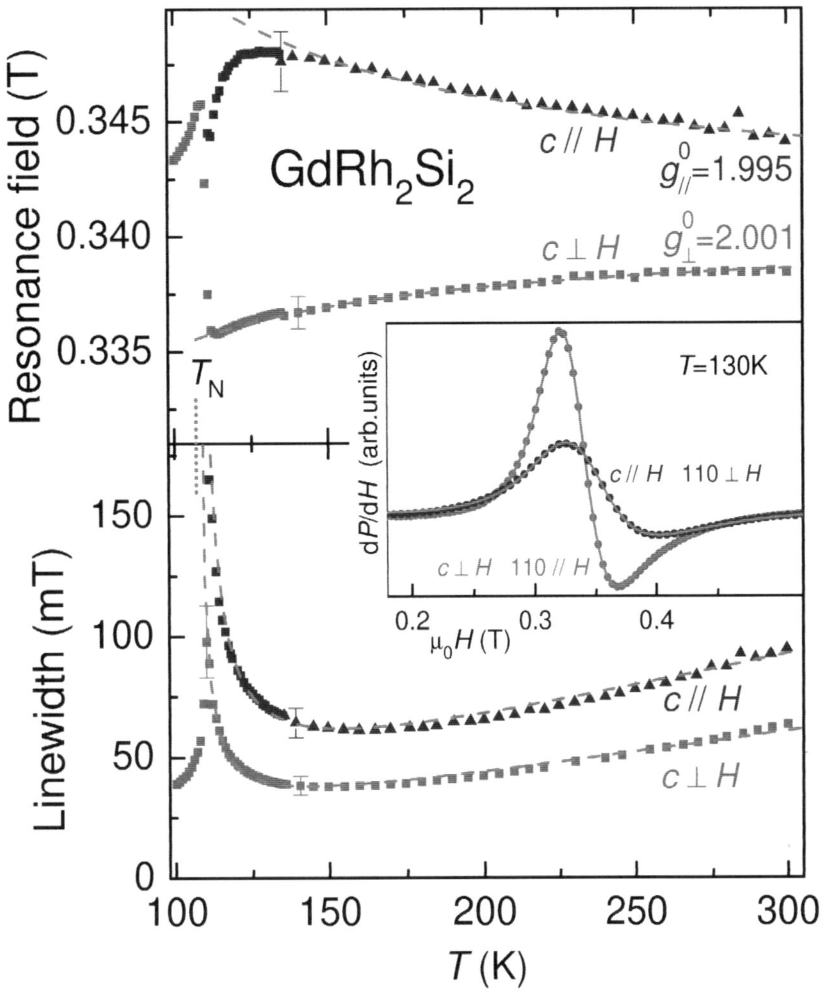

Figure 1. (Color online) Temperature dependence of the resonance field H_{res} and linewidth ΔH as obtained from fitting the X-band (9.4 GHz) spectra with Lorentzian shapes, see inset, solid lines. The dashed lines describe $H_{res}(T)$ with a weak anisotropy in the paramagnetic susceptibility (parameters see Eqs. 1 and main text) with high-T limits of the ionic g-values as indicated. The $\Delta H(T)$ data could be described by dashed lines according Eq.(2) with a dominant critical behavior for temperatures approaching $T_{\mathrm{N}} = 107$ K and with a linear dependence at elevated temperatures (parameters see main text).

$b_{\parallel} = 0.28$ mT/K, again in agreement with data of powdered samples [13]. As compared to the X-band data the paramagnetic linewidth of the Q-band data (measured for $c \perp H$ only) could be described with the same parameters of the critical part but with clearly different ones in the linear part: $b_{\perp}^{\mathrm{Q}} = 0.07$ mT/K and $\Delta H_0^{\mathrm{Q}} = 24$ mT.

3. Discussion and Conclusion

We have shown that at comparatively small energies (X-band, 9.4 GHz $\equiv$ 0.04 meV and Q-band, 34 GHz $\equiv$ 0.14 meV) the internal magnetic fields and the spin dynamics of $GdRh_2Si_2$ can be investigated by electron spin resonance of the Gd^{3+} spins. In the paramagnetic regime, $T > T_N = 107$ K, a common Gd^{3+} collective-mode resonance is observed. The ESR g-values follow nicely a temperature dependence as expected for anisotropic exchange-coupled paramagnets [11]. The spin dynamics is characterized by a coupled Gd^{3+} - conduction electron spin system and a linewidth divergence towards T_N determined by a slowing down of in-plane ferromagnetic fluctuations.

References

[1] Trovarelli O, Geibel C, Mederle S, Langhammer C, Grosche F M, Gegenwart P, Lang M, Sparn G and Steglich F 2000 *Phys. Rev. Lett.* **85** 626
[2] Quezel S, Rossat-Mignod J, Chevalier B, Lejay P and Etourneau J 1984 *Solid State Comm.* **49** 685
[3] Kliemt K and Krellner C 2015 *J. Crystal Growth* **419** 37
[4] Czjzek G, Oestreich V, Schmidt H, Latka K and Tomala K 1989 *J. Magn. Mag. Mat.* **79** 42
[5] Güttler M, Generalov A, Otrokov M M, Kummer K, Kliemt K, Fedorov A, Chikina A, Danzenbächer S, Schulz S, Chulkov E V, Koroteev Y M, Caroca-Canales N, Shi M, Radovic M, Geibel C, Laubschat C, Dudin P, Kim T K, Hoesch M, Krellner C and Vyalikh D V 2016 *Sci. Rep.* **6** 24254
[6] Elschner B and Loidl A 1997 *Electron-Spin Resonance on localized magnetic moments in metals* (*Handbook on the Physics and Chemistry of Rare Earths* vol 24) (Elsevier Science B.V.) chap 162, pp 221
[7] Kochelaev B I, Belov S I, Skvortsova A M, Kutuzov A S, Sichelschmidt J, Wykhoff J, Geibel C and Steglich F 2009 *Eur. Phys. J. B* **72** 485
[8] Kliemt K, Hofmann-Kliemt M, Kummer K, Yakhou-Harris F, Geibel C and Krellner C 2016 to be published
[9] Rauch D, Kraken M, Litterst F J, Süllow S, Luetkens H, Brando M, Förster T, Sichelschmidt J, Neubauer A, Pfleiderer C, Duncan W J and Grosche F M 2015 *Phys. Rev. B* **91** 174404
[10] Benner H and Boucher J P 1990 *Spin Dynamics in the Paramagnetic Regime: NMR and EPR in two-dimensional Magnets* vol 9 (Dordrecht, Boston, London: Kluwer Academic Publishers) pp 323
[11] Huber D L 2012 *Mod. Phys. Lett. B* **26** 1230021
[12] Note, that $\Theta_{||} = 9.2$ K and $\Theta_{\perp} = 6.5$ K would also fit the ESR data reasonably well. However, these values would not be consistent with the susceptibility data which are described by a *negative* Θ in $\chi \propto (T+\Theta)^{-1}$, indicating dominantly ferromagnetic exchange interactions.
[13] Kwapulińska E, Kaczmarska K and Szytuła A 1988 *J. Magn. Mag. Mat.* **73** 65
[14] Anderson P W and Weiss P R 1953 *Rev. Mod. Phys.* **25** 269
[15] Kötzler J and von Philipsborn H 1978 *Phys. Rev. Lett.* **40** 790

Interplay between hybridization gaps and antiferromagnetic gap in the hole-doped Kondo semiconductor $Ce(Os_{1-y}Re_y)_2Al_{10}$

Jo Kawabata[1], Toshikazu Ekino[2], Yoshihiro Yamada[1], Akira Sugimoto[2], Yuji Muro[3] and Toshiro Takabatake[1, 4,*]

[1] Department of Quantum Matter, Graduate School of Advanced Sciences of Matter, Hiroshima University, Higashi-Hiroshima 739-8530, Japan

[2] Graduate School of Integrated Arts and Sciences, Hiroshima University, Higashi-Hiroshima 739-8526, Japan

[3] Liberal Arts and Sciences, Faculty of Engineering, Toyama Prefectural University, Imizu, Toyama 939-0398, Japan

[4] Institute for Advanced Materials Research, Hiroshima University, Higashi-Hiroshima 739-8521, Japan

E-mail: takaba@hiroshima-u.ac.jp

Abstract. The Kondo semiconductor $CeOs_2Al_{10}$ undergoes an antiferromagnetic (AFM) order at an unexpectedly high temperature 28.5 K. We have performed break junction tunneling measurements for the hole-doped system $Ce(Os_{1-y}Re_y)_2Al_{10}$ ($y \leq 0.1$). The tunneling spectrum dI/dV for $y = 0$ displays successive openings of a hybridization gap V_1, an AFM gap V_{AF} and another hybridization gap V_2 in the density of states (DOS). On cooling from 36 K to T_N, both the gap value V_1 and the DOS at the Fermi level, E_F, decrease by 8% of the values at 36 K. This fact indicates that the development of short-range magnetic correlations reduces the *c-f* hybridization gap. For $y = 0.02$, a peak appears in dI/dV at $V = 0$ concurrently with the disappearance of V_2. With increasing y further, the in-gap states develop at E_F, in good agreement with the increase in the Sommerfeld coefficient of the heat capacity. Thereby, T_N, V_1 and V_{AF} decrease and disappear at $y = 0.05$. These facts provide compelling evidence that the presence of V_1 is necessary for the AFM order in $CeOs_2Al_{10}$.

1. Introduction

A family of cerium-based compounds CeT_2Al_{10} (T = Fe, Os and Ru) belongs to Kondo semiconductors. For example, the resistivity $\rho(T)$ in the Os compound shows a thermal activation-type temperature dependence at $30 < T < 80$ K. Nevertheless, this compound undergoes an antiferromagnetic (AFM) transition at rather high Néel temperature T_N of 28.5 K [1,2]. It has remained a mystery that the T_N with a small magnetic moment 0.3 μ_B/Ce is higher than T_N = 18 K for the Gd counterpart with 7 μ_B/Gd [3]. Below T_N, the slope of $\rho(T)$ increases abruptly, most likely by the formation of a superzone gap. The analysis of anisotropic magnetic susceptibility and the x-ray absorption spectra for single crystals determined the crystal-field ground state to be a Kramer's doublet dominated by $|J_z\rangle = |\pm 3/2\rangle$, where $|J_z\rangle$ is the *c*-axis component of the total angular momentum $J = 5/2$ [4].

The break junction tunneling spectroscopy (BJTS) is a powerful method to detect fine changes in the density of states (DOS) caused by AFM and charge-density-wave (CDW) transitions. The BJTS uses a crack prepared in a sample as an insulating barrier for the semiconductor-insulator-semiconductor (SIS) junction. Cracking the sample in a liquid-helium atmosphere provides a clean interface without any contamination and oxidation of reactive Ce based compounds. Furthermore, the SIS junction is not affected by the Seebeck effect so that symmetric tunneling spectra are observed [5]. The BJTS for the antiferromagnet $Ce(Fe_{0.95}Co_{0.05})_2$ showed peaks at $V = \pm 10$ mV below $T_N = 45$ K due to a superzone gap opening in the DOS at E_F [6]. The CDW gap in $CeTe_2$ was found to decrease when a short-range ferromagnetic order develops at low temperatures. This observation in $CeTe_2$ revealed the interplay between the CDW and magnetic order [7].

Recently, BJTS measurements for CeT_2Al_{10} (T = Fe, Os) have revealed openings of two gaps (Δ_1 = 150 meV, Δ_2 = 38 meV) in $CeFe_2Al_{10}$ and three gaps (Δ_1 = 100 meV, Δ_{AF} = 50 meV, Δ_2 = 25 meV) in $CeOs_2Al_{10}$ [8]. In the latter, the AFM gap Δ_{AF} develops below T_N. We pointed out that Δ_1 and Δ_2 are proportional to the Kondo temperature T_K, and the ratio Δ_1/Δ_2 is approximately 4 for the two compounds with T = Fe and Os. These relations, $\Delta \propto T_K$ and $\Delta_1/\Delta_2 = 4$, agree with those for the double *c-f* hybridization gaps calculated with the periodic Anderson model for a crystal-field ground state consisting of $|J_z\rangle = |\pm 3/2\rangle$ [9].

The 5*d* hole and electron doping effects in $Ce(Os_{1-y}Re_y)_2Al_{10}$ and $Ce(Os_{1-x}Ir_x)_2Al_{10}$ systems, respectively, have been studied by the measurements of magnetic properties, optical conductivity and muon spin relaxation (μSR) as well as neutron scattering [10-12]. It is found that the 4*f* state in $CeOs_2Al_{10}$ becomes more itinerant by doping of 5*d* holes, while the 4*f* state is localized by the doping of 5*d* electrons. The semiconducting increase in the $\rho(T)$ below 16 K changes to a metallic behavior at a small level of $y = 0.02$. For $y = 0.1$, $\rho(T)$ exhibits a broad maximum at around 100 K which is a characteristic of valence fluctuating Ce compounds. On the other hand, $\rho(T)$ for the electron doped sample with $x = 0.04$ shows a metallic behavior at low temperatures. For $x = 0.15$, $\rho(T)$ no longer shows the thermal activation behavior above T_N nor the increase below T_N. In both systems doped with 5*d* holes and electrons, the suppression of T_N is well correlated with that of the thermal activation energy Δ/k_B in $\rho(T)$. In contrast, the Sommerfeld coefficient γ in the specific heat increases sharply as x and y are increased. Therefore, we concluded that the presence of the hybridization gap V_1 is necessary for the AFM order in $CeOs_2Al_{10}$. Although the activation energy in $\rho(T)$ gives a crude estimation of the gap width, it does not provide us with the information on the temperature variation of the gap structure. Note that $\rho(T)$ depends on the scattering process and the DOS only at the Fermi level. It is important to observe how the three gaps V_1, V_{AF} and V_2 in $CeOs_2Al_{10}$ are changed by doping the 5*d* holes and electrons.

For this purpose, we have measured the BJTS on the system $Ce(Os_{1-y}Re_y)_2Al_{10}$ ($y \leq 0.1$). We have reported that the sample with $y = 0.02$ undergoes an AFM transition at 23 K, above which $\rho(T)$ still exhibits the activation-type temperature dependence. For $y = 0.05$, however, the AFM transition and the gap Δ/k_B in $\rho(T)$ disappear. Furthermore, $\rho(T)$ in $y = 0.1$ shows a characteristic temperature dependence with a broad maximum around 100 K [10].

2. Sample preparation and break-junction tunneling measurements

For BJTS measurements, polycrystalline samples of $Ce(Os_{1-y}Re_y)_2Al_{10}$ (y = 0, 0.02, 0.05 and 0.1) were prepared by arc melting and subsequent annealing in an evacuated quartz ampoule at 850 ℃ for 7 days [2]. The atomic composition was determined by electron-probe microanalysis, thereby the real compositions of Re were found to agree with the initial ones. In addition, a small amount of impurity phase Os_4Al_{13} has been detected. The midpoint of the jump in the specific heat has been taken as the AFM ordering temperature T_N [10].

The samples were shaped into a plate of 3×2×0.5 mm^3 for BJTS measurements. In order to crack it perpendicularly in the middle, we cut a groove into the surface. The plate was mounted on a flexible substrate, and an adjustable force was applied from its back to make a crack in a liquid-helium chamber. The spectra of dI/dV, where I and V represent the tunneling current and the bias voltage, respectively, have been recorded using a standard lock-in technique by applying V along the long axis of the sample [8]. The different thermal expansions between the sample and the substrate made the junction unstable at elevated temperatures. In fact, we could observe the spectra on heating just to 65 K.

3. Experimental results

The tunneling spectra dI/dV vs the bias voltage V for Ce(Os$_{1-y}$Re$_y$)$_2$Al$_{10}$ (y = 0, 0.02 and 0.05) at 4.4 K are shown in Fig. 1. The spectra are normalized by the values at V = –400 mV. The absolute value of dI/dV depends on the junction resistance R_J, which is determined by the thickness of insulating barrier [14]. In the high-bias range at V = –400 mV, the values of R_J are 20 – 10000 Ω at 4.4 K whose magnitude is much larger than the sample resistance 10 – 50 mΩ measured before breaking. This relation satisfies the condition to measure the voltage drop by the tunneling current in the insulating barrier. By measuring the spectra for many junctions, we confirmed that both the gap structures and the width $V^{\text{P-P}}$ hardly depend on R_J [15]. Such a result is in accordance with a principle of the tunneling spectroscopy.

There are three gap structures in the spectrum for y = 0. The peak structures at ± 200, 100 and 50 mV are denoted as V_1, V_{AF} and V_2, respectively, whose values agree with those obtained for single crystals [8]. The peak structures of V_{AF} change to a shoulder in the spectrum for y = 0.02, where a peak appears at V = 0. This peak structure develops in the spectrum for y = 0.05, where the gaps V_1 and V_{AF} disappear. These changes in the spectrum with y indicate that the 5d hole doping suppresses the hybridization gaps and induces an in-gap state at E_F. These behaviours are consistent with the gap structures in the DOS calculated on the basis of the periodic Anderson model which takes into a slight dispersion in the f-band in the orthorhombic Kondo semiconductors [16].

The temperature dependences of dI/dV vs V are shown in Fig. 2 for three selected compositions y = 0, 0.02 and 0.1. For y = 0, we present the data obtained with the single crystal [8] because the gap structures are clearer than that obtained with the polycrystal. We note that there is little difference in the gap structures between the two. The voltage difference between the shoulders at ±200 mV at 78.5 K is denoted as V_1. On cooling, the shoulders change to peaks. Below T_N, other peaks develop at ±100 mV, whose peak-to-peak voltage is denoted as the AFM gap V_{AF} = 200 mV. On further cooling below 16 K, additional peaks appear at ±50 mV, whose peak-to-peak voltage is denoted as V_2.

In the spectra for y = 0.02, the dI/dV at 40 K displays shoulders at $V_1/2$. The shoulders change to peaks at 35 K, and transform into shoulders again at T < 30 K. On cooling below 27 K (> T_N = 23 K), other shoulders appear at $V_{AF}/2$. At low temperatures, the dI/dV exhibits a cusp at V = 0, which develops

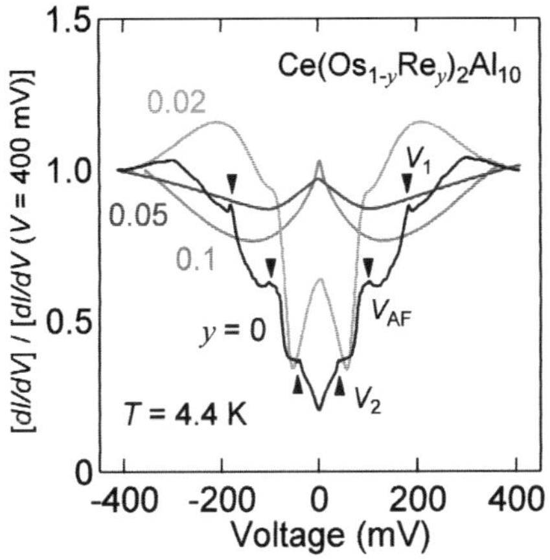

Figure 1. Tunneling conductance spectra dI/dV vs the bias voltage V normalized by the values at V = –400 mV for Ce(Os$_{1-y}$Re$_y$)$_2$Al$_{10}$ (y = 0, 0.02 and 0.05) at 4.4 K. The arrows on the spectrum for y = 0 indicate the structures V_1, V_{AF} and V_2, respectively.

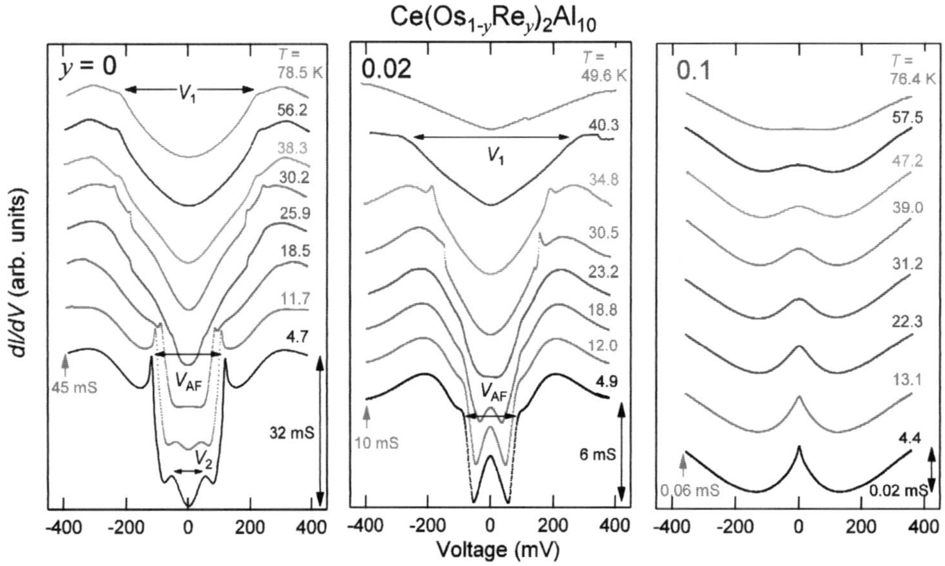

Figure 2. Temperature dependences of the tunneling conductance dI/dV vs V measured for break junctions of Ce(Os$_{1-y}$Re$_y$)$_2$Al$_{10}$. The spectra are shifted vertically for clarity. The widths between shoulders in the spectra are denoted as V_1, V_{AF} and V_2, respectively (see text).

into a peak on cooling. The spectra for $y = 0.1$ is characterized by one cusp at $V = 0$. The development of a cusp at $V = 0$ is a characteristic of a metallic heavy fermion state [17]. It indicates that the $4f$ state in $CeOs_2Al_{10}$ becomes more itinerant by $5d$ hole doping. With increasing y, in fact, the behavior of $\rho(T)$ changes from the Kondo semiconducting one to that for the valence fluctuating Ce compounds [10].

The temperature dependences of the gap widths $V^{P\text{-}P}$ (= V_1, V_{AF} and V_2) are plotted in Fig. 3(a). For $y = 0$, the black and blue data are obtained with a single crystal and a polycrystal, respectively, whose temperature variations are essentially same. The gap width V_1 for $y = 0$ decreases from 430 to 400 mV at around 36 K, which is still above $T_N = 28.5$ K. We recall that the CDW-like charge gap develops in $CeOs_2Al_{10}$ on cooling below 36 K [18]. Similarly for $y = 0.02$, V_1 decreases on cooling from 50 to 27 K (> $T_N = 23$ K). It is reminiscent of the CDW gap in $CeTe_2$ which decreases on cooling below 6.1 K, where a short-range ferromagnetic order develops [6]. By analogy, the reduction in V_1 occurring above T_N in Ce(Os$_{1-y}$Re$_y$)$_2$Al$_{10}$ ($y = 0$, 0.02) may be attributed to the onset of a short-range magnetic correlation. The gap V_{AF} for $y = 0$ increases gradually below T_N, whereas that for $y = 0.02$ seems to appear suddenly at 27 K. The gap V_2 for $y = 0$ appears below 16 K, but corresponding anomaly is absent for $y = 0.02$.

Figure 3(b) displays the temperature dependences of normalized zero bias conductance, $\{\mathrm{NZBC}\}^{1/2} = \{[dI/dV\,(V = 0)] / [dI/dV\,(V = -400\ \mathrm{mV})]\}^{1/2}$. We note that the NZBC is proportional to the square of the DOS at the Fermi level, $N(E_F)$. The NZBC for $y = 0$ gradually decreases and bends at around 36 K, which occurs simultaneously with the sudden decrease in V_1. It suggests that a part of Fermi surface is lost by the development of the short-range correlation. Besides, the significant change in the Fermi surface on cooling from 36 K manifests in the sharp decrease in the thermopower, which is very sensitive to the energy derivative of quasiparticle density of states at E_F, sharply decreases only along the b-axis [19]. For $y = 0.02$, the decrease in NZBC turns to an increase below 27 K due to the development of in-gap states at E_F. For $y = 0.1$ with no AFM transition, the NZBC increases monotonically on cooling.

Figure 4(a) displays the y dependences of gap widths V_1, V_{AF} and V_2 as well as $\{\mathrm{NZBC}\}^{1/2}$. The increase in $\{\mathrm{NZBC}\}^{1/2}$ with y is consistent with that in the γ value as shown in Fig. 4(b). The inverse correlation between T_N and γ in Fig. 4(b) indicates that the increase of $N(E_F)$ suppresses the AFM order.

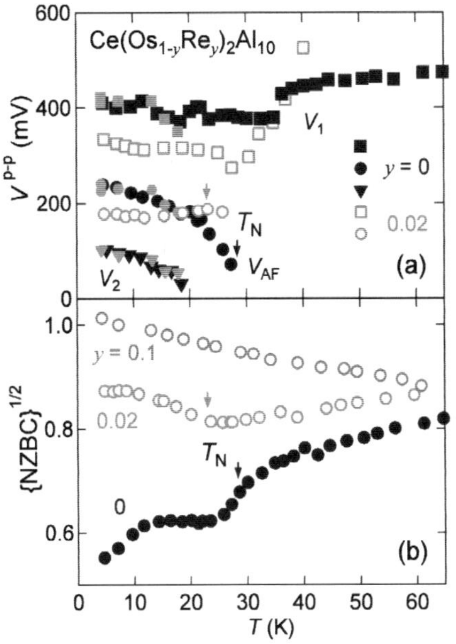

Figure 3. (a) Temperature dependences of gap widths V_1, V_{AF}, V_2 and (b) the square root of normalized zero bias conductance $\{NZBC\}^{1/2} = \{[dI/dV\ (V = 0)]\ /\ [dI/dV\ (V = -400\ mV)]\}^{1/2}$ for $Ce(Os_{1-y}Re_y)_2Al_{10}$ (y = 0, 0.02 and 0.1). For y = 0, the data in black and blue are obtained with a single crystal and a polycrystal, respectively. T_N denotes the AFM ordering temperature determined by the specific heat measurements.

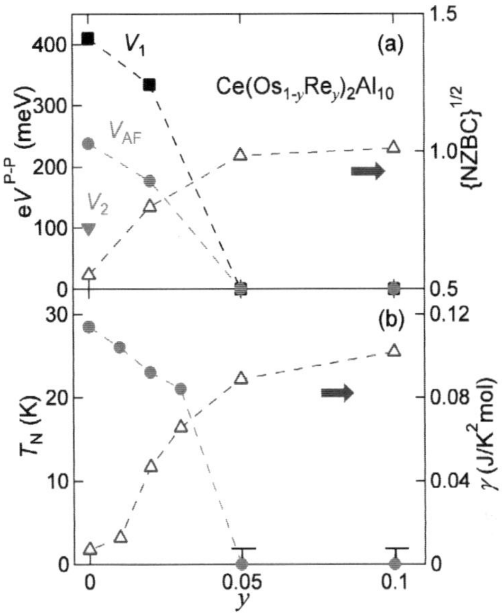

Figure 4. (a) Variations of gap widths V_1, V_{AF}, V_2 and the square root of normalized zero bias conductance $\{NZBC\}^{1/2}$ in the tunneling spectra, and (b) the Néel temperature T_N and the γ value in the specific heat as a function of y in $Ce(Os_{1-y}Re_y)_2Al_{10}$ [20].

Moreover, the T_N and the gap widths V_1 and V_{AF} are largely suppressed at a small hole-doping level y = 0.02, and all disappear at y = 0.05. Therefore, the AFM transition is strongly correlated with the hybridization gap V_1. This relation supports our previous argument that the presence of the hybridization gap V_1 is necessary for the unusual AFM order [10]. Here, let us compare the c-f hybridization gap V_1 with the spin gap which was observed by inelastic neutron scattering [12]. The spin gap exists at 11 meV in the undoped sample with y = 0. For y = 0.03, T_N is decreased to 21 K, but the spin gap still exists at the same energy 11meV while the excitation intensity is much reduced. On the other hand, the spin gap disappears in the 5d electrons doped systems $Ce(Os_{1.92}Ir_{0.08})_2Al_{10}$ with T_N = 21 K. It is highly required to examine by the BJTS wherever or not the hybridization gap V_1 is correlated with the spin gap.

4. Summary

We have performed the BJTS measurements for the $Ce(Os_{1-y}Re_y)_2Al_{10}$ ($y \leq 0.1$) to study how the two hybridization gaps V_1, V_2 and the AFM gap V_{AF} change by doping of the 5d holes in $CeOs_2Al_{10}$ with T_N = 28.5 K. For the undoped sample, both the hybridization gap V_1 and the magnitude of $N(E_F)$ decrease on cooling below 36 K. These reductions occurring above T_N are attributed to a short-range correlation. For y = 0.02, in-gap states develop at E_F with concomitant disappearance of V_2. With increasing y to 0.1, the in-gap states at E_F develops further. We found a strong correlation between the variations of T_N, V_1 and V_{AF}. Thus, we conclude that the presence of the hybridization gap V_1 above T_N is necessary for the unusual AFM order.

Acknowledgements
We acknowledge valuable discussions with K Umeo, T Onimaru, D T Adroja, S Kimura, T Yokoya, T Yamada and Y Ono. This work was partly supported by KAKENHI Grant Numbers Nos. 26400363, 16H01076 and 15J01007.

References

[1] Nishioka T, Kawamura Y, Takesaka T, Kobayashi R, Kato H, Matsumura M, Kodama K, Matsubayashi K and Uwatoko Y 2009 *J. Phys. Soc. Jpn.* **78** 123705
[2] Muro Y, Kajino J, Umeo K, Nishimoto K, Tamura R and Takabatake T 2010 *Phys. Rev. B* **81** 214401
[3] Muro Y, Kajino J, Onimaru T and Takabatake T 2011 *J. Phys. Soc. Jpn.* **80** SA021
[4] Strigari F *et al.* 2013 *Phys. Rev. B* **87** 125119
[5] Davydov D N, Kambe S, Jansen A G M, Wyder P, Wilson N, Lapertot G and Flouquet J 1997 *Phys. Rev. B* **55** R7299
[6] Fukuda H, Fujii H, Kamura H, Hasegawa Y, Ekino T, Kikugawa N, Suzuki T and Fujita T 2001 *Phys. Rev. B* **63** 054405
[7] Jung M H, Ekino T, Kwon Y S and Takabatake T 2000 *Phys. Rev. B* **63** 035101
[8] Kawabata J, Ekino T, Yamada Y, Sakai Y, Sugimoto A, Muro Y and Takabatake T 2015 *Phys. Rev. B* **92** 201113(R)
[9] Yamada T and Ono Y 2012 *Phys. Rev. B* **85** 165114
[10] Kawabata J, Takabatake T, Umeo K and Muro Y 2014 *Phys. Rev. B* **89** 094404
[11] Khalyavin D D, Adroja D T, Bhattacharyya A, Hillier A D, Manuel P, Strydom A M, Kawabata J and Takabatake T 2014 *Phys. Rev. B* **89** 064422
[12] Bhattacharyya A, Adroja D T, Strydom A M, Kawabata J, Takabatake T, Hillier A D, Sakai V G, Taylor J W and Smith R I 2014 *Phys. Rev. B* **90** 174422
[13] Ekino T, Takabatake T, Tanaka H and Fujii H 1995 *Phys. Rev. Let.* **75** 4262
[14] Wolf E L 2012 *Principles of Electron Tunneling Spectroscopy*, 2nd ed. (New York: Oxford University Press).
[15] Flachbart K, Gloos K, Konovalova E, Paderno Y, Reiffers M, Samuely P and Svec P 2001 *Phys. Rev. B* **64** 085104
[16] Riseborough P S and Magalhaes S G 2016 *J. Magn. Magn. Mat.* **400** 3
[17] Motoyama G, Ogawa S, Matsubayashi K, Fujiwara K, Miyoshi K, Nishigori S, Mutou T, Yamaguchi A, Sumiyama A and Uwatoko Y 2015 *Phys. Proc.* **75** 296
[18] Kimura S, Iizuka T, Miyazaki H, Irizawa A, Muro Y and Takabatake T 2011 *Phys. Rev. Lett.* **106** 056404
[19] Yamada Y, Kawabata J, Onimaru T and Takabatake T 2015 *J. Phys. Soc. Jpn.* **84** 084705

IOP Conf. Series: Journal of Physics: Conf. Series **807** (2017) 012009 doi:10.1088/1742-6596/807/1/012009

Zn-site Substitution Effect in $YbCo_2Zn_{20}$

Riki Kobayashi[1], Haruki Takamura[2], Yasuyuki Higa[2], Yoichi Ikeda[3,4], Kazuyuki Matsubayashi[5,6], Yoshiya Uwatoko[6], Hideki Yoshizawa[3], and Naofumi Aso[1]

[1]Faculty of Science, University of the Ryukyus, Nishihara, Okinawa 903-0213, Japan
[2]Graduate School of Engineering and Science, University of the Ryukyus, Nishihara, Okinawa 903-0213, Japan
[3]Neutron Science Laboratory, Institute for Solid State Physics, University of Tokyo, Tokai, Ibaraki 319-1106, Japan
[4]Institute for Materials Research, Tohoku University, Sendai 980-8577, Japan
[5]Graduate School of Informatics and Engineering, The University of Electro-Communications, Tokyo 182-8585, Japan
[6]Institute for Solid State Physics, University of Tokyo, Kashiwa, Chiba 277-8581, Japan

E-mail: kobariki@sci.u-ryukyu.ac.jp

Abstract.
We have investigated the substitution effect of $YbCo_2(Zn_{1-x}T_x)_{20}$ (T = Cu, Ga, and Cd) systems by using the experiments of X-ray powder diffraction (XRPD), specific heat, magnetic susceptibility, magnetization, and electrical resistivity in order to find out a material that approaches a quantum critical point by chemical pressure. The XRPD and electrical resistivity measurements clarify that the Cu-substitution makes the lattice constants shrink and keeps the magnetic electrical resistivity high, while the Ga- and the Cd-substitution show opposite relation of the Cu-substitution. However, we could not detect clear substitution effect in the specific heat, magnetic susceptibility, and magnetization measurements of Cu-substitution system within our experiments. It is necessary that to study the Cu-substitution samples that have higher x value at lower temperature.

Introduction

While the characteristic of heavy fermion ytterbium (Yb) compounds are similar to that of cerium (Ce) compounds due to the electron-hole analogy, there is a clear difference between two groups regarding the quantum critical behaviour, as reported in the study of β-$YbAlB_4$ and $YbRh_2Si_2$[1, 2]. This difference invokes fundamental interest to study the heavy fermion Yb compounds which are located near the quantum critical point (QCP).

A recently discovered $YbCo_2Zn_{20}$ is one of the most suitable material to investigate the QCP of Yb compounds. The Cubic $CeCr_2Al_{20}$-type ($Fd\bar{3}m$ space group) $YbCo_2Zn_{20}$, which has no magnetic order at ambient pressure, has extremely large value of electronic specific heat coefficient ($\gamma \simeq 8$ J/molK^2), and shows the field-induced order when the magnetic field only applies parallel to [111] direction[3, 4]. Especially, the existence of such a large γ value implies that the compound is located near the QCP, which means that we can easily access to the QCP of that. In fact, it was reported that an antiferromagnetic order was induced by the pressure of 1.5 GPa which is the lowest value in Yb compounds, except for β-$YbAlB_4$[5, 1]. Therefore, there is the possibility that chemical pressure also tunes the ground state to the magnetic one

Table 1. The actual value of x in three substitution systems estimated by ICP-AES.

nominal (%)	ICP-AES results (%)								
	Cu			Ga			Cd		
	C_{mag}	M, χ_{dc}	ρ_{mag}	C_{mag}	M, χ_{dc}	ρ_{mag}	C_{mag}	M, χ_{dc}	ρ_{mag}
2.5	1.9	1.9	1.9	3.3	3.2	3.2	1.4	1.3	1.4
5.0	2.5		2.5	6.0	6.0	6.0	2.5	2.4	2.2

and/or the QCP. While a large number of study of $YbCo_2Zn_{20}$ has been performed up to now, there is no information about the chemical pressure by substitution. Very recently, we have succeeded in growing single crystals where Co- and Zn-site of $YbCo_2Zn_{20}$ are substituted by several materials.

In this paper, we have focused on the substitution effect of Zn-site in $YbCo_2Zn_{20}$ in order to explore the material that approach QCP by chemical pressure. We have investigated $YbCo_2(Zn_{1-x}T_x)_{20}$ (T = Cu, Ga, and Gd) systems by using the measurements of the X-ray powder diffraction (XRPD), specific heat, magnetic susceptibility, magnetization, and electrical resistivity.

Experimental

Single crystals of $YbCo_2(Zn_{1-x}T_x)_{20}$ (T = Cu, Ga, and Gd) were grown by an Zn self-flux method. Materials where stoichiometric ratio of Yb : Co : Zn : T (T = Cu, Ga, and Cd) are 1 : 2 : $30(1-x)$: x ($x = 0.025$ and 0.05) were placed in an alumina crucible, sealed in a quartz ampule, heated up to 950 °C, and then cooled down to 750 °C at a rate of 1 °C/h. The Zn flux was removed by centrifugation. Almost obtained single crystals have a shining crystal-growth-plane perpendicular to the [111] direction.

In order to check the actual value of x, inductively coupled plasma atomic emission spectroscopy (ICP-AES) was performed. We picked up tiny pieces of the single crystals from the surrounding samples that was used for the experiments explained in the next paragraph. The actual values of x in three substitution systems which were estimated by ICP-AES are summarized in Table 1. The results of ICP-AES indicate that x are smaller than the nominal one, except for the Ga-substitution system.

All obtained samples were measured by using XRPD, specific heat, magnetic susceptibility, magnetization, and electrical resistivity. The XRPD measurements were performed at room temperature using single crystals powdered. The specific heat, magnetic susceptibility, magnetization, and electrical resistivity measurements were performed in the temperature range between 2.0 and 300 K by means of PPMS and SQUID magnetometer MPMS commercial equipment (Quantum Design).

Results and Discussion

Figure 1 (Left) indicates x dependence of the lattice constants for $YbCo_2(Zn_{1-x}T_x)_{20}$ (T = Cu, Ga, and Cd) systems. In Cu-substitution system, the lattice constant decreases with x, while it increases with x in Ga- and Cd-substitution systems. The results indicate the existence of positive chemical pressure in Cu-substitution system.

Figure 1 (Right) represents temperature dependence of the magnetic component of specific heat (C_{mag}) in the form of C_{mag}/T for $YbCo_2(Zn_{1-x}T_x)_{20}$ (T = Cu, Ga, and Cd) systems. The C_{mag} is obtained by subtracting the specific heat of nonmagnetic $LuCo_2Zn_{20}$ from that

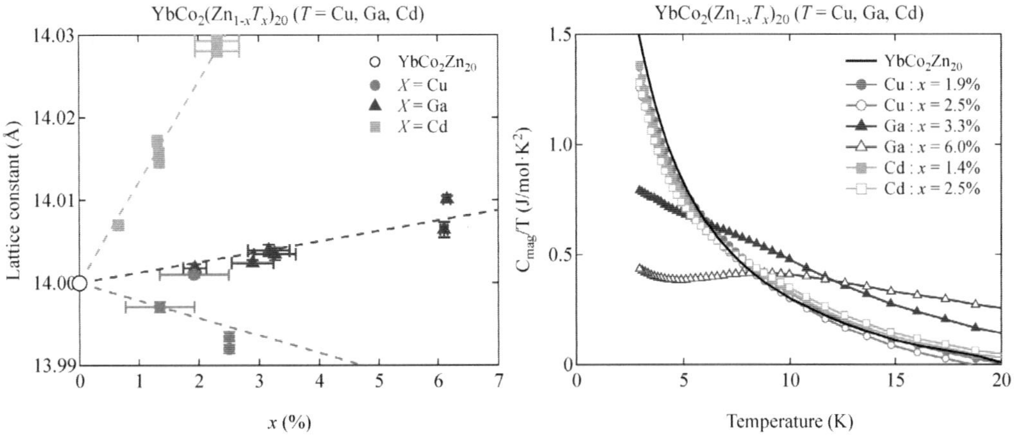

Figure 1. (Left) x dependence of the lattice constants for $YbCo_2(Zn_{1-x}T_x)_{20}$ (T = Cu, Ga, and Cd) systems. The open circle represents the lattice constant of the non-doped $YbCo_2Zn_{20}$. The filled circles, triangles, and squares indicate that of Cu-, Ga-, and Cd-substitution systems, respectively. (Right) Temperature dependence of the magnetic specific heat ($C_{\rm mag}$) in the form of $C_{\rm mag}/T$ for $YbCo_2(Zn_{1-x}T_x)_{20}$ (T = Cu, Ga, and Cd) systems. The solid line represents the $C_{\rm mag}$ of the non-doped $YbCo_2Zn_{20}$. The solid and open circles, triangles, and squares indicate that of $x = 1.9\%$ (Cu), 2.5% (Cu), 3.3% (Ga), 6.0% (Ga), 1.4% (Cd), 2.5% (Cd), respectively.

data. In this temperature region, the $C_{\rm mag}$ does not change by Cu- and Cd-substitution. The Ga-substitution, which has relatively high x value, suppresses the enhancement of $C_{\rm mag}$ at low temperature.

Figure 2 (Left) shows temperature dependence of the magnetic susceptibility ($\chi_{\rm dc}$) and the inverse magnetic susceptibility ($1/\chi_{\rm dc}$) for $YbCo_2(Zn_{1-x}T_x)_{20}$ (T = Cu, Ga, and Cd) systems. The applied magnetic field is 1.0 T parallel to [111]. All the data can be fitted well by Curie-Weiss law above 100 K, yielding the effective magnetic moment ($\mu_{\rm eff}$) and paramagnetic Curie-Weiss temperature ($\theta_{\rm p}$), which are summarized in Table 2. There is no remarkable change in $\mu_{\rm eff}$ and $\theta_{\rm p}$ for all the substitution systems. The effect of these substitution for $\chi_{\rm dc}$ is weak.

Figure 2 (Right) indicates isothermal magnetization curve for $YbCo_2(Zn_{1-x}T_x)_{20}$ (T = Cu, Ga, and Cd) systems at 2.0 K. The direction of applied magnetic field is parallel to [111]. The magnetization almost does not change by Cu- and Gd-substitution. The Ga-substitution in $x = 3.2$ and 6.0% suppress the magnetization. The magnetization shape of Cd-substitution in x =2.4 seems to slightly different to that of other substitutions. This difference may come from misalignment of the single crystal sample.

Figure 3 shows temperature dependence of the magnetic electrical resistivity ($\rho_{\rm mag}$) for $YbCo_2(Zn_{1-x}T_x)_{20}$ (T = Cu, Ga, and Cd) systems. The electrical current was applied parallel to [111]. The $\rho_{\rm mag}$ increases with decrease of temperature and then it makes the broad peak at $T_{\rho,\rm max}$ in the non-doped $YbCo_2Zn_{20}$. In three substitution systems however, the $T_{\rho,\rm max}$ disappears in this temperature region. The increase of x does not change the $\rho_{\rm mag}$ in Cu-substitution system, while it suppress $\rho_{\rm mag}$ in Ga- and Cd-substitution systems.

The Cu-substitution system which shows the positive chemical pressure and keeps $\rho_{\rm mag}$ high is the only candidate to approach QCP in three substitution systems. However, to measure samples having higher x value and/or to measure at lower temperature is required in order to

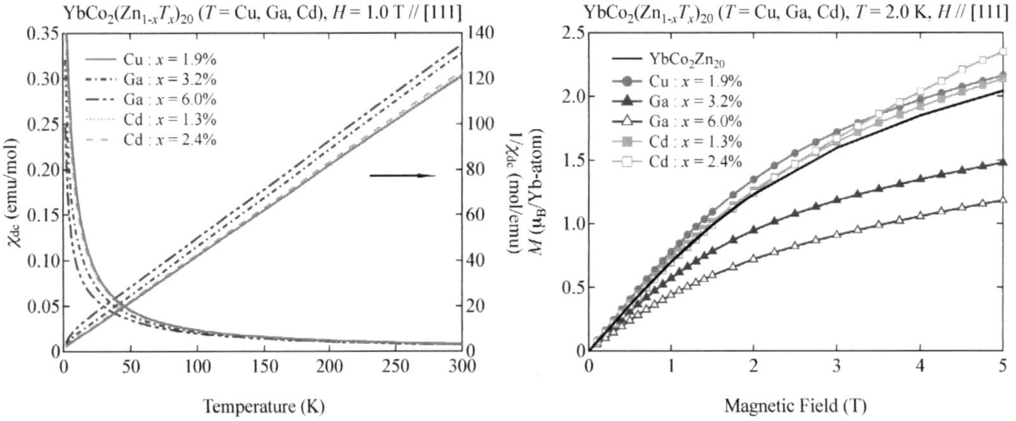

Figure 2. (Left) Temperature dependence of the magnetic susceptibility (χ_{dc}) and the inverse magnetic susceptibility ($1/\chi_{dc}$) for $YbCo_2(Zn_{1-x}T_x)_{20}$ (T = Cu, Ga, and Cd) systems. The solid, one-dot chain, and two-dot chain lines indicate that of x = 1.9% (Cu), 3.2% (Ga), 6.0% (Ga), 1.3% (Cd), and 2.4% (Cd), respectively. (Right) Isothermal magnetization curve for $YbCo_2(Zn_{1-x}T_x)_{20}$ (T = Cu, Ga, and Cd) systems at 2.0 K. The solid line represents that of the non-doped $YbCo_2Zn_{20}$. The filled circles indicates that of x = 1.9% (Cu), and the filled and open triangles and squares show that of x = 3.2% (Ga), 6.0% (Ga), 1.3% (Cd), and 2.4% (Cd), respectively.

Table 2. The effective moment (μ_{eff}) and paramagnetic Curie-Weiss temperature (θ_p) for three substitution systems. [1]These values refer to Takeuchi's results[6].

	$YbCo_2Zn_{20}$	Cu: 1.9%	Ga: 3.2%	Ga: 6.0%	Cd: 1.3%	Cd: 2.4%
μ_{eff} (μ_B)	4.27[1]	4.48(1)	4.35(1)	4.31(1)	4.49(1)	4.45(1)
θ_p (K)	-1.43[1]	-4.4(1)	-9.0(1)	-13(1)	-6.5(1)	-5.2(1)

know actual effect of Cu-substitution. There is no magnetic order at $x = 0 \sim 2.5\%$ in Cu-substitution system. The existence of chemical-pressure-induced magnetic order is one of the most strong evidence that the system crosses QCP. Therefore, to find that by measurement of higher x value samples is important. In addition, we could not detect clear substitution effect in C_{mag} and χ_{dc} measurements of Cu- and Cd-substitution systems within our experiments. These measurements at $T = 0.2 \sim 3$ K may clarify more detailed substitution effect.

Summary

We have investigated the substitution effect of $YbCo_2(Zn_{1-x}T_x)_{20}$ (T = Cu, Ga, and Cd) systems by the measurements of XRPD, specific heat, magnetic susceptibility, magnetization, and electrical resistivity. The Cu-substitution makes the lattice constants shrink and keeps the ρ_{mag} high, while Ga- and Cd-substitutions show opposite relation of Cu-substitution. We could not detect clear substitution effect in C_{mag}, χ_{dc}, and magnetization measurement of Cu-substitution system within our experiments, which imply that experiments at $T = 0.2 \sim 3$ K are

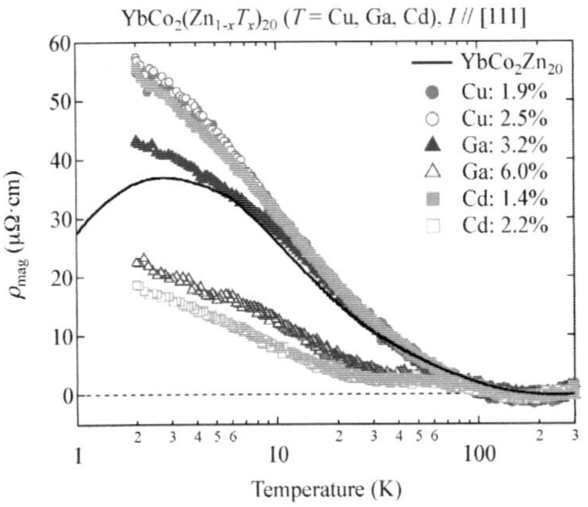

Figure 3. Temperature dependence of the magnetic electrical resistivity (ρ_{mag}) for $YbCo_2(Zn_{1-x}T_x)_{20}$ (T = Cu, Ga, and Cd) systems. The electrical current was applied parallel to [111]. The solid line shows that of the non-doped $YbCo_2Zn_{20}$. The filled and open circle, triangle, and square represent that of $x = 1.9\%$ (Cu), 2.5% (Cu), 3.2% (Ga), 6.0% (Ga), 1.4% (Cd), and 2.2% (Cd), respectively.

necessary to understand more detailed substitution effect. To find the chemical-pressure-induced magnetic order in the samples of higher x value is important to approach QCP.

Acknowledgments
We would like to thank Mr. H. Iwashita, T. Ohoka, and K. Kakazu for experimental support in early stage of the work. This work was supported by a Grant-in-Aid for Scientific Research C (Grant Number JP26400365). This work was carried out under the Visiting Researcher's Program of the Institute for Solid State Physics, the University of Tokyo.

References
[1] Matsumoto Y, Nakatsuji S, Kuga K, Karaki Y, Horie N, Shimura Y, Sakakibara T, Nevidomskyy A H and Coleman P 2011 *Science* **331** 316
[2] Custers J, Gegenwart P, Wilhelm H, Neumaier K, Tokiwa Y, Trovarelli O, Geibel C, Steglich F, Pépin C and Coleman P 2003 *Nature* **424** 524
[3] Torikachvili M S, Jia S, Mun E D, Hannahs S T, Black R C, Neils W K, Martien D, Bud'ko S L and Canfield P C 2007 *PNAS* **104** 9960-3
[4] Shimura Y, Sakakibara T, Yoshiuchi S, Honda F, Settai R and Ōnuki Y 2011 *J. Phys. Sci. Jpn.* **80** 073707
[5] Saiga Y, Matsubayashi K, Fujiwara T, Kosaka, M, Katano S, Hedo M, Matsumoto T and Uwatoko Y 2008 *J. Phys. Sci. Jpn.* **77** 053710
[6] Takeuchi T *et al.* 2011 J. Phys. Sci. Jpn. **80** 114703

Anomalous Antiferromagnetic Phase Diagram in $HoRu_2Al_{10}$

S Kamikawa[1*], I Ishii[1†], H Goto[1], K Takezawa[1], F Nakagawa[1], H Tanida[1], M Sera[1] and T Suzuki[1,2,3‡]

[1]Department of Quantum Matter, ADSM, Hiroshima University, Higashihiroshima, Hiroshima 739-8530, Japan

[2]Institute for Advanced Materials Research, Hiroshima University, Higashihiroshima, Hiroshima 739-8530, Japan

[3]Cryogenics and Instrumental Analysis Division, N-BARD, Hiroshima University, Higashihiroshima, Hiroshima 739-8526, Japan.

E-mail: *d140235@hiroshima-u.ac.jp, †ish@hiroshima-u.ac.jp, ‡tsuzuki@hiroshima-u.ac.jp

Abstract. To investigate an antiferromagnetic (AFM) phase transition of $HoRu_2Al_{10}$ with T_N= 5.2 K, we performed specific heat measurements on a single-crystalline sample under magnetic fields H. In H along the b- and c-axes, a sharp peak of the specific heat at T_N, indicating the phase transition, shifts to lower temperature with increasing H. From these measurements, we clarified a magnetic phase diagram for H // b and c above 2 K. In H // b, T_N decreases monotonically with increasing H, and then the phase boundary closes around 1.0 T, which is a usual behavior for an antiferromagnet. In H // c, although T_N shows a monotonic decrease with increasing H at low H, the phase boundary exhibits an inflection point around 4.0 T, suggesting that additional interaction are enhanced or induced by H // c in $HoRu_2Al_{10}$. From the stand point of fluctuation of quadrupole moments in isomorphic compound $HoFe_2Al_{10}$, we propose that the inflection point is originated from a quadrupolar instability.

1. Introduction

Ternary aluminides $LnTr_2Al_{10}$ (Ln = rare earth, Tr = Ru, Os, and Fe) with the orthorhombic $YbFe_2Al_{10}$-type structure (space group: $Cmcm$) have attracted many interests of magnetism. At the early stage, $CeRu_2Al_{10}$ has been studied intensively. It is the first example which shows an AFM ordering in the Kondo insulating state [1–4]. The AFM transition temperature of $CeRu_2Al_{10}$ is one hundred times higher than the expected temperature from the de Gennes law in $LnTr_2Al_{10}$ despite the strongly reduced Ce magnetic moment [3,4]. In addition, the ordering direction of the magnetic moment in the AFM state is along the magnetically hard axis [3,4]. The origin of the unusual AFM transition temperature and direction of the ordered moments has not been clarified. Recently, other $LnTr_2Al_{10}$ compounds have been also studied, such as an AFM transition with a valence instability [5], an incommensurate AFM ordering [6], successive magnetic phase transitions, and so on [7–11].

We focus on an AFM transition at T_N in $HoRu_2Al_{10}$ which was reported by Mizushima *et al.* [11]. A sharp peak of the specific heat C_p was observed at T_N. The temperature T dependencies of magnetic susceptibility along the all axes decrease below T_N, suggesting an

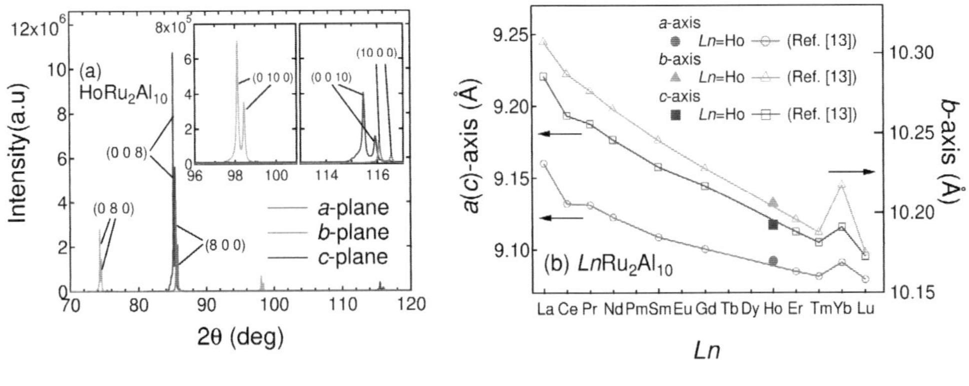

Figure 1. (a)XRD patterns of single-crystalline $HoRu_2Al_{10}$ for the all planes between $2\theta = 70$ and 120 degrees. The insets display the same data in an expanded scale between $2\theta = 96$ and 101 degrees, and 113 and 117 degrees, respectively. (b)The *Ln* dependences of lattice constants for all axes in $LnRu_2Al_{10}$ at room-temperature [13].

AFM transition at T_N in $HoRu_2Al_{10}$. The H dependence of the magnetization M along the c-axis at 2 K shows a metamagnetic transition at 0.8 T. In contrast, M along the a- and b-axes increase almost monotonically with increasing H up to 7 T. The results suggest that the magnetic moments of Ho are oriented toward the c-axis below T_N. As the Ho^{3+} ion in $HoRu_2Al_{10}$ is under the orthorhombic symmetry, degenerated 4*f*-electronic state of the free Ho^{3+} ion (total angular momentum $J = 8$) splits into seventeen singlets by the crystal electric field (CEF). In this situation, it is unlikely that there is a spin-degenerated state. In the present work, to investigate the AFM transition in $HoRu_2Al_{10}$, we performed specific heat measurements in H // b and c.

2. Experimental

By using Al self-flux method, we grew a single crystal of $HoRu_2Al_{10}$ and polycrystalline $LuRu_2Al_{10}$ which is a non-magnetic reference material. Specific heat under H is measured with a relaxation method by using the commercial PPMS system (Quantum Design). H is applied along the *b*- and *c*-axes between 2 and 10 K. Here, we note that the definition of the *b*- and *c*-axes are reverse between our and Mizushima's samples. To confirm the validity of our definition, we performed X-ray diffraction (XRD) experiments. The XRD patterns on our single-crystalline sample for the all planes between $2\theta = 70$ and 120 degrees are shown in Figure 1(a), where 2θ is the scattering angle. We observed Bragg peaks for the *a*-plane at 85.48, 85.74, 116.07, and 116.53 degrees, for the *b*-plane at 74.31, 74.52, 98.11, and 98.44 degrees, and for the *c*-plane at 85.12, 85.39, 115.46, and 115.91 degrees, respectively. In addition, we determined the lattice parameters of $HoRu_2Al_{10}$ ($a = 9.092$ Å, $b = 10.206$ Å, and $c = 9.117$ Å) by the Rietveld refinement of powder XRD patterns using the program RIETAN-FP [12]. The lattice parameters of $HoRu_2Al_{10}$ and $LnRu_2Al_{10}$ reported by Sera *et al.* [13] are shown by closed and open symbols in Figure 1(b), respectively. The lattice parameters of $HoRu_2Al_{10}$ are very close to the expected values from the lanthanide contraction. From the lattice parameters of $HoRu_2Al_{10}$, we estimated the Bragg peak positions in 2θ for all planes between 70 and 120 degrees, and found the peak position for (8 0 0) at 85.34 and 85.60 degrees, for (10 0 0) at 115.82 and 116.28 degrees, for (0 8 0) at 74.28 and 74.50 degrees, for (0 10 0) at 98.01 and 98.33 degrees, for (0 0 8) at 85.06 and 85.32 degrees, and for (0 0 10) at 115.32 and 115.78 degrees. Table 1 shows the relation between

Table 1. The relation between the plane indices and the Bragg peak positions in 2θ. We also show the calculated peak positions for each plane index obtained from the Rietveld refinement of powder XRD patterns using the program RIETAN-FP with the lattice parameters (a =9.092 Å, b =10.206 Å, and c = 9.117 Å) in $HoRu_2Al_{10}$ [12].

Plan index	Bragg peak position (deg)	
	Raw data	Calculation
(8 0 0)	85.48, 85.74	85.34, 85.60
(10 0 0)	116.07, 116.53	115.82, 116.28
(0 8 0)	74.31, 74.52	74.28, 74.50
(0 10 0)	98.11, 98.44	98.01, 98.33
(0 0 8)	85.12, 85.39	85.06, 85.32
(0 0 10)	115.46, 115.91	115.32, 115.78

the plane indices and the peak positions in 2θ. The observed peaks are corresponding to each plane index, respectively. This result indicates the validity of our definition for the crystal axis in $HoRu_2Al_{10}$. Therefore, we use our definition for the crystal axis.

3. Results and Discussion

Figure 2(a) shows the T dependencies of C_p of $HoRu_2Al_{10}$ and polycrystalline $LuRu_2Al_{10}$. The absolute value of C_p in $LuRu_2Al_{10}$ is almost the same as that in $HoRu_2Al_{10}$ above 60 K. We assumed $LuRu_2Al_{10}$ as a non-magnetic reference compound to $HoRu_2Al_{10}$. We subtracted C_p of $LuRu_2Al_{10}$ from that of $HoRu_2Al_{10}$ to estimate the magnetic specific heat C_m of $HoRu_2Al_{10}$. The T dependence of C_m and magnetic entropy S_m are shown in Figure 2(b). C_m exhibits a sharp peak at T_N, and a Schottky peak around 9 K. S_m reaches $1.26R\ln 2$(=7.26 J/K mol) at T_N, where R is the gas constant. S_m saturates around 90 K, suggesting that the CEF splitting in $HoRu_2Al_{10}$ is relatively small. The entropy release at T_N indicates that the AFM transition at T_N originates from the pseudo-degenerated state of singlets, though all the CEF states are

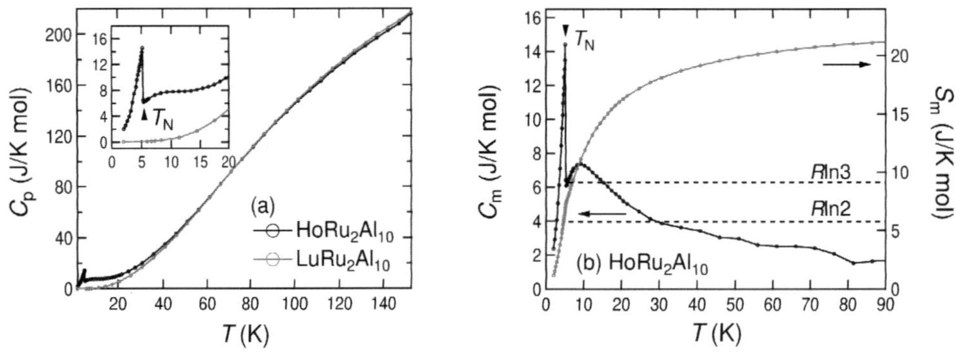

Figure 2. T dependencies of (a) the specific heat C_p of $HoRu_2Al_{10}$ and $LuRu_2Al_{10}$, and (b) the magnetic specific heat C_m and the magnetic entropy S_m of $HoRu_2Al_{10}$. The inset displays the same data in an expanded scale of C_p in $HoRu_2Al_{10}$ and $LuRu_2Al_{10}$ below 20 K.

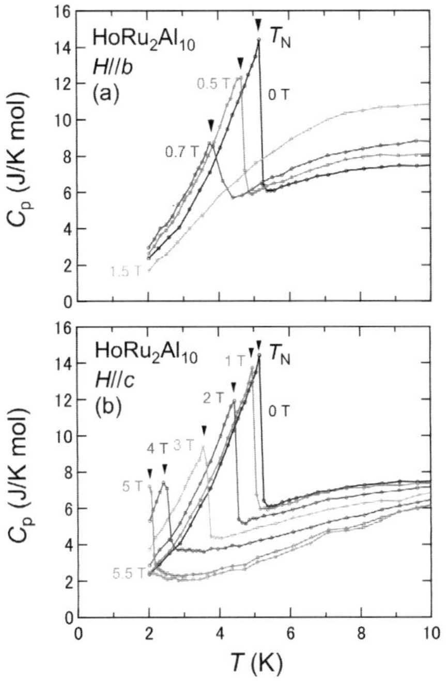

Figure 3. T dependencies of the specific heat C_p for (a) H // b and (b) H // c in $HoRu_2Al_{10}$.

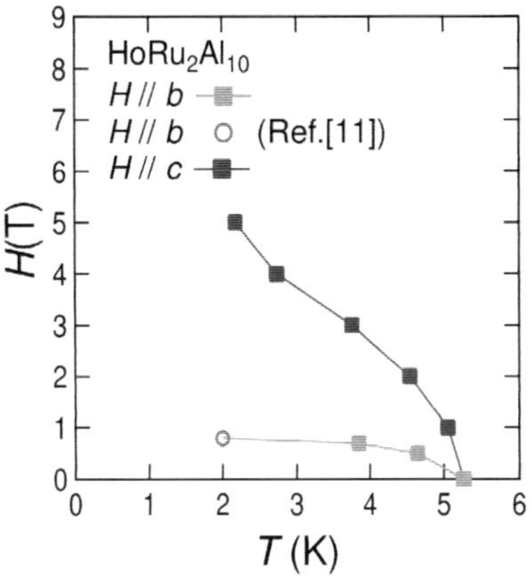

Figure 4. H-T diagram for H // b and c in $HoRu_2Al_{10}$. Closed and open symbols denote our and Mizushima's data [11], respectively.

singlets as mentioned above. Here, the pseudo-degenerated state means electronic states with energy separations which can be regarded as a degenerated state from the following viewpoint. If the system is in a condition at a certain temperature which has a higher energy scale than the energy separation, the energy separations may be ignored in the energy scale. In the case of $HoRu_2Al_{10}$, the entropy release at T_N is about Rln2, indicating that the CEF ground and first excited states seem to degenerate above T_N. Thus, we mention that the pseudo-degenerated state at T_N is a pseudo-doublet. The energy splitting between the CEF ground and first excited states is most likely less than T_N.

Figure 3(a) and (b) show the T dependencies of C_p in H // b and c, respectively. In H // b and c, the sharp peak of C_p shifts to lower T with increasing H, and then the peak is not observed above 0.7 and 5.5 T, respectively. The Schottky specific heat for H // b increases with increasing H. In contrast, the Schottky specific heat for H // c decreases. These behaviors are expected to originate from the change of the Schottky peak which depends on the energy change of the CEF states by the Zeeman effect. The increase of the Schottky specific heat for H // b suggests that energy separations between the CEF states become close. On the other hand, the decrease of the Schottky specific heat for H // c suggests that the energy separations spread. The H-T diagram for H //b and c obtained from the specific heat measurements are shown by closed symbols in Figure 4. We also plot the metamagnetic transition at 0.8 T in H //b at 2 K reported by Mizushima *et al.* [11] with a open symbol. The phase boundary for H //b closes around 1 T, which is a usual behavior originating from a competition between the Zeeman effect and spin interaction. In contrast to H // b, the phase boundary for H //c exhibits an inflection point around 3.0 K. This result suggests that additional interactions are enhanced or induced by H // c. We have reported that isomorphic compound $HoFe_2Al_{10}$ shows elastic softening originating from inter-level transition of quadrupole in the low energy range [14, 15]. Considering the result of $HoFe_2Al_{10}$, we assume that the inflection point of the H-T diagram for H // c is caused by enhancement of an instability due to the fluctuating quadrupole moments in H // c. To investigate the quadrupolar instability around T_N in $HoRu_2Al_{10}$, ultrasonic measurements in H along the all axes are in progress.

4. Conclusion

We performed specific heat measurements under H // b and c. The sharp peak of C_m is observed at T_N. S_m exceeds $R\ln2$ at T_N, indicating that the AFM transition at T_N originates from the pseudo-degenerated state of singlets. The peak of C_m shifts to lower T with increasing H // b and c. From the results, we clarified the H-T diagram of $HoRu_2Al_{10}$ for H // b and c above 2 K. In H // b, the phase boundary closes around 1 T. Meanwhile, in H // c, we found the inflection point of the phase boundary around 4 T, which is an anomalous behavior for antiferromagnet. Considering the result that fluctuation of quadrupole moments exist in isomorphic compound $HoFe_2Al_{10}$, we propose that the inflection point of the H-T diagram for H // c in $HoRu_2Al_{10}$ arises from the quadrupolar instability.

Acknowledgments

This work was supported by JSPS KAKENHI Grant Numbers 262870830A, 26800189, 24540376, 26800188, and 2624706001. This work was also supported by JSPS Core-to-Core Program, A. Advanced Research Networks.

References

[1] Strydom A M 2009 *Physica* B **404** 2981
[2] Nishioka T, Kawamura Y, Takesaka T, Kobayashi R, Kato H, Matsumura M, Kodama K, Matsubayashi K, and Uwatoko Y 2009 *J. Phys. Soc. Jpn.* **78** 123705
[3] Robert J, Mignot J M, André G, Nishioka T, Kobayashi R, Matsumura M, Tanida H, Tanaka D, and Sera M 2010 *Phys. Rev.* B **82** 100404(R)
[4] Khalyavin D D, Hillier A D, Adroja D T, Strydom A M, Manuel P, Chapon L C, Peratheepan P, Knight K, Deen P, Ritter C, Muro Y, and Takabatake T 2010 *Phys. Rev.* B **82** 100405(R)
[5] Peratheepan P, and Strydom A M 2015 *J. Phys. : Condens. Matter* **27** 095604
[6] Takai S, Matsumura T, Tanida H, and Sera M 2015 *Phys. Rev.* B **92** 174427
[7] Sera M, Nohara H, Nakamura M, Tanida H, Nishioka T, and Matsumura M 2013 *Phys. Rev.* B **88** 100404(R)
[8] Muro Y, Kajino J, Onimaru T, and Takabatake T 2011 *J. Phys. Soc. Jpn.* **80** SA021
[9] Kawamura Y, Tanimoto S, Nishioka T, Tanida H, Sera M, Matsubayashi K, Uwatoko Y, Kondo A, Kindo K, and Sekine C 2012 *J. Phys. : Conf. Series* **391** 012029
[10] Khandelwal A, Sharma V K, Chandra L S S, Singh M N, Sinha A K, and Chattopadhyay M K 2013 *Phys. Scr.* **88** 035706
[11] Mizushima T, Watanabe Y, Ejiri J, Kuwai T, and Isikawa Y 2015 *J. Phys. : Conf. Series* **592** 012051
[12] Izumi F and Momma K 2007 *Solid State Phenomena* vol 130 (Pfaffikon: Trans Tech Publication) p 15-20
[13] Sera M, Tanaka D, Tanida H, Moriyoshi C, Ogawa M, Kuroiwa Y, Nishioka T, Matsumura M, Kim J, Tsuji N, and Takata M 2013 *J. Phys. Soc. Jpn.* **82** 024603
[14] Kamikawa S, Ishii I, Noguchi Y, Goto H, Fujita T K, Nakagawa F, Tanida H, Sera M, and Suzuki T 2015 *Phys. Procedia* **75** 187
[15] Kamikawa S, Ishii I, Noguchi Y, Goto H, Fujita T K, Nakagawa F, Tanida H, Sera M, and Suzuki T (submitted)

Specific heat and electrical resistivity at magnetic fields in antiferromagnetic heavy fermion $CeAl_2$

T. Ebihara[1], M. Tsuchiya[1], Y. Saitoh[1], J. Jatmika[1], M. Tsujimoto[2], Y. Shimura[2], Y. Matsumoto[2] and S. Nakatsuji [2]

[1]Department of Physics, Shizuoka University, 836 Oya, Suruga Shizuoka Japan
[2]Institut for Solid State Physics, Kahsiwanoha, Kashiwa Chiba, Japan

E-mail: ebihara.takao@shizuoka.ac.jp

Abstract. We measured temperature dependence of electrical resistivity and the specific heat as a function of magnetic field up to 8 T in $CeAl_2$. At the metamagnetic transition around 6 T, the coefficient A ($\rho \propto AT^2$) and Sommerfeld coefficient reduce by 25 %. Such non drastic reduction implies magnetically insensitive phonon could enhance the A and γ in $CeAl_2$ even in low temperature.

1. Introduction

Cerium or Ytterbium based intermetallic compounds show heavy fermion, unconventional superconductivity, or non-Fermi liquid phenomena at the vicinity of quantum critical point. Such interesting properties emerge at much lower temperature than Debye temperature and are mainly caused by correlation between 4f and conduction electrons. The electronic states of Ce or Yb systems are tunable by alloying, pressure, or magnetic field

$CeAl_2$ is a prototype heavy fermion compound ordering antiferromagnetically at 4 K, which crystalizes into the Laves phase $MgCu_2$-type cubic structure with lattice parameter of 7.70 Å [1]. Below T_N at zero magnetic field, the Sommerfeld coefficient (γ) is 135 mJ/K^2 mol, which is about 40 times larger than γ=3.65 mJ/K^2 mol of non-f reference material $LaAl_2$ [2, 3]. $CeAl_2$ shows metamagnetic transition approximately at 6 T [6, 4, 5]. The Fermi surface is well clarified using de Haas-van Alphen effect above metamagnetic transition at 6 T. In the field induced ferromagnetic state, Fermi surface of $CeAl_2$ is closely similar to that of $LaAl_2$. The effective masses (m^*s) of $CeAl_2$ are in a range from 1.0 to 17 m_e [3, 7]. Comparing to effective masses of $LaAl_2$ between 0.2 and 1.7 m_e, the mass enhancement values range 5 to 10. Although the specific heat of $CeAl_2$ was measured at magnetic fields up to 5 T, the specific heat measurement is not performed in field induced ferromagnetic state [8]. Therefore, direct comparison between γ and m^* is not possible. The electrical resisitivity (ρ) shows $\rho \sim \rho_0 + AT^2$ behavior at low temperature in conventional heavy fermion system. The coefficient A is quadratically proportional to the density of sate at Fermi energy ($D(E_F)$) and thus square γ. Systematic measurements of electrical resistivity are not performed yet when crossing metamagnetic transition approximately at 6 T in $CeAl_2$, although detailed electrical resistivity measurements under high pressure were performed in low temperature [9]. Moreover in low temperature, power law of electrical resistivity deviates from T^2 as seen in Ref. [10]. $CeAl_2$ is an exception from conventional heavy fermion.

In this paper, we measured the specific heat and electrical resistivity with crossing the critical field approximately at 6 T for observing specific heat and deviation from standard power laws of electrical resistivity, and for determining the magnetic field dependence of γ and A in $CeAl_2$.

2. Experimental Methods

Single crystals of $CeAl_2$ were grown by the Czochralski pulling method. A part of single crystal was milled to fine powder for checking homogeneity, no unfavorable phase and correct composition (i.e. Ce:Al=1:2). Crystallographic directions were determined by X-ray back Laue method and samples were cut by spark cutter. Electrical resisitivity was measured by the standard four probe method in the condition of T>20 mK and H<10 T with dilution refrigerator at Institute for Solid State Physics (ISSP), the University of Tokyo. Specific heat measurement was performed by the relaxation method with dilution refrigerator also at ISSP.

3. Results and Discussion

3.1. Electrical Resistivity

We measured residual resistivity as a function of magnetic field for both $H\|$<100> and <110> axes at three temperatures of 100 mK (green), 410 mK (red) and 1000 mK (blue) up to 8 T shown in Figure 1 (a). In Fig. 1 (a), we indicated drops of ρ_0 by solid arrows for those axes. Drops for both $H\|$<100> and <110> axes are between 5 and 6 T, which reproduces critical magnetic fields determined by magnetization measurements [4, 5, 6]. The samples were polished for enhancing output voltage of measurements. Different amount of defects should be introduced to samples for $H\|$<100> and <110> by polishing. Thus, the zero field data are different from each other.

We measured the temperature dependence of electrical resistivity in low temperature as a function of magnetic field. Basically we need to subtract the electrical resistivity of $LaAl_2$ as the phonon part from electrical resistivity of $CeAl_2$. We estimated electrical resistivity of $LaAl_2$ using the Bloch-Grüneisen formula with taking into account the electrical resistivity of $LaAl_2$ seen in Ref. [9]. Electrical resistivity of normal metal is expressed as ρ=ρ_0+$\rho_{\mathrm{ph}}(T)$, where ρ_0 expresses residual resistivity and $\rho_{\mathrm{ph}}(T)$ means scattering by phonon. The ρ_{ph} follows the Bloch-Grüneisen formula in the form of $\rho_{\mathrm{ph}}=A'\,(T/\Theta_{\mathrm{D}})^5\int_0^{T/\Theta_{\mathrm{D}}}\frac{x^5}{(1-e^{-1})(e^x-1)}dx$, where A' is a constant term and Θ_{D} means Debye temperature. After we took into account the error bars when reading data from article, we regarded the electrical resistivity of $LaAl_2$ is nearly zero below 20 K. Thus, we show the electrical resistivity in H<110> and H<100> actually without subtracting electrical resistivity of $LaAl_2$ in temperature below 2 K. In magnetic material, electron-electron ($\rho_{\mathrm{e\text{-}e}}(T)$) and magnetic scattering parts ($\rho_{\mathrm{mag}}(T)$) are added to the form to be expressed as $\rho = \rho_0 + \rho_{\mathrm{ph}}(T) + \rho_{\mathrm{e\text{-}e}}(T) + \rho_{\mathrm{mag}}(T)$. The $\rho_{\mathrm{e\text{-}e}}(T)$ is proportional to AT^2 in conventional heavy fermion system.

We fitted the electrical resistivity at magnetic fields including zero field from base temperature to 0.4 K by $\rho = \rho_0 + AT^2$, where A is the coefficient of quadratic part of electrical resistivity based on the Landau-Fermi-liquid theory. Figure 1 (b) shows electrical resisitivity and its fitting curve at low temperature at zero magnetic field. The zero field date are the only exception that we could fit by quadratic function in electrical resistivity of $CeAl_2$. Thus, we used a fitting function as $\rho = \rho_0 + A^*T^{\mathrm{n}}$ at magnetic fields for both $H\|$<100> and <110> axes, where n is n^{th} power and A^* is the coefficient of nth power. We summarized n and A^* in Fig. 3 (a) and (b).

The values of n are almost constant near n=3, which is different from conventional heavy fermion showing n=2 as seen in Fig. 3 (a). Such deviation of n from two implies additional factor exists in low temperature electrical resistivity in $CeAl_2$. On the other hand, we could not

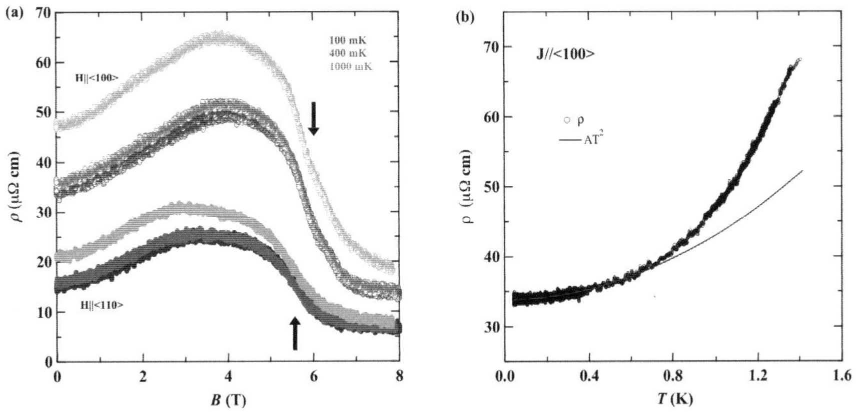

Figure 1. (a) Magnetic field dependence of residuar resistivity (ρ_0) at three temperatures. Open circles for $H\|$<100>, closed circles for $H\|$<110>, and arrows for indicating critical fields for both axes. (b) Temperature dependence of residual resistivity (ρ_0) at zero field. Black open circles show data points and a blue line represents quadratic fitting in a range from base temperature to 0.4 K.

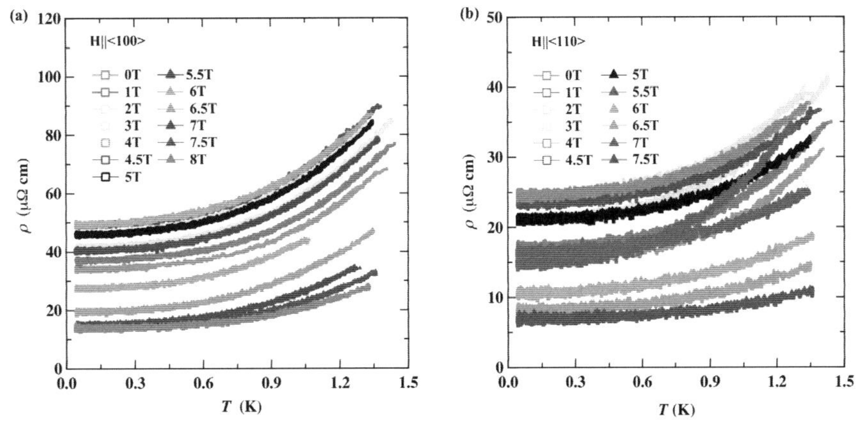

Figure 2. Magnetic field dependence of electrical resistivity when (a) $H\|$<100> and (b) $H\|$<110>.

see significant change in n-value at critical fields. The A^*-value should correspond to A-value, which is proportional to $D(E_F)$. There are shoulders near critical field seen in Fig. 3 (b).

3.2. Specific Heat

We performed specific heat measurements in $H\|$<110> and <111> in $CeAl_2$ shown in Fig. 4 (a) and (b). Below 0.2 K at magnetic fields, specific heat rapidly increases in magnetic field. The nuclear Schotkey anomaly causes this increase of specific heat. The specific heat is expressed as a function of $C(T) = C_{el} + C_{ph} + C_{nuc}$; $C_{el} = \gamma T$, $C_{ph} = \beta T^3$ and $C_{nuc} = \alpha T^{-2}$. At zero field, $C(T) = C_{el} + C_{ph}$ is used for fitting function because nuclear part does not contribute to the

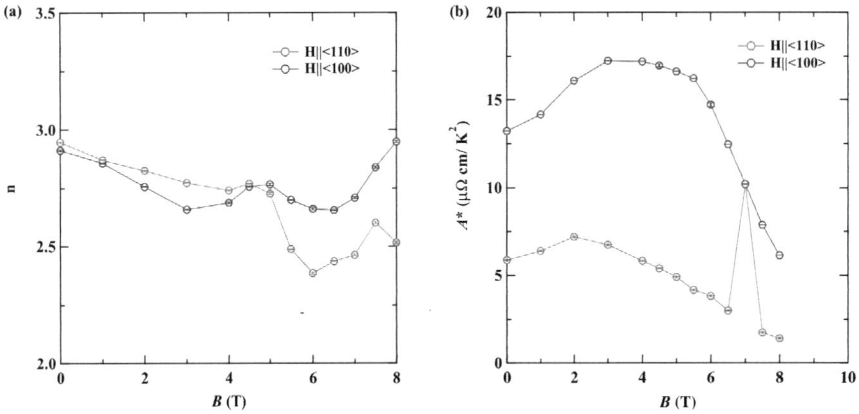

Figure 3. Magnetic field dependence of (a) n^{th} power and (b) coefficient A^* for $H\|$<100> and <110> in fitting function; $\rho = \rho_0 + A^*T^{\mathrm{n}}$.

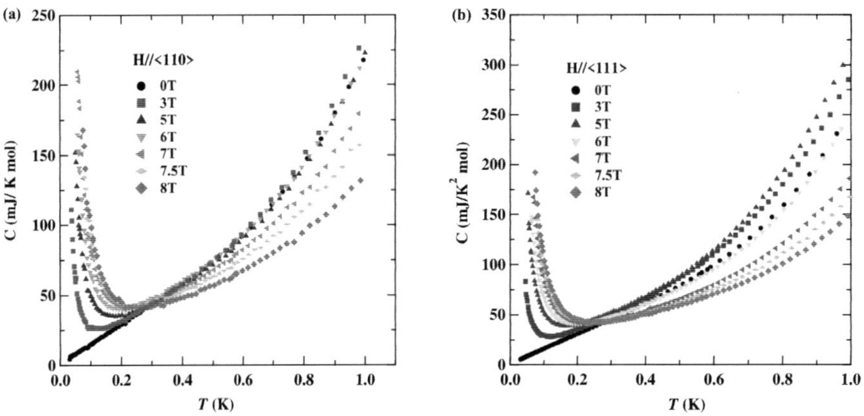

Figure 4. Magnetic field dependence of specific heat when (a) $H\|$<110> and (b) $H\|$<111>.

total specific heat $C(T)$. On the other hand, we use $C(T) = C_{\mathrm{el}} + C_{\mathrm{ph}} + C_{\mathrm{nuc}}$ for fitting data at magnetic fields with three parts. Both results at zero field and at magnetic fields are well fitted by the functions. Examples of fitting at 0 and 3 T are shown in Fig. 5. We determined γ and β as a function of magnetic field and depicted in Fig.6 (a) and (b). The γ and β reduce their values about 25 % and 70 % when crossing critical field about 6 T. This result infers there remains magnetically less sensitive enhancement factor in γ.

If we assume ideal condition such as spherical Fermi surface and electrons travel on the spherical Fermi surface, we can estimate the γ from m^*. Based on the free electron model, approximately 100 mJ/K^{2}mol is estimated as the γ-value from m^* about 16 m_{e} determined by de Haas-van Alphen above 6 T [3, 7]. Therefore, the less reduction of γ and m^* implies that magnetically less sensitive enhancement factor may exist around metamagnetic transition. At much higher fields in pulse magnet, Pricopi *et.al.* performed dHvA experiments to observe light effective mass about 3 m_{e} at 53 T in $CeAl_2$, which is almost the same as that in $LaAl_2$ [11, 12].

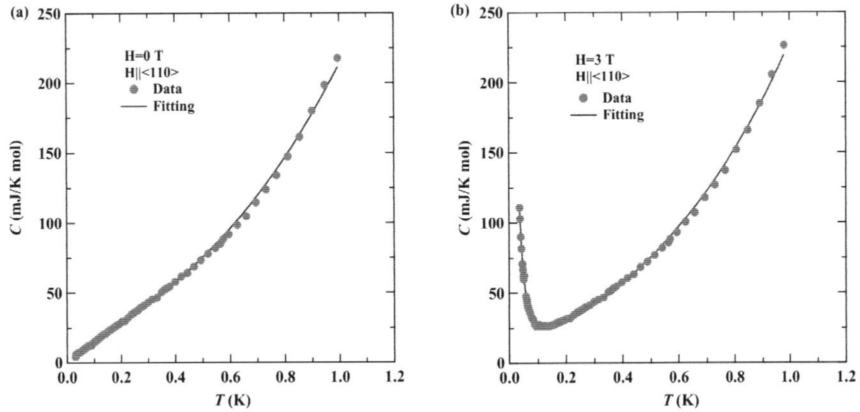

Figure 5. Fitting to the specific heat (a) at zero field and (b) at 3 T.

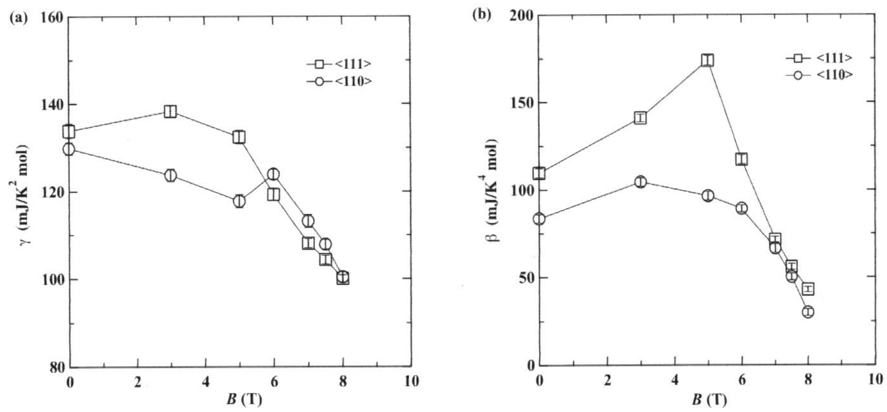

Figure 6. Magnetic field dependence of (a) electronic specific heat coefficient γ and (b) lattice specific heat coefficient β.

In conclusion, we measured electrical resistivity and specific heat at magnetic fields for observing magnetic field dependence of the coefficients A and γ. We reproduced T^3 dependence in electrical resistivity in low temperature. The γ-value reduces about 25 % when crossing metamagnetic transition, but the value maintain large γ-value as 100mJ/K^2 mol. Such non-drastic reduction implies magnetically insensitive phonon could enhance the A and γ in $CeAl_2$ even in low temperature. At much higher magnetic fields, we may observe an anomaly in specific heat, magnetostriction or magnetization measurements.

Acknowledgements
T. E acknowledges JSPS KAKENHI (Grant Number 16H04014). This work was carried out by the joint research in the Institute for Solid State Physics, the University of Tokyo. T. E. and M. T. express many thanks to Professor M. Koga for fruitful discussions.

References

[1] P.Villars and L.D.Calvert, *Pearson's Handbook of Crystallographic Data for Intermetallic Phases*, The Materials Information Society, Materials Park, OH 44073, 1991.
[2] H. Ambrüster and F. Steglich; Solid State Commun. **27** (1978) 873.
[3] M. Springford and P. H. P. Reinders; J. Mag. Mag. Mat. **76&77** (1988) 11.
[4] B. Barbara, M. F. Rossignol, J. X. Boucherle, J. Schweizer and J. L. Buevoz; J. Appl. Phys. **50** (1979) 2300.
[5] Y. Ōnuki, Y. Furukawa, and T. Komatsubara; J. Phys. Soc. Jpn. **53** (1984) 2734.
[6] Y. Saito, T. Ebihara, A. Matsuo, A. Kondo, K. Kindo; J. Phys. Soc. Conf. Proc. **3** (2014) 011025.
[7] P. H. P. Reinders and M. Springford; J. Mag. Mag. Mat. **79** (1989) 295.
[8] C. D. Bredl F. Steglich, K. D. Schotte; Z. Phys, B **29** (1978) 327.
[9] H. Miyagawa, G. Oomi, I. Satoh, T. Komatsubara, M. Hedo and Y. Uwatoko; Phys. Rev. B **78** (2008) 064403.
[10] G. Sparn, W. Lieke,U. Gottowick, F. Steglich and N. Grewe; J. Magn. Magn. Matt. **47&48** (1985) 521.
[11] L. Pricopi, E.G. Haanappel, S. Askénazy, N. Harrison, M. Bennett, P. Lajay A. Demuer and G. Lapertot; Physica B **294-295** (2001) 276.
[12] E.G. Haanappel, L. Pricopi, A. Demuer and P. Lajay; Physica B **259-261** (1999) 1081.

Elastic Softening in the Metallic Antiferromagnet YbCuGe

X J Xi[1], I Ishii[1], Y Noguchi[1], H Goto[1], S Kamikawa[1], K Araki[2], K Katoh[2] and T Suzuki[1,3,4]

[1]Department of Quantum Matter, ADSM, Hiroshima University, Higashihiroshima, Hiroshima 739-8530, Japan
[2]Department of Applied Physics, National Defense Academy, Yokosuka, Kanagawa 239-8686, Japan
[3]Institute for Advanced Materials Research, Hiroshima University, Higashihiroshima, Hiroshima 739-8530, Japan
[4]Cryogenics and Instrumental Analysis Division, N-BARD, Hiroshima University, Higashihiroshima, Hiroshima 739-8526, Japan

E-mail:d140698@hiroshima-u.ac.jp, tsuzuki@hiroshima-u.ac.jp, ish@hiroshima-u.ac.jp

Abstract. The Yb-based compound YbCuGe with a hexagonal structure is reported as a metallic antiferromagnet with T_N = 4.2 K. It is also proposed that this compound contains geometrical magnetic frustration due to the competition of exchange interaction within the quasi-two dimensional Yb plane of which Yb ions form a triangle. Previous research suggested that its magnetic anisotropy can be ascribed to the crystalline electric field (CEF) effect on a single Yb ion. In order to clarify the CEF effect in this compound, ultrasonic measurements and related theoretical fitting have been carried out. The transverse elastic modulus C_{44} exhibits a softening which begins at 120 K and stops at about 20 K. On the other hand, the longitudinal modulus C_{33} mode shows a continuous hardening before T_N. We carried out the theoretical strain-susceptibility calculation of C_{44} by using the CEF parameters reported by Katoh *et al.* According to the fitting results, it can be concluded that the softening at high temperature range can be explained by the quadrupole interaction between the ground and excited Kramers doublets under the hexagonal CEF.

1. Introduction

Yb-based compounds have recently attracted the interest of many researchers, especially the physical properties of a series of compounds Yb*T*Ge (*T* = transition metal) are well investigated [1-4]. Among them, the metallic compound YbCuGe with a hexagonal structure is reported as an antiferromagnet with a phase transition temperature of T_N = 4.2 K [5, 6]. In general, the Kondo interaction which makes a spin-singlet composed of *f*- and conduction-electrons may be suppressed by the spontaneous magnetization below phase transition temperature. However, comparing with some other compounds in this Yb*T*Ge series, such as YbPtGe and YbIrGe [1, 2], YbCuGe compound contains some different properties.

Above 100 K, as reported by Katoh *et al.*, the temperature dependence of the inverse magnetic susceptibility $1/\chi(T)$ of YbCuGe obeys the Curie-Weiss law. The obtained effective magnetic moment μ_{eff} is 4.27 μ_B for *H*//*a* and 4.43 μ_B for *H*//*c*, respectively. These values of μ_{eff} are close to the value of 4.53 μ_B expected for the free Yb^{3+} ion. It is considered that this magnetic anisotropy in the higher-temperature range is ascribed to the crystalline electric field (CEF) effect on a single Yb ion. The magnetic contribution $C_m(T)$ of the specific heat for YbCuGe shows a broad peak at 50 K. This can be ascribed to the Schottky contribution which is derived from four Kramer's doublets for the total angular momentum $\boldsymbol{J} = 7/2$ of the Yb^{3+} ion. Katoh *et al.* got the CEF parameters of $B_2^0 = -0.7$ K, $B_4^0 = 0.15$ K, and $B_4^3 = 0.5$ K in the first approximation which treats only the second- and fourth-order terms of the trigonal CEF model as the site symmetry of the Yb ions is trigonal. The first-, second-, and third-excitation energies of the CEF levels from the ground doublet were estimated to be 99 K, 162 K, and 219 K, respectively [6].

At low temperature, the peak of inverse magnetic susceptibility for $H//c$ indicates its antiferromagnetic ordering below 4.2 K. By the results of temperature dependence of the specific heat $C(T)$, the electronic specific heat coefficient γ was estimated as 1.35 mJ/(mol K^2). However, the magnetic entropy $S_m(T)$ at T_N is 28% of that of Rln2 which suggests that the slow release of magnetic entropy is probably ascribed to the spin fluctuation instead of the Kondo effect because of the small γ value. Moreover, it is also proposed that this compound contains geometrical magnetic frustration due to the competition of exchange interaction within the quasi-two dimensional Yb plane, which forms the triangle of Yb ions. All of these revealed that YbCuGe is a metallic antiferromagnet without heavy fermion character [6]. To investigate the phase transition at T_N and clarify the CEF effect in YbCuGe, we carried out ultrasonic measurements on a single-crystalline sample and related theoretical calculation was performed.

2. Experimental

Single crystal of YbCuGe was grown by the Bridgman method [5]. The elastic moduli C_{33} and C_{44} were measured as a function of the temperature T from 2 to 300 K using the phase comparison-type pulse echo method [7]. Ultrasound with the frequency of 10 or 30 MHz was generated and detected by $LiNbO_3$ transducers glued onto parallel surfaces of a sample. The elastic modulus C_{ii} was calculated using $C_{ii} = \rho v^2$ with a room-temperature mass density $\rho = 9.49$ g/cm^3, where v is the sound velocity in a sample. We estimated the absolute value of v at 4.2 K by using the relation $v = 2l/t$, where l is the sample length and t is the time interval between pulse echoes.

3. Results and discussion

Fig. 1 shows the T dependence of the longitudinal elastic modulus C_{33} from 2 K to 300 K, and C_{33} with the temperature range around phase transition is shown as the inset in Fig. 1. It can be seen that C_{33} increases monotonically with decreasing temperature down to T_N, indicating an elastic hardening characteristic before the phase transition. With further decreasing the temperature, C_{33} exhibits a step-like softening, and a peak is located at 4.2 K which is corresponding to the antiferromagnetic phase transition temperature as previously reported.

The variation of the transverse modulus C_{44} versus temperature for YbCuGe from 2 to 300 K is shown in Fig. 2. C_{44} exhibits hardening with decreasing T at higher temperatures, and then shows a softening which begins at 120 K. The softening slows down around 30 K and finally stops with further decreasing the temperature to 20 K. For the T below T_N, a similar step-like softening around T_N is observed in the transverse mode C_{44} as shown in the inset of Fig.2. We previously discussed such a step-like softening as a result of magnetostriction [8].

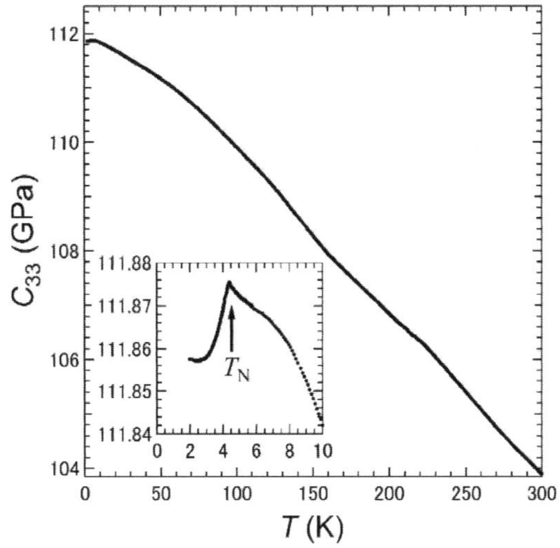

Figure 1: T dependence of the longitudinal elastic modulus C_{33}. The inset shows the T dependence of C_{33} below 10 K

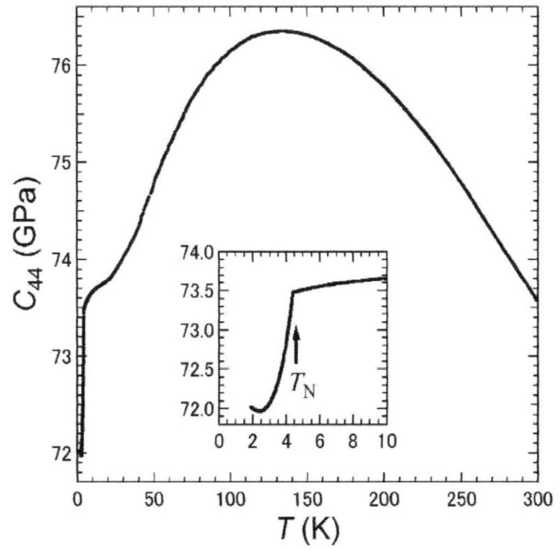

Figure 2: T dependence of the transverse elastic modulus C_{44}. The inset represents the T dependence of C_{44} below 10 K

We consider that the elastic softening of C_{44} may be due to the CEF effect. To obtain the detailed information, we carried out the theoretical fitting of C_{44} by using the CEF parameters reported by Katoh *et al.* as we mentioned above. We considered the effective Hamiltonian $H_{\rm eff}$ for the elastic modulus and the Hamiltonian of CEF as follows,

$$H_{eff} = H_{CEF} - gO_{yz}\varepsilon_{yz} - g'\langle O_{yz}\rangle O_{yz} \quad (1)$$

$$H_{CEF} = B_2^0 O_2^0 + B_4^0 O_4^0 + B_4^3 O_4^3 \quad (2)$$

where ε_{yz}, g, and g' are a strain, a strain-quadrupole coupling constant, and a quadrupole–quadrupole coupling constant, respectively. The T dependence of the elastic modulus, $C_{44}(T)$, is represented by the following equation [9],

$$C_{44}(T) = \frac{-N_0 g^2 \chi_s(T)}{1 - g'\chi_s(T)} + C_0(T) \qquad (3)$$

where χ_s is the so-called strain susceptibility and N_0 ($=1.85\times10^{28}$ m^{-3}) is the number density of Yb ions per unit volume at room temperature. To consider the background stiffness, based on the analysis for the similar materials with CEF effect before, here it is assumed as $C_0(T) = \mathrm{a}+\mathrm{b}T^2+\mathrm{c}T^4$ [8, 10, 11].

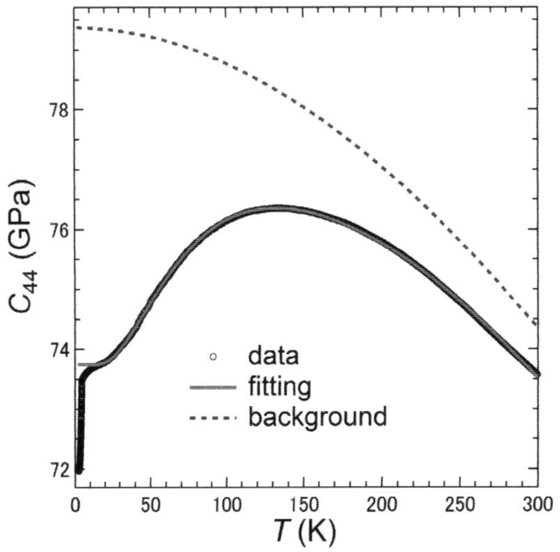

Figure 3: (Color online) T dependence of the elastic modulus C_{44}. The red solid and blue broken curves are the calculated curve and the background stiffness, respectively.

The fitting result of elastic modulus C_{44} is represented as the solid curve shown in Fig. 3. Here, we got the strain-quadrupole coupling constant, and the quadrupole–quadrupole coupling constant as $|g|$ = 237 K, and g'= -0.2 K, respectively. The theoretical fitting curve of C_{44} is matched well with the experimental data above 20 K, indicating that the softening in C_{44} originates from a quadrupole interaction. However, same as the other YbTGe system, the ground Kramers doublet of YbCuGe also contains no quadrupole degeneracy. This is very similar to the phenomenon we discovered in YbIrGe and YbPtGe which have a similar CEF effect as YbCuGe. Accordingly, comparing with the similar phenomenon we discovered in YbIrGe and YbPtGe, we conclude that the softening phenomenon of YbCuGe should also be due to the indirect quadrupole interaction between the ground doublet and the excited doublets [8, 10]. As shown in Fig.3, for the low temperature range (below 20 K) of C_{44}, the fitting curve shows a tendency to deviate from the experimental data which may arise from spin fluctuation of the phase transition.

However, the inverse magnetic susceptibility, especially for $H//a$, was not reproduced by this CEF parameters which are only considered the second- and fourth-order terms of Hamiltonian [6]. It means that it is necessary to consider the sixth-order terms of the CEF. To clarify the CEF level scheme more precise, ultrasonic measurements of other modes (C_{11} and C_{66}) are in progress. We are planning theoretical fitting of the elastic moduli by combining with the magnetic susceptibility and the Schottky specific heat of YbCuGe by using all terms of the trigonal CEF, and this work is in progress now.

4. Conclusion

Ultrasonic measurements have been performed on the single crystal of metallic antiferromagnet YbCuGe from 2 to 300 K. In order to study the CEF effect and the phase transition at T_N in this compound, the elastic moduli C_{33} and C_{44} were measured. A hardening feature above phase transition temperature is observed in longitudinal elastic modulus C_{33}. On the other hand, the transverse elastic modulus C_{44} shows elastic softening which begins at 120 K and disappears at 20 K with decreasing the temperature. The result of theoretical strain-susceptibility fitting reveals that the elastic softening phenomenon in YbCuGe originates from the indirect quadrupole interaction between the ground doublet and the excited doublets although all doublets have no quadrupole degeneracy.

5. Acknowledgement

This work was supported by JSPS KAKENHI Grant Numbers 262870830A, 26800189 and 2624706001. This work was also supported by JSPS Core-to-Core Program, A. Advanced Research Networks.

References

[1] Katoh K, Nakagawa S, Terui G and Ochiai A 2009 *J. Phys. Soc. Jpn.* **78** 104721
[2] Katoh K, Tsutsumi T, Yamada K, Terui G, Niide Y and Ochiai A 2005 *Physica B* **369** 81
[3] Katoh K, Mano Y, Nakano K, Terui G, Niide Y and Ochiai A 2004 *J. Magn. Magn. Mater.* **268** 212
[4] Enoki K, Hirose Y, Yoshiuchi S, Sugiyama K, Honda F, Takeuchi T, Yamamoto E, Haga Y, Hagiwara M, Kindo K, Settai R and Onuki Y 2012 *J. Phys. Soc. Jpn.* **81** SB056
[5] Heying B, Rodewald Ute Ch., Pöttgen R, Katoh K, Niide Y and Ochiai A 2005 *Monatshefte für Chemie* **136** 655
[6] Katoh K, Maeda M, Matsuda S and Ochiai A 2012 *J. Alloy Compd.* **520** 122
[7] Lüthi B, Bruls G, Thalmeier P, Wolf B, Finsterbusch D and Kouroudis I 1994 *J. Low Temp. Phys.* **95** 257
[8] Xi X J, Ishii I, Noguchi Y, Goto H, Kamikawa S, Araki K, Katoh K and Suzuki T 2015 *J. phys. Soc. Jpn.* **84** 124602
[9] Lüthi B 1980 *Dynamical Properties of Solids* ed G K Horton and A A Maradudin (North-Holland, Amsterdam) Chap. 4
[10] Ishii I, Noguchi Y, Kamikawa S, Goto H, Fujita T K, Katoh K and Suzuki T 2014 *J. phys. Soc. Jpn.* **83** 043601
[11] Nohara M, Suzuki T, Maeno Y, Fujita T, Tanaka I and Kojima H 1995 *Phys. Rev. B* **52** 570

Thermal conductivity of the Kondo semiconductor $CeRu_4Sn_6$

J Hänel[1], M Taupin[1], M Ikeda[1], V Martelli[1,2], P Tomeš[1], A Prokofiev[1], and S Paschen[1]

[1] Institute of Solid State Physics, Technische Universität Wien, Wiedner Hauptsr. 8-10, 1040 Wien, Austria

[2] Centro Brasileiro de Pesquisas Físicas, Rua Doutor Xavier Sigaud 150, CEP 22290-180, Brazil

E-mail: haenel@ifp.tuwien.ac.at

Abstract. We report measurements of the thermal conductivity κ on single crystalline $CeRu_4Sn_6$ in the temperature range between 80 mK and 80 K, along the main crystallographic directions. κ is phonon-dominated in the whole temperature range and is found to be essentially isotropic. At low temperatures the data are well approximated by $\kappa \propto T^2$, which is attributed to a predominant scattering of phonons on electrons. We describe the data with a Callaway fit in the whole temperature range giving good agreement at low and high temperatures.

1. Introduction

Kondo insulators or Kondo semiconductors (KSs) are strongly correlated electron systems where a narrow (pseudo)gap opens at low temperatures in the electronic density of states (DOS) at the Fermi level [1]. This is a result of the hybridization of localized f electrons with the conduction p or d electrons [2]. Even though known since decades [1], KSs remain at the forefront of condensed matter research. Recently it has been suggested that KSs may be candidates for strongly correlated materials with topologically protected surface states [3, 4, 5]. So far, experiments along these lines have focused on the cubic KS SmB_6 [6, 7].
Here $CeRu_4Sn_6$ comes into play. It has a tetragonal, non-centrosymmetric crystal structure ($I\bar{4}2m$) [8] and therefore can shed light on the role crystal symmetry plays. It was first synthesized in 1992 [9] and a development of a narrow gap was found in several physical properties [10, 11, 12]. Measurements on single crystals revealed large anisotropy in resistivity, specific heat, and thermopower [12, 13, 14]. Optical conductivity measurements and LDA+DMFT calculations attributed this to a gapping in the tetragonal plane while the out-of-plane direction stays weakly metallic [15].
Here we report on the thermal conductivity κ of $CeRu_4Sn_6$ single crystals, both along and perpendicular to the out-of-plane direction c, in the temperature range of 80 mK to 80 K.

2. Experimental

The polycrystalline starting material for the crystal growth was synthesized by melting first Ce and Ru, and subsequently the resulting Ce/Ru alloy with Sn in a horizontal water cooled copper boat using high-frequency heating. The purity of the starting materials was 99.99% for Ce (Ames

Lab) and Ru, and 99.9999% for Sn. The single crystal growth was performed by the floating zone melting technique using optical heating in a four-mirror furnace of Crystal Corporation. All steps were performed under Ar 99.9999% protective atmosphere after purging several times with Ar. Since $CeRu_4Sn_6$ melts incongruently we used a self-flux technique [12, 16].
The low-temperature (80 mK - 2 K) thermal conductivity was measured in a $^3He/^4He$ dilution refrigerator with a home-built setup employing a two-thermometer-one-heater technique using RuO_2 thick-film resistors as heater and thermometers. The measurements in the range 2 K to 80 K were carried out in a ^4He bath cryostat with thermocouples to measure the temperature gradient.

3. Physical properties

Figure 1 shows the thermal conductivity κ as a function of temperature T in the whole temperature range investigated, on both a linear (main panel) and a double-logarithmic (inset) scale. κ was measured along the out-of-plane direction c (called κ_c) and within the tetragonal plane, perpendicular to c ($\kappa_{\perp c}$). We calculated the electronic part of the thermal conductivity κ^{WF} from the electrical resistivity ρ [13] employing the Wiedemann-Franz law $\kappa^{\mathrm{WF}} = L_0 T/\rho$ with the Lorenz number $L_0 = 2.44 \times 10^{-8}\,\mathrm{W\Omega K^{-2}}$. κ is essentially isotropic, in marked contrast to other physical properties such as electrical resistivity [13], thermopower [14], and optical conductivity [15].
At low temperatures κ increases strongly with increasing temperature towards a maximum of roughly $\kappa_{\mathrm{max}} \approx 6.3\,\mathrm{Wm^{-1}K^{-1}}$ at $T_{\mathrm{max}} = 28\,\mathrm{K}$.

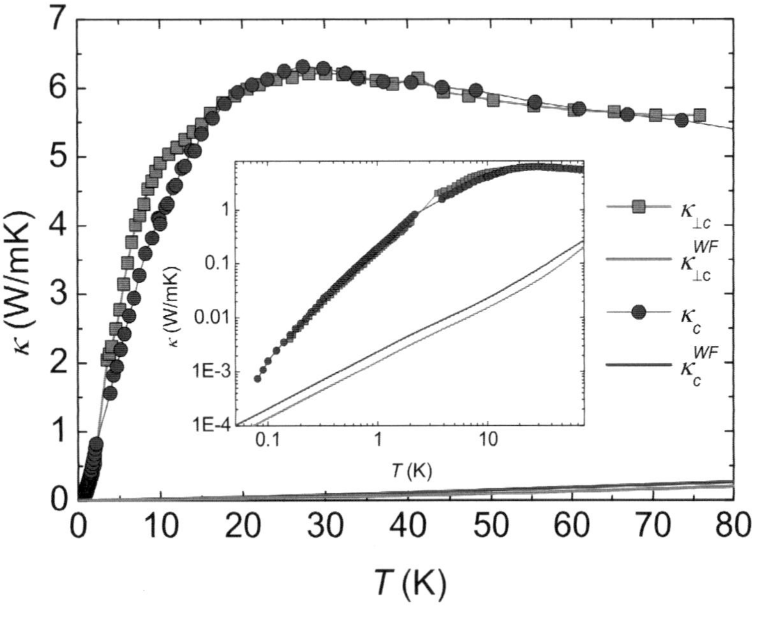

Figure 1. Thermal conductivity κ of $CeRu_4Sn_6$ as a function of temperature T measured parallel and perpendicular to the out-of-plane direction c. The electronic part of the thermal conductivity κ^{WF}, as obtained from resistivity ρ [13] via the Wiedemann-Franz law, is plotted as solid lines. The inset shows the same data on a double-logarithmic scale.

κ^{WF} is much lower than the total κ and thus κ is clearly phonon-dominated as expected in KSs. We estimate the phononic contribution κ_{ph} as $\kappa_{\mathrm{ph}} = \kappa - \kappa^{\mathrm{WF}}$. The result is plotted in

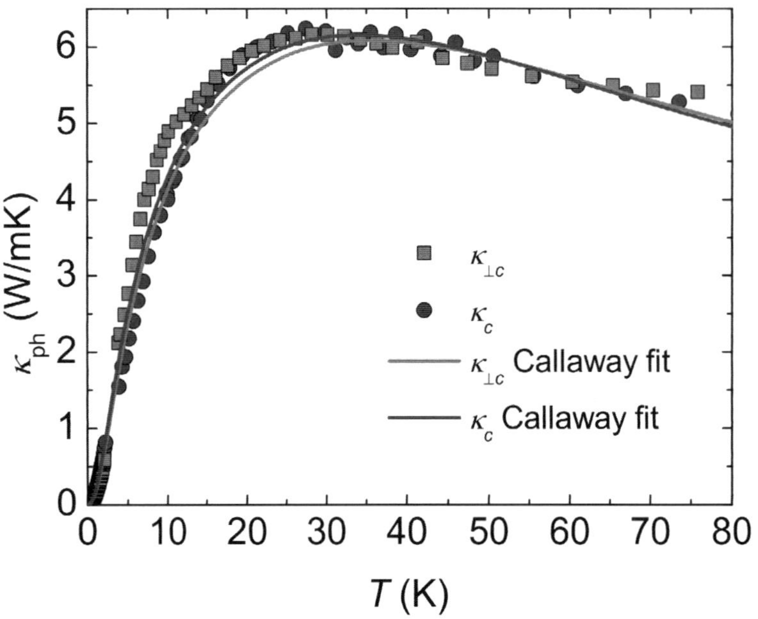

Figure 2. Phononic part of the thermal conductivity $\kappa_{\rm ph}$ of $CeRu_4Sn_6$ along and perpendicular to the c direction as a function of temperature T. Solid lines are Callaway fits to the data (see text).

Fig. 2, together with least-square fits to the Callaway model [17]

$$\kappa_{ph} = \frac{k_B^4 \cdot T^3}{2 \cdot \pi^2 \cdot v_g \cdot \hbar^3} \cdot \left[\int_0^{\frac{\theta_D}{T}} \tau \cdot \frac{x^4 \cdot e^x}{(e^x - 1)^2} \cdot dx \right] \quad , \tag{1}$$

where v_g is the sound velocity, θ_D is the Debye temperature, τ is the phonon relaxation rate, $x = \frac{\hbar\omega}{k_B T}$, and the other symbols have the usual meaning. Different scattering mechanisms influence τ according to the Matthiesen's rule

$$\frac{1}{\tau} = \sum_i \frac{1}{\tau_i} \quad . \tag{2}$$

The various scattering mechanisms have different dependencies on the frequency ω and subsequently lead to different temperature dependencies of $\kappa_{\rm ph}$. We have taken phonon-electron, -boundary, -defect and Umklapp scattering processes into account. Note that we have neglected N processes here which would lead to a correction term in Eqn. 1. However the influence of these processes is expected to be small in materials with complex crystal structure. Table 1 lists the results of the fits.

Figure 3 shows the same data as Fig. 2 on a double-logarithmic scale, which reveals the different scattering contributions most clearly. Straight lines indicate different power laws. The solid line shows T^3 behaviour typical for phonon-boundary scattering. The dashed line indicates T^2

Table 1. Fit parameters of the Callaway fits, where the scattering rates are phonon-electron scattering $\tau_{\rm pe}^{-1} = A \cdot \omega^2$, phonon-boundary scattering $\tau_{\rm b}^{-1} = B$, phonon-defect scattering $\tau_{\rm pd}^{-1} = C \cdot \omega^4$, and Umklapp scattering $\tau_{\rm U}^{-1} = D \cdot \omega^2 \cdot T \cdot exp(\frac{\theta_D}{-3T})$. The values in the table are the constants A, B, C, and D.

Direction	A (s)	B ($\rm s^{-1}$)	C ($\rm s^3$)	D ($\rm sK^{-1}$)
c	2.2×10^7	2.8×10^6	7250	9.16×10^4
$\perp c$	2.5×10^7	3.0×10^6	6878	9.06×10^4

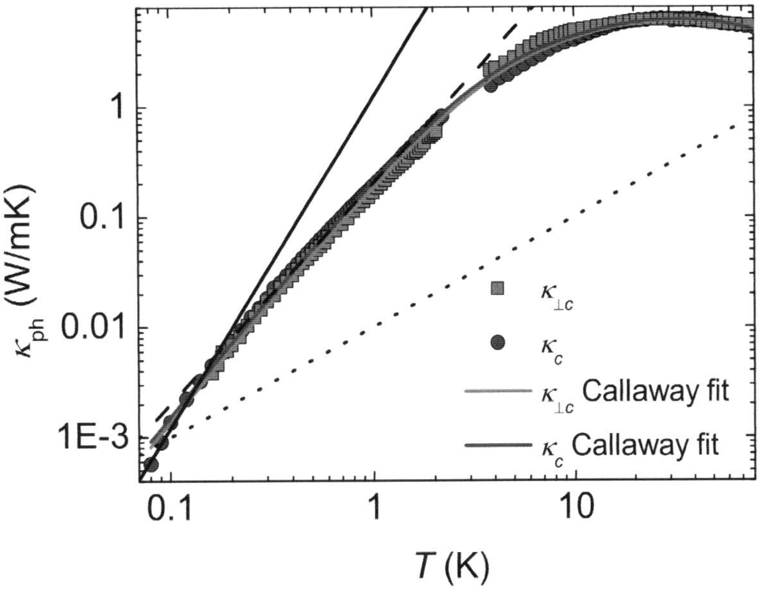

Figure 3. Phonon thermal conductivity $\kappa_{\rm ph}$ of $\rm CeRu_4Sn_6$ along and perpendicular to the c direction as a function of temperature T on a double-logarithmic scale. Solid red and blue lines are fits to the data according to the Callaway model [17]. Black lines indicate different power laws: T^3 (solid), T^2 (dashed), and linear-in-T (dotted).

behaviour which is seen for phonon-electron scattering, and the dotted line is linear in T.
The data follow the dashed line very closely in the range of 200 mK to 2 K. Thus the dominant scattering mechanism in this range is identified as phonon-electron scattering. At lower temperatures $\kappa_{\rm ph}$ shows a T^3 behaviour. The fit parameters of the Callaway fits give the boundary scattering rates $\tau_{\rm b,\perp c}^{-1} = 2.8 \times 10^6\,\rm s^{-1}$ and $\tau_{\rm b,c}^{-1} = 3.0 \times 10^6\,\rm s^{-1}$ for the measurement perpendicular and parallel to c, respectively. With a Debye temperature $\theta_D = 250\,\rm K$ [11] the sound velocity can be computed to be $v_g = 2316\,\rm m/s$. With this v_g the scattering times correspond to the mean free paths $l_{\perp c} = 0.83\,\rm mm$ and $l_c = 0.77\,\rm mm$. This is close to the smallest dimension of the samples and we therefore conclude that the phonons are predominantly scattered off the sample boundaries in this temperature range.
The onset of phonon-electron scattering, with the scattering rate $\tau_{\rm pe}^{-1}$, at higher temperatures leads to a further reduction of the thermal conductivity. According to Pippard's theory [18],

τ_{pe}^{-1} is proportional to the charge carrier concentration n. From Hall effect measurements n was determined to be isotropic at both high and low temperatures [19] in agreement with the isotropic behaviour of κ in these temperature ranges. However, at intermediate temperatures, signatures of the Kondo gap opening are anisotropic [15]. This may explain the small anisotropy seen in the thermal conductivity around 10 K (hump in $\kappa_{\perp c}$, see Fig. 2). As the gap opening is more pronounced for the plane perpendicular to c, the effect of electrons freezing out and thus disappearing as scattering partners for the phonons is more pronounced for $\kappa_{\perp c}$, which leads to slightly enhanced $\kappa_{\perp c}$ values. Similar anisotropies have been found in the KSs CeNiSn and CeRhSb and have likewise been attributed to anisotropies in the energy gap [20, 21, 22].

Above 2 K, κ_{ph} deviates from the T^2 law to lower values. Here scattering on point defects and U processes become the dominant effects. The values of the prefactors for these processes, as listed in Tab. 1, are similar to what was found in other systems [23]. However, for a quantitative discussion it is necessary to know the phonon band structure which is not available in this system. In addition, other aspects that play an important role in $CeRu_4Sn_6$ and are neglected in a simple Callaway model may come into play at elevated temperatures.

4. Conclusion

In summary we have investigated single-crystalline $CeRu_4Sn_6$ by means of thermal conductivity measurements in the temperature range of 80 mK to 80 K. The data are essentially isotropic in contrast to findings for other physical properties. The reason is that the thermal conductivity is dominated by its phonon contribution, even at the lowest temperature where the behaviour can be well explained by boundary and phonon-electron scattering. This gives further support to the previous conjecture that the strongly anisotropic electrical resistivity is mostly due to anisotropies in the charge carrier mobility, with the charge carrier concentration being essentially isotropic.

Acknowledgments

We gratefully acknowledge financial support from the Austrian Science Fund (FWF projects I 623-N16, W1243, and I 2535-N27) and from the European Research Council (ERC grant 227378). VM acknowledges the FAPERJ (Nota 10).

References

[1] Aeppli G and Fisk Z 1992 *Comments Condens. Matter Phys.* **16** 155
[2] Riseborough P S 2000 *Adv. Phys.* **49** 257
[3] Dzero M, Sun K, Galitski V and Coleman P 2010 *Phys. Rev. Lett.* **104** 106408
[4] Dzero M, Sun K, Coleman P and Galitski V 2012 *Phys. Rev. B* **85** 045130
[5] Alexandrov V, Dzero M and Coleman P 2013 *Phys. Rev. Lett.* **111** 226403
[6] Cooley J C, Mielke C H, Hults W L, Goettee J D, Honold M M, Modler R M, Lacerda A, Rickel D G and Smith J L 1999 *J. Supercond.* **12** 171
[7] Kim D J, Xia J and Fisk Z 2014 *Nat. Mater.* **13** 466
[8] Pöttgen R, Hoffmann R D, Sampathkumaran E V, Das I, Mosel B D and Müllmann R 1997 *J. Solid State Chem.* **134** 326
[9] Das I and Sampathkumaran E V 1992 *Phys. Rev. B* **46** 4250
[10] Strydom A M, Guo Z, Paschen S, Viennois R and Steglich F 2005 *Physica B: Condens. Matter* **359-361** 293
[11] Brüning E M, Brando M, Baenitz M, Bentien A, Strydom A M, Walstedt R E and Steglich F 2010 *Phys. Rev. B* **82** 125115
[12] Paschen S, Winkler H, Nezu T, Kriegisch M, Hilscher G, Custers J, Prokofiev A and Strydom A 2010 *J. Phys. Conf. Ser.* **200** 012156
[13] Winkler H, Lorenzer K A, Prokofiev A and Paschen S 2012 *J. Phys. Conf. Ser.* **391** 012077
[14] Hänel J, Winkler H, Ikeda M, Larrea J J, Martelli V, Prokofiev A, Bauer E and Paschen S 2014 *J. Electron. Mater.* **43** 2440
[15] Guritanu V, Wissgott P, Weig T, Winkler H, Sichelschmidt J, Scheffler M, Prokofiev A, Kimura S, Iizuka T, Strydom A M, Dressel M, Steglich F, Held K and Paschen S 2013 *Phys. Rev. B* **87** 115129

[16] Prokofiev A and Paschen S 2012 *Crystal Growth and Stoichiometry of Strongly Correlated Intermetallic Cerium Compounds* (InTech) chap 11, pp 263–284 Modern Aspects of Bulk Crystal and Thin Film Preparation
[17] Callaway J 1959 *Phys. Rev.* **113** 1046
[18] Pippard A B 1955 *Phil. Mag.* **46** 1104
[19] Winkler H 2013 *Physical properties of the anisotropic Kondo insulator* $CeRu_4Sn_6$ Ph.D. thesis Technische Universität Wien
[20] Paschen S, Wand B, Sparn G, Steglich F, Echizen Y and Takabatake T 2000 *Phys. Rev. B* **62** 14912
[21] Sera M, Kobayashi N, Yoshino T, Kobayashi K, Takabatake T, Nakamoto G and Fujii H 1997 *Phys. Rev. B* **55** 6421
[22] Isikawa Y, Mori K, Ogiso Y, Oyabe K and Sato K 1991 *J. Phys. Soc. Jpn.* **60** 2514
[23] Yan X, Bauer E, Rogl P and Paschen S 2013 *Phys. Rev. B* **87** 115206

Sb-NMR/NQR studies of heavy fermion system YbRhSb

Yasuki Kishimoto[1], Yosiki Awai[1], Hisashi Kotegawa[1], Hideki Tou[1*], Yuji Muro[2], Koji Nakamura[3], Masafumi Sera[3], and Toshiro Takabatake[3]

[1] *Department of Physics, Graduate School of Science, Kobe University, Kobe 657-8501, Japan*
[2] *Department of Intelligent Systems Design Engineering, Faculty of Engineering, Toyama Prefectural University, Imizu, Toyama 939-0398, Japan*
[3] *Department of Quantum Matters, AdSM, Hiroshima University, Higashi-Hiroshima, Hiroshima, Japan*

E-mail: `*:tou@crystal.kobe-u.ac.jp`

Abstract. We report a study of ^{121}Sb nuclear magnetic resonance (NMR) and nuclear quadrupole resonance (NQR) in a Yb-based heavy fermion system YbRhSb using a single crystal. Sharp ^{121}Sb-NMR lines were observed for each crystal axis for $B \| a, b, c$. From comparison with observed Sb-NMR spectra and numerical simulations, electric field gradient (EFG) parameters and Knight shift parameters are obtained at 10 K. The obtained quadrupole frequency, asymmetry parameter, isotropic Knight shift and anisotropic principal values of Knight shift tensors are ν_Q = 18.8 MHz, η = 0.2, $K_{iso} \approx$ 0.6 %, and $(K^x_{\rm ax}, K^y_{\rm ax}, K^z_{\rm ax}) \approx (-3.5, 4.1, -0.6)$ %, respectively. We also succeeded in observing ^{121}Sb-NQR signals at around 19.5 and 37 MHz, which are reproduced by the nuclear quadrupole Hamiltonian with ν_Q = 18.8 MHz and η = 0.2. We mention the relation between the anisotropy of the Knight shift and magnetic susceptibility. Our NMR/NQR results suggest that the weak ferromagnetism in YbRhSb is ascribed to a canted antiferromagnetic state.

1. Introduction

RTX type compounds (R = Ce, Yb, U; T = Co, Ni, Rh, X =Sn, Sb, Ge) with the orthorhombic ε-TiNiSi-type structure have been attracted attention because these compounds show various exotic ground states. For example, CeNiSn is well known as a Kondo semiconductor.[1] The isostructural U-based counterpart UCoGe shows coexistence between ferromagnetic state and superconductivity.[2]

Among them, Yb-based compounds with ε-TiNiSi-type structure shows unusual ferromagnetic properties. YbNiSn undergoes a ferromagnetic transition at $T_C \approx 5.6$ K with the ordered moment $\mu_s \approx 0.85\mu_B$/Yb.[3, 4] An electronic specific-heat coefficient is $\gamma \approx 300$ mJ/mol K below T_C, suggesting YbNiSn is classified into a class of heavy fermion ferromagnets.[5]

YbRhSb also reveals a ferromagnetic transition at $T_C \approx 2.7$ K.[6] The spontaneous moment is, however, unusually small of $\approx 3.0\times10^{-3}$ μ_B/Yb for the applied field (B) parallel to the crystal b-axis. A metamagnetic transition was observed at around 2 T and the magnetization reaches $1.4\mu_B$/Yb for $B\|a$-axis at 15 T. In addition, T_C decreases with the increase of magnetic field, differently from the behavior of typical ferromagnets. Moreover, a large electronic specific-heat coefficient is $\gamma \approx$370 mJ/mol K^2 in the ferromagnetic ordered state. However, it is strange that the largest spontaneous magnetization appears for the field direction parallel to the b axis below T_C, nevertheless the magnetic easy axis is the a axis in the paramagnetic state. These behaviors suggest that the weak ferromagnetism of YbRhSb is ascribed to a canted antiferromagnetic structure.[6]

In order to clarify microscopically the unusual weak ferromagnetic state, we carried out ^{121}Sb-nuclear magnetic resonance (NMR) and nuclear quadrupole resonance measurements on YbRhSb. Here we determine the ^{121}Sb-NMR/NQR parameters.

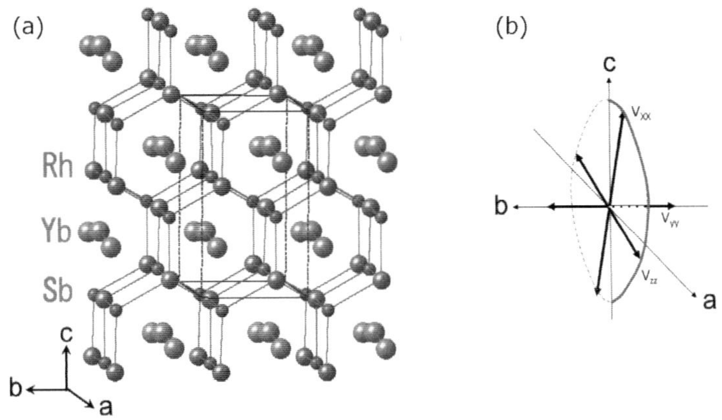

Figure 1. (a) Schematic crystal structure of YbRhSb. Unit cell is drown by blue dashed line. (b) Schematic figure of the principal axes of EFG tensors at Sb site.

2. Experimental

Single-crystalline samples of YbRhSb were grown by using the Bridgeman method. Stoichiometric amounts of the elements were sealed in a molybdenum crucible by arc welding under a purified argon atmosphere. The crucible was heated up to 1600 °C in a vacuum by using a tungsten mesh heater and slowly cooled. [6]

The powder x-ray-diffraction pattern showed that YbRhSb crystallizes with the orthorhombic ε-TiNiSi-type structure and the space group is $No.62, Pnma$.[6] The lattice constants is a = 7.004 Å, b = 4.492 Å, and c = 7.731 Å. The electron-probe microanalysis (EPMA) revealed the presence of an impurity phase $YbSb_2$ of less than 0.1% of the stoichiometric host phase. The single-crystal nature of the sample was also confirmed by the x-ray Laue diffraction method. The electrical resistivity, dc magnetic susceptibility measurements and specific heat measurements evidence a magnetic phase transition at 2.7 K.[6] The single-crystal sample was used for the NMR measurements, where the dimensions of the cuboidal sample were $L(a\text{-axis}) \times W(b) \times H(c) \approx 3 \times 3 \times 2$ mm^3.

^{121}Sb ($I = 5/2$, $^{121}\gamma_n/2\pi$ = 10.189 MHz/10 kOe, and $^{121}Q = -0.597 \times 10^{-28}$ m^2) NMR measurements for the single-crystal sample was carried out by using a conventional pulsed spectrometer. The ^{121}Sb nuclear magnetic resonance (NMR) spectrum was obtained by tracing the spin-echo intensity as a function of the magnetic field. The ^{121}Sb nuclear quadrupole resonance (NQR) spectrum was obtained by tracing the spin-echo intensity as a function of the frequency. A part of the obtained single-crystal sample was crushed into powder form with smaller grain size than the skin depth for NMR measurements to gain a signal to noise ratio at high temperatures.

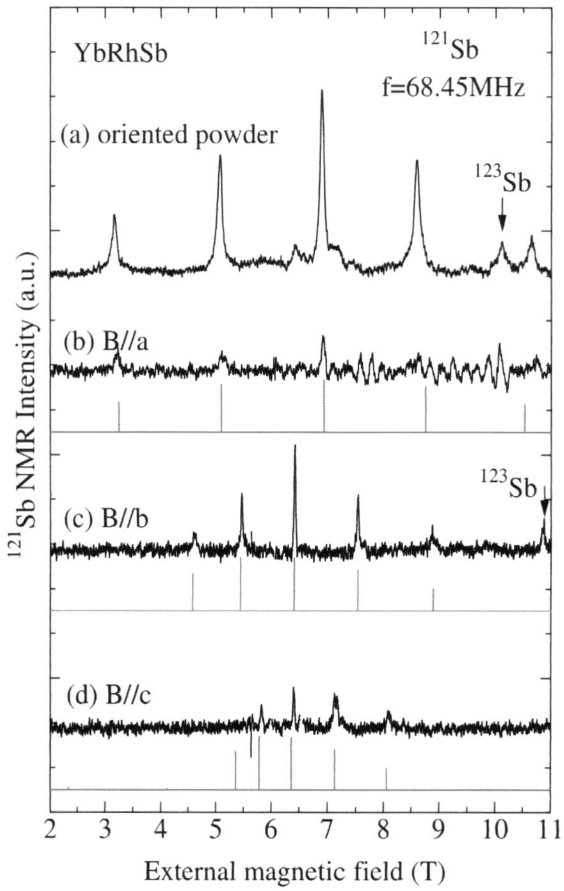

Figure 2. ^{121}Sb-NMR spectra for (a) oriented-powder, (b) $B\|a$, (c) $B\|b$, (d) $B\|c$-axes. Red solid lines are results of simulations for each field angle (see text).

3. Results and Discussions

Figure 2 shows the entire ^{121}Sb-NMR spectra for $B\|a, b$, and c directions observed at the fixed frequency $f = 68.45\,\mathrm{MHz}$ at $T = 10\,\mathrm{K}$. ^{121}Sb nucleus has the nuclear spin $I = 5/2$, then the ^{121}Sb-NMR line splits into five lines. In order to reproduce the observed NMR spectra, the resonance fields were calculated by an exact-diagonalization method for the 6×6 nuclear spin Hamiltonian matrix of ^{121}Sb nucleus. The Hamiltonian is written as,

$$\begin{aligned}\mathcal{H} &= -\gamma_{\mathrm{n}}\hbar I_z H_0 - \gamma_{\mathrm{n}}\hbar I_z H_0 K_{\mathrm{iso}} - \gamma_{\mathrm{n}}\hbar(K^z_{\mathrm{an}} I_z H_z + K^x_{\mathrm{an}} I_x H_x + K^y_{\mathrm{an}} I_y H_y) \\ &\quad + \frac{1}{6}h\nu_{\mathrm{Q}}\left[\left(3\cos^2\theta - \eta\cos 2\phi\sin^2\theta\right) I_z^2 + \frac{1}{4}\left\{3\sin^2\theta - \eta\cos 2\phi\left(1+\cos^2\theta\right)\right\}\left(I_+^2 + I_-^2\right)\right. \\ &\quad \left. + \frac{1}{4}\sin^2\theta\,(3+\eta\cos 2\phi)\,(I_+I_- + I_-I_+) - I\,(I+1)\right.\end{aligned}$$

$$-i\left(\frac{1}{2}\eta\sin 2\phi\cos\theta\left(I_+^2-I_-^2\right)+\frac{1}{4}\sin 2\theta\left(3+\eta\cos 2\phi\right)\left\{\left(I_+-I_-\right)I_z+I_z\left(I_+-I_-\right)\right\}\right)\Bigg]. \tag{1}$$

The first term is the Zeeman interaction, the second term is the hyperfine interaction described by the isotropic part of the Knight shift, K_{iso}, and the principal values of anisotropic Knight shift tensors, (K_{an}^x, K_{an}^y, K_{an}^z), where the shift tensors fulfills the relation of $K_{\mathrm{an}}^x+K_{\mathrm{an}}^y+K_{\mathrm{an}}^z=0$. The third one describes the nuclear quadrupole interaction. Here, h is the Planck constant and $\hbar\equiv h/2\pi$. ν_{Q} is the NQR frequency and is defined by $\nu_{\mathrm{Q}}\equiv 3eQV_{zz}/2hI(2I-1)$. η is the asymmetry parameter and is defined by $\eta\equiv\left(V_{xx}-V_{yy}\right)/V_{zz}$. V_{xx}, V_{yy}, and V_{zz} are the principal values of the electric field gradient (EFG) where $V_{zz}\geq V_{xx}\geq V_{yy}$. θ is the polar angle of the field direction with respect to V_{zz}, and ϕ is the azimuth angle of the field direction with respect to V_{xx}. For simplicity, we assumed that the principal axes of the EFG tensor coincide with those the Knight shift tensors.

In fact, the second order perturbation theory for the above nuclear Hamiltonian assuming $\nu_L\gg\nu_Q$, where $\nu_L=\frac{\gamma}{2\pi}H_0$, yields following equation for the transition frequency between mth and $m-1$th level:

$$\begin{aligned}h\nu_{m\leftrightarrow m-1} &= h(1+K_{iso})\nu_L+\frac{1}{2}h\left[\nu_Q\left(\frac{1}{2}-m\right)-K_{\mathrm{an}}^z\nu_L\right]\left(3\cos^2\theta-1\right)\\ &-\frac{1}{2}h\left[\nu_Q\eta\left(\frac{1}{2}-m\right)+\left(K_{\mathrm{an}}^y-K_{\mathrm{an}}^x\right)\nu_L\right]\cos 2\phi\sin^2\theta\\ &+\frac{\nu_Q^2}{12(1+K_{iso})\nu_L}\left[\frac{3}{2}\sin^2\theta\left\{(A+B)\cos^2\theta-B\right\}\right.\\ &\qquad\left.+\eta\cos 2\phi\sin^2\theta\left\{(A+B)\cos^2\theta+B\right\}\right]\\ &+\frac{\nu_Q^2}{72(1+K_{iso})\nu_L}\eta^2\left[A-(A+4B)\cos^2\theta-(A+B)\cos^2 2\phi(1-\cos^2\theta)^2\right], \end{aligned} \tag{2}$$

where $A=24m(m-1)-4I(I+1)+9$ and $B=\frac{1}{4}\left\{6m(m-1)-2I(I+1)+3\right\}$. Thus, the resonance frequency depends on H_0, ν_Q, η, K_{iso}, $K_{\mathrm{an}}^i(i=x,y,z)$, and field angle θ, ϕ with respect to the EFG principal axes. Then, the quadrupole frequency ν_Q should be slightly different from the peak positions of the observed NQR signals, as described later.

Red solid lines in Fig. 2 are the results of the simulations for each field direction with respect to the crystal axes. In order to reproduce the obtained NMR spectra, we obtained NMR/NQR parameters as follows: Isotropic and anisotropic NMR Knight shifts are $K_{iso}\approx 0.6$ % and $(K_{\mathrm{an}}^x, K_{\mathrm{an}}^y, K_{\mathrm{an}}^z)\approx(-3.5, 4.1, -0.6)$ %, respectively. NQR quadrupole frequency and asymmetry parameter are $\nu_Q=18.8$ MHz, $\eta=0.2$. Here, the angle-sets (θ,ϕ) are obtained to be $(7^\circ,0^\circ)$ for $B\|a$, $(90^\circ,7^\circ)$ for $B\|b$, and $(97^\circ,0^\circ)$ for $B\|c$. Thus we found that V_{yy} lies along b-axis, whereas respective V_{zz} and V_{xx} rotate from the a- and c-axes by $\sim 7^\circ$, as shown in the fig. 1(b). The reason why the principal EFG tensors do not coincide with the crystal axes is attributed to the lower symmetry of the crystal structure. In fact, the local symmetry at Sb site at Wycoff position 4c is .m., then only a mirror plane exists perpendicular to the b-axis. Thus, Sb $4c$ site does not have a local spatial symmetry.

We found from the anisotropic Knight shift that the maximum principal value of the Knight shift tensors is K_{an}^y for $B\|b$ axis, being incompatible with the susceptibility that the magnetic easy axis is a axis. This contradiction may be a possible origin for the weak ferromagnetism, i.e., a canted antiferromagnetic state, in YbRhSb. Furthermore, we should notice that the Wycoff position of the Yb site is same as that of the Sb site. That is, the local spatial symmetry

IOP Conf. Series: Journal of Physics: Conf. Series **807** (2017) 012014 doi:10.1088/1742-6596/807/1/012014

is broken at Yb site. In this case, a Dzyaloshinsky-Moriya type antisymmetric spin-orbit interaction[7] works and the alignment of the Yb magnetic moment might be forced to cant from the crystal axis. In fact, Dzyaloshinsky vector orients perpendicular to the *ac* plane (along *b* axis). This direction is consistent with the maximum principal direction of the Knight shift tensors. Therefore, we expect that the weak ferromagnetism in YbRhSb is arising from the lower symmetries of the Sb and Yb site. In any case, to check this scenario, measurements of the temperature dependences of both zero-field NMR and Knight shift are needed.

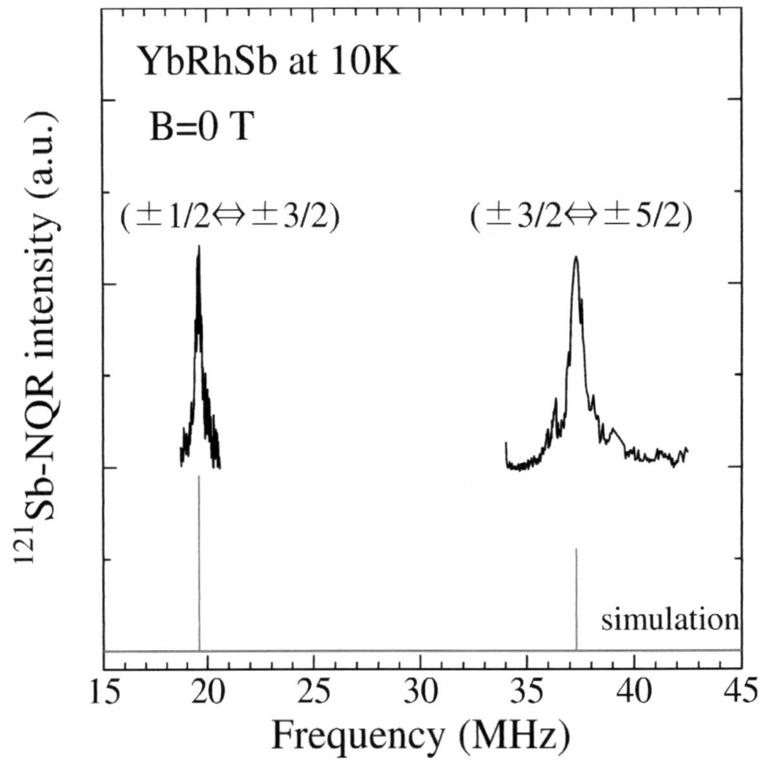

Figure 3. ^{121}Sb-NQR spectra of YbRhSb measured at 10 K. Respective resonance peaks at aroud ≈19.5 MHz and ≈37 MHz roughly correspond to the $(\pm\frac{1}{2} \leftrightarrow \pm\frac{1}{2})$ transition and $(\pm\frac{3}{2} \leftrightarrow \pm\frac{5}{2})$ transition.

We have succeeded in finding ^{121}Sb-NQR signals by using the NQR parameters. Figure 3 shows ^{121}Sb-NQR spectrum measured at 10 K. Red solid line is the result of the simulation with EFG parameters of $\nu_Q = 18.8$ MHz and $\eta = 0.2$. The good agreement between the experimental and simulation confirms that the obtained EFG parameters are valid.

4. Conclusions

We observed ^{121}Sb-NMR/NQR signals of a single crystal YbRhSb and determined ^{121}Sb NMR/NQR parameters. Notably, we succeeded in observing the Sb-NMR and NQR signals and determining EFG parameters, isotropic Knight shift and anisotropic Knight shift tensors at Sb site for the first time. Our NMR/NQR results suggest that the weak ferromagnetism in YbRhSb is ascribed to a canted antiferromagnetic state. The present result allows us to explore the origin of the weak ferromagnetism in YbRhSb by measuring temperature and field dependences of ^{121}Sb Knight shift and nuclear spin-lattice relaxation rate $^{121}(1/T_1)$.

5. Acknowledgments
This work has been supported by JSPS KAKENHI Grant Numbers 15H05885, (J-Physics), 26400359, and 15H05745, and partly supported by JSPS KAKENHI Grant Numbers, and 15H03689, 15K05882, and 15K21732.

References
[1] Takabatake T, Teshima F, Fujii H, Nishigori S, Suzuki T, Fujita T, Yamaguchi Y, Sakurai J, and Jaccard D, 1990 *Phys Rev. B* **41**, 9607.
[2] Aoki D, Huxley A, Ressouche E, Braithwaite D, Flouquet J, Brison J.P, Lhotel E, and Paulsen C, 2001 *Nature London* **413**, 613 .
[3] Bonville P, Bellot P, Hodges J.A., Imbert P, Jehanno G, Le Bras G, Hammann J, Leylekian L, Chevrier G, Thuery P, DOnofrio L, Hamzic A, and Barthelemy A,1992 *Physica B* **182**, 105.
[4] Kasaya M, Tani T, Kawate K, Mizushima T, Isikawa Y, and Sato K, 1991 *J. Phys. Soc. Jpn.* **60**, 3145.
[5] Drescher K, Abd-Elmeguid M.M, Micklitz H, and Sanchez J.P, 1996 *Phys. Rev. Lett.* **77**, 3228.
[6] Muro Y, Haizaki Y, Kim M.S., Umeo K, Tou H, Sera M, and Takabatake T, 2004 *Phys. Rev. B* **69**, 020401(R).
[7] Moriya T, in Magnetism I, edited by G.T. Rado and H. Suhl (*Academic*, New York, 1963) p. 85.

IOP Conf. Series: Journal of Physics: Conf. Series **807** (2017) 012015 doi:10.1088/1742-6596/807/1/012015

Magnetoresistance and Hall effect of antiferromagnetic uranium compound $URhIn_5$

Yoshinori Haga[1], Yuji Matsumoto[2], Jiří Pospíšil[1], Naoyuki Tateiwa[1], Etsuji Yamamoto[1], Tomoo Yamamura[3], Zachary Fisk[4]

[1]Advanced Science Research Center, Japan Atomic Energy Agency, Tokai, Ibaraki 319-1195, Japan
[2]Graduate School of Engineering, Nagoya Institute of Technology, Nagoya, 466-8555, Japan
[3]Institute for Materials Research, Tohoku University, Sendai, 980-8577, Japan
[4]Department of Physics and Astronomy, University of California, Irvine, California 92697, USA

E-mail: haga.yoshinori@jaea.go.jp

Abstract. Magnetoresistance and Hall effect in $URhIn_5$ have been measured across the antiferromagnetic transition temperature $T_N = 98$ K. In the paramagnetic state, Hall coefficient is small and temperature independent. Below the T_N, Hall coefficient suddenly changes to a negative value, suggesting that small light electron Fermi surface dominate electrical conductivity. The observation is consistent with small Fermi surfaces detected by previous de Haas-van Alphen experiment.

1. Introduction

Actinides and rare-earth compounds with the tetragonal $HoCoGa_5$-type structure (115 compounds) have extensively been studied because of the emergence of intriguing physical properties such as heavy fermion superconductivity and various types of magnetic ordering.[1] Substituting elements within the same structure enables us to study carrier doping and/or chemical pressure effects without changing lattice symmetry. In isostructural actinide compounds, uranium analogues can be formed as $UTGa_5$ where T stands for transition metal elements of group 8, 9 and 10. In contrast to rare-earth analogues which are formed with indium, those with uranium and other actinides are usually stabilized with gallium, reflecting the smaller atomic size of uranium than light rare-earths. $URhGa_5$ is known as a non-magnetic semietal where $5f$ electrons are itinerant to form conduction bands. Bulk physical properties and small Fermi surface pockets detected by the de Haas-van Alphen effect are consistent with semimetallic behavior.[2] It is interesting to note that the substitution of Rh by adjacent metals (Ru or Pd) alters the ground state: $URuGa_5$ is also non-magnetic, but expected to have large Fermi surfaces. $UPdGa_5$ is an antiferromagnet with large Fermi surfaces. The similar features are observed in the series of $UTGa_5$ with T = $3d$ (Fe, Co, Ni) and T = $5d$ (Os, Ir, Pt).[3] Together with the band structure calculations investigation, the substitution of the transition metal leads to a shift of the Fermi level in a relatively rigid band structure. [4]

Recently the first member of uranium 115 compounds having indium as a constituent elements, $URhIn_5$, has been synthesized in a single crystal form [5]. Unlike the isostructural and nonmagnetic $URhGa_5$, $URhIn_5$ orders antiferromagnetically with a significantly high transition temperature $T_N = 98$ K. More recently, a higher transition temperature 117 K has been reported for U_2RhIn_8 having the similar local structure.[6] Reflecting the larger atomic size of In than Ga, $URhIn_5$ has the larger interatomic distances, leaving an effectively large space for uranium atoms. Naively thinking, a larger atomic size leads to a localized state of $5f$ electrons with a magnetic ground state. The drastic difference of magnetic ground state also leads to a difference conduction band structure. In this paper, we report magnetoresistance and Hall effect measurements of $URhIn_5$.

2. Experimental

Single crystals of $URhIn_5$ have been grown from indium flux as described in ref. [1]. The sample was shaped in a rectangular bar for the transport measurements. Magnetoresistance and Hall resistivity were measured using an AC four-probe method in a temperature range between 2 and 300 K. Magnetic field up to 9 T was applied along the tetragonal c-axis, while the electrical current was applied along the a-axis.

3. Results and Analysis

Figure 1 shows the temperature dependence of magnetoresistance and Hall coefficient R_H of $URhIn_5$. In the paramagnetic state above 98 K, resistivity at 9 T coincides with that in 0 T, namely magnetoresistance is negligibly small. R_H in the paramagnetic state also shows temperature-independent and small values. The value is much smaller than that expected for a single carrier per unit cell volume (1.0×10^{-9} m^3/C). This is most likely due to multi-band effects. In general the conduction bands can involve many bands from different orbitals. The small values can arise from an accidental cancellation of the contributions from different bands. It should also be noted that the absence of the temperature dependence indicate that anomalous Hall effect often observed in magnetic actinide compounds is not significant in the paramagnetic state of $URhIn_5$.

In the antiferromagnetic state below $T_N = 98$ K, situations change drastically. With decreasing temperature, resistivity at 9 T gradually deviates from the 0 T values. Positive magnetoresistance is observed at low tempeartures below about 50 K. R_H shows an abrupt change at T_N. R_H shows negative values in the antiferromagnetic state and depends strongly on temperature. R_H decreases continuously with decreasing temperature below T_N, showing a minimum at 40 K and increase again with decreasing temperature. Negative sign indicates that the electron-like carrier mainly dominates electrical conduction. The R_H value at 40 K exceeds the value expected for a conventional one-carrier value, suggesting the smaller carrier number in the ordered state. This interpretation is consistent with the gap formation on the Fermi surface as suggested from a hump in the resistivity just below T_N. On the other hand, strong temperature dependence can be attributed to different contributions from bands with different characters.

To understand the complicated behavior in R_H, we assume a simple two-band compensated metal with a relatively small carrier number. In fact, at least one small Fermi surface pocket occupying a few per cent of Brillouin zone is found in the de Haas-van Alphen experiments.[7] In the analysis we assumed the contribution from anomalous Hall effect is small also in the antiferromagnetic state. Within the two-band model, the carrier number $n = 0.055$ per paramagnetic unit cell for the electron and hole bands was obtained in the antiferromagnetic state. This value is of the same order as the carrier number deduced from the Fermi surface volume obtained in the previous dHvA study. This result demonstrates the existence of small number of carrier in the antiferromagnetic state and reenforces the idea that a portion of the Fermi surface is gapped at T_N as suggested from resistivity behavior. Moreover, the present

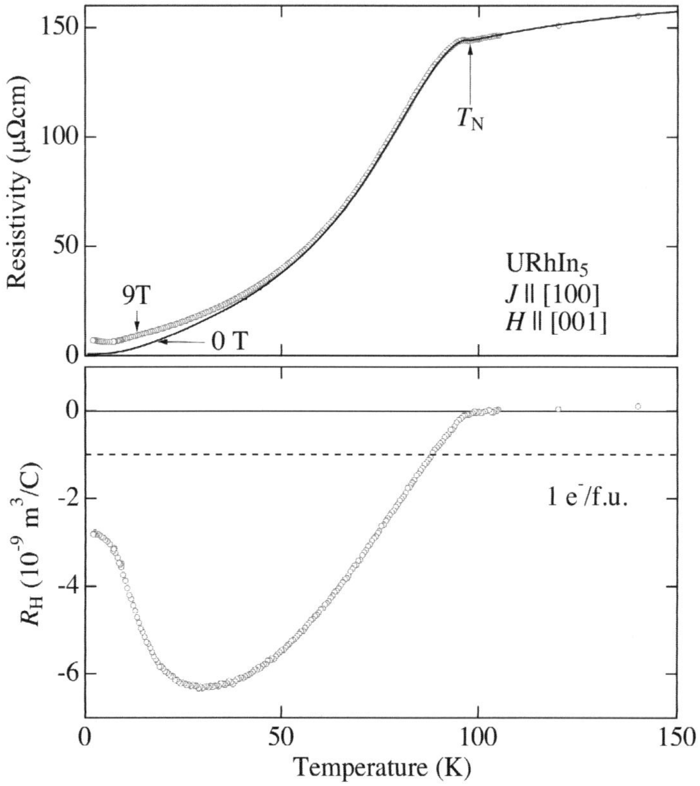

Figure 1. Upper panel: Temperature dependence of electrical resistivity in $URhIn_5$ measured at 0 T and 9 T. An arrow indicates the Néel temperarture. Lower panel: Temperature dependence of Hall coefficient in $URhIn_5$ at 9 T. Dashed line in lower panel corresponds to the Hall coefficient expected for one conduction electron per unit cell volume.

result denies the possible existence of larger Fermi surfaces which dHvA could not detect due to heavier mass and hence weak signal amplitudes.

We omit to analyze the data at $T < 30$ K, because our simple assumption of the compensated two-band picture might not be valid. Magnetoresistance follows H^2 above 30 K and the present the two-band analysis seems valid. However, a clear deviation from H^2 dependence is observed below 30 K. The detailed analyses of the field dependence are needed to understand the low-temperature behavior. The Hall resistivity also becomes field-dependent, most likely due to complicated multi-band effects beyond the simple two-band picture. It should be noted, however, that there are several experimental signatures suggesting a low temperature anomaly in the electronic state. Magnetic susceptibility shows an upturn below about 30 K[5]. Such an upturn would be attributed to the effect of paramagnetic impurities. However, the present anomaly is robust even in high magnetic field. Peculiar temperature dependence of the nuclear relaxation rate below about 30 K strongly suggest a recovery of density of states near Fermi level. [8] The analysis of the low temperature behavior to understand those anomalies is left for future study.

4. Summary

We investigated magnetoresistance and Hall effect on $URhIn_5$ both in paramagnetic and antiferromagnetic states. In contrast to the small Hall coefficient in the paramagnetic state, a significant negative Hall coefficient was observed in the antiferromagnetic state, indicating relatively small carrier number. The observation suggests that Fermi surface is partially gapped at the antiferromagnetic transition. The two-band analysis gave a carrier number of 5.5 % per formula unit, in agreement with the dHvA study.

References

[1] Thompson J D and Fisk Z 2011 *J. Phys. Soc. Jpn.* **81** 011002
[2] Ikeda S, Tokiwa Y, Okubo T, Haga Y, Yamamoto E, Inada Y, Settai R and Ōnuki Y 2002 *J. Nucl. Sci. Technol.* **Suppl. 3** 206
[3] Ikeda S, Tokiwa Y, MatsudaT D, Galatanu A, Yamamoto E, Nakamura A, Haga Y and Ōnuki Y 2005 *Physica B* **359-361** 1039
[4] Maehira T, Hotta T, Ueda K and Hasegawa A 2006 *New J. Phys.* **8** 24
[5] Matsumoto Y, Haga Y, Tateiwa N, Sakai H, Matsuda T D, Yamamoto E, Fisk Z 2013 *Phys. Rev. B* **88** 045120
[6] A. Bartha, M. Kratochvilova, M. Dusek, M. Divis, J. Custers, V. Sechovsky, J. Magn. Magn. Mater. **381**, 310 (2015).
[7] Matsumoto Y, Haga Y, Tateiwa N, Yamamoto E, Kimura N, Aoki H, Fisk Z 2014 *JPS Conf. Ser.* **3** 011097
[8] Sakai H, Kambe S, Tokunaga Y, Matsumoto Y, Tateiwa N, Haga Y, Fisk Z 2013 *Phys. Rev. B* **88** 045123

IOP Conf. Series: Journal of Physics: Conf. Series **807** (2017) 022001 doi:10.1088/1742-6596/807/2/022001

Ce 3p hard x-ray photoelectron spectroscopy study of the topological Kondo insulator $CeRu_4Sn_6$

M Sundermann[1], K Chen[1], Y Utsumi[2,3], Y-H Wu[4], K-D Tsuei[4], J Haenel[5], A Prokofiev[5], S Paschen[5], A Tanaka[6], L H Tjeng[2] and A Severing[1]

[1] Institute of Physics II, University of Cologne, Zülpicher Straße 77, 50937, Köln, Germany
[2] Max Planck Institute for Chemical Physics of Solids, Nöthnitzerstraße 40, 01187 Dresden, Germany
[3] Synchrotron SOLEIL, L'Orme des Merisiers, Saint-Aubin PB48, 91192 Gif-sur-Yvette, France
[4] National Synchrotron Radiation Reasearch Center, Hsinchu, Taiwan
[5] Institute of Solid State Physics, Vienna University of Technology, Austria
[6] Department of Quantum Matter, ADSM Hiroshima University, Japan

E-mail: sundermann@ph2.uni-koeln.de

Abstract. Bulk sensitive hard x-ray photoelectron spectroscopy data of the Ce 3p core level of $CeRu_4Sn_6$ are presented. Using a combination of full multiplet and configuration iteration model we were able to obtain an accurate lineshape analysis of the data, thereby taking into account correlations for the strong plasmon intensities. We conclude that $CeRu_4Sn_6$ is a moderately mixed valence compound with a weight of 8 % for the Ce f^0 configuration in the ground state.

1. Introduction

Kondo insulators are presently experiencing a renaissance since they are promising candidates for correlated materials with nontrivial topology [1, 2, 3, 4, 5, 6, 7]. Such predictions caused a flurry of experimental investigations on SmB_6 (see Ref. [7] and references therein) and more recently we suggested that also $CeRu_4Sn_6$ might be a good candidate for a strongly correlated material with topologically protected surface states [8].

$CeRu_4Sn_6$ is a tetragonal Kondo insulator where the hybridization of f and conduction electrons leads to the opening of a gap [9, 10, 11, 12, 13] at low temperature. Accordingly, the resistivity rises as temperature decreases, but saturates below $T \approx 10$ K. Our recent spectroscopic investigation of $CeRu_4Sn_6$ showed that it has a large Kondo temperature of about 170 K and a Γ_6 ground state crystal-field symmetry ($J_z = \pm 1/2$) [8] which is the favorable symmetry for strong hybridization and opening of a gap according to Dzero *et al.* [3, 4]. The opening of a gap in the presence of the $J_z = 1/2$ symmetry is supported by our band structure calculations and we inferred that $CeRu_4Sn_6$ must be topologically non trivial.

In that work [8] we had deduced the Kondo temperature from the temperature evolution of the $4f^0$ spectral weight in the L_3-edge absorption data where we had detected the $L_{\alpha 1}$ emission line. A difficulty of the analysis of the L-edge data in general is the assignment of the spectral weights, even when measured in the life time reduced partial fluorescence yield mode, since the line shape reflects the empty 5d density of states in the presence of the core hole. This uncertainty does not affect the temperature dependence but the absolute numbers for the spectral weights. In Ref. [8] we have been fitting empirical, identical line shapes to each f^n configuration and found an absolute f^0 occupation of $\approx 6\%$. Hard x-ray photoelectron spectroscopy (HAXPES) is a bulk sensitive technique [14, 15, 16, 17] and its data can be analyzed more quantitatively [18, 19] than that of x-ray absorption.

In cerium based Kondo systems the electron in the f shell hybridizes with the conduction electrons so that the resulting hybridized ground state is a mixed state that can be written as $\Psi_{\mathrm{GS}} = c_0 \mid f^0\rangle + c_1 \mid f^1\, \underline{v}\,\rangle + c_2 \mid f^2 \underline{\underline{v}}\,\rangle$ where the main component is given by the trivalent f^1 configuration with additional contributions of the divalent and tetravalent states (f^2 and f^0) with weights $w_n \equiv |c_n|^2$ $(n = 0,1,2)$. Here $\underline{v}$ $\left(\underline{\underline{v}}\right)$ denotes one (two) hole in the valence states. A core hole acts differently on the respective f^n configurations so that they are split in a PES and XAS experiment and the spectra exhibit three spectral weights that are related to the initial state f^n admixtures. However, due to strong final state effects *spectral* weights and f^n weights w_n in the ground state are not related 1:1. Here a configuration interaction model is required for converting between the two. Plasmons may complicate the situation when they cover up or distort the f^n spectral weights. In such cases the spectra have to be corrected for plasmons by performing a full multiplet calculation that contains the plasmons as part of the lineshape (see Analysis). Below we present HAXPES data of $CeRu_4Sn_6$ at $T \leq 90\,K$ that have been analyzed quantitatively with the combination of a full multiplet and configuration interaction calculation.

2. Experiment & Analysis

Single crystals were grown by the self-flux floating-zone melting method using optical heating in a four-mirror furnace [20]. The samples are single grain and single phase, as shown by Laue and x-ray diffraction, scanning electron microscopy, and energy dispersive x-ray spectroscopy.

Hard x-ray core level photoemission spectra are performed at the Taiwan beamline BL12XU at the SPring-8 synchrotron radiation facility in Japan. The high incident energy of 6.47 keV ensures bulk sensitivity. The photo electrons were detected by a MB Scientific A-1HE analyzer with a pass energy of 200 meV in vertical geometry for the Ce $3d$ test measurement and in horizontal geometry [21] for the Ce $3p$ and Sn $3p$, yielding an instrumental FWHM of about 1 eV. The Fermi energy was determined by a Au thin film measured with same conditions. The samples were cooled down and cleaved in ultra high vacuum of $\approx 10^{-8}$ mbar and were then immediately transferred to the measurement position (pressure $<3\cdot 10^{-10}$ mbar and $T<90\,K$). The quality of the sample and sample surface was verified by performing long energy scans that verify the absence of extra peaks (Fermi energy to 1500 eV binding energy).

We have analyzed the HAXPES data with a combination of full multiplet and configuration interaction model. We use the XTLS 9.0 code [22] for these calculations with input parameters from atomic calculations using Cowans atomic code [23]. The code uses a configuration interaction model based on a single non-dispersive valence state [24] combined with a full multiplet calculation. The plasmon parameters are determined from emission lines that are not subject to the configuration interaction, here the Sn $3p$, so that, once the overall lineshape of the multiplet is adjusted, the configuration interaction model only depends on four parameters, Coulomb exchange interaction between the f electrons (U_{ff}) and between the f electrons and core hole (U_{fc}), the effective f-electron binding energy ϵ_f and the isotropic hybridization V_{eff}. These parameters are uniquely defined by the intensity ratios and energy differences of the three valence states in the spectra. More details about the simulation can be found in Ref. [25] and [26].

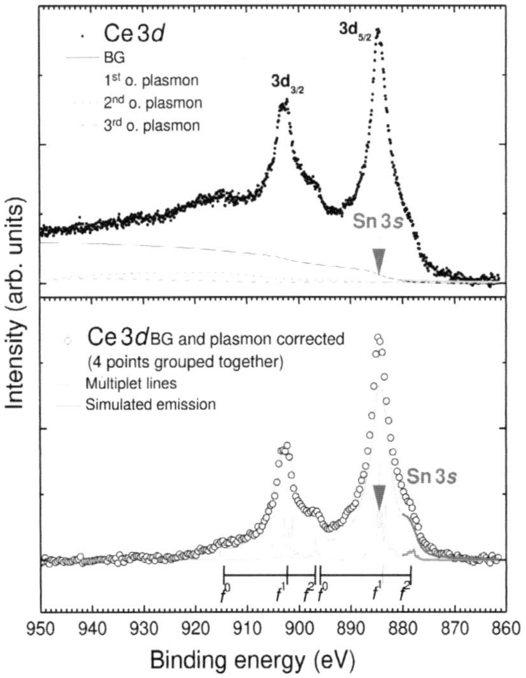

Figure 1. HAXPES data of Ce $3d$ core level emission. Top: data (black dots), Shirley-type background (black line), and plasmon intensities (green lines). Bottom: consistency check of the fit to Ce $3p$ data. in Fig. 2. Background and plasmon corrected data (black circles), multiplet structure from configuration interaction model with parameters as obtained from fit to Ce $3p$ in Fig. 2 (orange line) and total calculated Ce $3d$ emission intensity (red line). The discrepancy is due to the Sn $3s$. The bottom ruler indicates the position of the emission for the different $4f$ occupancies.

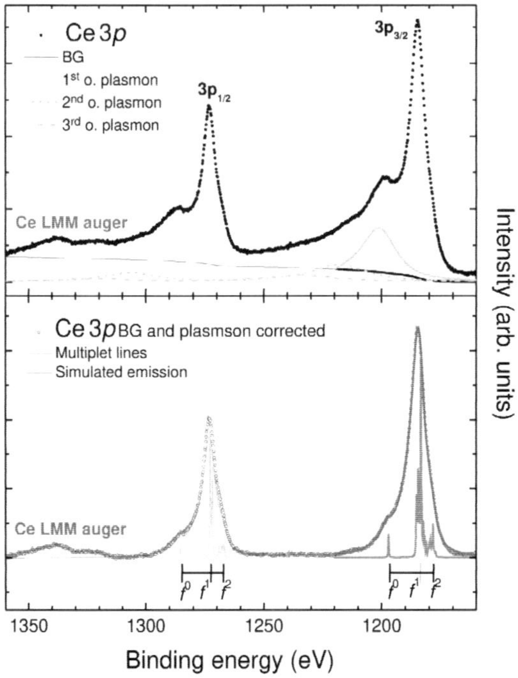

Figure 2. HAXPES data of Ce $3p$ core level emission. Top: data (black dots), Shirley-type background (black line), and plasmon intensities (green lines). Bottom: fit to background and plasmon corrected data (black circles), multiplet structure from optimized configuration interaction model (orange line) and total calculated Ce $3p$ emission intensity. The bottom ruler indicates the position of the emission for the different $4f$ occupancies.

3. Results

For the rare earth the $3d$ core level spectrum is typically used for determining the valence state. In Fig. 1 (top) the Ce $3d$ spectrum of $CeRu_4Sn_6$ is shown. There are two sets of emission lines due to the $3d$ spin-orbit splitting ($3d_{3/2}$ and $3d_{5/2}$). However, in $CeRu_4Sn_6$ the Sn $3s$ emission appears at 885 eV binding energy (see red arrow) so that it severely distorts the spectral weights. We therefore measured the Ce $3p$ emission line which is shown in Fig. 2 (top). The advantage of the Ce $3p$ is also that the two spin-orbit contributions do not overlap because the spin-orbit coupling of the Ce $3p$ electrons is much stronger and the resulting splitting of the Ce $3p_{1/2}$ and Ce $3p_{3/2}$ much larger than for Ce $3d$ (note the different energy scales of Fig. 1 and 2). A drawback is the stronger intrinsic lifetime broadening due to the higher binding energy. The expected averaged energy positions of the f^n configurations are marked by the ruler at the bottom of Fig. 1 and 2, respectively.

Before analyzing the Ce data further we look at the Sn $3p$ emission lines which are also

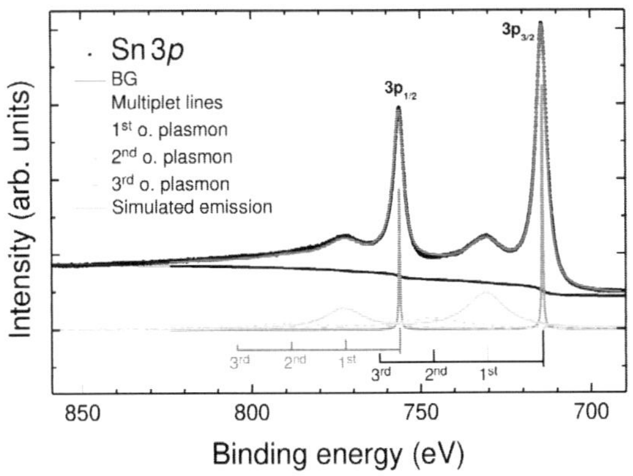

Figure 3. HAXPES data of the Sn $3p$ core level emission. Data (black dots), Shirley-type background (black line), main emission lines (orange line) and fitted plasmon intensities (green lines). The red line is the sum of the calculated contributions.

spin orbit split ($3p_{1/2}$ and $3p_{3/2}$) (see Fig. 3 (top)). The multiplet structure consists of a single emission line and Sn is not affected by the configuration interaction so that the additional humps at about 16 eV distance from the main emission lines are identified as strong plasmons. These plasmons have about the same energy distance from the main emission line as the expected f^0 contributions so that these plasmon intensities have to be taken into account when analyzing the Ce $3p$ spectra. After subtracting a Shirley-type background the plasmon parameters (energy positions, intensity ratios, and widths of 1st, 2nd, and 3rd order plasmons) are determined form the Sn $3p$ emission. In Fig. 3 (top) the black line shows the Shirley-type background and the green lines the plasmon contributions (solid 1st order, dashed 2nd order, and dotted 3rd order). The orange line is the multiplet line broadened with a Gaussian FWHM=1 eV and Lorentzian FWHM=3.1 eV as well as a Mahan lineshape with energy cut-off ξ=8 eV and asymmetry factor α=0.1, and the red line is the total fit. The plasmon parameters are listed in Table 1.

We now use these plasmon parameters for describing the Ce $3p$ multiplet structure i.e. we attach the plasmon of this shape and energy to each emission line [25, 26]. In Fig. 2 (top) the Shirley-type background (black) and resulting plasmon intensities (green) are shown. In the bottom part the background and plasmon corrected data are shown. There is clearly a strong shoulder at the energy position where we expect the f^0 spectral weight. The combination of multiplet and configuration interaction calculation yields the multiplet structure (orange). For the fit the multiplet emission lines are broadened with a Gaussian FWHM=1 eV, a Lorentzian FWHM=3.8 eV and a Mahan lineshape with energy cut-off ξ=8 eV and asymmetry factor α=0.4. The red line is the total fit. The resulting weights w_n of the f^n configurations in the initial state and the resulting configuration interaction parameters are listed in Table 2. The resulting f^0 weight (w_0) to the ground state amounts to 8%. The strong Lorentzian broadening and overlap between f^1 and f^2 limits the precise separation between theses contribution so that we give the sum of w_1 and w_2. When fitting the main emphasis has been put on the Ce $3p_{3/2}$ part of the data since it is not affected by the Auger intensities.

As a consistency check the parameters from the Ce $3p$ emission data were used for Ce $3d$. The black and green lines in Fig. 1 (top) refer to the Shirley-type background and the plasmon intensities, using the parameters from Table 1. The bottom panel of Fig. 1 shows the background an plasmon corrected data (4 energy intervals grouped together), the multiplet structure (orange) resulting from the multiplet calculation with the configuration interactions parameters from Table 2 and the calculated emission (red line). The multiplet lines of the Ce $3d$ are broadened with a Gaussian FWHM=1 eV, a Lorentzian FWHM=1.6 eV and a Mahan lineshape with energy cut-off ξ=8 eV and asymmetry factor α=0.3. The Ce $3d_{3/2}$ part is quite well described while the Ce $3d_{5/2}$ shows some discrepancy between calculation and data. The latter was expected because of the contribution of Sn $3s$. Note, that the Ce $3d_{3/2}$ is still affected by the plasmon of

Table 1. Properties of the plasmons in $CeRu_4Sn_6$.

Order of plasmon	1	2	3	4
Scaling factor	0.5	0.25	0.13	0.06
Energy shift (eV)	16.1	32.2	48.3	64.4
Lorentzian FWHM (eV)	10.0	20.0	30.0	40.0

Table 2. Optimized configuration interaction parameters (U_{ff}, U_{fc}, V_{eff}, ϵ_f) and resulting f^n weights w_n in the ground state in %.

U_{ff}(eV)	U_{fc}(eV)	ϵ_f(eV)	V_{eff}(eV)	w_0(%)	w_1+w_2(%)
9.1(9)	10.6(9)	-2.5(3)	0.26(2)	8(2)	92(2)

the Sn 3s.

The present experiment and analysis of the Ce 3p core level emission data finds 8% of f^0 configuration in the initial state. The value is slightly larger than in our previous L-edge absorption study [8] since in the present analysis final state effects have been taken into account. Comparing results of the same type of analysis, we can state that hybridization and occupation of f^0 of $CeRu_4Sn_6$ are only slightly larger than in $CeRu_2Si_2$, a magnetically non-ordering, metallic heavy fermion compound. However, ground state symmetries, that also have an impact on the efficiency of hybridization are not taken into account in such a comparison.

We summarize that the Ce 3p core level is an alternative to the Ce 3d for determining valencies with photoemission and the present work provides a more quantitative value for the absolute f^0 weight in the ground state of $CeRu_4Sn_6$, thus confirming its moderate intermediate valency.

4. Acknowledgment

K.C. M.S. and A.S. are grateful for support from the German funding agency DFG through Project 600575. J.H. and S.P. are grateful for the support from the Austrian Science Fund FWF through the project I2535-N27 and doctoral school W1243.

References

[1] Dzero M, Sun K, Galitski V and Coleman P 2010 *Phys. Rev. Lett.* **104**(10) 106408 http://link.aps.org/doi/10.1103/PhysRevLett.104.106408
[2] Takimoto T 2011 *Journal of the Physical Society of Japan* **80** 123710 http://dx.doi.org/10.1143/JPSJ.80.123710
[3] Dzero M, Sun K, Coleman P and Galitski V 2012 *Phys. Rev. B* **85**(4) 045130 http://link.aps.org/doi/10.1103/PhysRevB.85.045130
[4] Dzero M and Galitski V 2013 *Journal of Experimental and Theoretical Physics* **117** 499–507 ISSN 1090-6509 http://dx.doi.org/10.1134/S1063776113110083
[5] Alexandrov V, Dzero M and Coleman P 2013 *Phys. Rev. Lett.* **111**(22) 226403 http://link.aps.org/doi/10.1103/PhysRevLett.111.226403
[6] Lu F, Zhao J, Weng H, Fang Z and Dai X 2013 *Phys. Rev. Lett.* **110**(9) 096401 http://link.aps.org/doi/10.1103/PhysRevLett.110.096401
[7] Dzero M, Xia J, Galitski V and Coleman P 2016 *Annual Review of Condensed Matter Physics* **7** 249–280 http://dx.doi.org/10.1146/annurev-conmatphys-031214-014749
[8] Sundermann M, Strigari F, Willers T, Winkler H, Prokofiev A, Ablett J, Rueff J P, Schmitz D, Weschke E, Moretti Sala M, Al-Zein A, Tanaka A, Haverkort M, Kasinathan D, Tjeng L, Paschen S and Severing A 2015 *Scientific Reports* **5** 17937 http://dx.doi.org/10.1038/srep17937
[9] Das I and Sampathkumaran E V 1992 *Phys. Rev. B* **46**(7) 4250–4252 http://link.aps.org/doi/10.1103/PhysRevB.46.4250
[10] Strydom A, Guo Z, Paschen S, Viennois R and Steglich F 2005 *Physica B: Condensed Matter* **359361** 293 –

295 ISSN 0921-4526 proceedings of the International Conference on Strongly Correlated Electron Systems http://www.sciencedirect.com/science/article/pii/S092145260500116X

[11] Paschen S, Winkler H, Nezu T, Kriegisch M, Hilscher G, Custers J, Prokofiev A and Strydom A 2010 *Journal of Physics: Conference Series* **200** 012156 http://stacks.iop.org/1742-6596/200/i=1/a=012156

[12] Brüning E M, Brando M, Baenitz M, Bentien A, Strydom A M, Walstedt R E and Steglich F 2010 *Phys. Rev. B* **82**(12) 125115 http://link.aps.org/doi/10.1103/PhysRevB.82.125115

[13] Wissgott P and Held K 2016 *The European Physical Journal B* **89** 1–7 ISSN 1434-6036 http://dx.doi.org/10.1140/epjb/e2015-60753-5

[14] Tanuma S, Powell C and Penn D 2011 *Surface and Interface Analysis* **43** 689–713 ISSN 1096-9918 http://dx.doi.org/10.1002/sia.3522

[15] Braicovich L, Brookes N B, Dallera C, Salvietti M and Olcese G L 1997 *Phys. Rev. B* **56**(23) 15047–15055 http://dx.doi.org/10.1103/PhysRevB.56.15047

[16] Laubschat C, Weschke E, Holtz C, Domke M, Strebel O and Kaindl G 1990 *Phys. Rev. Lett.* **65**(13) 1639–1642 http://dx.doi.org/10.1103/PhysRevLett.65.1639

[17] Sekiyama A, Iwasaki T, Matsuda K, Saitoh Y, Onuki Y and Suga S 2000 *Nature* **403** 396 – 398 http://dx.doi.org/10.1038/35000140

[18] Allen J, Oh S, Gunnarsson O, Schönhammer K, Maple M, Torikachvili M and Lindau I 1986 *Advances in Physics* **35** 275–316 http://dx.doi.org/10.1080/00018738600101901

[19] Hüfner S 1992 *Zeitschrift für Physik B Condensed Matter* **86** 241–246 ISSN 1431-584X http://dx.doi.org/10.1007/BF01313831

[20] Prokofiev A and Paschen S 2012 *Modern Aspects of Bulk Crystal and Thin Film Preparation* (InTech) chap Crystal Growth and Stoichiometry of Strongly Correlated Intermetallic Cerium Compounds http://www.intechopen.com/books/modern-aspects-of-bulk-crystal-and-thin-film-preparation/crystal-growth-and-stoichiometry-of-strongly-correlated-intermetallic-cerium-compounds

[21] Weinen J, Koehte T C, Agrestini S, Kasinathan D, Strigari F, Haupricht T, Liao Y F, Tsuei K D and Tjeng L H 2016 *Hard X-ray Photoelectron Spectroscopy (HAXPES)* (Springer International Publishing) chap 11 Springer Series in Surface Sciences 59 ISBN 978-3-319-24041-1 http://link.aps.org/doi/10.1007/978-3-319-24043-5

[22] Tanaka A and Jo T 1994 *J. Phys. Soc. Jpn.* **63** 2788–2807 http://dx.doi.org/10.1143/JPSJ.63.2788

[23] Cowan R D 1981 *The Theory of Atomic Structure and Spectra* (Berkeley: University of Calivornia Press)

[24] Imer J M and Wuilloud E 1987 *Zeitschrift für Physik B Cond. Mat.* **66** 153–160 ISSN 0722-3277 http://dx.doi.org/10.1007/BF01311650

[25] Strigari F, Sundermann M, Muro Y, Yutani K, Takabatake T, Tsuei K D, Liao Y, Tanaka A, Thalmeier P, Haverkort M, Tjeng L and Severing A 2015 *Journal of Electron Spectroscopy and Related Phenomena* **199** 56–63 ISSN 0368-2048 http://www.sciencedirect.com/science/article/pii/S036820481500016X

[26] Sundermann M, Strigari F, Willers T, Weinen J, Liao Y, Tsuei K D, Hiraoka N, Ishii H, Yamaoka H, Mizuki J, Zekko Y, Bauer E, Sarrao J, Thompson J, Lejay P, Muro Y, Yutani K, Takabatake T, Tanaka A, Hollmann N, Tjeng L and Severing A 2016 *Journal of Electron Spectroscopy and Related Phenomena* **209** 1 – 8 ISSN 0368-2048 http://www.sciencedirect.com/science/article/pii/S0368204816300123

IOP Conf. Series: Journal of Physics: Conf. Series **807** (2017) 022002 doi:10.1088/1742-6596/807/2/022002

Uniaxial pressure effects on the unusual antiferromagnetic transition in the Kondo semiconductor $CeOs_2Al_{10}$

Kyosuke Hayashi[1], Kazunori Umeo[1,2,*], Yoshihiro Yamada[1], Jo Kawabata[1], Yuji Muro[4], and Toshiro Takabatake[1,3]

[1] Graduate School of Advanced Sciences of Matter, Hiroshima University, Higashi-Hiroshima, Hiroshima 739-8530, Japan

[2] Cryogenics and Instrumental Analysis Division, N-BARD, Hiroshima University, Higashi-Hiroshima, Hiroshima 739-8526, Japan

[3] Institute for Advanced Materials Research, Hiroshima University, Higashi-Hiroshima, Hiroshima 739-8530, Japan

[4] Department of Intelligent Systems Design Engineering, Faculty of Engineering, Toyama Prefectural University, Imizu, Toyama 939-0398, Japan

E-mail: kumeo@sci.hiroshima-u.ac.jp

Abstract. We have studied the relation between the unusual antiferromagnetic (AFM) ordering and anisotropic *c-f* hybridization in $CeOs_2Al_{10}$ by measuring the magnetic susceptibility χ under uniaxial pressures P up to 0.45 GPa. For $P \,//\, B \,//\, a$ and $P \,//\, B \,//\, c$, the AFM ordering temperature T_N = 28.5 K hardly changes with increasing P, while the value of χ decreases and the maximal temperature T_χ^{max} of $\chi(T)$ shifts to higher temperatures. These findings suggest that the intensity of *c-f* hybridization is enhanced along the pressurized *a* and *c* axes. On the other hand, for $P \,//\, b$, T_N increases up to 30 K at 0.15 GPa, while the value of χ remains unchanged. This observation for $P \,//\, b$ suggests that the net intensity of *c-f* hybridization decreases, while the *c-f* hybridization along the *b*-axis hardly changes. We discuss these results in terms of the atomic displacements of Os, Al(3) and Al(4) under uniaxial pressures.

1. Introduction

Cerium based intermetallic compounds display rich physical properties, such as, heavy-fermion state, unusual magnetically ordered state, and semiconducting state, which arise from the hybridization between the conduction electrons and localized 4*f* electrons, namely, *c-f* hybridization. Among these compounds, CeNiSn and CeRhSb, so called Kondo semiconductors, possess an anisotropic (pseudo) gap in the density of states at the Fermi level due to the anisotropic *c-f* hybridization [1]. In fact, applying uniaxial pressure on CeNiSn along the orthorhombic *a*-axis enhances the *c-f* hybridization, increasing the gap magnitude [2].

The conventional Kondo semiconductors did not exhibit magnetic ordering because of the strong *c-f* hybridization. However, it has been established that CeT_2Al_{10} (T = Ru, Os) exhibit both Kondo semiconducting behavior and antiferromagnetic (AFM) orders at T_N = 27.3 and 28.5 K, respectively [3-8]. Both compounds crystallize in the orthorhombic $YbFe_2Al_{10}$-type structure, in which Ce and T atoms form a zigzag chain along the *c*-axis as shown in Fig. 1. The easy magnetization axis is the *a*-axis in the paramagnetic state, while the AFM ordered moment μ_{AFM} is oriented parallel to the *c*-axis. Although the magnitudes of μ_{AFM} of 0.3~0.4μ_B/Ce in CeT_2Al_{10} are smaller than 7μ_B/Gd for GdT_2Al_{10}, T_N's in the Ce compounds are higher than those in the Gd counterparts, respectively [9,10]. An optical conductivity study has suggested that the unusual AFM order is induced by the formation of charge density wave occurring along the *b*-axis [11]. In spite of intensive works, the origin of the AFM order remains unsolved yet.

The analyses of the magnetic susceptibility by crystal field model and of polarization-dependent soft x-ray absorption spectra of CeT_2Al_{10} have revealed that the Ce 4*f* wave function lies in the *b*-*c* plane [12,13]. It suggested that the hybridization between the Ce 4*f* electrons and Al 3*p* electrons is important for the unusual AFM. Furthermore, for $CeRu_2Al_{10-y}Si_y$, the T_N and the absolute value of paramagnetic Curie temperature $|\theta_P|$ decrease with increasing *y*. The decreases in T_N and $|\theta_P|$ with *y* coincide with those reported in the 4*d* electron doped system $Ce(Ru_{1-x}Rh_x)_2Al_{10}$ with respect to the number of doped electrons per formula unit. This coincidence indicates that the Ru 4*d*- and Al 3*p*- electrons in $CeRu_2Al_{10}$ play the equivalent role in the unusual AFM order [14,15]. Namely, the hybridization of the Ce 4*f* electrons with the 4*d*-3*p* hybridized conduction band is essential in AFM order in $CeRu_2Al_{10}$[16]. On the other hand, the variations of lattice parameters in the series of LnT_2Al_{10} (Ln = lanthanides, T = Fe, Ru, Os) suggested that the intensity of the *c*-*f* hybridization is stronger in the *a*-*c* plane than that along the *b*-axis [10,17].

In order to examine the relation between the unusual AFM order and *c*-*f* hybridization, the pressure effects on the resistivity and magnetic susceptibility χ of $CeOs_2Al_{10}$ have been investigated previously [18]. The T_N of $CeOs_2Al_{10}$ reached a maximum of 29 K at 0.7 GPa, and suddenly disappeared at P_c = 2.5 GPa. On the other hand, the maximal temperature in $\chi(T)$, which is proportional to the Kondo temperature, increased linearly with pressure. However, hydrostatic pressure studies provided little information of the role of the anisotropic *c*-*f* hybridization in the unusual AFM order, because the contraction under hydrostatic pressures is almost isotropic for $P < P_c$ [19]. In order to obtain clearer information, we have measured the magnetization under the uniaxial pressures along the principal axes.

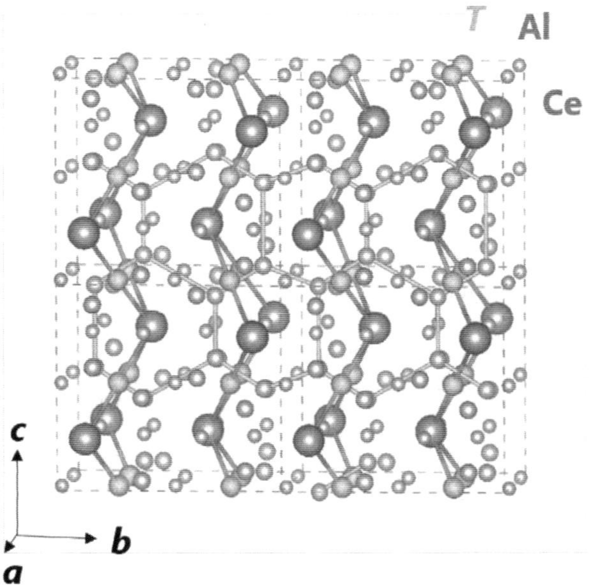

Figure 1. Crystal structure of CeT_2Al_{10} [11].

2. Experiment

Single-crystalline samples of $CeOs_2Al_{10}$ were grown using an Al self-flux method as described previously [6]. Powder x-ray diffraction patterns of the samples confirmed the $YbFe_2Al_{10}$-type structure. The samples were cut into plates of the width of 0.5 ~ 0.8 mm with the masses of 8 ~ 15 mg for *P* // *a* and *P* // *c* and 25 mg for *P* // *b*, respectively. The sample plate was sandwiched between ZrO_2 pistons in the homemade pressure cell. The magnetization was measured by using a commercial superconducting quantum interference device magnetometer (Quantum Design MPMS) at a magnetic field $B = 4$ T in the temperature range from 2 to 300 K. The direction of the uniaxial pressure *P* is always parallel to *B*. The pressure was determined by the known pressure dependence of the superconducting transition temperature of tin.

3. Results and Discussion

Figure 2 shows the temperature dependence of the magnetic susceptibility χ of $CeOs_2Al_{10}$. The values of χ for *P* // *a* and *P* // *c* decrease with pressure. On the other hand, the value of χ for *P* // *b* hardly changes for $P \leq 0.15$ GPa, while it slightly decreases for $P \geq 0.2$ GPa as shown in the inset of Fig. 2 (b). Furthermore, maximal temperatures T_χ^{max} of $\chi(T)$ increase from 43 K at 0 GPa up to 48 K at *P* // *a* = 0.45 GPa and to 50 K at *P* // *c* = 0.31 GPa.

In order to examine the pressure dependence of T_N for *P* // *a*, we plot the $d(\chi T)/dT$ vs *T* in Figure 3 (a). The peak temperatures of $d(\chi T)/dT$ do not change up to 0.45 GPa. Moreover, for *P* // *c*, the anomaly in $\chi(T)$ due to the AFM order also hardly changes as shown in Fig. 3 (b). However, as shown in Fig. 4, for *P* // *b*, the T_N increases up to 0.15 GPa and decreases for $P > 0.15$ GPa.

Figure 5 shows the dependences of T_N of $CeOs_2Al_{10}$ on uniaxial pressures and hydrostatic pressure. The inset displays the pressure dependence of T_χ^{max}. For *P* // *b*, T_N increases up to 30 K at 0.15 GPa,

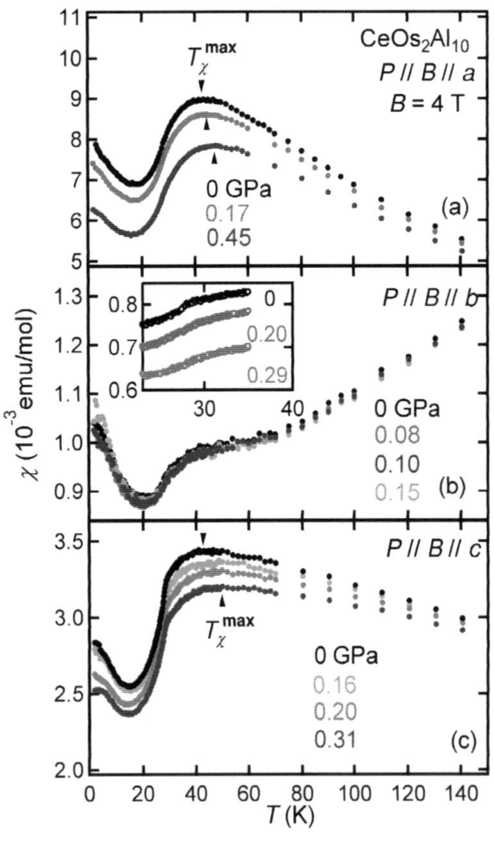

Figure 2. Temperature dependence of the magnetic susceptibility χ of $CeOs_2Al_{10}$ under uniaxial pressures. The inset in Fig. 2 (b) shows the shifted data at around T_N for *P* // *b* > 0.2 GPa.

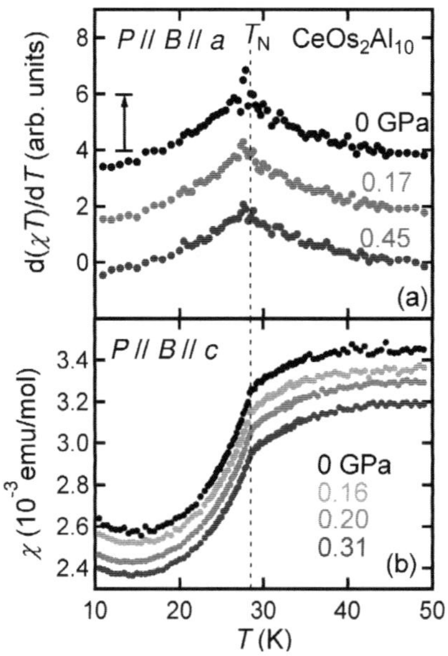

Figure 3. Temperature dependence of the $d(\chi T)/dT$ for *P* // *B* // *a* (a) and $\chi(T)$ for *P* // *B* // *c* (b) of $CeOs_2Al_{10}$.

while T_N does not change for P // a and P // c. This maximum of T_N = 30 K is higher than T_N = 29 K under the hydrostatic pressure of 0.7 GPa. Furthermore, the values of dT_χ^{max}/dP for P // a and P // c are 11 K/GPa and 23 K/GPa, respectively, which are larger than those of 7 K/GPa and 11 K/GPa for χ_a and χ_c under the hydrostatic pressure [18].

We obtained two important results: (i) for P // a and P // c, T_N does not change, although the *c-f* hybridizations along the pressurized direction become stronger. (ii) for P // b, T_N rises, whereas the *c-f* hybridizations along *b*-axis hardly changes. We discuss these results by considering the atomic displacement under uniaxial pressures as shown in Fig. 6. Here, we consider the hybridization among the Os 5*d*, Al(3), Al(4) 3*p* and Ce 4*f* charge cloud , because the 4*f* charge is distributed mainly in the *b*-*c* plane [12,13]. For simplicity, the Al(3) is not drawn in Fig. 6 (b – d), because Al(3) seems to move to the same direction of Al(4). First, we discuss (i). Applying pressure along the *a*-axis causes the extension of the lattice along the *b*-axis and *c*-axis. Here, we will consider the atomic displacement in the *a-c* plane, because the intensity of *c-f* hybridization along the *b*-axis is insensitive to the change in the *b*-axial length. One expects that each atom move along the arrows as shown in Fig. 6 (b). In this situation, the Os atoms approach to the 4 *f* charge cloud and the Al atoms move away from the Ce atom to the directions of arrows. The, the *c-f* hybridization in the *a-c* plane may hardly change, because the change of distance between the Os 5*d* - Al 3*p* hybridized charge cloud and the Ce 4*f* charge cloud is small. For P // c, as shown in Fig. 6 (c), Os and Al move to opposite directions to that for P // a. Thus, the *c-f* hybridization in the *a-c* plane also may be constant. Therefore, T_N hardly changes for P // a and P // c. On the other hand, for P // b, both Os and Al would move away from the 4 *f* charge cloud as shown in Fig. 6 (d). Thus, the hybridization between the Os 5*d* - Al 3*p* hybridized charge cloud and Ce 4*f* charge cloud is weakened, which results in the rise in T_N.

As we noted above, the hybridization between the Os 5*d* - Al(3), Al(4) 3*p* electrons and Ce 4*f* electron should play an important role in the unusual AFM order in $CeOs_2Al_{10}$. To confirm this scenario, we plan to measure the magnetization in the configuration $P \perp B$.

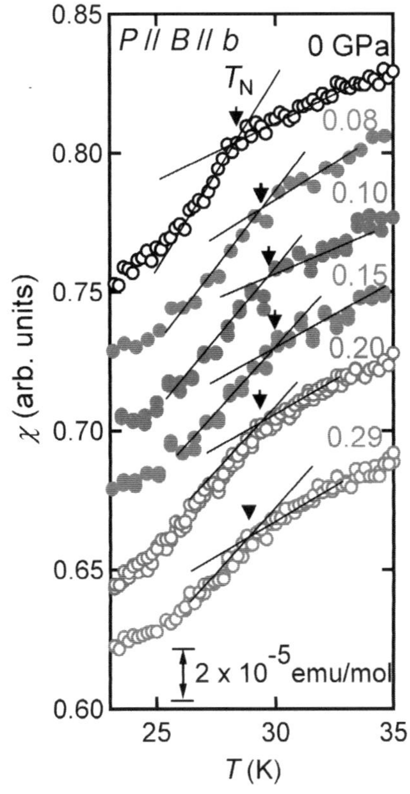

Figure 4. $\chi(T)$ for P // B // b. T_N was defined as the cross-point of the two lines as depicted in the figure.

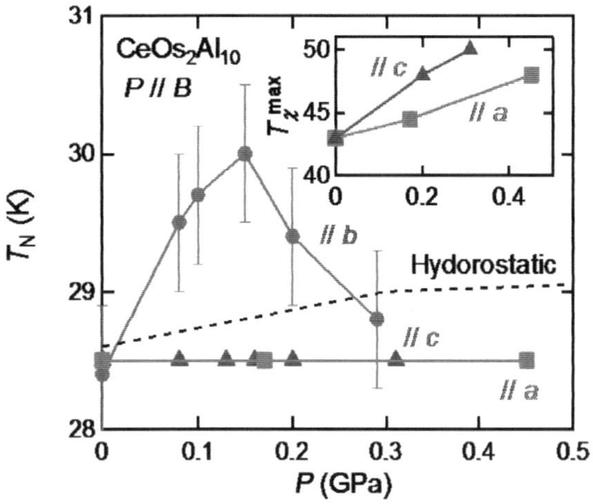

Figure 5. Pressure dependence of T_N of $CeOs_2Al_{10}$ for uniaxial and hydrostatic pressures [18]. The inset shows the pressure dependence of the maximal temperature of $\chi(T)$.

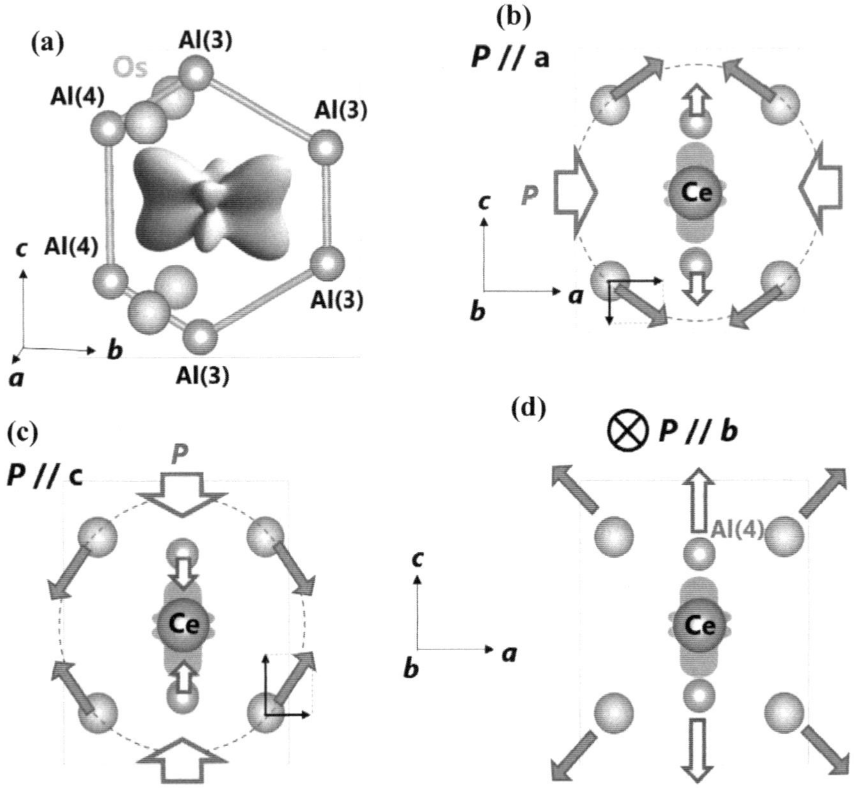

Figure 6. (a) Charge distribution of Ce 4*f* electrons in $CeOs_2Al_{10}$ [12,13]. (b – d) Displacements of Os and Al (4) under uniaxial pressures *P* // *a* (b), *P* // *c* (c) and *P* // *b* (d), respectively, viewed from the *b*-direction.

Acknowledgements
The magnetization measurements under uniaxial pressures were performed at N-BARD, Hiroshima University. This work was partly supported by KAKENHI Grant Numbers Nos. 25400375, 26400363 and 16H01076.

References

[1] Takabatake T, Teshima F, Fujii H, Nishigori S, Suzuki T, Fujita T, Yamaguhci Y, Sakurai J and Jaccard D 1990 *Phys. Rev. B* **41** 9607
[2] Umeo K, Igaue T, Chyono H, Echizen Y, Takabatake T, Kosaka M and Uwatoko Y 1999 *Phys. Rev. B* **60** R6957
[3] Thiede V M T, Edel T and Jeitschko W 1998 *J. Mater Chem.* **8** 125
[4] Muro Y, Motoya K, Saiga Y and Takabatake T 2010 *J. Phys. Conf. Ser.* **200** 012136
[5] Nishioka T, Kawamura Y, Takesawa T, Kobayashi R, Kato H, Matsumura M, Kodama K, Matsubayashi K and Uwatoko Y 2009 *J. Phys. Soc. Jpn.* **78** 123705
[6] Muro Y, Umeo K, Nishimoto K, Tamura R and Takabatake T 2010 *Phys. Rev. B* **81** 214401
[7] Khalyavin D D, Hillier A D, Adroja D T, Strydom A M, Manuel P, Chapon L C, Peratheepan P, Knight K, Deen P, Ritter C, Muro Y and Takabatake T 2010 *Phys. Rev. B* **82** 100405 (R)
[8] Kato H, Kobayashi R, Takesaka T, Nishioka T, Matsumura M, Kaneko K and Metoki N 2011

J. Phys. Soc. Jpn. **80** 073701

[9] Morrison G, Haldolaarachchige N, P Young D and Chan J Y 2012 *Condens. Matter* 356002

[10] Muro Y, Kajino J, Onimaru T and Takabatake T 2011 *J. Phys. Soc. Jpn.* **80** SA021

[11] Kimura S, Iizuka T, Miyazaki H, Hajiri T, Matsunami M, Mori T, Irizawa A, Muro Y, Kajino J and Takabatake T 2011 *Phys. Rev. B* **84** 165125

[12] Yutani K, Muro Y, Kajino J, Sato T J and Takabatake T 2012 *J. Phys. Conf. Ser.* **391** 012070

[13] Strigari F, Willers T, Muro Y, Yutani K, Takabatake T, Hu Z, Agrestini S, Kuo C Y, Chin Y Y, Lin H J, Pi T W, Chen C T, Weschke E, Schierle E, Tanaka A, Haverkort M W, Tjeng L H and Severing A 2013 *Phys. Rev. B* **87** 125119

[14] Kobayashi R, Ogane Y, Hirai D, Nishioka T, Matsumura M, Kawamura Y, Matsubayashi K, Uwatoko Y, Tanida H and Sera M 2013 *J. Phys. Soc. Jpn.* **82** 093702

[15] Muro Y, Hida K, Fukuhara T, Kawabata J, Yutani K and Takabatake T 2014 *JPS Conf. Proc.* **3** 012017

[16] Hayashi K, Muro Y, Fukuhara T, Kawabata J, Kuwai T and Takabatake T 2016 *J. Phys. Soc. Jpn.* **85** 034714

[17] Sera M, Tanaka D, Tanida H, Moriyoshi C, Ogawa M, Kuroiwa Y, Nishioka T, Matsumura M, Kim J, Tsuji N and Tanaka M 2013 *J. Phys. Soc. Jpn.* **82** 024603

[18] Umeo K, Ohsuka T, Muro Y, Kajino J and Takabatake T 2011 *J. Phys. Soc. Jpn.* **80** 064709

[19] Kawamura Y, Hayashi J, Takeda K, Sekine C, Tanida H, Sera M and Nishioka T 2016 *J. Phys. Soc. Jpn.* **85** 044601

IOP Conf. Series: Journal of Physics: Conf. Series **807** (2017) 022003 doi:10.1088/1742-6596/807/2/022003

Dilution effects on the antiferromagnetic Kondo semiconductor $CeOs_2Al_{10}$

Y Okada[1], J Kawabata[1], Y Yamada[1], Y Muro[3] and T Takabatake[1,2]

[1] Graduate School of Advanced Sciences of Matter, [2] Institute for Advanced Materials Research, Hiroshima University, Higashi-Hiroshima 739-8530, Japan
[3] Faculty of Engineering, Toyama Prefectural University, Izumi, Toyama 939-0398, Japan

E-mail: takaba@hiroshima-u.ac.jp

Abstract. We have studied the effects of dilution of Ce sublattice on the unusual antiferromagnetic (AFM) order in the Kondo semiconductor $CeOs_2Al_{10}$ at 28.5 K by the magnetic, transport and specific-heat measurements of single crystals of $Ce_{1-z}La_zOs_2Al_{10}$. The effective magnetic moment and paramagnetic Curie temperature hardly change with z up to 0.5, indicating that the 4*f* state remains unchanged at high temperatures. The suppression of the Néel temperature T_N is much weaker than that in 5*d* hole doped system, $Ce(Os_{1-y}Re_y)_2Al_{10}$. Therefore, the AFM interaction is robust against the violation of the coherent Ce sublattice. The activation energy in the resistivity decreases in parallel with T_N, confirming the argument that the presence of the *c*-*f* hybridization gap is a requisite for the unusual AFM order in this system.

1. Introduction

In the so-called Kondo semiconductors with renormalized gaps at the Fermi level, the ground states remain nonmagnetic due to the strong hybridization of the 4*f* states and conduction bands [1]. However, recently found Kondo semiconductors CeT_2Al_{10} (T = Ru, Os) crystallizing in the orthorhombic $YbFe_2Al_{10}$-type structure order antiferromagnetically (AFM) at rather high temperature T_N = 28 K [2-4]. In order to elucidate the mechanism for the unusual AFM order, we have studied the effects of electron- and hole-doping on the magnetic and transport properties [5,6], electron tunneling [7] and spin-gap formation [8] in $Ce(Os_{1-y}M_y)_2Al_{10}$ (M = Re, Ir). These studies have revealed that 5*d* hole doping with Re substitution for Os moves $CeOs_2Al_{10}$ to a valence fluctuation regime while 5*d* electron doping with Ir substitution moves to a localized Kondo regime. Furthermore, the 5*d* hole doping suppresses the T_N more drastically than the 5*d* electron doping, whereby the suppressions of T_N are well correlated with those of the activation energy in the electrical resistivity. Because the activation energy is a measure of the hybridization gap, this correlation suggested that the hybridization gap is necessary for the unusual AFM order in this system.

The effect of dilution of the Ce 4*f* sublattice on the physical properties of $CeRu_2Al_{10}$ was studied by substituting La for Ce [9-11]. It was expected that the La substitution expands the lattice and therefore the 4*f* state would be more localized. As z is increased from 0 to 1 in $Ce_{1-z}La_zRu_2Al_{10}$, the lattice parameters a and c increased by 0.25% and the b parameter increased by 0.15%. However, the magnetic susceptibility along the easy a axis for T > 100 K hardly changed with z up to z = 0.7 where the AFM order disappeared. Therefore, little change occurred in the effective magnetic moment per Ce atom and paramagnetic Curie temperature. The rather robust AFM order against dilution is in contrast with the strong suppression of the AFM order observed in $Ce(Os_{1-x}M_x)_2Al_{10}$ for M = Re and Ir [7]. In order to understand the contrasting responses to the substitution in the T sublattice and that in the Ce sublattice, we have studied the dilution effect on the AFM order in $Ce_{1-z}La_zOs_2Al_{10}$. Another issue in $Ce_{1-z}La_zRu_2Al_{10}$ is the reorientation of the ordered magnetic moment μ_{AF} from $//c$ to $//b$, which occurs at $z \approx 0.07$ [11]. In this work, we have examined whether such reorientation occurs in the Os system $Ce_{1-z}La_zOs_2Al_{10}$.

2. Experimental

Single crystals of $Ce_{1-z}La_zOs_2Al_{10}$ were grown by the Al self-flux method as previously reported [4]. The chemical compositions were determined by the wave-length dispersive electron-probe microanalysis. The compositions of La (z) in the crystals were found to slightly deviate from the initial ones (Z). For example, starting with the compositions Z = 0.05, 0.10, 0.20, 0.30 and 0.5, the actual values of z were 0.043, 0.096, 0.15, 0.24 and 0.51, respectively. A small amount of impurity $OsAl_4$ was detected. The x-ray diffraction analysis on the powdered samples indicated that a and c parameters increase linearly by 0.3% and the b parameter increases by 0.2% as z increases from 0 to 1.0.

Using the samples cut along the principal axes, we measured the electrical resistivity ρ, magnetic susceptibility χ and specific heat C in the temperature range from 2 to 300 K. The measurement of $\rho(T)$ was performed by the ac four-terminal method. Along the three principal axes, $\rho(T)$ was measured with bar shaped samples longer than 0.4 mm. A SQUID magnetometer (Quantum Design MPMS) was used for $\chi(T)$ measurements. The magnetization up to 14 T was measured by the extraction method using the Quantum Design PPMS. The measurement of $C(T)$ was performed by the thermal relaxation method on the PPMS.

3. Results and Discussion

The magnetic susceptibility χ of $Ce_{1-z}La_zOs_2Al_{10}$ along the a axis is shown in figure 1 as $1/\chi$ vs temperature T. The value is normalized per Ce mol. The Curie Weiss fit to the data above 200 K yielded the effective magnetic moment μ_{eff} which stays at 2.62 – 2.66 μ_B/Ce for $0 \leq z \leq 0.51$. The absolute values of the paramagnetic Curie temperature Θ_p stays at 22 – 30 K, as is shown in the inset of figure 1. Both μ_{eff} and Θ_p are insensitive to the La substitution, which are in common with those observed for $Ce_{1-z}La_zRu_2Al_{10}$ [9-11]. This result indicates that the $4f$ hole doping does not alter the c-f hybridization at high temperatures. It is in contrast with the strong effect of $5d$ hole doping by Re substitution for Os, which shifts the system to the valence fluctuation regime as is evidenced by the large increase in $|\Theta_p|$ (see the inset of figure 1) [5]. This opposite effect is not due to the volume effect because the unit cell volume increases by a similar ratio for both substitutions, La for Ce and Re for Os. In the latter, the a

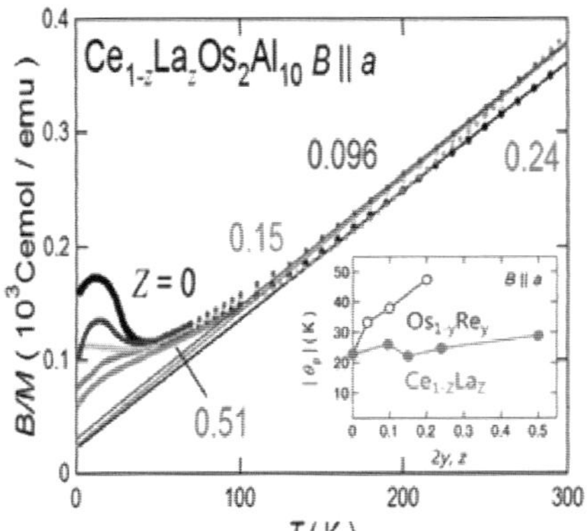

Figure 1. Temperature dependence of the inverse magnetic susceptibility $1/\chi$ measured in $B//a$ = 1 T for $Ce_{1-z}La_zOs_2Al_{10}$. The inset shows the absolute values of the paramagnetic Curie temperature $|\Theta_p|$ for $B//a$ as a function of the doped hole concentration in $Ce_{1-z}La_zOs_2Al_{10}$ and $Ce(Os_{1-y}Re_y)_2Al_{10}$ [5].

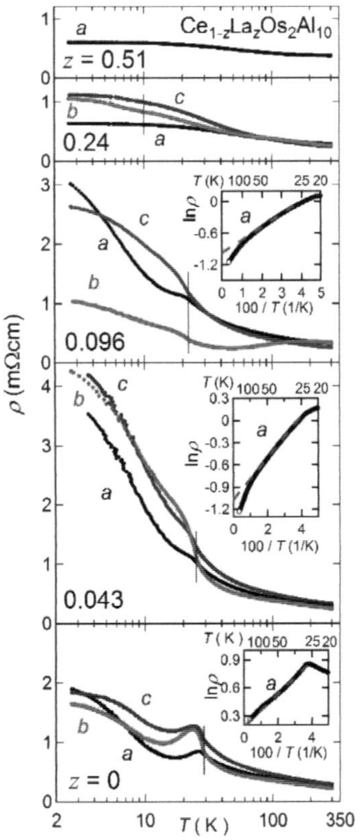

Figure 2. Temperature dependence of the electrical resistivity $\rho(T)$ along the principal axes for $Ce_{1-z}La_zOs_2Al_{10}$. The inset is the Arrhenius plot showing the thermal activation behavior of $\rho(T)$ along the *a* axis.

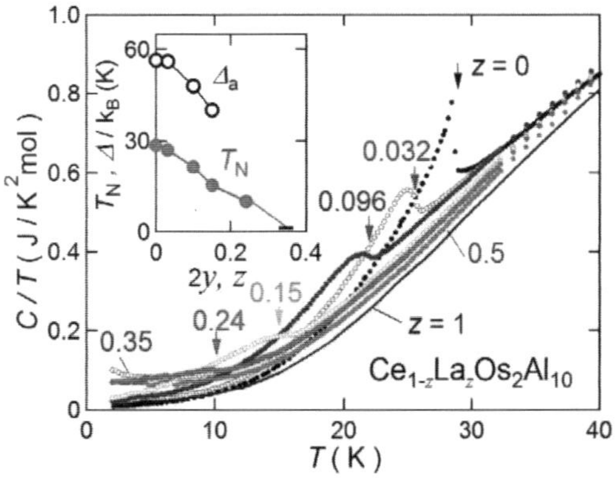

Figure 3. Temperature dependence of the specific heat divided by temperature C/T for $Ce_{1-z}La_zOs_2Al_{10}$. The inset shows the thermal activation energy in the resistivity along the *a* axis Δ_a and antiferromagnetic ordering temperature T_N as a function of *z*.

and *c* parameters enlarge by 0.2% and the *b* parameter enlarges by 0.3% as the Re content increases from 0 to 0.5 [5], whose elongations are comparable in the former as described in the previous section. We note here that the χ_b and χ_c at $T > 50$ K are essentially unchanged with *z* but the drop in χ_b and χ_c at $T < T_N$ for $z = 0$ disappeared for $z > 0.24$.

The presence of the hybridization gap in $CeOs_2Al_{10}$ manifests in the thermal activation type behaviour of resistivity, $\rho(T) = \rho_0 \exp(\Delta/2k_BT)$ [4]. The results of $\rho(T)$ along the three principal axes for $Ce_{1-z}La_zOs_2Al_{10}$ are presented in figure 2 together with the Arrhenius plot of $\rho(I\ //a)$ in the inset. The vertical lines denote T_N's determined by the specific heat measurement as will be described below. For $z = 0$, the thermal activation behaviour is observed in two regions, $5 < T < 16$ K and $30 < T < 80$ K. The increase in $\rho(T)$ at $T < T_N$ becomes stronger for $z = 0.043$ but is weakened as *z* is further increased to 0.096. However, the thermal activation behaviour above T_N remains in $\rho(T)$ along the *a* axis as shown in the inset. For $z = 0.24$, $\rho(T)$'s no longer follow the activation type form. We note here that the magnitudes of $\rho(T{=}2.5\ \text{K})$ for $z = 0.043$ and 0.096 are several times larger than those of the corresponding sample $Ce_{1-z}La_zRu_2Al_{10}$ [10]. The larger enhancement in the Os system than in the Ru system is consistent with the smaller value of the Sommerfeld coefficient in the former, as shown below.

Figure 3 shows the temperature dependence of C/T for $Ce_{1-z}La_zOs_2Al_{10}$. The midpoint of the jump in C/T was taken as T_N, which is 28.5 K for $z = 0$. With increasing *z*, the jump at the AFM order shifts to low temperatures and becomes broader. A weak anomaly is noticeable at 10 K for $z = 0.24$ but no jump is observed down to 2 K for $z = 0.35$. The Sommerfeld coefficient γ was estimated by extrapolating

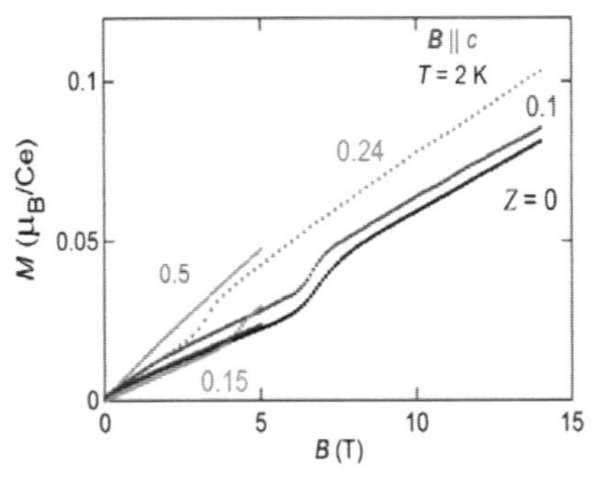

Figure 4. Magnetization curves $M(B)$ for the field B //c at 2 K for $Ce_{1-z}La_zOs_2Al_{10}$.

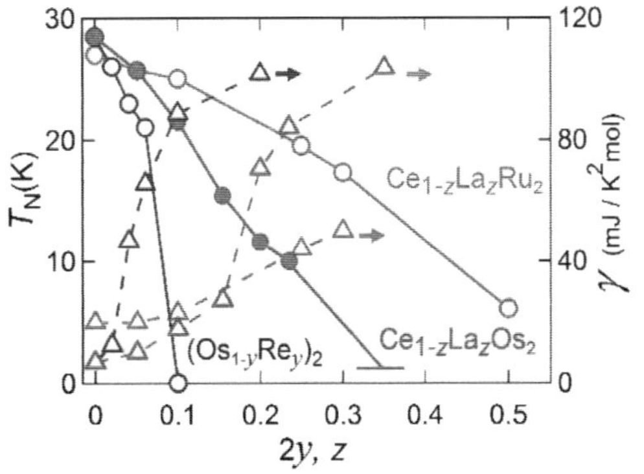

Figure 5. Néel temperature T_N and Sommerfeld coefficient of the electronic specific heat γ as a function of z and $2y$ in $Ce_{1-z}La_zOs_2Al_{10}$, $Ce_{1-z}La_zRu_2Al_{10}$ [9] and $Ce(Os_{1-y}Re_y)_2Al_{10}$ [5].

the C/T vs T^2 plot to $T = 0$. We discuss later the dependence of γ on z in comparison with other doped systems. The dependences of T_N and the thermal activation energy Δ_a in the resistivity along the a axis are plotted in the inset of figure 3. It is noteworthy that the decrease in T_N follows that of Δ_a. This relation supports our argument that the hybridization gap is a requisite for the unusual AFM order, which has been derived from our systematic investigation of $Ce(Os_{1-x}M_x)_2Al_{10}$ for M = Re and Ir [5,6].

As mentioned in the introduction, the direction of AFM ordered moments μ_{AF} in $Ce_{1-z}La_zRu_2Al_{10}$ changes from //c to //b as z is increased to 0.07 [10]. In order to examine whether such reorientation occurs in $Ce_{1-z}La_zOs_2Al_{10}$, we have measured the magnetization curves $M(B)$ up to 14 T at 2 K. The results of $M(B)$ for B//c are shown in figure 4. The metamagnetic anomaly at 6.5 T for z = 0 is the manifestation of the spin flop transition from μ_{AF}//c to $\mu_{AF}\perp c$ [4,5]. When z increases to 0.24, the metamagnetic anomaly still exists at B//c = 2.5 T, indicating the direction of μ_{AF} to be kept along the c axis. This result is consistent with the AFM structure for z = 0.1 with μ_{AF} //c = $0.23\mu_B$ which was determined by neutron diffraction experiments [12]. We note that the $M(B)$ curves for B//a and B//b did not show any anomaly up to 14 T (not shown).

Figure 5 shows the variations of T_N and the γ value as a function of doping amount for the three systems: 4f hole doped systems $Ce_{1-z}La_zOs_2Al_{10}$ and $Ce_{1-z}La_zRu_2Al_{10}$ [9-11] and 5d hole doped system $Ce(Os_{1-y}Re_y)_2Al_{10}$ [5]. Between the two La substituted systems, the degree of suppression of T_N for the Os compound is stronger than that for the Ru compound. However, it is much milder when compared with that of Re substituted system. This is well correlated with the gradual increases in the γ value as a function of z in $Ce_{1-z}La_zOs_2Al_{10}$ and $Ce_{1-z}La_zRu_2Al_{10}$ which are slower than that as a function of $2y$ in $Ce(Os_{1-y}Re_y)_2Al_{10}$. Therefore, we conclude that the gapped electronic state in $CeOs_2Al_{10}$ below T_N is much weakly affected by 4f hole doping compared with the 5d hole doping.

4. Summary

We have studied the effect of dilution of the 4f sublattice on the unusual AFM order in the Kondo semiconductor $CeOs_2Al_{10}$ by measuring the magnetic, thermal and transport properties of $Ce_{1-z}La_zOs_2Al_{10}$. The measurements indicate that the 4f hole doping hardly change both the Ce valence state and the c-f hybridization strength. The AFM order is mildly suppressed by the doping of 4f holes, in contrast with the strong suppression caused by the 5*d* hole doping. The contrasting effects provide a strong constraint for the theory to explain the mechanism of the unusual AFM order in the presence of the hybridization gap in the Kondo semiconductor.

Acknowledgements
We acknowledge valuable discussions with K Umeo, T Onimaru, D T Adroja, S Kimura, T Yokoya. We thank H Tanida and M Sera for providing us with the specific heat data of $Ce_{1-z}La_zRu_2Al_{10}$. This work was partly supported by JSPS KAKENHI Grant Numbers Nos. 26400363, 16H01076, 15K05180 and 15J01007.

References
[1] Takabatake T, Teshima F, Fujii H, Nishigori S, Suzuki T, Fujita T,Yamaguchi Y, Sakurai J and Jaccard D 1990 *Phys. Rev. B* **41**, 9607
[2] Strydom A M 2009 *Physica B* **404** 2981
[3] Nishioka T, Kawamura Y, Takesaka T, Kobayashi R, Kato H, Matsumura M, Kodama K, Matsubayashi K and Uwatoko Y 2009 *J. Phys. Soc. Jpn.* **78** 123705
[4] Muro Y, Kajino J, Umeo K, Nishimoto K, Tamura R and Takabatake T 2010 *Phys. Rev. B* **81** 214401
[5] Kawabata J, Takabatake T, Umeo K and Muro Y 2014 *Phys. Rev. B* **89** 094404
[6] Yamada Y, Kawabata J, Onimaru T and Takabatake T 2015 *J. Phys. Soc. Jpn.* **84** 084705
[7] Kawabata J, Ekino T, Yamada Y, Sakai Y, Sugimoto A, Muro Y and Takabatake T 2015 *Phys. Rev. B* **92** 201113(R)
[8] Bhattacharyya A, Adroja D T, Strydom A M, Kawabata J, Takabatake T, Hillier A D, Sakai V G, Taylor J W and Smith R I 2014 *Phys. Rev. B* **90** 174422
[9] Tanida H, Tanaka D, Sera M, Moriyoshi C, Kuroiwa Y, Takesaka T, Nishioka T, Kato H and Matsumura M 2010 *J. Phys. Soc. Jpn.* **79** 043708
[10] Oogane Y, Kawamura Y, Nishioka T, Kato H, Matsumura M, Yamamoto Y, Kodama K, Tanida H and Sera M 2012 *J. Phys.: Conf. Ser.* **400** 032073
[11] Yoshida K, Okubo R, Tanida H, Matsumura T, Sera M, Nishioka T, Matsumura M, Moriyoshi C and Kuroiwa Y 2015 *Phys. Rev. B* **91** 235124
[12] Adroja D T *et al.* 2015 *Phys. Rev. B* **92** 094425

Pressure Evolution of Characteristic Electronic States in $EuRh_2Si_2$ and $EuNi_2Ge_2$

Fuminori Honda[1], Keigo Okauchi[2], Ai Nakamura[1], Dai Aoki[1], Hiromu Akamine[3], Yousuke Ashitomi[3], Masato Hedo[4], Takao Nakama[4], and Yoshichika Ōnuki[4]

[1]Institute for Materials Research, Tohoku University, Oarai, Ibaraki 311-1313, Japan
[2]Department of Quantum Science and Energy Engineering, Graduate School of Engineering, Tohoku University, Sendai, 980-8579, Japan
[3]Graduate School of Engineering and Science, University of the Ryukyus, Nishihara, Okinawa 903-0213, Japan
[4]Faculty of Science, University of the Ryukyus, Nishihara, Okinawa, 903-0213, Japan

E-mail: honda@imr.tohoku.ac.jp

Abstract. $EuRh_2Si_2$ and $EuNi_2Ge_2$ are known to reveal the valence transition at pressure of about 1 and 2 GPa, respectively, from a divalent state (Eu^{2+}) to a nearly Eu-trivalent state ($Eu^{3-\delta'}$). We have succeeded in growing single crystals of $EuRh_2Si_2$ by the Bridgman method and $EuNi_2Ge_2$ by the In-flux method. In order to clarify the transition of electronic properties from Eu^{2+} to $Eu^{3-\delta'}$ in detail, we carried out electrical resistivity measurements under pressure. $EuRh_2Si_2$ indicates a remarkable first-order valence transition in the pressure range from 1 to 2 GPa with a sharp hysteresis in the electrical resistivity. At 2.12 GPa, the temperature dependence of the electrical resistivity exhibit a broad peak, implying a moderate heavy-fermion state such as $EuPd_2Si_2$ and $EuIr_2Si_2$. While, $EuNi_2Ge_2$ indicates a first-order valence transition in the pressure range from 2 to 3.22 GPa with a hysteresis in the electrical resistivity. At 3.60 GPa, the electrical resistivity of $EuNi_2Ge_2$ also shows a characteristic behavior for a moderate heavy-fermion compound. Pressure - temperature phase diagrams for both compounds are constructed.

1. Introduction

Eu atoms in a compound exhibit two kinds of valence states with Eu^{2+} ($4f^7 6s^2$) and Eu^{3+} ($4f^6 5d^1 6s^2$). Since Eu^{2+} is magnetic, Eu-compounds with Eu^{2+} state order magnetically at low temperatures due to the Ruderman–Kittel–Kasuya–Yosida interaction. The magnitude of the magnetic moment of Eu^{2+} ion is the same as that of the corresponding Gd^{3+} electronic state ($J = S = 7/2$, $L = 0$), where J is a total angular momentum, S is a spin angular momentum, and L is an orbital angular momentum. On the contrary, the electronic state of Eu^{3+} is non-magnetic, since J becomes zero as a result of $S = L = 3$. Vast majority of Eu compounds, Eu ions are in the divalent electronic state and orders magnetically. Compounds with Eu-trivalent state are small in number compared to the Eu-divalent compounds. It should be noted that the valence state of the Eu ion is changed by external parameters such as element substitution (chemical pressure), temperature, magnetic field, and pressure, owing to the differences in the ionic sizes and magnetic properties between the Eu^{2+} and Eu^{3+} ions.

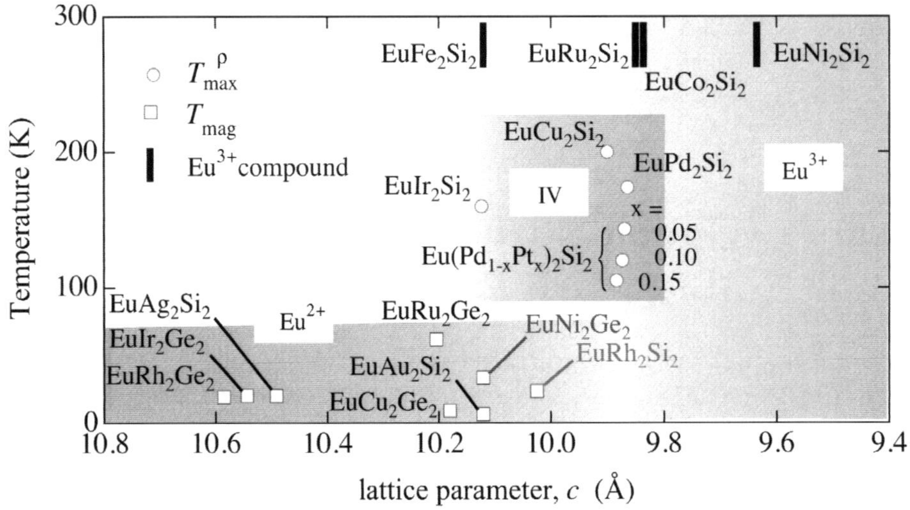

Figure 1. Electronic properties of EuT_2X_2 with $ThCr_2Si_2$-type tetragonal structure as a function of the lattice parameter along the c-axis. Open square (□) indicates magnetic ordering temperature, open circle (○) indicates a temperature where the electrical resistivity shows broad maximum, and black bar indicates the c-axis lattice parameter of trivalent Eu compounds.

One of the most systematically studied Eu compounds is a EuT_2X_2 (T: transition metal, X: Si, Ge) system with the $ThCr_2Si_2$-type tetragonal structure (space group: $I4/mmm$, No. 139). The rich variety of electronic properties have been reported such as magnetic ordering, moderate heavy fermion, intermediate valence, and valence transition from more than 15 EuT_2X_2 compounds and their element substituted systems. For example, in $EuFe_2Si_2$ and $EuAg_2Si_2$, Eu ions are trivalent and divalent, respectively, [1] and $EuPd_2Si_2$ is an intermediate valence compound indicating a significant temperature dependent valence change below 200 K. [2] It is characteristic that the temperature dependence of the electrical resistivity indicates broad maximum at T_{max}^{ρ} where the steep valence change occurs. In order to overview the electronic properties of $ThCr_2Si_2$-type EuT_2X_2 compounds, we tentatively plotted the magnetic ordering temperature T_{mag} with Eu^{2+} compounds and T_{max}^{ρ} of intermediate valence (IV) compounds with respect to the lattice parameter along the c-axis, as shown in Fig. 1.

From a brief look, the compounds with longer c-axis lattice parameter order magnetically, and several compounds with $c \simeq 10.1$ Å reveal intermediate valence state. In this plot, $EuRh_2Si_2$ and $EuNi_2Ge_2$ stands around the boarder between magnetically ordered state (Eu^{2+}) and intermediate valence state. Actually, it is reported that $EuRh_2Si_2$ and $EuNi_2Ge_2$ indicate pressure–induced valence transition at 1 and 2 GPa, respectively. [3, 4]

Eu electronic state of $EuRh_2Si_2$ is divalent at ambient pressure and orders antiferromagnetically below a Néel temperature $T_N = 23.5$ K. [5] With increasing pressure P, the electronic state is changed abruptly at the critical pressure $P_v \simeq 1$ GPa, revealing the first order valence transition. [3] Namely, the temperature dependence of the magnetic susceptibility at 1.00 GPa indicates a hysteresis in the cooling and warming processes. The Eu electronic state at 1.00 GPa is nearly divalent, $Eu^{2+\delta}$ ($\delta \ll 1$), at temperatures higher than the characteristic temperature T_v ($\simeq 38$ K), while it becomes nearly trivalent, $Eu^{3-\delta'}$ ($\delta' < 1$) below T_v. Here, T_v is called the valence transition temperature.

As for $EuNi_2Ge_2$, Eu electronic state is also divalent at ambient pressure and orders

IOP Conf. Series: Journal of Physics: Conf. Series **807** (2017) 022004 doi:10.1088/1742-6596/807/4/022004

antiferromagnetically below $T_N = 30$ K. [6] The electronic state is changed abruptly at the critical pressure $P_c \simeq 2$ GPa, revealing the first-order phase transition. [4, 7, 8] The temperature dependences of the electrical resistivity and thermoelectric power at 2.3 GPa indicate a clear hysteresis in the temperature cooling and warming processes. [8] It is also reported that the antiferromagnetic transition is also observed at 2.3 GPa.

Although it is reported that there are quite a few EuT_2X_2 compounds exhibiting valence transition under pressure or element substitution, most of works were done using polycrystalline samples. Recently, we have succeeded in growing single crystals of $EuRh_2Si_2$ by the Bridgman method and $EuNi_2Ge_2$ by the In-flux method. In order to clarify furthermore the valence transition in EuT_2X_2 system, we have carried out the electrical resistivity experiments under high pressures using single crystal samples.

2. Experimental Procedure

Single crystals of $EuRh_2Si_2$ were grown by the Bridgman method. Details for the single crystal growth were described elsewhere. [9] A typical example of the as-grown single crystal of $EuRh_2Si_2$ is shown in the inset of Fig. 2. Each single crystals are small in size, less than 1 mm in length, but we obtained a number of such crystals. The brilliant plane corresponds to the tetragonal (001) plane, which is a characteristic shape of $ThCr_2Si_2$-type single crystals. While, single crystals of $EuNi_2Ge_2$ was grown by the In-flux method. A detailed sample preparation procedure is described elsewhere. [10]

The specific heat was measured by a thermal relaxation method in the temperature range from 2 to 150 K by the physical property measurement system (PPMS, Quantum Design Co. Ltd.). The electrical resistivity was measured by the conventional four-probe AC method in the temperature range from 3 to 300 K at ambient and under high pressures. Pressure was applied by the Bridgman-anvil-type cell using tungsten-carbide WC opposed anvils with Daphne-7373 oil as a pressure transmitting medium. Pressure inside the pressure cell was determined from the superconducting transition of Pb placed next to the sample.

3. Results and Discussion

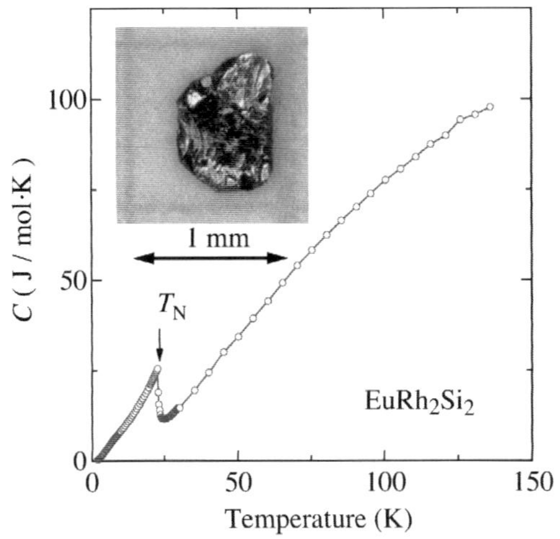

Figure 2. Temperature dependence of the specific heat of $EuRh_2Si_2$ single crystal. Inset shows a photograph of a typical single crystal.

Figure 2 shows the temperature dependence of the specific heat C of $EuRh_2Si_2$ single crystal. Reflecting the antiferromagnetic ordering at $T_N = 23.4$ K, a sharp λ-type anomaly is observed, as indicated by an arrow. A broad hump around 7 K in the specific heat curve is due to a Schottky

IOP Conf. Series: Journal of Physics: Conf. Series **807** (2017) 022004 doi:10.1088/1742-6596/807/4/022004

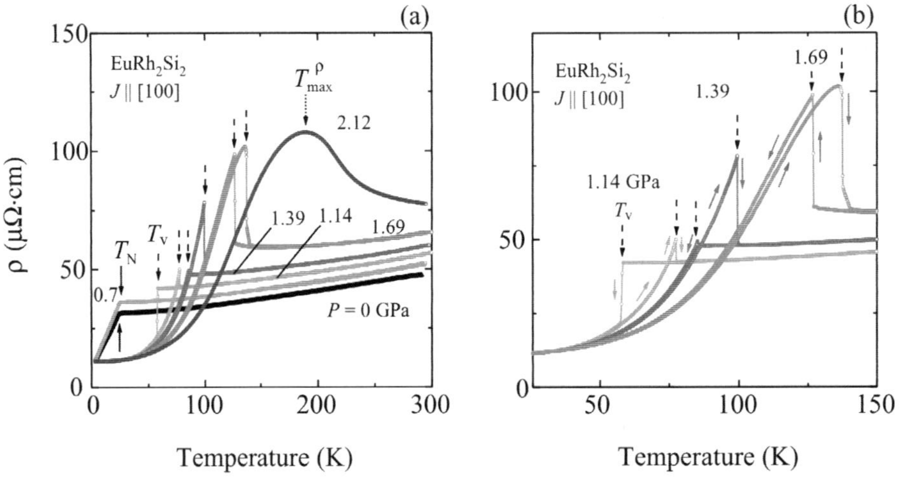

Figure 3. Temperature dependences of the electrical resistivities of $EuRh_2Si_2$ for the current along the [100] direction under several pressures in the temperature range (a) 0 - 300 K and (b) 25 - 150 K.

peak of the 4f-spins with $S = 7/2$, where the Zeeman splitting of the degenerated spins is realized due to a large magnetic exchange field in the magnetically ordered Eu compound, as described in $EuGa_4$. [11]

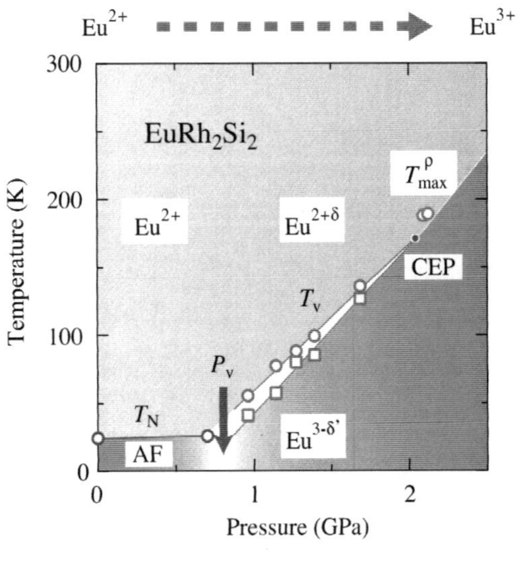

Figure 4. Pressure – temperature phase diagram in $EuRh_2Si_2$, where T_N is the Néel temperature, T_v the valence transition temperature, and T^{ρ}_{max} the temperature showing a peak of the resistivity curve. "AF" means antiferromagnetic phase and the critical end point "CEP" of valence transition is marked as a closed circle (●).

The temperature dependences of the electrical resistivity $\rho(T)$ of $EuRh_2Si_2$ under pressures up to 2.12 GPa are displayed in Fig. 3(a). Magnified plot in $\rho(T)$ curves for the results obtained $1.1 < P < 1.7$ GPa is shown in Fig. 3(b) for clarity. At 0 and 0.7 GPa, $\rho(T)$ shows a kink at around 25 K reflecting the antiferromagnetic ordering, which is consistent with the data obtained by polycrystalline samples. [3] Further increasing pressure, the $\rho(T)$ curves of single crystal show drastic changes compared to the results obtained by polycrystalline samples. Thus, the present resistivity data on a single crystal indicate characteristic features accompanied by the valence transition. At 1.14 GPa, $\rho(T)$ suddenly drops at 58 K with decreasing temperature, on the other hand, $\rho(T)$ shows spike-like peak at 77 K with increasing temperature. Such prominent and sharp hysteresis behaviors are seen at 1.39 and 1.69 GPa.

Notice that the valence transition in the warming process, namely from the moderate heavy fermion state to the nearly divalent state, possesses a large drop of the electrical resistivity under P = 1.39 and 1.69 GPa, as shown in Fig. 3(b). Furthermore, with increasing pressure, the transition from the nearly divalent state to the moderate heavy fermion state in the cooling process indicates a sharp and significantly large jump of the resistivity and decreases nearly quadratically with decreasing temperature, as shown in Fig. 3(b) at T_v = 126.9 K under 1.69 GPa. Note that the $\rho(T)$ curve at 1.69 GPa resembles that obtained in $Eu(Rh_{0.5}Ir_{0.5})_2Si_2$. [12]

On the other hand, the $\rho(T)$ curve at P = 2.12 GPa is the same between cooling and warming processes, *i.e.* no hysteresis behavior was observed. A broad resistivity peak at around 190 K under P = 2.12 GPa is characteristic, which is discussed later. Figure 4 shows the pressure–temperature phase diagram in $EuRh_2Si_2$, where T_N and T_v are the Néel temperature and the valence transition temperature, and T^{ρ}_{max} is the temperature showing a peak of the resistivity curve. "AF" means the antiferromagnetic phase and the critical end point "CEP" of the first order valence transition is marked the closed circle (●). Critical pressure P_v, where the antiferromagnetism is suppressed, is below 1 GPa which is consistent with the data obtained on polycrystals. Hysteresis width with respect to the temperature decreases with increasing pressure and varnishes around 2 GPa.

Figure 5 shows the temperature dependence of the specific heat C of $EuNi_2Ge_2$ single crystal. A λ-type anomaly is observed at the antiferromagnetic ordering temperature T_N = 33.5 K, as indicated by an arrow. A broad hump due to a Schottky peak of the $4f$-spins, which is even more clearer than that of $EuRh_2Si_2$, around 10 K is observed.

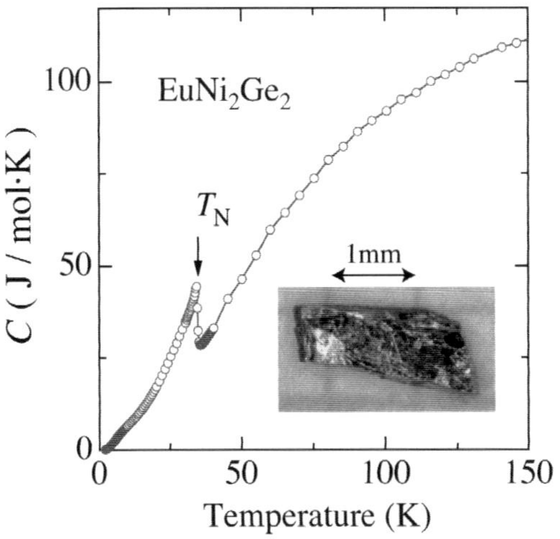

Figure 5. Temperature dependence of the specific heat of $EuNi_2Ge_2$ single crystal. Inset shows a photograph of a typical single crystal.

The temperature dependences of the electrical resistivity $\rho(T)$ below and above 2 GPa are displayed in Figs. 6(a) and 6(b), respectively. Note that the resistivity is vertically shifted in Fig. 6(b) for clarity. Below 2 GPa, electrical resistivity increases monotonically with increasing pressure. The Néel temperature is also gradually increases with a rate of 6.3 K/GPa. Critical pressure P_v, where the antiferromagnetism varnishes, is about 2.1 GPa in $EuNi_2Ge_2$, which is practically consistent with the previous studies. [4, 7, 8] At 2.17 GPa, $\rho(T)$ curve shows a clear hysteresis around 60 K, indicating a first order valence transition. $\rho(T)$ drops with two-steps

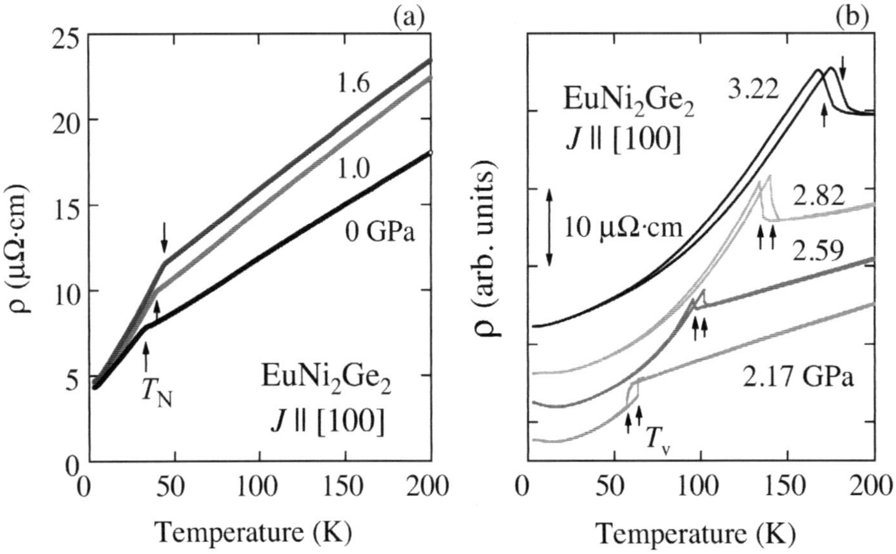

Figure 6. Temperature dependences of the electrical resistivities of $EuNi_2Ge_2$ for the current along the [100] direction under (a) lower pressure below 1.6 GPa and (b) higher pressure region $2 < P < 3.5$ GPa.

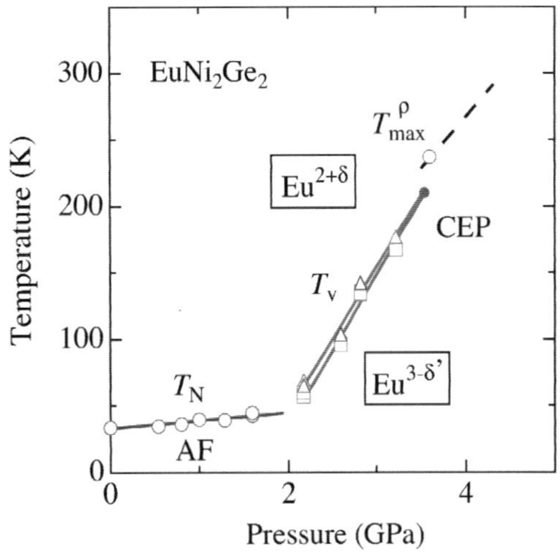

Figure 7. Pressure – temperature phase diagram in $EuNi_2Ge_2$, where T_N is the Néel temperature, T_v the valence transition temperature, and T^{ρ}_{max} the temperature showing a peak of the resistivity curve. "AF" means antiferromagnetic phase and the critical end point "CEP" of valence transition is marked as a closed circle (●).

at T_v. So far, it is not clear that the two-step decrease corresponds to the two different valence transitions or due to extrinsic reason(s) such as inhomogeneity of sample itself and/or pressure. Nevertheless, it was previously reported that there were two-step decrease of $\rho(T)$ at 2.3 GPa measured on the polycrystalline sample of $EuNi_2Ge_2$. [8] Further pressure experiments are in progress in order to clarify the electronic properties around P_v in detail. The $\rho(T)$ anomaly changed spike-like one, above 2.59 GPa and the sharp spike-like $\rho(T)$ becomes broadened for further increasing pressure. As discussed later, $\rho(T)$ at 3.60 GPa shows single broad peak without hysteresis, implying that the electronic state becomes a moderate heavy fermion state.

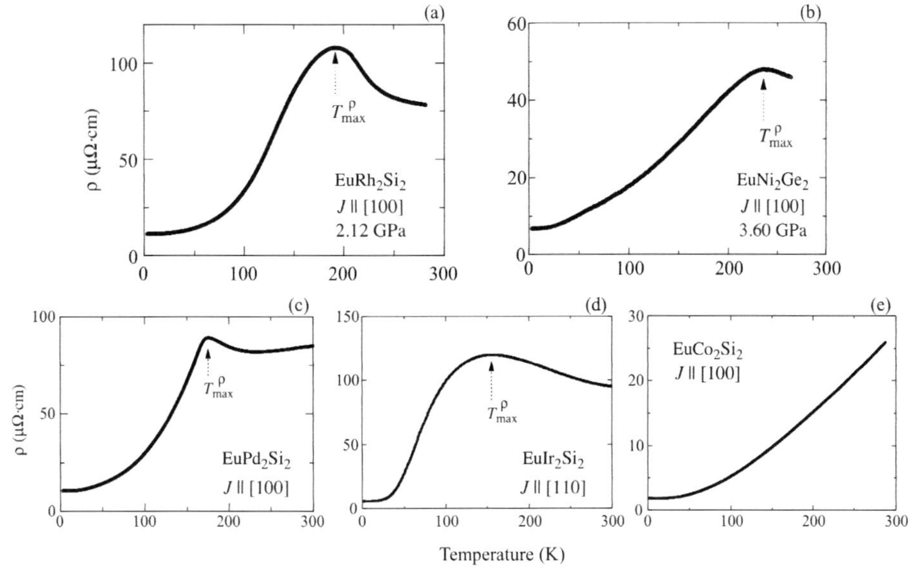

Figure 8. Temperature dependences of the electrical resistivities in (a) $EuRh_2Si_2$ at 2.12 GPa, (b) $EuNi_2Ge_2$ at 3.60 GPa, (c) $EuPd_2Si_2$, (d) $EuIr_2Si_2$, and (e) $EuCo_2Si_2$

Figure 7 shows the constructed pressure – temperature phase diagram in $EuNi_2Ge_2$ from the present single crystal study. The obtained pressure – temperature phase diagrams on $EuRh_2Si_2$ and $EuNi_2Ge_2$ are quite identical. Magnetic ordering state with Eu^{2+} is changed to the nearly trivalent state through the first order valence transition, critical end point of the valence transition, and moderate heavy fermion state. The first order valence transition is characterized by a sudden change of electrical resistivity and electrical resistivity indicates a spike-like shape as approaching to the critical end point. Moderate heavy fermion state, characterized by a gradual valence change with respect to the temperature, is accompanied by a broad peak of the resistivity. This pressure – temperature phase diagram can be a universal phase diagram in EuT_2X_2 and might be applicable to the other Eu compounds.

It is interesting to compare the present electrical resistivities in the moderate heavy-fermion state in $EuRh_2Si_2$ and $EuNi_2Ge_2$ with the resistivities of $EuPd_2Si_2$, $EuIr_2Si_2$, and $EuCo_2Si_2$ [13, 12, 14, 15], as shown in Fig. 8. $EuPd_2Si_2$ and $EuIr_2Si_2$ are well-known valence cross-over compounds. As shown in Fig. 8(c) and 8(d), the valence crossover is observed around T^{ρ}_{max} = 170 and 150 K in $EuPd_2Si_2$ and $EuIr_2Si_2$, respectively, as the broad peak in the temperature dependence of the electrical resistivity. [12] From the Mössbauer and photoemission spectroscopic experiments [13, 16, 12], it was clarified that the valence of $EuPd_2Si_2$ is 2.22 at room temperature, starts to increase continuously from 200 K, and gradually saturates below 100 K, reaching a value of 2.74 at 20 K. In the case of $EuIr_2Si_2$, the Eu valence of 2.3 at 290 K gradually increases and reaches a value of 2.8 at 4.2 K, with a slightly steeper increase around 150 K. [17] On the other hand, Eu valence in $EuCo_2Si_2$ is nearly trivalent (Eu^{3+}), where the electrical resistivity shows normal metallic behavior. [18] From these results, it can be concluded that the electronic states in $EuRh_2Si_2$ at 2.12 GPa and $EuNi_2Ge_2$ at 3.60 GPa are a moderate heavy fermion state with gradual change of Eu valence with respect to the temperature. Actually, $\rho(T)$ in $EuRh_2Si_2$ shows normal metallic behavior around 5 GPa, which is similar to $\rho(T)$ in $EuCo_2Si_2$. [9, 14] It is also expected that the $\rho(T)$ curve in $EuNi_2Ge_2$ turns into that of $EuCo_2Si_2$

under higher pressure than 3.60 GPa.

4. Summary

We have succeeded in growing single crystals of $EuRh_2Si_2$ and $EuNi_2Ge_2$ and studied electronic properties under high pressure. In $EuRh_2Si_2$, antiferromagnetism is suppressed below 1 GPa and first order valence transition was observed between 1 and 2 GPa. Critical end point of the first order valence transition is estimated as $P_{CEP} \simeq 2.05$ GPa and $T_{CEP} \simeq 170$ K. In $EuNi_2Ge_2$, antiferromagnetism is suppressed around 2 GPa and first order valence transition was observed between 2.17 and 3.22 GPa. Critical end point in $EuNi_2Ge_2$ is estimated as $P_{CEP} \simeq 3.5$ GPa and $T_{CEP} \simeq 210$ K. Corresponding pressure – temperature (P–T) phase diagrams of both $EuRh_2Si_2$ and $EuNi_2Ge_2$ are constructed. The obtained P–T phase diagram might be a universal phase diagram of EuT_2X_2 compounds.

Acknowledgments

This work was supported by JSPS KAKENHI Grant Numbers 15K05156 and 15H05884.

[1] Bauminger E *et al.* 1974 *J. Phys. Colloq.* **35** C6-61
[2] Nagarajan R *et al.* 1981 *Phys. Lett.* A **81** 397
[3] Mitsuda A *et al.* 2012 *J. Phys. Soc. Jpn.* **81** 023709
[4] Hesse H J *et al.* 1997 *J. Alloys Compd.* **246** 220
[5] Hossain Z *et al.* 2001 *J. Alloys Compd.* **323-324** 396
[6] Bud'ko S L *et al.* 1999 *J. Magn. Magn. Mater.* **205** 53
[7] Takeshita N *et al.* 2000 *Science and Technology of High Pressure, Proc. of AIRAPT-17*, (eds. Manghnani M H, Nelis W J, and Nicol M F, Universities Press, Hyderabad, India) 445
[8] Nakamura A *et al.* 2012 *J. Phys.: Conf. Ser.* **400** 032106
[9] Honda F *et al.* 2015 *J. Phys. Soc. Jpn.* **85** 063701
[10] Honda F *et al.* 2016 *to be published*
[11] Nakamura A *et al.* 2013 *J. Phys. Soc. Jpn.* **82** 104703
[12] Seiro S and Geibel C 2011 *J. Phys.: Condens. Matter* **23** 375601
[13] Sampathkumaran E V *et al.* 1981 *J. Phys.* C: *Solid State Phys.* **14** L237
[14] Ōunce Y *et al.* 2016 *Philosophical Magazine* (2016) to be published. (presented in this conference)
[15] Ōnuki Y *et al.* 2015 *J. Phys.: Conf. Ser.* **592** 012049
[16] Mimura K *et al.* 2011*Jpn. J. Appl. Phys.* **50** 05FD03.
[17] Chevalier B *et al.* 1986 *J. Phys.* C: *Solid State Phys.* **19** 4521
[18] Seiro S, *et al.* 2013 *Phys. Status Solidi* B **250** 621

Anisotropic Thermal Expansion of α-$YbAlB_4$

Yosuke Matsumoto[1], K. Kuga[1], T. Tomita[1], R. Küchler[2] and S. Nakatsuji[1]

Institute for Solid State Physics, University of Tokyo, Kashiwa 277-8581, Japan
Max Planck Institute for Chemical Physics of Solids, Dresden D-01187, Germany

E-mail: matsumoto@issp.u-tokyo.ac.jp

Abstract. Thermal expansion measurements were performed for α-$YbAlB_4$ using powder X-ray diffraction and our high resolution capacitance dilatometry. Highly anisotropic thermal expansion, which is much larger along the c-axis compared to the a- and b-axes, was observed. In addition, high resolution capacitance dilatometry revealed negative thermal expansion appears below 90 K only along the a-axis, not along the c-axis. The observation is consistent with anisotropic hybridization stronger in the ab-plane, suggesting the formation of anisotripic heavy fermion well below the valence fluctuation temperature scale of 200 $\sim$ 300 K. Furthermore, linear thermal expansion coefficient along the a-axis starts to decrease below 115 K. The characteristic temperature scale of 90 - 115 K may correspond to the crystal electric field excited level. The in-plane negative thermal expansion can be regarded as a development of anisotropic hybridization due to $|J_z = \pm 5/2 >$ crystal electric field ground states which becomes dominant on cooling below the excited states.

1. Introduction

In f-electron intermetallics, Quantum criticality (QC) is one of the subjects which has been extensively studied as a source of novel quantum phenomena, such as unconventional superconductivities and non-Fermi liquid behavior (NFL)[1, 2]. So far, the most of the studies have been restricted to magnetic QC, which arises at a magnetic quantum critical point (QCP) in the Kondo lattice systems with integer valence.

On the other hand, in the case of mixed valence compounds, local f moments are expected to be screened by conduction electrons at their relatively higher Kondo temperatures. Therefore, heavy fermion (HF) and QC behavior have not been expected in this class of materials. In sharp contrast, the first Yb-based HF superconductor β-$YbAlB_4$ provides a unique example of a HF formation and a QC in the strongly mixed valence state [3, 4, 5, 6]. Furthermore, the QC cannot be described by the standard theory for the spin-density-wave instability[7, 8, 9]. The diverging magnetic susceptibility along the c-axis exhibits the T/B scaling in the wide temperature (T) and magnetic field (B) region spanning 3 $\sim$ 4 orders of magnitude[5], indicating the QC emerges without tuning any control parameter. In addition, recent resistivity measurements under pressure revealed the anomalous metallic state extends over a finite pressure range up to 0.4 GPa, suggesting a formation of an anomalous metallic phase[10].

β-$YbAlB_4$ has a locally isostructral polymorph α-$YbAlB_4$, which is also strongly mixed valent. These have an orthorhombic crystal structure with space groups $Pbam$ and $Cmmm$ for α- and β-$YbAlB_4$, respectively [11]. The Yb valence estimated by a hard x-ray photoemission spectroscopy is +2.73 for α-$YbAlB_4$ and +2.75 for β-$YbAlB_4$ at 20 K [6]. The valence fluctuation temperature scale was estimated to be $\sim$ 200 - 300 K for both compounds[5, 12, 13]. Nevertheless, these two systems exhibit a HF behavior with a characteristic temperature scale of $\sim$ 8 K, which is far lower than the valence fluctuation scale [5]. This is quite unusual because Pauli paramagnetism is usually expected in the mixed valence compounds below the valence fluctuation temperature scale. The small temperature scale of $\sim$ 8 K for the anomalous HF state may indicate that α-$YbAlB_4$ is also close to a QCP although it has a Fermi liquid (FL) ground state at zero field in contrast to β-$YbAlB_4$ [12].

Another remarkable feature for both systems is the anisotropic hybridization between conduction and f electrons (c-f hybridization). From the argument based on the local symmetry of the Yb site, the crystal field ground doublet of both α- and β-$YbAlB_4$ is suggested to be made solely of $|J_z = \pm 5/2>$ [14]. In this case, the c-f hybridization is expected to be highly anisotropic and have a node along the c-axis[15]. Indeed, it was already suggested experimentally from the resistivity measurements of α-$YbAlB_4$ [12]. The resistivity is highly anisotropic and the one in the ab-plane is 10 times larger than the one along the c-axis. In addition, A coefficients of T^2 behavior of the resistivity, which were determined below 0.24 K, are also highly anisotropic with $A_{ab}/A_c \sim 10$, suggesting anisotropic effective mass with $m_{ab}/m_c \sim 3$. Consistently, transverse magnetoresistivity also exhibits a large anisotropy [16]. These are consistent with the hybridization node along the c-axis. Interestingly, it has been pointed out that the anisotropic c-f hybridization plays an important role in the formation of HF state under the strong valence fluctuation and the novel QC found in β-$YbAlB_4$ [15, 17]. On the other hand, a critical valence fluctuation has been also suggested as the origin of the QC [18, 19]. Therefore, the role of the anisotropic c-f hybridization is still controversial.

In order to further clarify the anisotropic nature of the system, here, we performed thermal expansion measurements using powder X-ray diffraction and capacitance dilatometry. While the temperature dependence of the lattice parameters were determined with a resolution of $\Delta L/L \sim 2\times10^{-4}$ by the X-ray diffraction, the measurements using the capacitance dilatometer along the a- and c- axes were performed with a high resolution of $\Delta L/L \sim 2\times10^{-7}$. Interestingly, we found highly anisotropic thermal expansion, which is much larger along the c-axis compared to the a- and b-axes. In addition, high resolution capacitance dilatometry revealed a negative thermal expansion appears below 90 K only along the a-axis, not along the c-axis. The observation is consistent with anisotropic hybridization stronger in the ab-plane, suggesting anisotripic heavy fermion formation well below the valence fluctuation temperature scale of 200 - 300 K. Furthermore, the linear thermal expansion coefficients start to change at 115 K on cooling. The characteristic temperature scales of 90 - 115 K may correspond to the crystal electric field (CEF) excited level, below which the anisotropic hybridization is expected to develop as $|J_z = \pm 5/2>$ states become dominant.

2. Experimental

We used high purity single crystals of α-$YbAlB_4$ with typical RRR (Residual Resistivity Ratio) ranging from 20 to 110 grown by a flux method [11]. Powder X-ray diffraction measurements were carried out from room temperature down to 16 K. Single crystaline samples were grind into fine powder and measured using a commercial X-ray diffraction system (Rigaku RINT2100) in the angle range between $10 \leq 2\theta \leq 90°$. Diffraction patterns were analyzed to determine the lattice constant using the Rietveld analysis program PDXL (Rigaku).

High resolution measurements using capacitance dilatometer [20] were performed along the a- and c-axes at temperature range from 2 K to room temperature with a resolution of $\Delta L/L \sim 2\times10^{-7}$. The measured single crystal has a dimension of $\sim$ 1 mm for both along the a- and c-axes. In order to subtract the cell background effect, we also measured a high purity Cu piece with a thickness of 1.0 mm.

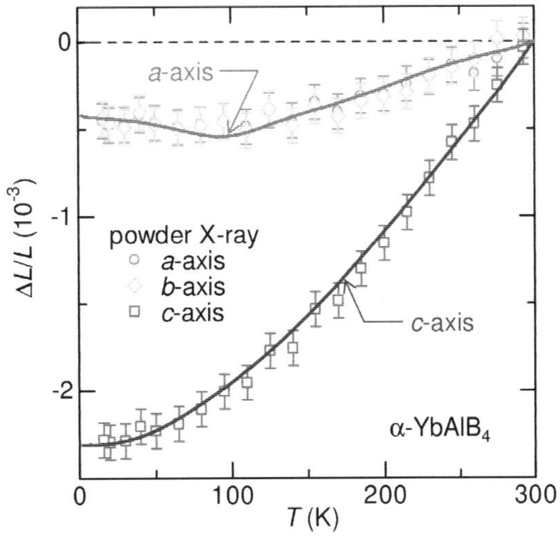

Figure 1. Linear thermal expansions $\Delta L/L$ of α-$YbAlB_4$ measured from room temperature (300 K). Open symbols are the data obtained from the X-ray diffraction measurements. Solid lines indicate the data obtained from capacitance dilatometry.

3. Results and Discussions

First we present the temperature dependence of the linear thermal expansions $\Delta L/L$ measured from room temperature in Figure 1. The data obtained from the X-ray diffraction and those from capacitance diratometer are consistent with each other within the experimental accuracy. As it is clearly seen from the figure, $\Delta L/L$ is highly anisotropic. $\Delta L/L$ along the c-axis is much larger than those along the a- and b- axes. This may come from much harder covalent bond in the in-plane boron network than the c-axis aluminum and ytterbium chain. On the other hand, the X-ray diffraction data indicate that the in-plane anisotropy is small, *i.e.* , $\Delta L/L$ along the a- and b- axes are roughly overlapping to each other.

Remarkably, $\Delta L/L$ along the a-axis, measured by capacitance dilatometer, clearly shows negative thermal expansion below 90 K, *i.e.* , the lattice expand along the a-axis on cooling. In the case of Yb-based intermetallics, negative thermal expansion can be associated with a valence change due to Kondo effect. This is because the ionic radius of Yb^{2+} ($4f^{14}$) is larger than that of Yb $^{3+}$ ($4f^{13}$). If the valence of the system changes through Kondo effect from Yb $^{3+}$ towards Yb^{2+} on cooling, the lattice will expand correspondingly. This negative thermal expansion is observed in Yb-based Kondo lattice systems such as $YbIr_2Zn_{20}$ and $YbCu_2Si_2$ [21]. Therefore, the negative thermal expansion observed in α-$YbAlB_4$ can also be regarded to arise from valence change from Yb $^{3+}$ towards Yb^{2+} due to hybridization developing in the low temperatures.

One important feature here is that the negative thermal expansion is only observed along the a-axis, not along the c-axis. This is consistent with the anisotropic hybridization larger in the ab-plane and clearly suggests the formation of anisotropic heavy fermion in α-$YbAlB_4$. Note that, although thermal expansion along the b-axis has not been measured using capacitance dilatometer, it is also expected to exhibit the same negative thermal expansion. This is because the in-plane resistivity does not exhibit anisotropy [12] and, furthermore, no in-plane large anisotropy has been expected in the c-f hybridization theoretically [15]. The small in-plane anisotropy in $\Delta L/L$ by X-ray diffraction is also consistent with this expectation.

In order to discuss the anisotropic thermal expansion in detail, we present the temperature

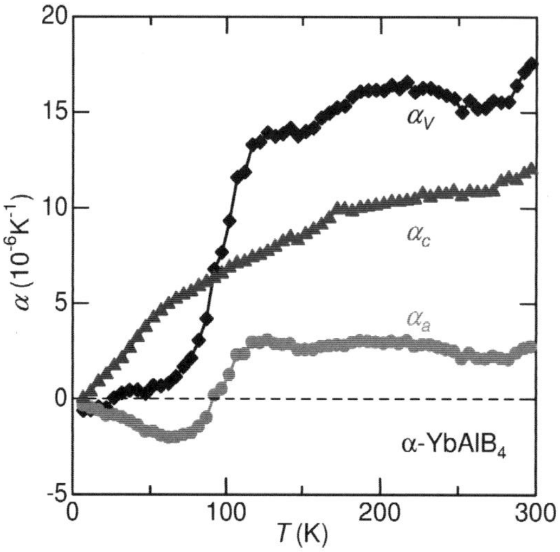

Figure 2. Temperature dependence of the linear thermal expansion coefficient along the a-axis, α_a, and that along the c-axis, α_c. The volume thermal expansion coefficient $\alpha_V \equiv 2\alpha_a + \alpha_c$ is also shown.

dependence of the linear and volume thermal expansion coefficients in Figure 2. Here, the linear thermal expansion coefficient α is defined as $\alpha \equiv (1/L)(dL/dT)$ and estimated from the data by capacitance dilatometry. The volume thermal expansion coefficient α_V is given as $\alpha_V \equiv (1/V)(dV/dT) = \alpha_a + \alpha_b + \alpha_c \approx 2\alpha_a + \alpha_c$, neglecting in-plane anisotropy. α_i denotes α along the i-axis. While α_c exhibits a monotonic decrease on cooling, α_a exhibits almost temperature independent behavior at higher temperatures and then starts to decrease below 115 K and becomes negative below 90 K. On the other hand, α_V does not exhibit clearly negative value beyond experimental accuracy, while it also starts to decrease below 115 K. Interestingly, α_c compensates the negative contribution coming from α_a, which results in the very small values of α_V close to zero below $\sim$ 60 K.

The characteristic temperature scale of 115 K (or 90 K) may correspond to the CEF excited state. As an excited state, $|J_z = \pm 1/2> + \gamma |J_z = \mp 3/2>$ with $\gamma \sim 0.28$ has been suggested at the excited level of 80 K for β-YbAlB$_4$ [14], which well reproduces the anisotropic magnetic susceptibility at the temperature range from 10 to 300 K. Although α-YbAlB$_4$ was not mentioned in their work, considering the quite similar local symmetry of the Yb site and almost identical magnetic susceptibility above $T \sim 8$ K, one can expect quite similar CEF levels for both α- and β-YbAlB$_4$. As it is already mentioned, the anisotropic hybridization is supposed to arise from CEF ground state of $|J_z = \pm 5/2>$. If the system is cooled below the CEF excited level, the anisotropic hybridization will develop as $|J_z = \pm 5/2>$ become dominant, which may results in anisotropic thermal expansion. If this is a case, considering that α_a starts to decrease already at 115 K, the CEF excited level should be a bit higher than this temperature. To confirm this, neutron inelastic scattering measurements are highly desired. Note that it is quite unusual to assume well defined CEF levels far below the valence fluctuation temperature scale. This must be closely related to one of the central issue in α- and β-YbAlB$_4$, why HF and QC behaviors appear even under the strong valence fluctuation or why local moments survive under the strong valence fluctuation. Another point we should note is that the negative thermal expansion is expected also at the valence fluctuation temperature scale of 200 - 300 K, since the Yb valence is changing in this temperature range. However, this $4f$ electron contribution may be masked by a large lattice contribution at high temperatures. In order to discuss how the $4f$ electron contribution evolves on cooling, we need to subtract the lattice contribution using non-magnetic α-LuAlB$_4$, which is an important future issue.

Finally, the volume thermal expansion α_V is supposed to be proportional to the specific heat. Therefore, in the lowest temperatures, α_V/T in α-YbAlB$_4$ should be constant corresponding to its Fermi liquid ground state. Unfortunately, the resolution and the temperature range of the current measurements are not enough to see this behavior of the moderately heavy fermions with $\gamma \sim 130$ mJ/K^{2}mol [12]. The measurements with a better resolution of $\Delta L/L \sim 1 \times 10^{-8}$ down to much lower temperature of 0.1 K is now going to be measured.

4. Conclusion

In this work, the anisotropic nature of α-YbAlB$_4$ was investigated by thermal expansion measurements. Using powder X-ray diffraction and a high-resolution capacitance dilatometry, we revealed that α-YbAlB$_4$ exhibits highly anisotropic thermal expansion, where the one along the c-axis is much larger than those along the a- and b-axes. This is possibly due to much harder covalent bond in the in-plane boron network than the c-axis aluminum and ytterbium chain. More interestingly, high resolution capacitance dilatometry, which were done along the a- and c-axis, revealed a negative thermal expansion appears below 90 K only along the a-axis, not along the c-axis. The observation suggests anisotropic hybridization, which develops stronger in the ab-plane well below the valence fluctuation temperature scale of 200 - 300 K. One can further expect anisotropic Fermi liquid formation in the low temperature limit. On the other hand, the linear thermal expansion coefficient along the a-axis starts to decrease below 115 K. The characteristic temperature scale of 90 - 115 K may correspond to the CEF excited level. In this case, the in-plane negative thermal expansion can be regarded as a development of anisotropic hybridization due to $|J_z = \pm 5/2>$ crystal electric field states which becomes dominant on cooling below the excited state.

Acknowledgments

This work was partially supported by CREST of JST and Grants-in-Aid (Nos. 25800190, 15K13515 and 16H04011) from the Japanese Society for the Promotion of Science (JSPS), by Grants-in-Aids for Scientific Research on Innovative Areas (15H05882, 15H05883) of the Ministry of Education, Culture, Sports, Science and Technology, Japan.

References

[1] v Löhneysen H, Rosch A, Vojta M and Wölfle P 2007 *Rev. Mod. Phys.* **79** 1015–1075
[2] Gegenwart P, Si Q and Steglich F 2008 *Nature Phys.* **4** 186–197
[3] Nakatsuji S, Kuga K, Machida Y, Tayama T, Sakakibara T, Karaki Y, Ishimoto H, Yonezawa S, Maeno Y, Pearson E, Lonzarich G G, Balicas L, Lee H and Fisk Z 2008 *Nature Phys.* **4** 603–607
[4] Kuga K, Karaki Y, Matsumoto Y, Machida Y and Nakatsuji S 2008 *Phys. Rev. Lett.* **101** 137004
[5] Matsumoto Y, Nakatsuji S, Kuga K, Karaki Y, Horie N, Shimura Y, Sakakibara T, Nevidomskyy A H and Coleman P 2011 *Science* **331** 316–319
[6] Okawa M, Matsunami M, Ishizaka K, Eguchi R, Taguchi M, Chainani A, Takata Y, Yabashi M, Tamasaku K, Nishino Y, Ishikawa T, Kuga K, Horie N, Nakatsuji S and Shin S 2010 *Phys. Rev. Lett.* **104** 247201
[7] Hertz J A 1976 *Phys. Rev. B* **14** 1165–1184
[8] Moriya T 1985 *Spin Fluctuations in Itinerant Electron Magnetism* (Berlin: Springer)
[9] Millis A J 1993 *Phys. Rev. B* **48** 7183–7196
[10] Tomita T, Kuga K, Uwatoko Y, Coleman P and Nakatsuji S 2015 *Science* **349** 506–509
[11] Macaluso R T, Nakatsuji S, Kuga K, Thomas E L, Machida Y, Maeno Y, Fisk Z and Chan J Y 2007 *Chem. Mater.* **19** 1918–1922
[12] Matsumoto Y, Kuga K, Tomita T, Karaki Y and Nakatsuji S 2011 *Phys. Rev. B* **84** 125126
[13] Matsuda Y H, Nakamura T, Kuga K, Nakatsuji S, Michimura S, Inami T, Kawamura N and Mizumaki M 2013 *J. Korean Phys. Soc.* **62** 125126

[14] Nevidomskyy A H and Coleman P 2009 *Phys. Rev. Lett.* **102** 077202
[15] Ramires A, Coleman P, Nevidomskyy A H and Tsvelik A M 2012 *Phys. Rev. Lett.* **109** 176404
[16] Matsumoto Y, Hong J, Kuga K and Nakatsuji S 2015 *Journal of Physics: Conference Series* **592** 012086
[17] O'Farrell E C T, Matsumoto Y and Nakatsuji S 2012 *Phys. Rev. Lett.* **109** 176405
[18] Watanabe S and Miyake K 2010 *Phys. Rev. Lett.* **105** 186403
[19] Watanabe S and Miyake K 2010 *J. Phys. Soc. Jpn.* **83** 103708
[20] Küchler R, Bauer T, Brando M and Steglich F 2012 *Review of Scientific Instruments* **83** 095102
[21] Takeuchi T, Yasui S, Toda M, Matsushita M, Yoshiuchi S, Ohya M, Katayama K, Hirose Y, Yoshidtani N, Honda F, Sugiyama K, Hagiwara M, Kindo K, Yamamoto E, Haga Y, Tanaka T, Kubo Y, Settai R and Ōnuki Y 2010 *J. Phys. Soc. Jpn.* **79** 064609

Local moments in the heterogeneous electronic state of Cd-substituted $CeCoIn_5$: NQR relaxation rates

H. Sakai[1], F. Ronning[2], T. Hattori[1], Y. Tokunaga[1], S. Kambe[1], J.-X. Zhu[3], N. Wakeham[2], H. Yasuoka[2], E. D. Bauer[2], and J. D. Thompson[2]

[1] Advanced Science Research Center, Japan Atomic Energy Agency, Tokai, Ibaraki, 319-1195, JAPAN
[2] Condensed Matter and Magnet Science, Los Alamos National Laboratory, Los Alamos, New Mexico 87545, USA
[3] Theoretical Division, Los Alamos National Laboratory, Los Alamos, New Mexico 87545, USA

E-mail: sakai.hironori@jaea.go.jp

Abstract. We have used nuclear quadrupole resonance (NQR) to probe microscopically the response of a prototypical quantum critical metal $CeCoIn_5$ to substitutions of small amounts of Cd for In. Approximately half of the Cd substituents induce local Ce moments in their close proximity, as observed by site-dependent longitudinal nuclear spin relaxation rates $1/T_1$. To reaffirm that localized f moments are induced around the Cd substituents, we find a Gaussian spin-echo decay rate $1/T_{2G}$ of transverse nuclear spin relaxation. Further, T_1T/T_{2G}^2 for the NQR subpeak is found to be proportional to temperatures, again indicating local moments fluctuations around the Cd substituents, while that for the NQR main peak shows a $T^{0.7}$-dependence. The latter temperature dependence is close to 0.75 in pure $CeCoIn_5$ and indicates that the bulk electronic state is located close to a two dimensional quantum critical instability.

1. Introduction

Quantum critical points (QCPs) are an excellent organizing principle for the non-Fermi liquid (NFL) behavior observed in many systems and a source of fluctuations from which unconventional superconductivity (SC) emerges. $CeCoIn_5$ is an unconventional superconductor that belongs to the $CeTIn_5$ (T=Co, Rh, Ir) family, and is known to lie in close proximity to an AFM QCP at ambient pressure [1, 2]. A small amount of Cd ($5s^2$) substitution for In ($5s^25p$), induces long-range AFM order [3]. For ~1% Cd substitution the AFM propagation vector is $\boldsymbol{Q} = (\frac{1}{2}, \frac{1}{2}, \frac{1}{2})$ and the ordered moment is ~0.7 μ_B/Ce [4, 5]. Applying pressure suppresses the AFM; however, the fluctuations and the signatures in electrical transport properties at this pressure (P)-induced AFM QCP are absent [6]. The SC dome for ~1% Cd substitution remains broad in the T-P plane as in the pure $CeCoIn_5$ [7]. Recently we reported a qualitative difference in local electronic environments around Sn and Cd substituents that were probed by nuclear quadrupole resonance (NQR). While Sn substitutions introduce a rather homogeneous electronic state suppressing the AFM fluctuations, Cd substitutions produce an electronic heterogeneous state where Cd substituents enhance AFM behavior by inducing unscreened localized moments in their immediate vicinity [8]. At the same time, the bulk electronic state far from the Cd substituents is the same as in the pure $CeCoIn_5$.

In this paper, the localized f nature in the vicinity of Cd substituents in $CeCoIn_5$ is further investigated by measuring the longitudinal and transverse nuclear spin relaxation rates.

2. Experimental

Single crystals of $CeCoIn_5$ and $CeCoIn_{4.925}Cd_{0.075}$ (1.5 % Cd-doped) were grown by the self-flux method [9]. Small crystals for zero-field NQR studies were selected from the same batch used for magnetic susceptibility, electrical resistivity and specific heat measurements. The AFM transition temperature T_N for the 1.5% Cd-doped sample is 3.7 K [8]. $CeCoIn_5$ forms in the tetragonal $HoCoGa_5$-type structure ($P4/mmm$) with two crystallographically inequivalent In sites denoted as In(1) ($4/mmm$) and In(2)($2mm$). Microprobe measurement using wavelength dispersive spectroscopy and extended x-ray absorption fine-structure (EXAFS) observe a homogeneous distribution of Cd with a preferential occupation on the In(1) site [10]. Consequently, the fractional occupancy for the substituted atoms on these In(1) and In(2) sites is roughly 50/50 [10]. NQR measurements were carried out using a phase-coherent, pulsed spectrometer. Using the conventional definition of the electric field gradient (EFG), the NQR frequency ν_Q is defined as $3e^2qQ/\{2I(2I-1)h\}$ by the nuclear quadrupolar moment eQ and the principal component of the EFG tensor $eq \equiv V_{ZZ}$. For a non-axial EFG, the asymmetry parameter is defined as $\eta \equiv \{|V_{YY}| - |V_{XX}|\}/|V_{ZZ}|$ ($|V_{XX}| \leq |V_{YY}| \leq |V_{ZZ}|$). The ^{115}In NQR spectrum for $CeCoIn_5$ consists of 4 lines for In(1) sites with an equal separation of ν_Q=8.17 MHz and η=0, and 4 unequally-separated lines for In(2) sites whose positions give ν_Q=15.5 MHz and η=0.39 [11], by diagonalization of the electric quadrupole Hamiltonian $\mathcal{H}_Q = (h\nu_Q/6)\{3I_z^2 - I(I+1) + (\eta/2)(I_+^2 + I_-^2)\}$.

The spin-echo inversion recovery method was used to determine the longitudinal nuclear spin relaxation rates $1/T_1$. The nuclear magnetization $M(t)$ was recorded as a function of time t after the inversion rf pulse, always together with recording $M(\infty)$ which is the echo intensity without an inversion pulse. The inversion pulse width was adopted to be $\sim$20 μsec as long as possible. To evaluate T_1, each $R(t)$ was fit by the recovery function for the $4\nu_Q$ transition line of $I = 9/2$ with a single T_1: $R(t) \propto (4/33)\exp(-3t/T_1) + (80/143)\exp(-10t/T_1) + (49/165)\exp(-21t/T_1) + (16/715)\exp(-36t/T_1))$, which is derived assuming an EFG $\eta = 0$. In the case of finite η, the recovery function was calculated by numerical diagonalization of the master equation for nuclear relaxation.[12] T_1 did not change significantly if a finite η of 0.01–0.1 was assumed in these cases.

The transverse nuclear spin relaxation rates $1/T_2$ were derived by fitting the echo decay $E(2\tau)$ to $E(2\tau) = E(0)\exp(-2\tau/T_{2L})\exp(-\frac{1}{2}(2\tau/T_{2G})^2)$ with τ being the separation time between excitation and refocusing pulses. In a strongly correlated metal, $1/T_2$ contains the T_1-process which is represented by Lorentz-type decay on a scale T_{2L} and an indirect (nuclear) spin-spin coupling by Gaussian-type decay with a time scale T_{2G}.

3. Results and Discussions

Figure 1 shows a summary of our findings in the 1.5% Cd-doped $CeCoIn_5$ reported in Ref. [8]. As seen in Fig. 1(a), the $4\nu_Q$ line for In(1) splits into a main peak (A) and two satellite peaks labeled B, and C. The split In(1) spectrum for the Cd case is consistent with previous reports [5]. The resolved line splitting means that the local EFGs for the A, B, and C peaks are discretely distributed for sites near Cd. We can estimate η for A, B, and C sites for Cd substitutions to be 0.005, 0.09, and 0.12, respectively. For spectral assignment, EFGs have been calculated by density functional theory (DFT) assuming 2.5% Cd-doping in a $2 \times 2 \times 2$ supercell [8]. These calculations reproduce the main NQR spectral features, e.g., the In(1) NQR lines are discrete. From a comparison with the calculations, B and C lines can be assigned to the nearest neighbor In(1) sites for Cd substituted on In(2) (hereafter, Cd(2) dopants) and In(1) (Cd(1) dopants) sites, respectively. The respective peak dependences of the longitudinal nuclear spin relaxation rates $1/T_1$ for the In(1) sites, as seen in Fig. 1(b), show a striking enhancement of $1/T_1$ just

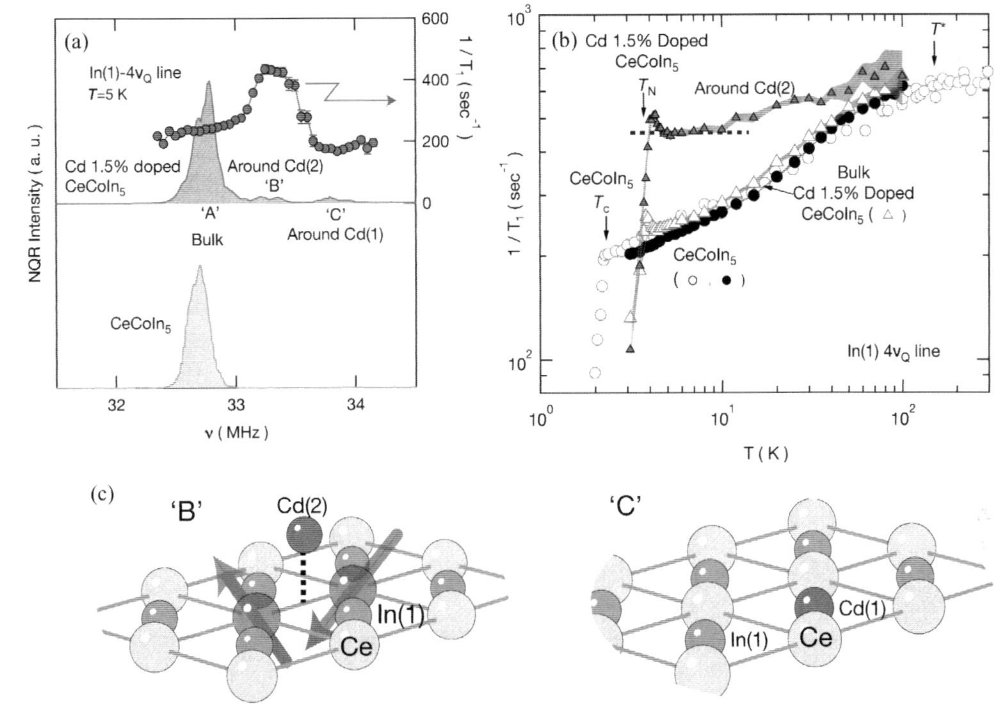

Figure 1. (a) ^{115}In(1) NQR spectra for the $4\nu_Q$ transition in $CeCoIn_5$ and Cd 1.5% doped $CeCoIn_5$ at 5 K. The frequency dependence of longitudinal nuclear spin relaxation rate $1/T_1$ is superimposed. (b) Temperature dependence of $1/T_1$ for In(1) NQR in $CeCoIn_5$ and 1.5% Cd-doped $CeCoIn_5$. The shaded area on each point reflects error associated with the measurement of $1/T_1$. The data for $CeCoIn_5$ are also plotted for a wide temperature range (from Ref. [13]). (c) Schematic illustrations for the local environment for the Cd dopants associated with 'B' and 'C' peaks in Fig. 1(a).

on the B peak. In general, $1/T_1$ is a measure of spin fluctuations, since $1/T_1$ is proportional to $T\sum_{\boldsymbol{q}} \mathrm{Im}\chi_\perp(\boldsymbol{q},\omega_0)$ where $\mathrm{Im}\chi_\perp(\boldsymbol{q},\omega_0)$ is imaginary part of dynamical spin susceptibility, $\boldsymbol{q}$ is propagation wave vector, and ω_0 is an NQR frequency. It should be noted that no enhancement of $1/T_1$ on the C peak indicates that antiferromagnetic fluctuations are not enhanced around Cd(1) dopants. The temperature dependence of $1/T_1$ for the B peak is significantly enhanced for temperatures well below T^* and less temperature dependent. Indeed, $1/T_1$ for peak B is nearly temperature independent below ~10 K, as indicated by the dotted line in Fig. 1(b), although it shows a critical slowing down of $1/T_1$ just above T_N. This constant $1/T_1$ behavior indicates that the $4f$ spin fluctuations behave as if localized. On the other hand, $1/T_1$ for the bulk region in 1.5% Cd-doped $CeCoIn_5$ is nearly equivalent to that in $CeCoIn_5$. These results strongly suggest that a small group of Ce sites neighboring Cd(2) dopants remains unscreened by conduction electrons, but the majority of the Cd-doped sample is in the same electronic state as in undoped $CeCoIn_5$. Such local environments around Cd(1) and Cd(2) dopants are schematically illustrated in Fig. 1(c). Interestingly, the Cd(1) dopants do not induce local moments, proving anisotropic p-f hybridization paths between Ce-In(1) and Ce-In(2) [14], which is also experimentally suggested by the anisotropic Grüneisen parameters in $CeCoIn_5$[15].

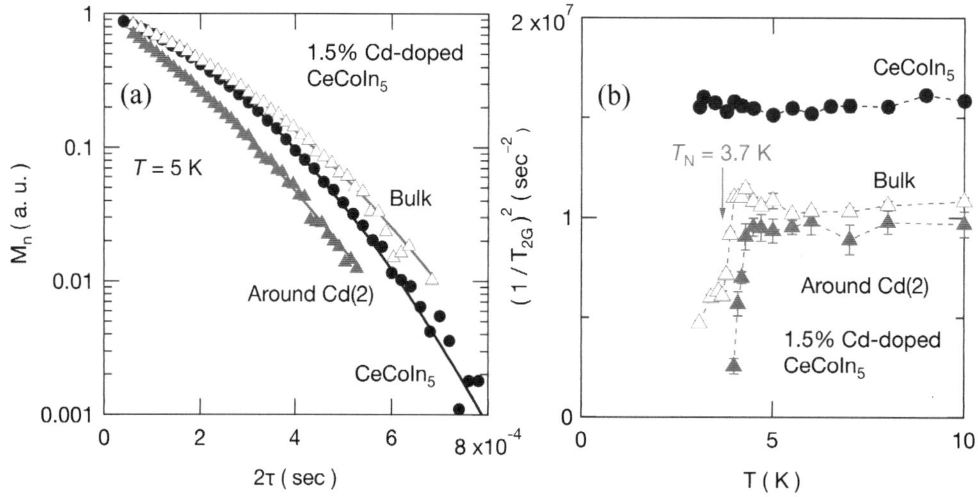

Figure 2. (a) Echo decay curves at 5 K for In(1)-NQR in $\mathrm{CeCoIn_5}$ and 1.5% Cd-doped $\mathrm{CeCoIn_5}$. (b) Temperature dependence of $1/T_{2G}^2$ derived from the echo decays.

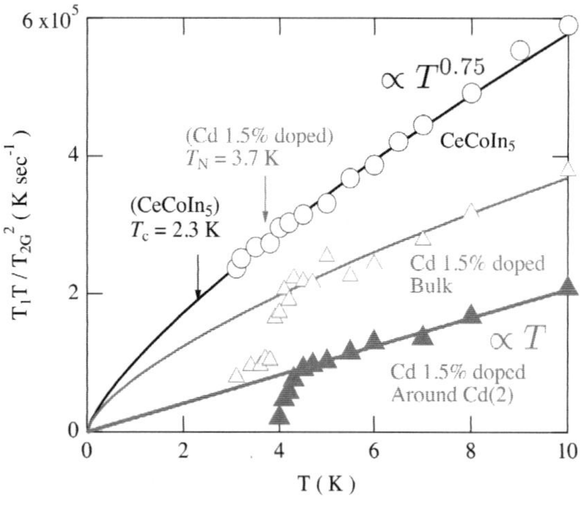

Figure 3. Temperature dependence of T_1T/T_{2G}^2 for In(1) NQR in $\mathrm{CeCoIn_5}$ and 1.5% Cd-doped $\mathrm{CeCoIn_5}$. The solid curves represent fits to cT^n.

For further confirmation of the local moments around Cd(2) dopants, we have measured the transverse nuclear spin relaxation rates $1/T_2$, which provide a measure of the magnetic correlation length ξ. As shown in Fig. 2(a), the obtained echo decay curves on a semi-log scale are convex upward, indicating finite Gaussian components. Since the T-dependence and orders of magnitude of the obtained $1/T_{2L}$ are consistent with $1/T_1$ values from NQR [8], our estimates of $1/T_{2G}$ are reasonable. As shown in Fig. 2(b), $1/T_{2G}^2 \propto \int \mathrm{Re}\chi^2(\boldsymbol{q},0)d\boldsymbol{q}$ [16, 17] is nearly T-independent below ~10 K except for a decrease just above T_N in Cd-doped $\mathrm{CeCoIn_5}$. In the paramagnetic state for a strongly correlated system, the static scaling law $\mathrm{Re}\chi(q,0) \sim \xi^2 g(q\xi,0)$ holds with a scaling function g such that $(1/T_{2G}^2) \sim \xi(T)$ [18]. $\xi(T \to 0)/a$ for pure $\mathrm{CeCoIn_5}$ has been already estimated to be about two at the low temperature [19, 2]. The bulk T-independent values in Cd-doped crystals are reduced roughly by a factor of 1/2 to 2/3 relative to pure $\mathrm{CeCoIn_5}$. Therefore, we deduce that Cd doping reduces ξ/a to approximately unity.

Recently, the quantity of T_1T/T_{2G}^2 has been proposed to be a measure for identifying the quantum criticality class for heavy fermion materials [20, 21]. The technical validity of this method has been originally discussed in high-T_c cuprates [22, 23]. The T-dependences of T_1T/T_{2G}^2 are discussed in Ref. [20]: it becomes (i) constant for 3D spin density wave (SDW) criticality, (ii) proportional to T in the case of localized moments, and (iii) proportional to T^α ($0 < \alpha < 1$) for 2D quantum criticality. Indeed, as shown in Fig. 3, T_1T/T_{2G}^2 for undoped $CeCoIn_5$ is found to be proportional to $T^{0.75\pm0.02}$, which is consistent with quasi-2D quantum criticality in this material. Similarly, T_1T/T_{2G}^2 for the bulk electronic state in 1.5% Cd-doped $CeCoIn_5$ shows a $T^{0.7\pm0.05}$-dependence in the temperature range $\sim 4 < T < 10$ K, although a deviation just above T_N is seen due to a critical slowing down. On the other hand, T_1T/T_{2G}^2 for the electronic state around Cd(2) dopants is proportional to T for $4.5 < T < 10$ K, again indicating the surrounding spin dynamics are dominated by localized $4f$ electrons.

4. Summary

Since NQR measurements are a powerful microscopic tool, the site-to-site relaxation rates reveal the heterogeneous electronic state in 1.5% Cd-doped $CeCoIn_5$: the majority of which is an undisturbed quantum critical metal, but the minority is localized $4f$ states confined around Cd(2) dopants. Further NQR investigations about the heterogeneous electronic states under pressures are ongoing.

5. Acknowledgments

Work at Los Alamos National Laboratory was performed under the auspices of the US Department of Energy, Office of Basic Energy Sciences, Division of Materials Sciences and Engineering. Work in Japan was partly supported by the Reimei Research Program of JAEA. section.

References

[1] Petrovic C, Pagliuso P G, Hundley M F, Movshovich R, Sarrao J L, Thompson J D, Fisk Z and Monthoux P 2001 *J. Phys. :Condens. Matter* **13** L337
[2] Sakai H, Brown S E, Baek S H, Ronning F, Bauer E D and Thompson J D 2011 *Phys. Rev. Lett.* **107** 137001
[3] Pham L D, Park T, Maquilon S, Thompson J D and Fisk Z 2006 *Phys. Rev. Lett.* **97** 056404
[4] Nicklas M, Stockert O, Park T, Habicht K, Kiefer K, Pham L D, Thompson J D, Fisk Z and Steglich F 2007 *Phys. Rev. B* **76** 052401
[5] Urbano R R, Young B L, Curro N J, Thompson J D, Pham L D and Fisk Z 2007 *Phys. Rev. Lett.* **99** 146402
[6] Seo S, Lu X, Zhu J X, Urbano R R, Curro N, Bauer E D, Sidorov V A, Pham L D, Park T, Fisk Z and Thompson J D 2014 *Nature Physics* **10** 120
[7] Sidorov V A, Nicklas M, Pagliuso P G, Sarrao J L, Bang Y, Balatsky A V and Thompson J D 2002 *Phys. Rev. Lett.* **89** 157004
[8] Sakai H, Ronning F, Zhu J X, Wakeham N, Yasuoka H, Tokunaga Y, Kambe S, Bauer E D and Thompson J D 2015 *Phys. Rev. B* **92** 121105(R)
[9] Tokiwa Y, Movshovich R, Ronning F, Bauer E D, Papin P, Bianchi A D, Rauscher J F, Kauzlarich S M and Fisk Z 2008 *Phys. Rev. Lett.* **101** 037001
[10] Booth C H, Bauer E D, Bianchi A D, Ronning F, Thompson J D, Sarrao J L, Cho J Y, Chan J Y, Capan C and Fisk Z 2009 *Phys. Rev. B* **79** 144519
[11] Kohori Y, Yamato Y, Iwamoto Y, Kohara T, Bauer E D, Maple M B and Sarrao J L 2001 *Phys. Rev. B* **64** 134526
[12] Chepin J and J H Ross J 1991 *J. Phys. :Condens. Matter* **3** 8103
[13] Yashima M, Kawasaki S, Kawasaki Y, q Zheng G, Kitaoka Y, Shishido H, Settai R, Haga Y and Ōnuki Y 2004 *J. Phys. Soc. Jpn.* **73** 2073
[14] Haule K, Yee C H and Kim K 2010 *Phys. Rev. B* **81** 195107
[15] Oeschler N, Gegenwart P, Lang M, Movshovich R, Sarrao J L, Thompson J D and Steglich F 2003 *Phys. Rev. Lett.* **91** 076402
[16] Pennington C H and Slichter C P 1991 *Phys. Rev. Lett.* **66** 381
[17] Walstedt R E and Cheong S W 1995 *Phys. Rev. B* **51** 3163

[18] Kambe S, Sakai H, Tokunaga Y, Matsuda T D, Haga Y, Chudo H and Walstedt R E 2009 *Phys. Rev. Lett.* **102** 037208
[19] Stock C, Broholm C, Hudis J, Kang H J and Petrovic C 2008 *Phys. Rev. Lett.* **100** 087001
[20] Kambe S, Sakai H, Tokunaga Y, Matsuda T D, Haga Y, Yasuoka H, Chudo H and Walstedt R E 2010 *J. Phys.: Conf. Ser.* **200** 012076
[21] Kambe S, Sakai H, Tokunaga Y, Chudo H and Walstedt R E 2010 *physica status solidi (b)* **247** 520
[22] Itoh Y, Yasuoka H, Fujiwara Y, Ueda Y, Machi T, Tomeno I, Tai K, Koshizuka N and Tanaka S 1992 *J. Phys. Soc. Jpn.* **61** 1287
[23] Curro N J, Imai T, Slichter C P and Dabrowski B 1997 *Phys. Rev. B* **56** 877

Second-order Structural Transition in $(Ca_{0.5}Sr_{0.5})_3Rh_4Sn_{13}$

Y W Cheung[1], Y J Hu[1], S K Goh[1,*], K Kaneko[2], S Tsutsui[3], P W Logg[4], F M Grosche[4], H Kanagawa[5], Y Tanioku[5], M Imai[5], T Matsumoto[5], K Yoshimura[5]

[1] Department of Physics, The Chinese University of Hong Kong, Shatin N.T., Hong Kong, China
[2] Materials Sciences Research Center, Japan Atomic Energy Agency, Tokai, Naka, Ibaraki 319-1195, Japan
[3] Japan Synchrotron Radiation Research Institute (JASRI), SPring-8, Sayo, Hyogo 679-5198, Japan
[4] Cavendish Laboratory, University of Cambridge, J. J. Thomson Avenue, Cambridge CB3 0HE, United Kingdom
[5] Department of Chemistry, Graduate School of Science, Kyoto University, Kyoto 606-8502, Japan

E-mail: skgoh@phy.cuhk.edu.hk

Abstract. $(Ca_{0.5}Sr_{0.5})_3Rh_4Sn_{13}$ is a member of the substitution series $(Ca_xSr_{1-x})_3Rh_4Sn_{13}$ which has recently been argued to feature a structural quantum critical point at $x_c = 0.9$. In the stoichiometric compound $Sr_3Rh_4Sn_{13}$, the structural transition at $T^* \approx 138$ K has been shown to be second-order. Moving towards x_c, we examine the character of the structural transition in $(Ca_{0.5}Sr_{0.5})_3Rh_4Sn_{13}$ (*i.e.* $x = 0.5$, $T^* \approx 55$ K) using electrical resistivity, heat capacity and X-ray scattering. The absence of the thermal hysteresis in specific heat around T^*, and the continuous evolution of the superlattice reflection detected by X-ray diffraction are consistent with the scenario that the structural transition associated with a modulation vector $\boldsymbol{q} = (0.5\ 0.5\ 0)$ in $(Ca_{0.5}Sr_{0.5})_3Rh_4Sn_{13}$ remains second-order on approaching the quantum critical point.

1. Introduction

$Sr_3Rh_4Sn_{13}$ is a conventional s-wave superconductor with a critical transition temperature (T_c) of 4.7 K [1, 2, 3]. In addition to the superconducting transition, another second-order phase transition takes place at T^*=138 K. This high temperature transition has been established to be a structural phase transition between the $Pm\bar{3}n$ and $I\bar{4}3d$ space groups above and below T^*, respectively. T^* can be suppressed either by applying pressure or by calcium substitution, *i.e.* by forming a $(Ca_xSr_{1-x})_3Rh_4Sn_{13}$ substitution series [1, 3]. If the structural transition remains second order as $T^* \to 0$, a structural quantum critical point can be realized [1, 4].

In the substitution series $(Ca_xSr_{1-x})_3Rh_4Sn_{13}$, T^* can be completely suppressed solely by fine tuning the calcium content, and it has been established that $T^* \to 0$ when $x = x_c \approx 0.9$ [1]. In the vicinity of x_c, several important observations are noted: (a) Debye temperature is a minimum, (b) T_c is a maximum, and (c) strong-coupling superconductivity can be stabilized. All these features can be nicely understood in the framework of structural quantum criticality, and

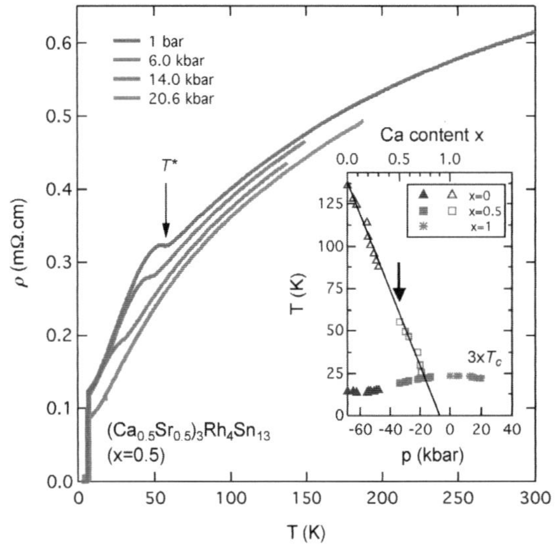

Figure 1. Temperature dependence of the electrical resistivity measured at different pressures. The arrow indicates T^* at ambient pressure. Inset: Universal temperature-pressure phase diagram constructed using $Sr_3Rh_4Sn_{13}$, $(Ca_{0.5}Sr_{0.5})_3Rh_4Sn_{13}$, and $Ca_3Rh_4Sn_{13}$. For details, see reference [1]. $Sr_3Rh_4Sn_{13}$ is located at the leftmost edge of the phase diagram. The location of $(Ca_{0.5}Sr_{0.5})_3Rh_4Sn_{13}$ is shown by the arrow.

they thus provide strong support for the identification of x_c as a quantum critical point [1, 3]. In $Sr_3Rh_4Sn_{13}$ (x=0), specific heat data unambiguously proved that the structural transition is second-order [1, 3]. In this article, we move closer towards x_c, and demonstrate that the T^* transition remains second-order in $(Ca_{0.5}Sr_{0.5})_3Rh_4Sn_{13}$ ($x = 0.5$) using our latest x-ray diffraction data.

2. Experimental Details

Single crystals of $(Ca_{0.5}Sr_{0.5})_3Rh_4Sn_{13}$ were grown using the Sn-flux method as described elsewhere [5]. Electrical resistivity and heat capacity were measured using Quantum Design's Physical Properties Measurement System. High pressure electrical resistivity was measured using a piston-cylinder pressure cell with Daphne 7373 as the pressure transmitting medium. Superlattice reflection was measured at BL35XU of SPring-8, Japan with a Si(11 11 11) backscattering analyzer on the two-theta arm, which helps to reduce the background. The X-ray energy is 21.747 keV.

3. Results and Discussion

The inset to Figure 1 displays the universal temperature-pressure phase diagram constructed earlier [1], with the arrow marking the position of $(Ca_{0.5}Sr_{0.5})_3Rh_4Sn_{13}$. It is therefore more than halfway between $x = 0$ and x_c. The main panel of Figure 1 shows the temperature dependence of the electrical resistivity for $(Ca_{0.5}Sr_{0.5})_3Rh_4Sn_{13}$ at ambient pressure, 6.0 kbar, 14.0 kbar and 20.6 kbar. In addition to the superconducting transition at low temperature, the structural transition at $T^* \sim 55$ K can be identified as the hump in the electrical resistivity. When pressure is applied, T^* decreases rapidly, reaching 46.9 K and 29.7 K at 6.0 kbar and 14.0 kbar, respectively. At 20.6 kbar, the T^* feature is no longer visible in the electrical resistivity. This confirms the trend that pressure moves the system towards the right hand side of the phase diagram.

In $Sr_3Rh_4Sn_{13}$, a pronounced lambda-like jump was detected at T^* [1, 6], without any thermal hysteresis [1]. In the inset to Figure 2, we display the specific heat data of $(Ca_{0.5}Sr_{0.5})_3Rh_4Sn_{13}$ around T^* collected on warming and on cooling. The specific heat anomaly is significantly weaker here compared with the case of $Sr_3Rh_4Sn_{13}$. However, the absence of thermal hysteresis

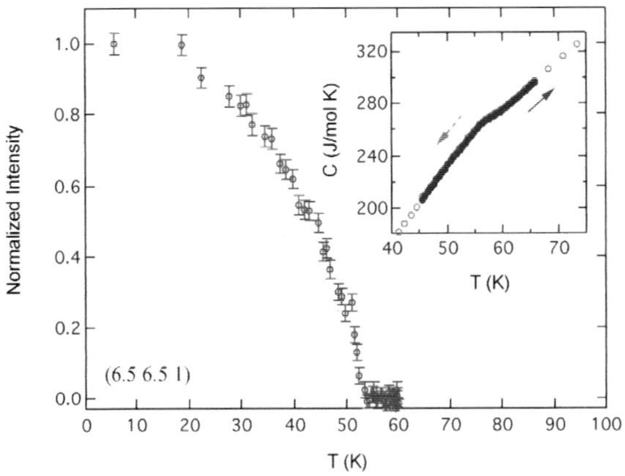

Figure 2. Temperature evolution of the X-ray diffraction intensity associated with the superlattice reflection with $\boldsymbol{q}$ = (0.5 0.5 0). (Inset) Temperature dependence of the specific heat in $(Ca_{0.5}Sr_{0.5})_3Rh_4Sn_{13}$ around T^*, collected on cooling (open symbols) and on warming (closed symbols).

is still clear. In the main panel of Figure 2, we present the temperature evolution of the elastic X-ray scattering intensity for $\boldsymbol{Q}$ = (6.5 6.5 1.0). The additional reflection at $\boldsymbol{Q} = \boldsymbol{k}+\boldsymbol{q}$ below T^*, where $\boldsymbol{k}$ = (6.0 6.0 1.0) corresponds to the Bragg spots in the high temperature ($T > T^*$) phase, is consistent with a modulation vector with $\boldsymbol{q}$ = (0.5 0.5 0) (or the **M** point). T^* determined from the onset of the superlattice reflection is slightly below that obtained from the specific heat data. This difference can be attributed to a slight variation in the calcium content between different samples. From the slope $\mathrm{d}T^*/\mathrm{d}x$ (*c.f.* Figure 1 and Ref. [1]), a variation of 0.03 in x can cause a variation of up to 5 K in T^*. Finally, the gradual and continuous increase of the diffraction intensities on cooling below T^* is consistent with the scenario of a second-order phase transition.

4. Conclusion

In summary, we have studied single crystals of $(Ca_{0.5}Sr_{0.5})_3Rh_4Sn_{13}$ using electrical resistivity, heat capacity and X-ray diffraction. From the absence of the thermal hysteresis around T^* in the heat capacity data and the continous evolution of the superlattice peak intensity at (6.5 6.5 1), we conclude that the structural transition in $(Ca_{0.5}Sr_{0.5})_3Rh_4Sn_{13}$ is a second-order phase transition with $\boldsymbol{q}$ = (0.5 0.5 0). Similar conclusion has recently been reached in related compounds $La_3Co_4Sn_{13}$ [7], $Sr_3Ir_4Sn_{13}$ [6], $Sr_3Rh_4Sn_{13}$ [6] and $Ca_3Ir_4Sn_{13}$ [8] from x-ray/neutron diffraction. These data highlights the similarity across different 3-4-13 families in which the notion of structural quantum criticality has been discussed [1, 4, 9].

Acknowledgments

The authors acknowledge financial support from Research Grant Council (ECS/24300214), The National Natural Science Foundation of China (No. 11504310), CUHK Direct Grant (No. 3132719), EPSRC of the UK (EP/K012894/1), Grant-in-Aids for Scientific Research (B) (No. 22350029 and 16H04131), and (C) (No. 24540336), and for Challenging Exploratory Research (No. 15K14170) from Japan Society for the Promotion of Science. This experiment was carried out under the approval of JASRI (Proposal No. 2013B1095, 2015B1294 and 2016A1160).

References

[1] Goh S K, Tompsett D A, Saines P J, Chang H C, Matsumoto T, Imai M, Yoshimura K and Grosche F M 2015 *Phys. Rev. Lett.* **114**(9) 097002
[2] Kuo C N, Tseng C W, Wang C M, Wang C Y, Chen Y R, Wang L M, Lin C F, Wu K K, Kuo Y K and Lue C S 2015 *Phys. Rev. B* **91**(16) 165141

[3] Yu W C, Cheung Y W, Saines P J, Imai M, Matsumoto T, Michioka C, Yoshimura K and Goh S K 2015 *Phys. Rev. Lett.* **115**(20) 207003
[4] Klintberg L E, Goh S K, Alireza P L, Saines P J, Tompsett D A, Logg P W, Yang J, Chen B, Yoshimura K and Grosche F M 2012 *Phys. Rev. Lett.* **109**(23) 237008
[5] Yang J, Chen B, Michioka C and Yoshimura K 2010 *J. Phys. Soc. Jpn.* **79** 113705
[6] Lue C S, Kuo C N, Tseng C W, Wu K K, Liang Y H, Du C H and Kuo Y K 2016 *Phys. Rev. B* **93**(24) 245119
[7] Cheung Y W, Zhang J Z, Zhu J Y, Yu W C, Hu Y J, Wang D G, Otomo Y, Iwasa K, Kaneko K, Imai M, Kanagawa H, Yoshimura K and Goh S K 2016 *Phys. Rev. B* **93**(24) 241112(R)
[8] Mazzone D G, Gerber S, Gavilano J L, Sibille R, Medarde M, Delley B, Ramakrishnan M, Neugebauer M, Regnault L P, Chernyshov D, Piovano A, Fernández-Díaz T M, Keller L, Cervellino A, Pomjakushina E, Conder K and Kenzelmann M 2015 *Phys. Rev. B* **92**(2) 024101
[9] Ślebarski A, Fijałkowski M, Maśka M M, Mierzejewski M, White B D and Maple M B 2014 *Phys. Rev. B* **89**(12) 125111

CePdAl - a Kondo lattice with partial frustration

Veronika Fritsch[1], Stefan Lucas[2], Zita Huesges[2], Akito Sakai[1], Wolfram Kittler[3], Christian Taubenheim[3], Sarah Woitschach[2], Bjørn Pedersen[4], Kai Grube[5], Burkhard Schmidt[2], Philipp Gegenwart[1], Oliver Stockert[2], and Hilbert v. Löhneysen[3,5]

[1]Experimental Physics VI, Center for Electronic Correlations and Magnetism, Institute of Physics, University of Augsburg, 86135 Augsburg, Germany
[2]Max Planck Institute for Chemical Physics of Solids, 01187 Dresden, Germany
[3]Karlsruhe Institute of Technology, Physikalisches Institut, 76131 Karlsruhe, Germany
[4]Heinz Maier-Leibnitz Zentrum (MLZ), Technische Universität München, 85748 Garching, Germany
[5]Karlsruhe Institute of Technology, Institut für Festkörperphysik, 76131 Karlsruhe, Germany

E-mail: veronika.fritsch@physik.uni-augsburg.de

Abstract. Magnetic frustration, which is well-defined in insulating systems with localized magnetic moments, yields exotic ground states like spin ices, spin glasses, or spin liquids. In metals magnetic frustration is less well defined because of the incipient delocalization of magnetic moments by the interaction with conduction electrons, viz., the Kondo effect. Hence, the Kondo effect and magnetic frustration are antithetic phenomena. Here we present experimental data of electrical resistivity, magnetization, specific heat and neutron diffraction on CePdAl, which is one of the rare examples of a geometrically frustrated Kondo lattice, demonstrating that the combination of Kondo effect and magnetic frustration leads to an unusual ground state.

1. Introduction

Geometric frustration can prevent magnetic ordering despite the presence of a sizable magnetic exchange interaction. In insulating systems this leads to exotic ground states like spin ices [1], spin glasses [2] or spin liquids [3]. The existing theoretical concepts to describe these states involve localized magnetic moments and usually their nearest and next-nearest neighbor exchange interactions without considering the charge degrees of freedom [4]. In metals the situation is different due to the presence of conduction electrons coupling to the local moments. Firstly, especially in f-electron systems, the indirect long-range Ruderman-Kittel-Kasuya-Yosida (RKKY) interaction renders the description by nearest and next-nearest neighbor exchange not in general sufficient. In $4f$-electron systems partial frustration is found in a few materials where one part of the magnetic moments forms long-range order, the other part remains disordered down to lowest temperature. Examples are systems like $R\mathrm{InCu_4}$ with R = Gd, Tb, Dy, Ho, Er and Tm [5, 6, 7] or CePdAl to be presented here [8]. Secondly, due to the Kondo effect local moments can be screened by the conduction electrons, leading to a non-magnetic ground state. Magnetic frustration and Kondo effect can in this respect be regarded as being antithetic: The Kondo effect, yielding a delocalization of the magnetic moments due to virtual transitions of $4f$ electrons to the Fermi level, is not beneficial for the formation of a frustrated state. On the other hand, magnetic exchange interactions between the local moments can result in a breakdown of

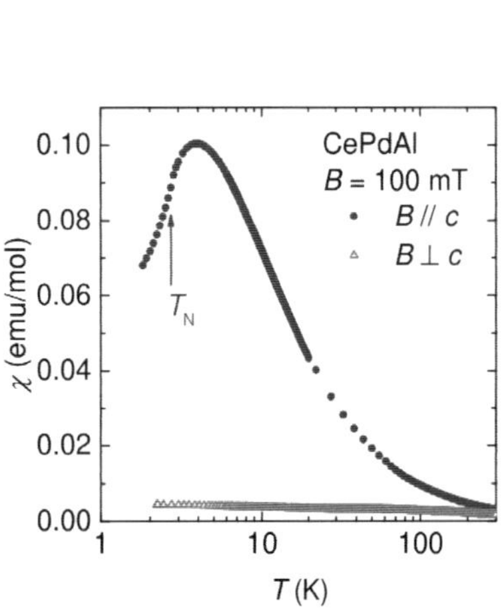

Figure 1. DC susceptibility $\chi = M/B$ versus temperature T in an external magnetic field $B = 100$ mT aligned parallel (blue circles) and perpendicular (red triangles) to the crystallographic c axis. The maximum of $\chi(T)$ indicates antiferromagnetic order, while the maximum of $d\chi/dT$ yields the Neél temperature T_N [13].

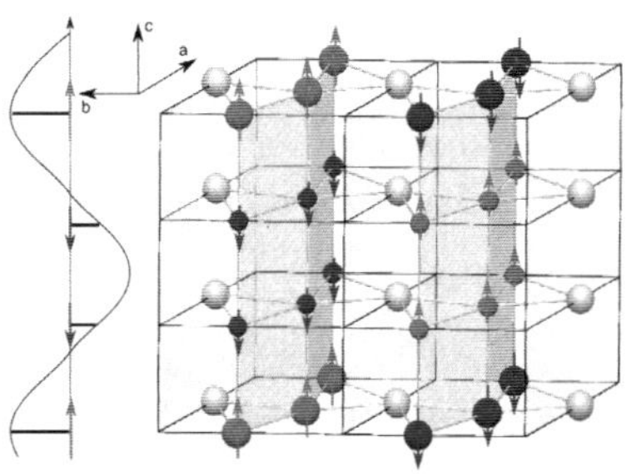

Figure 2. Magnetic structure of CePdAl: two-thirds of the Ce moments form corrugated antiferromagnetic planes in the ac-plane with a sine-like modulation along the c axis, which are separated by planes of frustrated moments (shown in yellow). The sketch neglects the small incommensuration τ^* along the c direction

Kondo screening [9]. Hence, the frustration parameter, the ratio between Curie-Weiss and Néel temperature as defined for insulating systems, is no longer useful in the presence of the Kondo effect.

CePdAl is a partially frustrated Kondo lattice with a Kondo temperature $T^* \approx 6$ K [8, 10, 11]. Therefore it provides the possibility to investigate the interplay between magnetic frustration and Kondo physics. Here we present measurements of the magnetization, the electrical resistivity and the specific heat as well as experimental data from neutron scattering experiments indicating that the frustrated moments in CePdAl remain fluctuating down to lowest temperature [12] and affect the long-range order and magnetic excitations of the non-frustrated moments.

CePdAl crystallizes in the hexagonal ZrNiAl structure. The magnetic moments are located at the cerium sites, which form distorted kagomé planes stacked exactly on top of each other. At $T_N = 2.7$ K antiferromagnetic order sets in. It is an Ising system with the magnetic moments aligned along the c axis, as shown by the susceptibility measurements presented in Fig. 1. These data are in nice agreement with previously published susceptibility data by Isikawa et al. [14], who attributed the magnetocrystalline anisotropy to the effect of the crystal electric field. The magnetic structure derived from powder neutron diffraction with a wave vector $\boldsymbol{q} = \left(\frac{1}{2}, 0, \frac{1}{3} + \tau^*\right)$ [8] is visualized in Fig. 2. Within the basal plane, two thirds of the cerium moments form ferromagnetic chains, which are coupled antiferromagnetically and separated by the frustrated moments. Thus in contrast to the crystallographic structure with a single cerium site, the magnetic structure below T_N of CePdAl features three inequivalent cerium sites.

Due to the geometrical frustration present in the ab plane in this compound one third of the magnetic moments does not participate in the long-range magnetic order [8]. These moments form a rectangular lattice in the ac plane (shown in yellow in Fig. 2), separated from each other

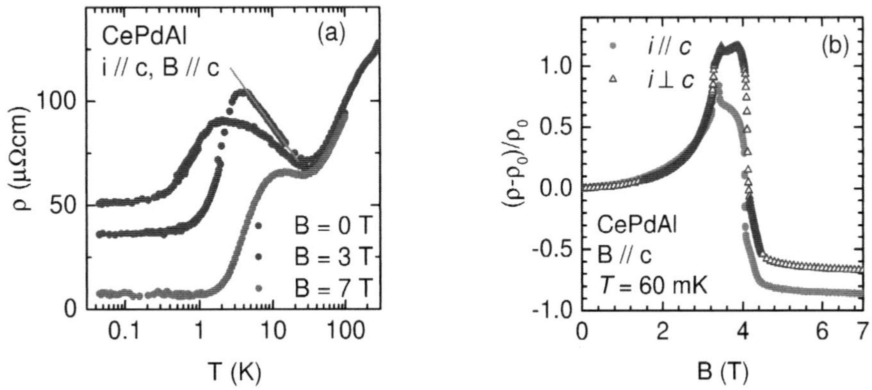

Figure 3. (a) Resistivity ρ vs. temperature T of CePdAl with the current i applied along the c axis in zero field (blue symbols) and an external magnetic field aligned parallel to the c axis with $B = 3\,\mathrm{T}$ (purple symbols) and $B = 7\,\mathrm{T}$ (green symbols). The red line is a guide to eye, showing the logarithmic increase of the resistivity related to the Kondo effect. (b) Field dependence of the resistivity at $T = 60$ mK with the magnetic field aligned parallel to the c axis and the current parallel (red circles) and perpendicular (blue triangles) to the applied field.

by antiferromagnetically ordered corrugated planes (grey shaded in Fig. 2) [15].

2. Experimental Details

A single crystal of CePdAl was grown by the Czochralski method [16, 17]. A Physical Properties Measurement System (PPMS, Quantum Design) was used to obtain the specific heat and resistivity data in the temperature range between 350 mK and room temperature. Specific-heat and resistivity measurements at lower temperatures were performed in a dilution refrigerator. The electrical resistivity was measured with a standard four-contact method employing a LR700 resistance bridge, for the specific heat measurements a standard heat-pulse technique was used. The magnetization measurements at $T = 0.5$ and 5 K were performed in a Magnetic Properties Measurement System (MPMS, Quantum Design) equipped with a SQUID Magnetometer in magnetic fields up to $B = 7\,\mathrm{T}$. The magnetization data at $T = 1.6$ K and the dc susceptibility between 1.6 K and room temperature were measured in a Vibrating Sample Magnetometer (VSM, Oxford Instruments) in magnetic fields up to $B = 12$ T. Neutron scattering was performed at the instrument RESI (FRM2, Munich) with a neutron wavelength $\lambda = 1.03\,\text{Å}$ in a ^{3}He cryostat.

3. Results and Discussion

The electrical resistivity of CePdAl is shown in Fig. 3 (a) in the temperature range between 40 mK and 300 K in zero field (blue symbols) and an external magnetic field applied parallel to the c axis with $B = 3\,\mathrm{T}$ (purple symbols) and $B = 7\,\mathrm{T}$ (green symbols). In zero field a logarithmic increase of $\rho(T)$ (illustrated by the red line) is observed when lowering the temperature, characteristic for the presence of the Kondo effect. At $T_{max} \approx 4\,\mathrm{K}$ the resistivity passes through a maximum and then drops steeply due to the onset of lattice coherence and antiferromagnetic order. A minimum around $T \approx 30\,\mathrm{K}$ in the thermopower [18] confirms the presence of the Kondo effect in this material. An external magnetic field of $B = 3\,\mathrm{T}$ lowers and broadens the maximum in $\rho(T)$ due to the lower Néel temperature $T_{\mathrm{N}} = 1.7\,\mathrm{K}$, while the low-$T$ resistivity increases by roughly 30% in $B = 3$ T.

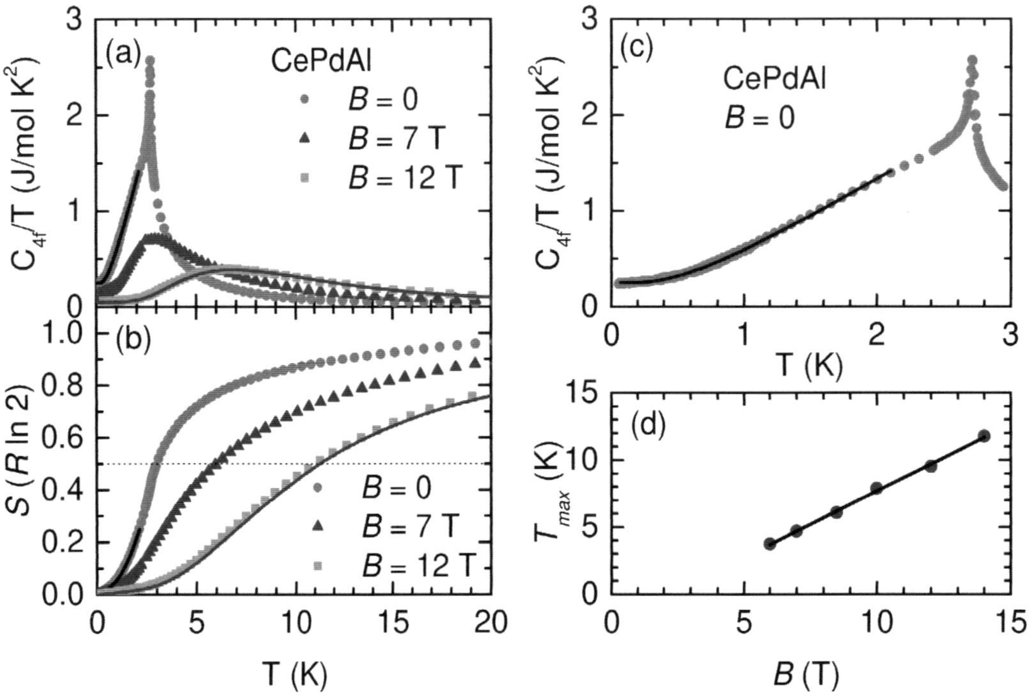

Figure 4. (a) Specific heat of CePdAl in the representation C/T vs T in zero field (red circles) and in an external magnetic field along the easy axis $B = 7\,\mathrm{T}$ (purple triangles) and $B = 12\,\mathrm{T}$ (orange squares). The black line is a fit to the zero-field data as described in the text. The blue line is a fit with the Kondo model after ref. [19], see text for details. (b) Corresponding magnetic entropies $S(T)$. The blue line is a fit with the Kondo model as used in panel (a). (c) Enlarged presentation of the zero-field specific heat and the corresponding fit. (d) Temperatures T_{max} of the maxima of the Schottky anomaly in CePdAl vs magnetic field B.

In an external magnetic field $B = 7\,\mathrm{T}$ the resistivity at the lowest attainable temperature drops to $7.5\,\mu\Omega\,\mathrm{cm}$ with the room temperature resistivity of $126\,\mu\Omega\,\mathrm{cm}$ yielding a residual-resistivity ratio $RRR = 17$, which indicates a satisfying sample quality. Of course one could argue that magnetic impurities would be suppressed in an external magnetic field as well, however that would usually happen at significantly lower fields. The non-monotonic low-T $\rho(T, B)$ is confirmed when considering the field dependence of the isothermal resistivity at lowest temperature, as presented in Fig. 3 (b). The resistivity increases with increasing field with peaks at 3.4 T and 3.9 T (the latter being a shoulder for $i//B$) indicating magnetic transitions in nice agreement with the $\rho(B)$ data by Goto *et al.* [10], who also suggested that these transitions indicate the successive alignment of the in zero field frustrated Ce moments with increasing magnetic field.

In order to obtain insight into the magnetic low-energy states, we analyzed the specific heat of CePdAl in zero field down to very low temperatures. Figure 4 (a) shows the $4f$-electron contribution C_{4f} to the specific heat of of CePdAl in zero field, $B = 7\,\mathrm{T}$ and $B = 12\,\mathrm{T}$ with $B//c$. C_{4f} was obtained by subtracting the specific heat of the non-magnetic parent compound LuPdAl as well as the nuclear contribution $\propto 1/T^2$ due to the electric quadrupolar moments of Pd and Al calculated from high-field data [20]. The black line, better visible in panel (c) of

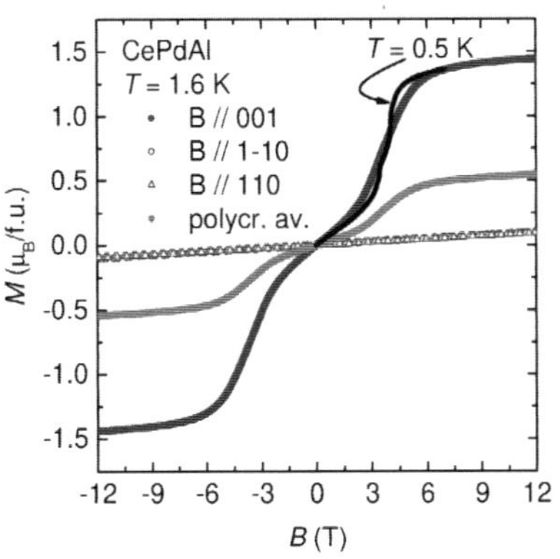

Figure 5. Magnetization of CePdAl at $T = 1.6$ K with the magnetic field aligned parallel to the [001], [110] and $[1\bar{1}0]$ direction. The red curve is the polycrystalline average of all three directions. The black line shows the magnetization M of CePdAl at $T = 0.5$ K with the magnetic field aligned parallel to the [001].

Fig. 4 is a fit of the zero field data with

$$C_{4f} = \gamma T + b \cdot T^2 \exp\left(-\frac{\Delta}{k_B T}\right). \quad (1)$$

Here γT represents the electronic linear contribution to the specific heat with $\gamma = 250\,\mathrm{mJ/molK^2}$, which we attribute to the frustrated Ce moments. The second term $b \cdot T^2 \exp\left(-\frac{\Delta}{k_B T}\right)$ suggests the presence of gapped spin waves with $\Delta/k_B = 920\,\mathrm{mK}$ in addition to the low-energy excitations due to frustrated Ce moments also found in previous NMR measurements [12]. The observed T^2 behavior is reminiscent of two-dimensional antiferromagnetic spin waves [21], in agreement with the corrugated antiferromagnetic planes visualized in Fig. 2.

The energy levels of the Ce are split into three doublets due to the local $m2m$ symmetry of the Ce ion. The crystal field levels were found at $\Delta/k_\mathrm{B} = 244$ and $510\,\mathrm{K}$ by inelastic neutron scattering [11]. The ground state wave function is an almost pure $\left|\frac{5}{2}\right\rangle$ doublet state, in line with the Ising anisotropy shown in the magnetization and susceptibility (see figs. 1 and 5). Thus at low temperatures we can assume an effective spin-$\frac{1}{2}$ system with an enhanced g-factor $g = 4.2$ to account for $J = \frac{5}{2}$ [22]. The magnetic entropy calculated from the specific heat is shown in Fig. 4 (b). For the zero-field measurement at the Néel temperature $T_\mathrm{N} = 2.7\,\mathrm{K}$ less than 50% of the expected entropy of $R \ln 2$ is gained, which is another hint to the frustration present in CePdAl.

As demonstrated by the specific heat data in $B = 7$ and 12 T a Schottky anomaly evolves in magnetic fields. Fig. 4 (d) shows the temperature T_{max} of the maxima in C_{4f} of this Schottky anomaly in magnetic fields larger than 6 T. The linear field dependence confirms that it is indeed a Schottky anomaly. From the slope of the data we estimate the Zeeman splitting of the levels and the involved magnetic moment, the latter being $1.77\,\mu_\mathrm{B}$, which is in fair agreement with the ordered moment of $1.6\,\mu_\mathrm{B}$ found by Dönni *et al.* [8] and compares well to $1.83\,\mu_\mathrm{B}$ found by Prokěs *et al.* [23].

The specific heat in $B = 12\,\mathrm{T}$ can be described very nicely with the single-ion resonance-level Kondo model [19]. The solid blue lines in Fig. 4 (a) and (b) are fits with the resulting fit parameters as follows: Kondo temperature $T^* = 3.23 \pm 0.03\,\mathrm{K}$ and Zeeman splitting $\Delta E/k_\mathrm{B} = 22.9\,\mathrm{K}$, where the latter nearly perfectly agrees with $T_{max} = 9.5$ K when considering $k_B \cdot T_{max} = 0.42 \cdot \Delta E$. In lower fields the emerging correlation effects compromise this fit by producing an additional background, which, however, does not shift T_{max}. Thus, although the magnetization data (see Fig. 5 and ref. [10]), suggest that in an external field $B = 7\,\mathrm{T}$ aligned parallel to the c-axis the system is already in the paramagnetic state the specific heat

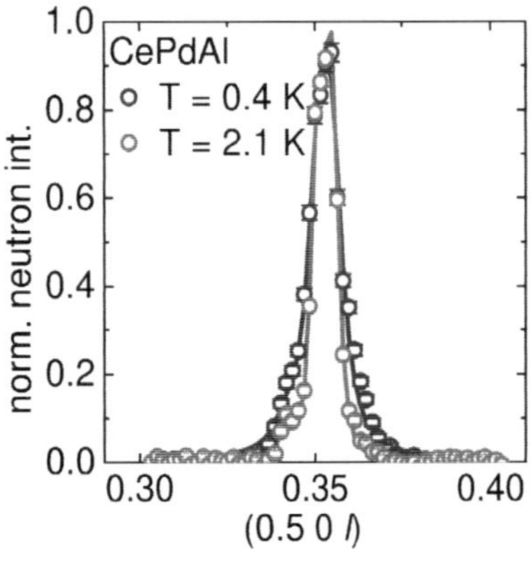

Figure 6. Scans along [001] across a magnetic Bragg (0.5 0 0.35) peak at $T = 2.1$ and 0.4 K.

in $B = 7$ T cannot be described with the single-ion resonance-level Kondo model. The presence of correlations between the magnetic moments, even in regimes beyond the long-range ordered state, is in line with the recent results of Prokěs *et. al* [24], who found that the three magnetic cerium sites still carry different magnetic moments at $T = 100\,\text{mK}$, $p = 0.85\,\text{GPa}$ and in $B = 9\,\text{T}$.

From the Sommerfeld coefficient γ and the value of the magnetic susceptibility χ at low temperatures the Wilson ratio can be calculated [25]. The Sommerfeld coefficient was determined from the value of C_{4f}/T extrapolated to zero temperature to be $\gamma = 250$, 134 and 57 mJ/mol K^2 in $B = 0$, 7 and 12 T. The magnetization, as presented in Fig. 5, was measured at $T = 1.6$ K in magnetic fields up to $B = 12$ T aligned along the easy axis [001] and perpendicular along the hard axes [110] and $[1\bar{1}0]$. The saturation moment at $B = 12\,\text{T}$ is $\mu = 1.44\,\mu_\text{B}$/f.u., roughly in line with the ordered moment $\mu \approx 1.6\,\mu_\text{B}$/f.u. found in previous neutron diffraction experiments [8] and our estimations from the specific heat data discussed above. In a high magnetic field with all moments aligned a saturation moment $m = 2.1\,\mu_\text{B}$ is expected for free Ce^{3+}. The significant reduction of our measured data is a further fingerprint of the Kondo screening effects in CePdAl. For $B \perp c$ no transitions are found and M/B is constant in field. From ref. [14] we know that M/B is temperature-independent below $T \approx 3$ K as well, so we can assume $\chi = dM/dB = 0.00712\,\mu_\text{B}/\text{mol}\,\text{T}$ for all temperatures below 3 K and fields applied along the hard axis. For $B = 0$ we use our data taken at $T = 0.5$ K, which are in agreement with the data reported previously by Goto *et al.* [10]. For the magnetic field aligned parallel to the easy axis the data at $T = 0.5$ and 1.6 K above $B = 6$ K converge, thus we can use the data from our measurement at $T = 1.6$ K for $B = 7$ and 12 T. For the calculation of the Wilson ratio we take the polycrystalline average of these data. The resulting values are $\chi = 0.039$, 0.019 and 0.0078 μ_B/mol T in $B = 0$, 7 and 12 T. The Wilson ratios calculated thereof are $R_W = 1.44$, 1.18 and 1.17 in $B = 0$, 7 and 12 T, which is close to the value for free electrons in all fields. In a phenomenological "two-component model", which was for example used previously for the analysis of $CeCu_{6-x}Au_x$ and $CeAl_2$, the specific heat at low temperatures is separated into two parts: on the one hand magnetic excitations of the antiferromagnetic state, rapidly vanishing with decreasing temperature, and on the other hand heavy fermion excitations [26, 27]. In analogy one could argue that in zero field only the one third of frustrated, disordered moments contributes to γ, which would result in a corrected Sommerfeld coefficient $\gamma' = 0.750$ J/mol K^2 and a corrected Wilson ratio $R'_W = 4.32$. These values would qualify CePdAl as a strongly correlated heavy fermion system. In higher fields of 7 and 12 T this argument is not valid anymore, here the magnetic moments contribute equally to γ, resulting in a Wilson ratio as expected for normal metals.

In order to take a closer look on the magnetic ground state of CePdAl we performed elastic neutron scattering experiments. Figure 6 displays scans across a magnetic Bragg position of CePdAl at $T = 2.1$ and 0.4 K. The peak width of the magnetic Bragg peak increases towards lower temperatures, even well below T_N. This measured width corresponds to antiferromagnetic domains with a size of the order of 200 Å at $T = 0.4$ K. Such a behavior is in marked contrast to usual magnetically ordered systems with true long-range order at lowest temperatures. There must exist an effect limiting the size of the antiferromagnetic domains towards lower temperatures. A possible origin of this behavior are the magnetically frustrated moments in CePdAl. Although no other long-range order is detected, one can assume that the frustrated moments - bearing a non-zero moment - couple to the ordered moments perturbing the long-range ordered structure resulting in a reduction of the domain size.

4. Conclusion

Our results show clearly that magnetic frustration and the Kondo effect are present in CePdAl simultaneously. Our data suggest a bipartite system: on the one hand the antiferromagnetically ordered part, on the other hand the frustrated part. However, these subsystems are not completely independent, as evidenced by the broadening of the magnetic Bragg peaks towards lower temperatures. Furthermore the missing entropy in magnetic fields above the magnetic transitions points to the presence of magnetic correlations and possibly magnetic frustration in areas beyond the long-range ordered regime. Overall CePdAl offers an excellent opportunity to explore the interplay between magnetic frustration and Kondo physics.

Acknowledgements

This work was supported by the Deutsche Forschungsgemeinschaft through FOR 960, the Helmholtz Association through VI 521 and JSPS Postdoctoral Fellow for Research Abroad.

References

[1] Bramwell S T and Gingras M P 2001 *Science* **294** 1495–1501
[2] Mydosh J A 1993 *Spin glasses: an experimental introduction* (London, Washington DC: Taylor & Francis)
[3] Balents L 2010 *Nature* **464** 199–208
[4] Ramirez A P 1994 *Annu. Rev. Mater. Sci.* **24** 453–480
[5] Nakamura H, Ito K, Wada H and Shiga M 1993 *Physica B* **186-188** 633–635
[6] Fritsch V, Hemberger J, Büttgen N, Scheidt E W, Krug von Nidda H A, Loidl A and Tsurkan V 2004 *Phys. Rev. Lett.* **92** 116401
[7] Fritsch V, Thompson J D, Sarrao J L, Krug von Nidda H A, Eremina R M and Loidl A 2006 *Phys. Rev. B* **73** 094413
[8] Dönni A, Ehlers G, Maletta H, Fischer P, Kitazawa H and Zolliker M 1996 *J. Phys.: Cond. Mat.* **8** 11213–11229
[9] Senthil T, Vojta M and Sachdev S 2004 *Phys. Rev. B* **69** 035111
[10] Goto T, Hane S, Umeo K, Takabatake T and Isikawa Y 2002 *J. Phys. Chem. Sol.* **63** 1159
[11] Woitschach S, Stockert O, Koza M M, Fritsch V, v Löhneysen H and Steglich F 2013 *phys. stat. sol. (b)* **250** 468–471
[12] Oyamada A, Maegawa S, Nishiyama M, Kitazawa H and Isikawa Y 2008 *Phys. Rev. B* **77** 064432
[13] Fisher M E 1962 *Philosophical Magazine* **7** 1731 – 1743
[14] Isikawa Y, Mizushima T, Fukushima N, Kuwai T, Sakurai J and Kitzawa H 1996 *J. Phys. Soc. Jpn.* **65** Suppl. B 117
[15] Fritsch V, Bagrets N, Goll G, Kittler W, Wolf M J, Grube K, Huang C L and Löhneysen H v 2014 *Phys. Rev. B* **89** 054416
[16] Abell J S 1989 *Handbook of the physics and chemistry of rare earths* vol 12 (North-Holland) chap Preparation and crystal growth of rare-earth elements and intermetallic compounds, pp 1–51
[17] High-purity rare-earth metals acquired from Materials Preparation Center, Ames Laboratory, US DOE Basic Energy Sciences, Ames, IA, USA, <http://www.mpc.ameslab.gov>
[18] Huo D, Kuwai T, Mizushima T, Isikawa Y and Sakurai J 2002 *Physica B: Condensed Matter* **312 & 313** 232–234 ISSN 0921-4526 the International Conference on Strongly Correlated Electron Systems
[19] Schotte K and Schotte U 1975 *Physics Letters A* **55** 38–40

[20] Sakai A, Lucas S, Gegenwart P, Stockert O, v Löhneysen H and Fritsch V 2016 Quantum criticality in geometrically frustrated $CePd_{1-x}Ni_xAl$. to be published
[21] Ramirez A P, Espinosa G P and Cooper A S 1990 *Phys. Rev. Lett.* **64** 2070–2073
[22] Abragam A and Bleaney B 2012 *Electron paramagnetic resonance of transition ions* (Oxford University Press) page 33
[23] Prokeš K, Manuel P, Adroja D, Kitazawa H, Goto T and Isikawa Y 2006 *Physica B* **385** 359–362
[24] Prokeš K, Hartwig S, Stunault A, Isikawa Y and Stockert O 2015 Probing magnetism in cepdal under multi-extreme conditions using polarized neutrons *Journal of Physics: Conference Series* vol 592 (IOP Publishing) p 012082
[25] Hewson A C 1997 *The Kondo problem to heavy fermions* vol 2 (Cambridge university press)
[26] Schlager H G, Schröder A, Welsch M and v Löhneysen H 1993 *J. Low Temp. Phys.* **90** 181–204
[27] Bredl C, Steglich F and Schotte K 1978 *Zeitschrift für Physik B Condensed Matter* **29** 327–340

Local Kondo entanglement and symmetry protected local criticality in the two-impurity Kondo problem

Yuxiang Li, Xiao-Yong Feng, and Jianhui Dai
Department of Physics, Hangzhou Normal University, Hangzhou 310036, China
E-mail: daijh@hznu.edu.cn

Abstract.
The local Kondo entanglement is defined as the concurrence of a short-ranged Kondo singlet state consisting of a localized magnetic moment and a nearby conduction electron. We derive the entanglement phase diagram of the Rasul-Schlottmann model, the effective spin-only Hamiltonian for the two-impurity Kondo model in the numerical renormalization group approach. We show that the local Kondo entanglement vanishes exactly at the two-impurity Kondo critical point, associated concomitantly with a jump in the inter-impurity entanglement. We discuss how to generalize this result to a Kondo lattice model preserving the same enhanced spin symmetry.

1. Introduction
The two-impurity Kondo model (TIKM)

$$H_{TIKM} = H_0 + J_K[\vec{S}_1 \cdot \vec{s}_c(1) + \vec{S}_2 \cdot \vec{s}_c(2)] \tag{1}$$

is a prototype model system to understand the competition between Kondo effect and the Ruderman-Kittel-Kasuya-Yosida (RKKY) interaction in strongly correlated electron systems[1, 2, 3]. Here, H_0 represents the conduction electron bath; $(\vec{S}_1, \vec{S}_2)$ the impurity spins localized at the sites $\vec{R}_1$ and $\vec{R}_2$; $(\vec{s}_c(1), \vec{s}_c(2))$ the conduction electron spins at the sites $\vec{r}_1$ and $\vec{r}_2$. Usually, the sites $\vec{R}_1$ and $\vec{r}_1$ (or $\vec{R}_2$ and $\vec{r}_2$) are the nearby sites in the real space so that J_K (> 0) describes the short-ranged Kondo coupling. For a metallic bath with the band width D and finite density of states at the Fermi level ρ_F, the system involves two distinct energy scales: the single-ion Kondo temperature $T_K \sim D\exp(-\frac{1}{2J_K\rho_F})$ and the inter-impurity RKKY interaction J_R. Thus the TIKM has two stable fixed points: the strong Kondo coupling limit $(J_R \to 0)$ where individual impurity spins are completely quenched by Kondo effect, and the strong RKKY interaction limit $(T_K \to 0)$ where the two impurity spins are locked into an antiferromagnetic singlet[1, 4].

The primary focus in the two-impurity problem is the situation when T_K and J_R are comparable to each other. In this intermediate parameter regime the physical properties crossover from one limit to another[4, 5]. Of particular interesting is the case when the crossover is sharpened leading to a phase transition. Indeed, when the bath has a hidden particle-hole symmetry at half-filling, an unstable interacting fixed point near $J_R/T_K \approx 2.2$ was revealed by the numerical renormalization group[4] and conformal field theory studies[6]. It was soon

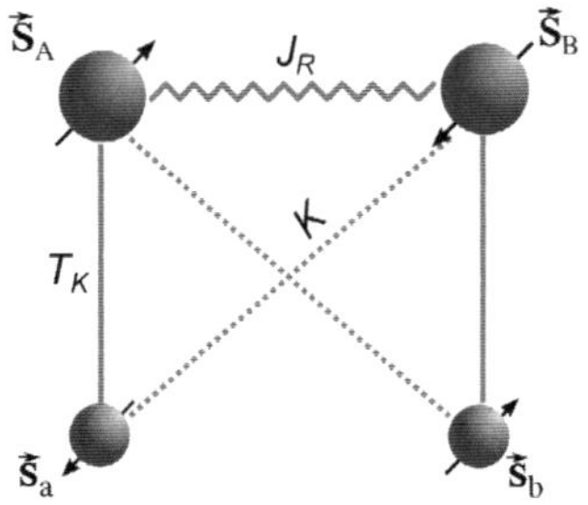

Figure 1. (color online) The cartoon picture of the RS model: the local magnetic moments and the conduction electron spins are denoted by big blue circled and small green circled arrows, respectively. T_K, J_R, and K represent the single-ion Kondo energy scale, the RKKY interaction, and the interaction induced cross Kondo coupling.

pointed out by Rasul and Schlottmann[7] that most of the singular features associated with the unstable fixed point can be understood phenomenologically by an *spin-only* effective model involving two impurity spins $\vec{S}_A$, $\vec{S}_B$, and two conduction electron spins $\vec{s}_c(a)$, $\vec{s}_c(b)$, as shown in Fig.1. The Rasul-Schlottmann (RS) Hamiltonian is given by

$$H_{RS} = T_K[\vec{S}_A \cdot \vec{s}_c(a) + \vec{S}_B \cdot \vec{s}_c(b)] + J_R\vec{S}_A \cdot \vec{S}_B + K[\vec{S}_A \cdot \vec{s}_c(b) + \vec{S}_B \cdot \vec{s}_c(a)]. \quad (2)$$

This model can be regarded as the effective Hamiltonian of the original TIKM at the half-filling where only the spin degrees of freedom are relevant while the charge degrees of freedom are frozen [1, 6, 8]. The J_R describes the intersite RKKY interaction energy, T_K the splitting between the Kondo singlet and the spin triplet states, while the cross-coupling K represents the interaction-induced Kondo frustration as a result of many-body process.

It is known that the two-impurity Kondo critical point may be replaced by a crossover in some circumstances within different numerical approaches[8, 9, 10, 11], while with this symmetry a groundstate degeneracy at the critical point is observed [8]. Such long-standing discrepancy between various numerical approaches has been ultimately resolved by the Natural Orbital Renormalization Group approach, confirming the link of the two-impurity Kondo critical point to the hidden particle-hole symmetry[12]. Compatible with all these observations, the RS model indeed exhibits an enhanced degeneracy at a special point $P : (K/T_K = 1, J_R/T_K = 2)$ corresponding to the critical point[7]. More recently, we have thoroughly reexamined the RS model (in an unpublished arxiv preprint [13]) from the quantum entanglement perspective. In this proceedings paper, we shall briefly illustrate the main results of this work by highlighting the concept of *local Kondo entanglement*. We also discuss how to generalize these results to a Kondo lattice model preserving the same enhanced spin symmetry.

2. Solutions of the RS model

For our purpose we assume that all the interactions involved are positive and spin-$SU(2)$ invariant. Because Eq. (2) is also invariant under the simultaneous permutations between ($\vec{S}_A$, $\vec{S}_B$) and (T_K, K), the regime with strong frustration corresponds to $K = T_K$. Hence we only need to consider the situation $0 \leq K/T_K \leq 1$. In the following we set $T_K = 1$ and (J_R, K) are tunable independently.

The eigenstates of H_{RS} are solved based on a complete set of the conventional basis $|S^z(A), s_c^z(a); S^z(B), s_c^z(b)\rangle$ and labeled by the total spin S, its z-component S_z, and the parity (with respect to permutations of the two local moments) [13]. Because we consider T_K, J_R, and

K are all antiferromagnetic, the groundstate is among the mixed states of the two singlets of even parity. They are denoted by (each up to a normalization factor)

$$\begin{aligned} |\Psi_{(0^+)}\rangle &= |\Downarrow\downarrow;\Uparrow\uparrow\rangle + |\Uparrow\uparrow;\Downarrow\downarrow\rangle + \frac{\Delta - J_R + 2}{J_R - 2K}(|\Downarrow\uparrow;\Downarrow\uparrow\rangle + |\Uparrow\downarrow;\Uparrow\downarrow\rangle) \\ &+ \frac{-\Delta + 2K - 2}{J_R - 2K}(|\Downarrow\uparrow;\Uparrow\downarrow\rangle + |\Uparrow\downarrow;\Downarrow\uparrow\rangle) \end{aligned} \tag{3}$$

and

$$\begin{aligned} |\Psi'_{(0^+)}\rangle &= |\Downarrow\downarrow;\Uparrow\uparrow\rangle + |\Uparrow\uparrow;\Downarrow\downarrow\rangle + \frac{-\Delta - J_R + 2}{J_R - 2K}(|\Downarrow\uparrow;\Downarrow\uparrow\rangle + |\Uparrow\downarrow;\Uparrow\downarrow\rangle) \\ &+ \frac{\Delta + 2K - 2}{J_R - 2K}(|\Downarrow\uparrow;\Uparrow\downarrow\rangle + |\Uparrow\downarrow;\Downarrow\uparrow\rangle) \end{aligned} \tag{4}$$

The corresponding eigen energies are

$$E_{(0^+)} = E_{(0)} - \frac{1}{2}\Delta, \quad E'_{(0^+)} = E_{(0)} + \frac{1}{2}\Delta, \tag{5}$$

with $E_{(0)} = -J_R/4 - K/2 - 1/2$ and $\Delta = \sqrt{(1 + K - J_R)^2 + 3(K-1)^2}$.

It is apparent that the groundstate is the singlet Ψ_{0^+}, and at the point P (i.e., $K = T_K = J_R/2$) the repulsive level spacing (energy gap) $\Delta = 0$ so that the singlet $\Psi_{0'+}$ is degenerate with Ψ_{0^+}. Notice that one of the odd triplet Ψ_{1^-} is also degenerate with Ψ_{0^+} for $K = 1, J_R \geq 2$[13]. Hence the model indeed shows a strong frustration along the line $K = 1$ and exhibits an enlarged symmetry at P[14]. In particular, the correct forms of wavefunctions across P can be approached by taking either limits ($K = 1, J_R = 2 - \epsilon$) and ($K = 1, J_R = 2 + \epsilon$), $\epsilon \to 0^+$. It readily reveals a discontinuity in $|\Psi_{(0^+)}\rangle$.

3. Entanglement phase diagram

One of the conventional entanglement quantities in impurity spin systems is the single-impurity Kondo entanglement (SIKE), defined as the entanglement between a local moment, say $\vec{S}(A)$, and the rest of the system, denoted by $\tilde{A}$. It is usually measured by the von Neumann entropy $\mathcal{E}_{SIKE} = -Tr_{(A)}\{\hat{\rho}_{imp}(A)\log_2 \hat{\rho}_{imp}(A)\}$, where $\hat{\rho}_{imp}(A)$ is the reduced density matrix $\hat{\rho}_{imp}(A) = Tr_{\tilde{A}}\hat{\rho}$. Here, $\hat{\rho} = |\Psi_G\rangle\langle\Psi_G|$ is the density matrix for the groundstate (in our case $|\Psi_G\rangle = |\Psi_{(0^+)}\rangle$) of the whole system. It is straightforwardly seen that $\mathcal{E}_{SIKE} = 1$ due to the $SU(2)$-spin invariance, indicating a maximal entanglement between a local moment and the reminder of the total system. Another frequently used entanglement quantity in the TIKM is the entanglement of the two local moments with conduction electron bath, called the two-impurity Kondo entanglement (TIKE). This entanglement is determined by the reduced density matrix of the two impurities, $\hat{\rho}_{imp}(AB) = Tr_{(c)}\hat{\rho}$, with $Tr_{(c)}$ indicating trace over the Hilbert subspace spanned by conduction electrons. It can be quantified by the von Neumann entropy[15]

$$\mathcal{E}_{TIKE} = -p_s \log_2 p_s - (1 - p_s)\log_2 \frac{1 - p_s}{3}. \tag{6}$$

Here, $p_s = \frac{1}{4} - f_{AB}$ is the fidelity of the spin singlet within the reduced two impurity state, and $f_{AB} = \langle\Psi_{0^+}|\vec{S}(A)\cdot\vec{S}(B)|\Psi_{0^+}\rangle$ is the spin-spin correlation function on the groundstate. In our case,

$$f_{AB} = \frac{1}{2}\frac{1 + K - J_R}{\sqrt{(J_R - K - 1)^2 + 3(K-1)^2}} - \frac{1}{4}. \tag{7}$$

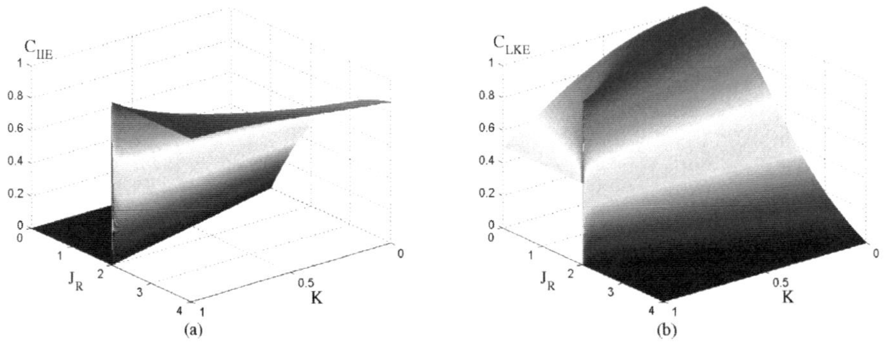

Figure 2. (color online) The concurrence of the IIE (a) and the LKE (b) as functions of J_R and K. The single-ion Kondo energy scale is taken as unit $T_K = 1$.

We find that $\mathcal{E}_{TIKE}$ is not only a smoothly varying function of f_{AB}, but also a smooth function of K and J_R without detectable feature across the point P[16].

Obviously, both SIKE and TIKE cannot capture the variation of Kondo effect across the transition point. In Ref.[15], the inter-impurity entanglement (IIE) is introduced. The IIE is measured by the concurrence or negativity, $\mathcal{C}_{IIE}$, and its evaluation is also related to the reduced two-impurity density matrix $\hat{\rho}_{imp}(AB)$. The concurrence can be expressed by[15, 16]

$$\mathcal{C}_{IIE} = \max\{-2f_{AB} - 1/2, 0\}. \tag{8}$$

The result is plotted in Fig.2(a). For fixed $K < 1$, $\mathcal{C}_{IIE}$ increases continuously with J_R. For $K = 1$, $\mathcal{C}_{IIE}$ shows a sudden increase from zero to unity when J_R goes across $J_R = 2$, indicative of transition to the inter-impurity singlet state.

But IIE itself is not a decisive quantity for the Kondo effect breakdown. We now turn to such a quantity by introducing the local Kondo entanglement (LKE)[13], i.e., the entanglement between a local moment, say $\vec{S}(A)$, and the conduction electron at its nearest neighbor site, $\vec{s}_c(a)$. The LKE differs to the conventional impurity entanglements as it involves only a spatially short-ranged Kondo pair as one can clearly see from the original Kondo coupling in Eq.(1). Similar to IIE, the LKE can be evaluated by the concurrence or negativity, via the corresponding reduced density matrix $\hat{\rho}_{LK}(Aa) = Tr_{(\tilde{A}\tilde{a})}\hat{\rho}$, with $Tr_{(\tilde{A}\tilde{a})}$ indicating the trace in the Hilbert space except the subspace spanned by $\vec{S}_A$ and $\vec{s}_c(a)$. Thus we have

$$\mathcal{C}_{LKE} = \max\{-2f_{Aa} - 1/2, 0\}, \tag{9}$$

where $f_{Aa} = \langle\Psi_{0+}|\vec{S}(A)\cdot\vec{s}_c(a)|\Psi_{0+}\rangle$ is the correlation function of the local Kondo singlet state,

$$f_{Aa} = \frac{1}{4}\frac{J_R + 2K - 4}{\sqrt{(J_R - K - 1)^2 + 3(K-1)^2}} - \frac{1}{4}. \tag{10}$$

The result of $\mathcal{C}_{LKE}$ is plotted in Fig.2(b). Interestingly, $\mathcal{C}_{LKE}$ develops a maximum for $0 < K < 1$, $J_R < 2$ and decreases monotonically for $J_R > 2$. But along the line $K = 1$ it shows a sudden suppression for $J_R > 2$.

An entanglement phase diagram in terms of K and J_R is drawn in Fig.3, where three different phases divided by the lines $J_R = K + 1$ and $J_R = 4 - 2K$ are indicated: the IIE phase

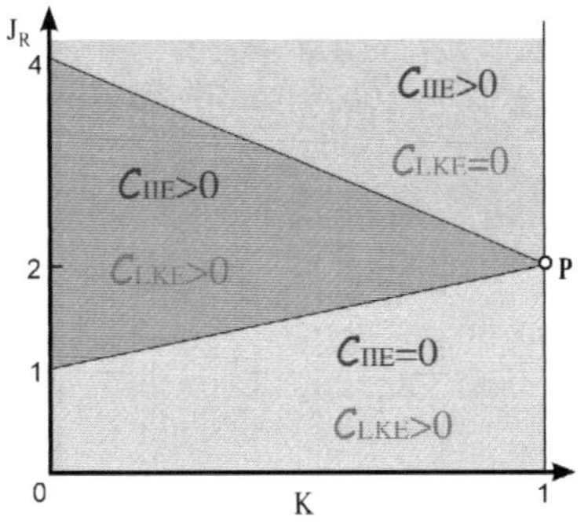

Figure 3. (color online) Entanglement phase diagram: there are three distinct phases divided by the lines $\mathcal{C}_{LKE} = 0$ and $\mathcal{C}_{IIE} = 0$. These two lines intersect at the singular point: $P : (K = 1, J_R = 2)$.

($\mathcal{C}_{IIE} > 0$,$\mathcal{C}_{LKE} = 0$), the LKE phase ($\mathcal{C}_{LKE} > 0$,$\mathcal{C}_{IIE} > 0$), and the co-existence phase ($\mathcal{C}_{IIE} > 0$, $\mathcal{C}_{LKE} > 0$). It is important to understand that the Kondo screening takes place only in the regions with non-zero LKE, i.e., the LKE and co-existence phases. In the co-existence phase $K + 1 < J_R \leq 4 - 2K$, one has $\frac{\sqrt{3}}{2} \leq \mathcal{C}_{IIE} + \mathcal{C}_{LKE} < 1$, so only one of the entanglements could be maximized or violate the Bell inequality[16]. Moreover, the IIE and LKE phases could contact only at the point P: by increasing J_R across P along the strong frustration line $K = 1$, f_{AB} has a sudden drop from $1/4$ to $-1/2$, f_{Aa} has a jump from $-1/2$ to 0. Or, $\mathcal{C}_{IIE}$ has a jump from 0 to 1 while $\mathcal{C}_{LKE}$ has a sudden drop from $1/2$ to 0.

Therefore, together with the discontinuity of the wavefunction $|\Psi_{(0+)}\rangle$ mentioned previously, the sudden changes along the line $K = 1$ in the IIE and LKE do evidence a phase transition companied by the breakdown of Kondo effect. Of course, a true second order phase transition usually requires a continuous suppression of the order parameter. The discontinuity in the IIE or LKE (as an order parameter here) may be either due to the simplicity of the present model or due to the finite degrees of freedom associated with the local components[4, 7, 12].

4. Extension to the lattice case: discussions

It is interesting to discuss implications of the present results in more generic Kondo lattice models or in heavy fermion metals where the magnetic quantum phase transitions may be influenced by the variation of Kondo effect. A local quantum phase transition is proposed in the generic Kondo lattice phase diagram where the criticality is associated with a critical breakdown of the collective Kondo effect[17, 18]. Near the critical point, the Hall constant shows a discontinuous jump due to the reconstruction of the Fermi surface across the critical point[17, 19]. Experimental evidence for this scenario comes from several prototypes of heavy fermion metals including $YbRh_2Si_2$[20] and CeNiAsO[21] where the observed Hall constant exhibits a sudden change accompanying the magnetic phase transition.

Although this scenario could be naturally understood based on the Kondo entanglement picture, a lattice model Hamiltonian with exact solutions showing the Kondo entanglement breakdown transition is still missing. Based on above considerations, the inter-impurity antiferromagnetic singlet discussed in the TIKM evolves into the antiferromagneic ordered state in the lattice case, leading to the transition from the paramagnetic to antiferromangeitc phases.

IOP Conf. Series: Journal of Physics: Conf. Series **807** (2017) 032004 doi:10.1088/1742-6596/807/3/032004

So we can extend the RS model to the following dimerized Kondo lattice model

$$\begin{aligned} H_{DKL} &= \sum_j T_K[\vec{S}_{A,j}\cdot\vec{s}_c(a,j)+\vec{S}_{B,j}\cdot\vec{s}_c(b,j)]+J_R\vec{S}_{A,j}\cdot[\vec{S}_{B,j-1}+\vec{S}_{B,j}] \\ &+ \sum_j K\vec{S}_{A,j}\cdot[\vec{s}_c(b,j-1)+\vec{s}_c(b,j)]+K\vec{S}_{B,j}\cdot[\vec{s}_c(a,j)+\vec{s}_c(a,j+1)]. \end{aligned} \tag{11}$$

Apparently, each building block of this lattice model is the four spin RS model. Thus the lattice model shares the same symmetry in addition to the transitional invariance. Generally, in the regime with relatively small K, the Kondo and inter-local-moment singlets can co-exist in the intermediate regime of J_R. But the co-existence regime diminishes with increasing K and a direct Kondo singlet breakdown transition takes place when the Kondo coupling is maximally frustrated (at $K = T_K$). In the Kondo lattice phase diagram this condition should correspond to the regime with strong geometric frustrations and quantum fluctuations but no spin liquid phase sets in[18]. Because the lattice model respects the enhanced spin symmetry as that of the RS model, we expect that the groundstate of this model is degenerated at the critical point mentioned above. The LKE breakdown is thus naturally expected at the critical point. Therefore, the LKE is an appropriate measure of the local quantum criticality in Kondo lattice systems. The detailed results of the dimerized Kondo lattice model Eq.(11) will be reported in the further coming work.

Acknowledgement
The authors would like to thank Professors Lu Z Y, Si Q and Steglich F for helpful discussions. This work was supported in part by the NSF of China under Grants No. 11304071 and No. 11474082.

References

[1] Jayaprakash C, Krishnamurthy H R and Wilkins J W 1981 *Phys. Rev. Lett.* **47** 737
[2] Hewson A C 1993 *The Kondo Problem to Heavy Fermions* (Cambridge University Press, Cambridge, England)
[3] Coleman P 2007 in *Handbook of Magnetism and Advanced Magnetic Materials* vol 1 (Wiley) p 95
[4] Jones B A and Varma C M 1987 *Phys. Rev. Lett.* **58** 843; Jones B A, Varma C M and Wilkins J W 1988 *Phys. Rev. Lett.* **61** 125
[5] Jones B A and Varma C M 1989 *Phys. Rev.* B **40** 324; Jones B A, Kotliar B G and Millis A J 1989 *Phys. Rev.* B **39** 3415
[6] Affleck I and Ludwig A W W 1992 *Phys. Rev. Lett.* **68** 1046
[7] Rasul J W and Schlottmann P 1989 *Phys. Rev. Lett.* **62** 1701; Schlottmann P and Rasul J W 1990 *Physica* B **163** 544
[8] Affleck I, Ludwig A W W and Jones B A 1995 *Phys. Rev.* B **52** 9528
[9] Sakai O, Shimizu Y and Kasuya T 1990 *Solid State Commun.* **75** 81
[10] Fye R M 1994 *Phys. Rev. Lett.* **72** 916
[11] Silva J B *et al* 1996 *Phys. Rev. Lett.* **76** 275
[12] He R Q, Dai J and Lu Z Y 2015 *Phys. Rev.* **91** 155140
[13] Li Y, Feng X Y and Dai J 2015 *arXiv*:1503.05091 (unpublished)
[14] In the conformal field theory and non-Abelian bosonization approaches the enlarged hidden $SO(7)$ symmetry at the non-trivial critical point is identified[8].
[15] Cho S Y and Mckanzie R H 2006 *Phys. Rev.* A **73** 012109
[16] The extended introduction to the LKE in generic multiple-impurity Kondo model, the list of all eigenstates and eigenvalues, the proofs of discontinuity of the groundstate wavefunction and an entanglement summation rule, and the results of the conventional TIKE and the fidelities of the Kondo singlet or the inter-impurity AFM states of the groundstate could be found in our unpublished preprint [13].
[17] Si Q, Rabello S, Ingersent K and Smith L J 2001 *Nature* **413** 804
[18] Si Q 2006 *Physica B: Conden. Mat.* **378** 23
[19] Coleman P, Pepin C, Si Q and Ramazashvili R 2001 *J. Phys.* C **13** R723
[20] Paschen S *et al* 2004 *Nature* **432** 881
[21] Luo Y K *et al* 2013 *Nature Mat.* **13** 777

Characterization of $YbNi_4(P_{1-x}As_x)_2$, $x = 0, 0.2$ single crystals grown by Czochralski method

K Kliemt and C Krellner

Institute of Physics, Goethe-University Frankfurt, Max-von-Laue Stasse 1, 60438 Frankfurt, Germany

E-mail: kliemt@physik.uni-frankfurt.de

Abstract. We have investigated large single crystals of $YbNi_4P_2$ that were grown from a levitating melt by the Czochralski method. The new samples facilitate the determination of the absolute values of the electrical resistivity. Phase pure polycrystalline samples of the non-magnetic reference $LuNi_4P_2$ were prepared and the electrical resistivity was measured. Furthermore we have grown a single crystal of the As substituted compound $YbNi_4(P_{1-x}As_x)_2$, $x = 0.2$ and investigated the homogenity of the As distribution.

1. Introduction

$YbNi_4P_2$ has a Curie temperature of T_C= 0.17 K [1]. This is one of the lowest Curie temperatures ever observed among stoichiometric compounds [2]. The Curie temperature can be decreased even further in the substitution series $YbNi_4(P_{1-x}As_x)_2$. At $x \approx 0.1$ the rare case of ferromagnetic quantum critical point (FM QCP) occurs [3]. $YbNi_4P_2$ was found to be a Kondo lattice system with a Kondo temperature of $T_K \approx 8$ K revealed from specific heat measurements on powder samples. The occurence of coherent Kondo scattering was observed in the resistivity which steeply decreases below 20 K [1]. Resistivity measurements from 2 - 300 K on small Bridgman grown $YbNi_4P_2$ single crystals show an anisotropy between the $j \parallel c$ and the $j \perp c$-direction, which was explained by the presence of the crystalline electric field [4]. A further measurement on single crystals gave absolute values at 300 K of 90 $\mu\Omega$cm for $j \parallel c$ and 160 $\mu\Omega$cm for $j \perp c$-direction [3]. Until now, all reported resistivity measurements on single crystals were performed on crystals grown by the Bridgman method. This method yields thin needle shaped crystals of about $2 - 3$ mm along the [001]−direction and of $0.5 - 0.7$ mm in the perpendicular extension.

Here, we have investigated the electrical transport properties down to 1.8 K of single crystal samples which were cut from one large crystal. This single crystal had a conical shape with a diameter of $3 - 9$ mm and a length of 14 mm. It belongs to a new generation of single crystals grown from a levitating melt by the Czochralski method [5].

In the past, samples of the substitution series $YbNi_4(P_{1-x}As_x)_2$ with As-contents up to $x = 0.13$ [3] have been investigated and the material was studied close to the ferromagnetic QCP. Until now, the T-x phase diagram is incomplete due to a lack of samples with $x > 0.13$. To observe the emergence of Fermi liquid behaviour in the system, samples with these higher As concentrations are required. Using a seed from the new crystal generation, we have grown a single crystal of the substituted compound $YbNi_4(P_{1-x}As_x)_2$ with $x = 0.2$ with 16 mm in length and $3-6$ mm in diameter. It turned out that the As distribution is homogeneous.

2. Experimental

An $YbNi_4P_2$ single crystal was grown by the Czochralski method from a Ni-P self-flux as described in [5]. The crystal structure was confirmed by powder X-ray diffraction (Cu K_α). The chemical composition was checked by energy-dispersive X-ray spectroscopy (EDX). Upon investigating a polished surface by scanning electron microscopy and polarization microscopy an excellent homogenity of this surface and the absence of Ni-P flux-inclusions was found. The single crystal was oriented using the X-ray Laue back-scattering technique. From this crystal, samples for the electrical transport measurement were cut using a spark erosion device. The samples had initial dimensions of $\approx$ 3 mm x 0.5 mm x 0.15 mm and were oriented along the crystallographic main symmetry directions [100], [110] and [001]. Electrical transport measurements were performed between 1.8 and 300 K using the ACT option of a commercial PPMS (Quantum Design). The Czochralski method was also applied to grow a single crystal from a melt where 20 at% of P was replaced by As. The crystal structure and the chemical composition were investigated by powder X-ray diffraction (PXRD) and EDX, respectively. Phase pure polycrystalline samples of the non-magnetic reference compound $LuNi_4P_2$ were prepared according to [6, 7]. PXRD measurements confirmed the $P4_2/mnm$ tetragonal structure with lattice parameters a = 7.047(6) Å and c = 3.583(4) Å, which is in agreement with the data published for polycrystalline samples [6]. PXRD data were refined using the General Structure Analysis System (GSAS) [8, 9].

3. Electrical resistivity $\rho(T)$ of $YbNi_4P_2$

For a tetragonal material Ohm's law reads

$$\vec{j} = \begin{pmatrix} \sigma_a & & \\ & \sigma_a & \\ & & \sigma_c \end{pmatrix} \vec{E}, \qquad \rho = \frac{1}{\sigma} \tag{1}$$

with σ_a beeing the isotropic conductivity in the basal plane and σ_c the conductivity along the c-direction for the electrical field $\vec{E}$. The coordinates are choosen such that the z-axis is oriented along the crystallographic c-direction. Our Czochralski grown $YbNi_4P_2$ single crystals reproduce the temperature dependence of the reported data as shown in the main part of Fig. 1 and with them it is also possible to resolve the temperature dependence of the resistivity ρ_{100} and ρ_{110} of the in-plane directions. As expected according to Eqn. 1, the resistivity is isotropic in the a-a plane. We determined the absolute values at room temperature applying

$$\rho(T) = \frac{ab}{lI} U(T) \tag{2}$$

with the width a, the thickness b of the sample, the contact distance l and the applied constant current I and measured voltage U. For $j \parallel 001$ we determined the resistivity $\rho_{001}(300\text{K}) = 60\mu\Omega\text{cm}$, for $j \perp 001$ we found $\rho_{110}(300\text{K}) = 70\mu\Omega\text{cm}$. The absolute value determined for the [100]-direction agrees with that of the [110]-direction. For comparison, the resistivity of the non-magnetic reference $LuNi_4P_2$ was measured and is shown in the inset of Fig. 1. For the determination of the absolute values, the measurement device (U, I) as well as the geometry of the samples (a, b, l) have to be considered as sources for uncertainties. In the resistivity measurement, the main contribution to the uncertainty comes from the geometry factors $\Delta a/a \simeq 3\%$, $\Delta b/b \simeq 3\%$ and $\Delta l/l \simeq 1\%$. The uncertainty of the voltage $\Delta U/U$ is between 1% and 3% for different temperatures. The uncertainty of the current [10] with $\Delta I/I \simeq 5 \cdot 10^{-4}\%$ is neglegible.

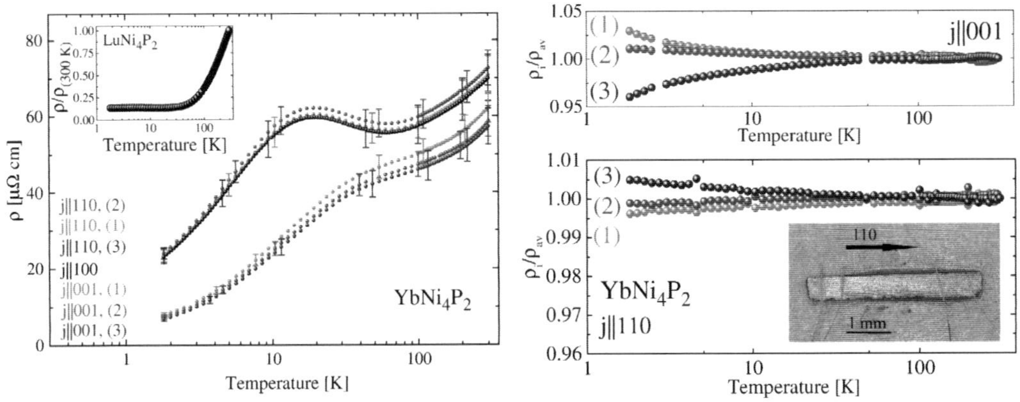

Figure 1. *Left:* Electrical resistivity ρ as a function of temperature from 1.8K to 300K measured with current j perpendicular ($j \parallel 100$ and $j \parallel 110$) and parallel to the crystallographic c-axis. The inset shows the electrical resistivity measured on a polycrystalline sample of $LuNi_4P_2$. The resistivity ratio was determined to be $\rho_{\text{poly}}(300\text{K})/\rho_{\text{poly}}(1.8\text{K}) = 7.4$. *Right:* Ratio of the normalized electrical resistivity ρ_i of measurement i divided by the average resitivity ρ_{av} for the respective current direction as a function of temperature. The samples were thinned down in two steps from 0.15 (i=1, curve (1)) to 0.09 mm (i=3, curve (3)). In the inset, an oriented sample with the four platinum wire contacts prepared for the measurement is shown.

The aim of our optimization of the crystal growth process is to minimize the crystal defects. Indispensable is the use of high-purity starting materials. The contamination of the melt by crucible materials has to be avoided. Low growth rates avoid flux inclusions. Adjusting the temperature during the growth experiment reduces the evaporation of the elements and therefore a shift in the stoichiometry. At lowest temperatures, the scattering on crystal defects is the only contribution to the resistivity. Therefore, the resistivity ratio $\text{RR}_{1.8\text{K}} = \rho(300\text{K})/\rho(1.8\text{K})$ is an indicator for the amount of crystal defects.

For 6 different samples, which were cut from the same single crystal, different $\text{RR}^{001}_{1.8\text{K}}$ values between 11 and 17 were found. For the purest sample, we determined the absolute resistivity value to be $\rho_{001}(1.8\text{K}) = 3.5\ \mu\Omega\text{cm}$. In the other case, if the current is perpendicular to the c-direction, $\text{RR}^{110}_{1.8\text{K}}$ and $\text{RR}^{100}_{1.8\text{K}}$, respectively, spread less for the 6 examined samples. The values are between 3.0 and 3.07. The absolute value is $\rho_{110}(1.8\text{K}) = 22.8\ \mu\Omega\text{cm}$. The large variation in $\text{RR}^{001}_{1.8\text{K}}$ can be caused by a misalignment of the four contacts for the current and voltage measurement. During the growth small-angle grain boundaries can evolve in .a single crystal. They lead to the distorsion of the lattice and enhance the residual resistivity. A further analysis by Laue method is required to clarify the origin of the variation of $\text{RR}^{001}_{1.8\text{K}}$.

A common technique to improve the crystallinity of materials is annealing. An experiment at 800° C in an inert atmosphere for 5 days yields an increase of 25% of $\text{RR}^{001}_{1.8\text{K}}$. Due to the large anisotropy for the different current directions, it is not clear if this improvement is the result of the annealing procedure or was caused by a slight misalignment of the contacts during the resistivity measurement. For further annealing studies with different parameters, well cut samples with a precise orientation are necessary. We investigated the influence of grinding and polishing to the residual resistivity ratio. Two samples cut from the single crystal were oriented along the [110] and the [001]-direction. They were thinned down in two steps from 0.15 to 0.09 mm and the respective temperature dependence of the resistivity ρ_1, ρ_2, ρ_3 (curves (1-3)) was measured and is shown on the left hand side of Fig. 1. We determined from our data for each current direction the ratio ρ_i/ρ_{av} with $\rho_{av} = (\rho_1 \cdot \rho_2 \cdot \rho_3)^{1/3}$ to investigate if our polishing procedure causes stress and strain in the samples and therefore influences the sample quality. The right part of Fig. 1 shows that $RR_{1.8K}$ slightly varies between the different thicknesses labelled by (1), (2), (3). For the current parallel to the c-direction, the polishing diminishes the value of $\rho_{001}(1.8K)$. For the perpendicular direction, the opposite effect occurs. This result might be caused by the anisotropy of $RR_{1.8K}$. In the thinner sample the ideal current flow without a perpendicular component is approached. That means that the current is confined to the [001] or the [110]-direction and the respective perpendicular current component is reduced upon thinning down the sample. For the [001]-direction ρ_i becomes smaller for increasing i leading to a larger $RR_{1.8K}$ and for the [110]-direction ρ_i becomes larger leading to a smaller $RR_{1.8K}$ value. This investigation also demonstrates that the polishing procedure itself seems not to enlarge the number of crystal defects.

4. The substituted compound $YbNi_4(P_{1-x}As_x)_2$, $x = 0.2$

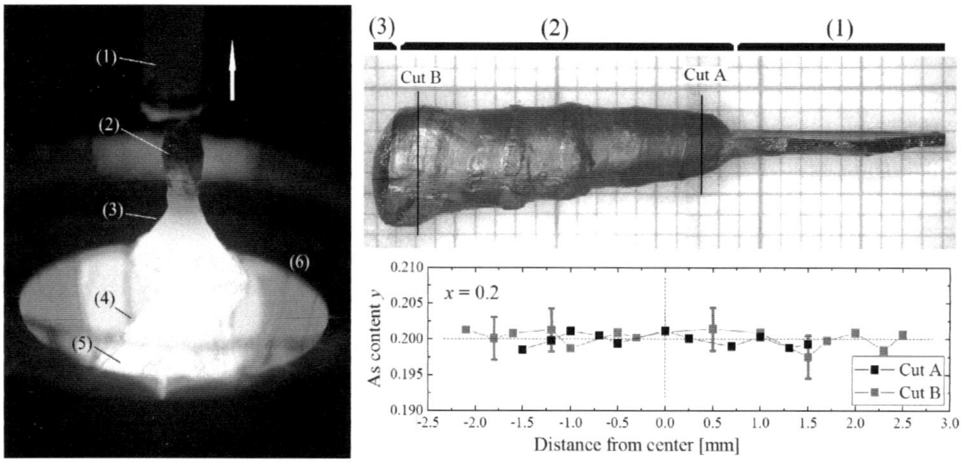

Figure 2. *Left:* The $YbNi_4(P_{1-x}As_x)_2$, $x = 0, 0.2$ crystals were grown from a levitating melt. The arrow mark the pulling direction of the seed (2) in the seed holder (1). The meniscus (4) marks the border between the grown sample (3) and melt (5) levitating in a cold copper crucible (6). The heating power is provided via a radio-frequency induction coil. The evaporation of the volatile elements was slowed down by applying an Ar pressure of 20 bar in the growth chamber; *Right, upper part:* The as grown sample with the nominal composition of $YbNi_4(P_{0.8}As_{0.2})_2$. The single crystal seed (1) consisted of $YbNi_4P_2$ and was oriented along the [001]-direction. The determination of the As content of region (2) of the grown sample is shown in the lower part of the figure. The growth was terminated with a faster growth velocity leading to the incorporation of flux in part (3) of the sample; *Right, lower part:* The real As content y in the grown sample along two radial lines, A,B (black and grey symbols) was determined using EDX in dependence of the distance from the center of the sample.

The growth of the substituted compound by the Czochralski method Fig. 2, was done similarily to that reported in [5]. The liquidus temperature is measured during the growth experiments. Within the accuracy of the pyrometer $\Delta T \approx \pm 50K$, no change compared to the stoichiometric compound was detected. The growth was started at about 1400°C using an $YbNi_4P_2$ single crystal as the seed. After seeding, the sample was pulled with a rate of 0.3 mm/h. Such low growth velocities are necessary to obtain crystals without flux inclusions.

The distribution coefficient of As in the system $\kappa = c_l^{\mathrm{As}}/c_s^{\mathrm{As}}$ with the concentrations of As in the melt c_l^{As} and in the solid c_s^{As}, respectively, is not known and does not necessarily have to be exactly one [11]. A deviation from $\kappa = 1$ would lead to the enrichment or depletion of As in the melt and an inhomogeneous As distribution in the grown crystal. The As distribution in the single crystal was investigated by EDX, and showed that the initial As:P ratio of the melt of 1:4 can be found all over in our sample except at the first part which is connected to the seed. The seed consisted of $YbNi_4P_2$ and over a length of about 2.5 mm the crystal structure included more and more As and the lattice adapted to the new lattice constants. The analysis along radial lines (Cut A,B) on polished surfaces shown in Fig. 2, yields a homogeneous As concentration from the center to the surface of the crystal. A part of the $YbNi_4(P_{1-x}As_x)_2$, $x = 0.2$ sample was investigated by PXRD and the structure refinement yields an enlargement of the unit cell of $\Delta V/V = 0.0138$ compared to $x = 0$. The lattice constants of the tetragonal $P4_2/mnm$ structure are a = 7.0803(5)Å and c = 3.6131(2)Å.

5. Summary

The electrical resistivity was measured on Czochralski grown $YbNi_4P_2$ single crystals from 1.8 to 300 K and the absolute values at room temperature were determined to be 60 $\mu\Omega$cm for j parallel to the c-direction and 70 $\mu\Omega$cm perpendicular. Comparing with the measurement on Bridgman grown samples [3], the data reproduce the temperature dependence of the resistivity but the absolute values differ. The resistivity ratio $RR_{1.8K}$ is strongly anisotropic. For j parallel to the c-direction we found value between 11 and 17 for different samples. In the perpendicular directions, it is almost sample independent with values about 3. Polycrystalline $LuNi_4P_2$ samples, exhibit the resistivity characteristics of a non-magnetic metal. A single crystal of the substituted compound $YbNi_4(P_{1-x}As_x)_2$ with $x = 0.2$ was grown. The phosphorous to arsenic ratio in the melt is maintained in the crystal. The As distribution is homogeneous throughout the whole sample. The enlargement of the unit cell is $\Delta V/V = 0.0138$ compared to $x = 0$.

Acknowledgments

We thank W. Assmus and F. Ritter for valuable discussions and K.-D. Luther for technical support. The authors grateful acknowledge support by the DFG via project KR 3831/4-1.

References

[1] C. Krellner, S. Lausberg, A. Steppke, M. Brando, L. Pedrero, H. Pfau, S. Tencé, H. Rosner, F. Steglich, and C. Geibel. *New J. Phys.*, 13:103014, 2011.
[2] M. Brando, D. Belitz, F.M. Grosche, and T.R. Kirkpatrick. *Rev. Mod. Phys.*, 2016 (accepted).
[3] A. Steppke, R. Küchler, S. Lausberg, E. Lengyel, L. Steinke, R. Borth, T. Lühmann, C. Krellner, M. Nicklas, C. Geibel, F. Steglich, and M. Brando. *Science*, 339:933, 2013.
[4] C. Krellner and C. Geibel. *J. Phys. Conf. Ser.*, 391:012032, 2012.
[5] K. Kliemt and C. Krellner. *J. Cryst. Growth*, 449:129 – 133, 2016.
[6] S. I. Chikhrij, S. V. Orishchin, and Y. B. Kuz'ma. *Dopov Akad Nauk A*, 7:79, 1986.
[7] W. Jeitschko, F. Terbüchte, E.J. Reinbold, P.G. Pollmeier, and T. Vomhof. *J. Less. Common Met.*, 161:125, 1990.
[8] A.C. Larson and R.B. Von Dreele. *Los Alamos National Laboratory Report LAUR*, page 86, 2000.
[9] B. H. Toby. *J. Appl. Crystallogr.*, 34:210, 2001.
[10] QuantumDesign. *Electro-transport Application note, Physical Property Measurement System*, 2016.
[11] K.-T. Wilke and J. Bohm. *Kristallzüchtung*. VEB Deutscher Verlag der Wissenschaften, Berlin, 1988.

Site-selective NMR measurements in single crystal $PrNb_2Al_{20}$

Tetsuro Kubo[1,*], Hisashi Kotegawa[1], Hideki Tou[1,†], Hisatomo Harima[1], Ryuji Higashinaka[2], Akihiro Nakama[2], Yuji Aoki[2], and Hideyuki Sato[2]

[1]Department of Physics, Graduate School of Science, Kobe University, 1-1 Rokkodai, Kobe, Hyogo 657-8501, Japan

[2]Department of Physics, Graduate School of Science and Engineering, Tokyo Metropolitan University, 1-1 Minamiohsawa, Hachioji, Tokyo 192-0397, Japan

E-mail: *tkubo@stu.kobe-u.ac.jp †tou@crystal.kobe-u.ac.jp

Abstract. We report the result and analysis of the nuclear magnetic resonance (NMR) measurements at the magnetic field of 8.57 T and temperature of 80 K using a single crystal $PrNb_2Al_{20}$. Combining the field angle dependence of NMR spectra, numerical simulation, and band calculation, we deconvoluted the NMR signals from the Nb, Al $48f$, and Al $96g$ sites. Unfortunately, the overlapping of NMR lines prevents us to extract the NMR signals from the Al $16c$ site. However, the obtained nuclear quadrupole resonance (NQR) parameters for the Nb and Al $48f$ sites are consistent with the previous reports [T. Kubo *et al.*, 2014 *JPS Conf. Proc.* **3** 012031; 2015 *J. Phys.: Conf. Ser.* **592** 012093; 2016 *J. Phys. Conf.: Ser.* **683** 012015] and for the Al $96g$ site are refined to be $\nu_Q = 0.98$ MHz and $\eta = 0.46$.

1. Introduction

PrT_2X_{20} system (T = Ti, V, Nb, Rh, Ir, X = Al, Zn) which crystallizes in a cubic $CeCr_2Al_{20}$-type structure with the space group $Fd\bar{3}m$ has been studied intensively because of their intriguing behaviors at low temperatures such as superconductivity, quadrupole ordering, and non-Fermi liquid (NFL) behaviors [1, 2, 3, 4, 5]. In this system, the $4f$ electronic ground state of Pr^{3+} ion is considered to be the non-Kramers' Γ_3 doublet which has no magnetic dipole but has two electric quadrupoles O_{20} and O_{22}, and one magnetic octupole T_{xyz}. These multipole degrees of freedom are believed to play an important role in exotic ground states in these materials at low temperatures. Among them, $PrNb_2Al_{20}$ shows NFL behaviors such as T-linear resistivity below 2 K and logarithmic divergent specific heat divided by temperature below 10 K, in which the relationship between these behaviors and quadrupole degrees of freedom has been discussed [1].

To clarify the importance of multipole degrees of freedom, site-selective nuclear magnetic resonance (NMR) measurements are desirable since NMR can obtain the microscopic electronic information from static and dynamical viewpoints based on the local symmetry at the ligand sites of Pr ions. The NMR spectrum of $PrNb_2Al_{20}$ becomes complex because of its crystal structure. Thus, it is important to deconvolute the complex NMR lines since the microscopic electronic state at the ligand sites directly reflects the essential information of coupling between conduction electrons and multipoles through the bond symmetry between the Pr and ligand sites [6, 7, 8, 9, 10]. In this paper, we report the results of ^{93}Nb- and ^{27}Al-NMR measurements using a single crystal and the analysis of the NMR lines.

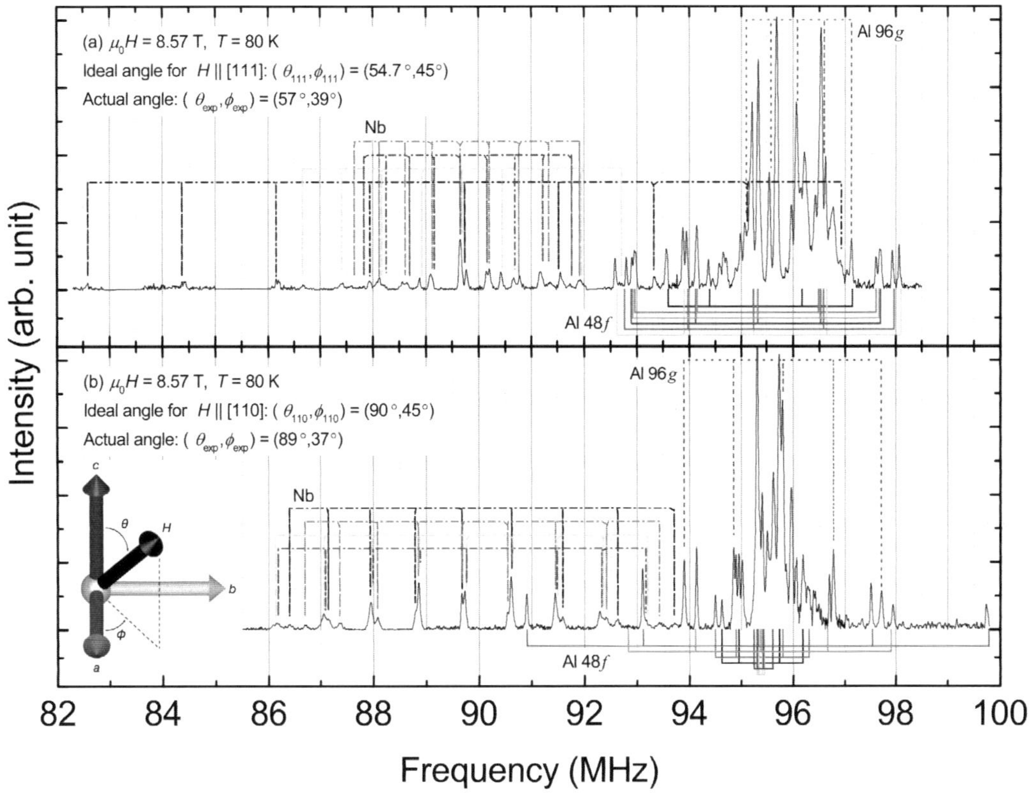

Figure 1. NMR lines with the applied magnetic field parallel to one of the (a) ⟨111⟩ and (b) ⟨110⟩ directions at 8.57 T, and at 80 K. Colored dash-dotted, solid, and dashed lines are the results of the simulation for Nb, Al 48f, and Al 96g sites, respectively. As shown in the inset of Fig. 1(b), θ and ϕ denote the field angle from the crystal c-axis and in the ab-plane, respectively. Both the ideal and actual angles are presented. The details of the simulation are described in the text.

2. Experimental Procedure

The single crystal sample of $PrNb_2Al_{20}$ was prepared by the Al-flux method. Details of the sample preparation method were reported in the previous paper [1]. The field angle dependence of NMR spectra is measured by using a 17 T superconducting magnet and a phase-coherent pulsed spectrometer. The angle between the applied field and crystal axes is controlled by a uniaxial goniometer. ^{93}Nb- (nuclear spin $I = 9/2$, gyromagnetic ratio $\gamma/2\pi = 10.421$ MHz/T, nuclear quadrupole moment $Q = -0.32 \times 10^{-28}$ m^2) and ^{27}Al- ($I = 5/2$, $\gamma/2\pi = 11.094$ MHz/T, $Q = 0.1466 \times 10^{-28}$ m^2) [11, 12] NMR lines are obtained by summing up the fast Fourier transformed spin echo signals obtained with a step of 5 kHz at a fixed magnetic field $\mu_0 H \approx 8.57$ T.

3. Results and Discussion

Figure 1 shows ^{93}Nb- and ^{27}Al-NMR spectra obtained at the fixed magnetic field 8.57 T applied parallel to one of the (a) ⟨111⟩ and (b) ⟨110⟩ axes, and at a temperature of 80 K. Each of the resonance lines are as sharp as about 30 kHz. There are one Nb site and three crystallographically inequivalent Al sites (16c, 48f, and 96g) in $PrNb_2Al_{20}$. Due to the local symmetry at the ligand sites which is lower than cubic as described below, these sites split further under the magnetic field. In order to deconvolute the obtained lines into the above sites, we carried out the simulation numerically solving the $(2I+1)\times(2I+1)$ nuclear spin hamiltonian $\mathcal{H}$ by an exact diagonalization method. Such a hamiltonian can be written as follows:

$$\mathcal{H} = \mathcal{H}_Z + \mathcal{H}_Q + \mathcal{H}_{hf}, \tag{1}$$

$$\mathcal{H}_Z = -\gamma_n \hbar \boldsymbol{I} \cdot \boldsymbol{H} \tag{2}$$

$$\mathcal{H}_Q = \frac{eQ}{6I(2I-1)} \sum_{i,j} V_{ij} \left[\frac{3}{2}(I_i I_j + I_j I_i) - \delta_{ij} I(I+1) \right] \tag{3}$$

$$= \frac{1}{6} h\nu_Q \left[(3I_Z^2 - I(I+1) + \frac{\eta}{2}\left(I_+^2 + I_-^2\right) \right] \tag{4}$$

$$\mathcal{H}_{hf} = -\gamma_n \hbar \boldsymbol{I} \left(\tilde{D} + \tilde{K} \right) \boldsymbol{H} \tag{5}$$

where h is the Planck constant ($\hbar \equiv h/2\pi$) and e is the elementary charge. The first term in Eq. (1) is a Zeeman interaction between the nuclear spin $\boldsymbol{I}$ and applied field $\boldsymbol{H}$, and is expressed in Eq. (2). The second term is a nuclear quadrupole interaction between the nuclear spin and electric field gradient (EFG) created by the surrounding charge distribution at the nuclear site, and is expressed in Eq. (3) . EFG is characterized by the second rank EFG tensor $\tilde{V}$ ($V_{ij} \equiv \partial^2 V/\partial x_i^2$ ($x_i = x, y, z$)). In the principal coordinate system, $\tilde{V}$ is diagonalized and Eq. (3) can be rewritten as Eq. (4), where $\nu_Q \equiv 3eQ/2hI(2I-1)$ is the nuclear quadrupole frequency, $\eta \equiv (V_{XX} - V_{YY})/V_{ZZ}$ is the asymmetry parameter, and V_{XX}, V_{YY}, and V_{ZZ} are the principal values of $\tilde{V}$ which satisfy the relation $|V_{ZZ}| \geq |V_{YY}| \geq |V_{XX}|$ and $V_{XX}+V_{YY}+V_{ZZ} = 0$. The third term is a hyperfine interaction between the nuclear spin and electron spins, which is represented by the second rank classical dipole field tensor $\tilde{D}$ and Knight shift tensor $\tilde{K}$, and is expressed in Eq. (5). $\tilde{D}$ describes the classical dipole interaction between the nuclear spin and Pr $4f$ magnetic dipoles. $\tilde{K}$ has the information of c-f hybridization and is decomposed into the isotropic part K_{iso} and the anisotropic part $\tilde{K}_{ani}$. $\tilde{D}$ and $\tilde{K}_{ani}$ are traceless and are designated by their principal values (D_1, D_2, D_3) and (K_1, K_2, K_3) that satisfy the relation of $|D_3| \geq |D_2| \geq |D_1|$ and $|K_3| \geq |K_2| \geq |K_1|$.

In order to calculate NMR lines, we have to know the principal values and the principal axis directions of $\tilde{V}$, $\tilde{D}$, and $\tilde{K}$. The principal values of $\tilde{D}$ are calculated by the lattice sum of the classical dipole field $\boldsymbol{H}_D = \sum_i \left[-\boldsymbol{\mu}_i/r_i^3 + (\boldsymbol{\mu}_i \cdot \boldsymbol{r}_i)\boldsymbol{r}_i/r_i^5 \right]$, where $\mu_i = g_J \mu_B \boldsymbol{J}_i$ is the $4f$ electronic magnetic dipole in which g_J, μ_B, and $\boldsymbol{r}_i$ are the Landé g-factor, the Bohr magneton, and the distance between the site considered and ith Pr site, respectively. The magnitude of magnetic dipole is estimated to be $\approx 0.2\,\mu_B$ at 80 K from the crystalline electric field analysis [10]. ν_Q and η for Nb, Al $48f$, and Al $96g$ sites are previously determined [13, 14, 15]. The principal values of $\tilde{K}$ are treated as fitting parameters.

Next, we consider the principal axis directions. Generally, these directions do not coincide with the crystal axes or with each other due to the local symmetry at the ligand sites. In such a case, the analysis taking into account of the local symmetry is previously reported [9, 15, 16, 17, 18, 19]. Shown in Fig. 2 is the geometry of neighboring Nb, Al $48f$, and Al $96g$ atoms around a Pr atom. The local symmetry axis directions at these ligand sites are also represented. For the Nb site with local symmetry of $.\bar{3}m$, the maximum principal axis direction coincides with the local threefold rotation axis. As shown in Fig. 2(a) with the crystal axis directions, this axis is parallel to [111] direction for the Nb site at (1/2, 1/2, 1/2). $\tilde{V}$, $\tilde{D}$, and $\tilde{K}$ are axially symmetric around this axis. The Al $16c$ site has the local symmetry of $.\bar{3}m$, thus the principal axis directions are same with those for the Nb site. For the Al $48f$ site at which local symmetry is $2.mm$, there are two mirror planes. For example, the Al $48f$ site at $(1/8, 1/8, z)$ have the mirror planes (110) and $(1\bar{1}0)$. In this case, the principal axis directions are constrained in these mirror planes but the maximum principal axis direction cannot be determined beforehand. For the Al $96g$ site at which local symmetry is $..m$, there are only one mirror plane which is parallel to one of the {110} plane. In this case, one of the three principal axes are perpendicular to the plane and the other two axes are in the plane. Since the principal axis directions in the plane cannot be predetermined, we calculated $\tilde{V}$ for each sites by a band calculation in local density approximation for the isostructural $LaNb_2Al_{20}$. The details of the calculation method are described in the previous paper [20]. As shown in Fig. 2(b) with the crystal axis

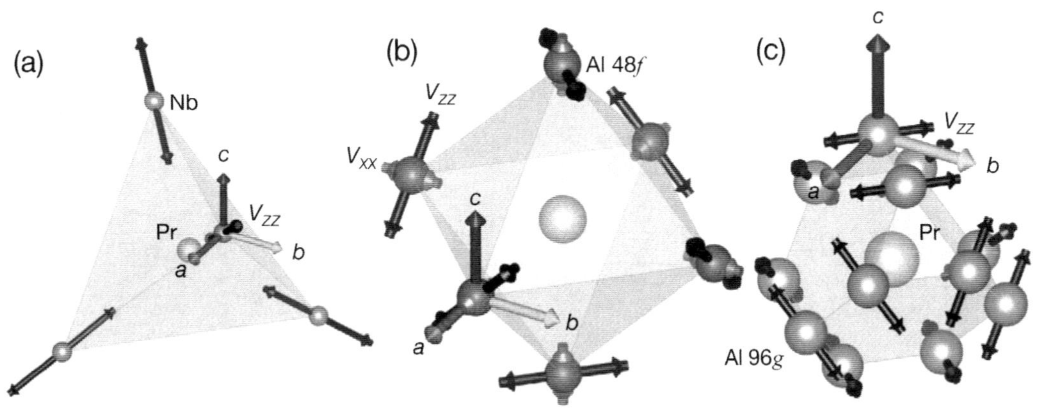

Figure 2. (a) Pr and surrounding Nb atoms with crystal axes and local symmetry axis at Nb sites. Red, light green, and blue arrows denote crystal a, b, and c axes, respectively. Black arrows indicate local threefold rotation axes, which coincide with the maximum principal axis directions of $\tilde{V}$, $\tilde{K}$, and $\tilde{D}$ in this case. (b) Pr and nearest-neighboring Al 48f atoms with the crystal and EFG principal axes. Black and gray arrows indicate the maximum (V_{ZZ}) and minimum (V_{XX}) principal axis directions of the EFG tensor from the band calculation, respectively. (c) Pr and nearest-neighboring Al 96g atoms with the crystal and EFG principal axes. Black arrows indicate V_{ZZ} directions from the band calculation. These figures were drawn with VESTA 3 [21].

$H \parallel \langle 111 \rangle$	NMR sites	NMR lines
Nb	2	18
Al 16c	2	10
Al 48f	2	10
Al 96g	3	15

$H \parallel \langle 110 \rangle$	NMR sites	NMR lines
Nb	2	18
Al 16c	2	10
Al 48f	3	15
Al 96g	4	20

Table 1. (Left) The number of NMR sites and corresponding number of NMR lines for $H \parallel \langle 111 \rangle$ case. (Right) The number of NMR sites and corresponding number of NMR lines for $H \parallel \langle 110 \rangle$ case.

directions, V_{ZZ} and V_{XX} directions for the Al 48f site at $(x, 1/8, 1/8)$ are obtained to be [011] and [100] directions, respectively. Figure 2(c) shows the crystal axis and V_{ZZ} directions for the Al 96g site. For the Al 96g site at (x, x, z), the V_{ZZ} direction is calculated to be parallel to the $[1\bar{1}0]$ direction.

Using the principal values and principal axis directions, we can obtain each of the tensor components in the crystal coordinate by an appropriate frame transformation $\tilde{T}_c = R\tilde{T}_p R^{-1}$, where $\tilde{T}_p$ and $\tilde{T}_c$ are the representations of a tensor in its principal and crystal coordinate, respectively, and R is a frame transformation matrix between these coordinates. In addition, tensor components are transformed from one site to another under symmetry operations such as 120° rotation around one of the $\langle 111 \rangle$ axes and mirror operation for one of the $\{110\}$ planes.

Based on the above considerations, we analyzed the NMR lines. In Table 1, we summarized the number of the NMR sites and corresponding number of NMR lines for two magnetic field directions. As shown in Fig. 1, the number of the observed NMR lines are larger than expected number of NMR lines for each of the field directions. This is due to a misalignment of the sample, which causes further splitting of the NMR sites. Such a misalignment is inevitable since we use a uniaxial goniometer in the present measurement. For general field direction, we have four Nb, six Al 48f, and twelve Al 96g sites.

First, we determined the exact field angle by calculating the ^{93}Nb-NMR lines because the principal axis directions of $\tilde{V}$, $\tilde{D}$, and $\tilde{K}$ at Nb sites are uniquely determined as discussed above. We defined the

	ν_Q (MHz) / η (exp.)	ν_Q (MHz) / η (cal.)	V_{ZZ} (cal.)
Nb	1.82 / 0	1.238 / 0	$\parallel \langle 111 \rangle$
Al 16c	– / –	0.715 / 0	$\parallel \langle 111 \rangle$
Al 48f	2.28 / 0.17	2.427 / 0.191	$\parallel \langle 110 \rangle$
Al 96g	0.98 / 0.46	1.005 / 0.348	$\parallel \langle 110 \rangle$

	(D_1, D_2, D_3) (%) (calc.)	D_3	(K_{iso}, K_1, K_2, K_3) (%) (exp.)	K_3
Nb	$(0.018, 0.018, -0.036)$	$\parallel \langle 111 \rangle$	$(0.43, -0.08, -0.08, 0.16)$	$\parallel \langle 111 \rangle$
Al 16c	$(-0.132, -0.132, 0.264)$	$\parallel \langle 111 \rangle$	–	–
Al 48f	$(0.005, 0.025, -0.030)$	$\parallel \langle 110 \rangle$	$(0.25, 0.03, -0.09, 0.12)$	$\parallel \langle 1\bar{1}0 \rangle$
Al 96g	$(-0.026, -0.081, 0.107)$	$\perp \langle 110 \rangle$	–	–

Table 2. (Upper) The experimental and calculated NQR parameters for each of the ligand sites. (Lower) The calculated dipole field tensor and experimentally determined Knight shift tensor with D_3 and K_3 directions, respectively. The magnitude of D_i's and K_i's is given as the relative values for the Zeeman frequency $\gamma_n H/2\pi$. As described in the text, NMR signals from the Al 16c site cannot be deconvoluted due to the overlapping of NMR lines.

field angle θ to be the polar angle of H from c-axis and ϕ to be the azimuthal angle of H in ab-plane as shown in the inset of Fig. 1. When we applied the magnetic field parallel to the [111] direction, we have $(\theta_{111}, \phi_{111}) = (54.74^\circ, 45^\circ)$. Similarly, when we applied the magnetic field parallel to the [110] direction, we have $(\theta_{110}, \phi_{110}) = (90^\circ, 45^\circ)$. As shown by the dash-dotted lines in Fig. 1, we could assigned ^{93}Nb-NMR lines to the four Nb sites. From the fitting of these lines, we obtained $(\theta_{exp}, \phi_{exp} = 57^\circ, 39^\circ)$ for Fig. 1(a) and $(\theta_{exp}, \phi_{exp} = 89^\circ, 37^\circ)$ for Fig. 1(b). Obtained NQR and Knight shift parameters are summarized in Table 2 with the result of the band calculation. The obtained values are consistent with the previous reports [13, 14, 15]. Therefore, we could determined the field angle for two magnetic field direcitons.

Next, we analyzed the ^{27}Al-NMR lines by using the obtained field angles. NMR lines from Al 16c and 96g sites overlap strongly each other so that, in general, it is difficult to deconvolute them. Nevertheless, we could assigned NMR lines from one of the Al 96g site since these lines could be clearly separated from another Al lines as shown by the dashed lines in Fig. 1. The obtained NQR parameters (ν_Q = 0.98 MHz, η = 0.46) are consistent with the previous NMR and NQR result [15] and the V_{ZZ} direction corresponds to the predetermined one by the band calculation. Fore more precise deconvolution of the Al 16c and 96g lines, the NMR measurement using a double-axis goniometer and the NQR measurement below 1.5 MHz are needed.

Finally, we discuss about the ^{27}Al-NMR lines from the Al 48f site. As shown by the solid lines in Fig. 1, all of the ^{27}Al-NMR lines from the Al 48f site are well reproduced by the simulation and the obtained NQR parameters (ν_Q = 2.28 MHz, η = 0.17) are consistent with the previous reports [14, 15]. The EFG principal axis directions at the Al 48f site correspond to the calculated ones, thus the validity of the band calculation is confirmed. In addition, we could determined $\tilde{K}$ at the Al 48f site as shown in Table 2.

By using these assignment of the ligand sites in $PrNb_2Al_{20}$, we may bring out the information of multipole fluctuations. Measurements of the temperature and field dependences of the Knight shift and NMR relaxation rate using single crystal $PrNb_2Al_{20}$ are ongoing and the results will be presented elsewhere.

4. Summary

We measured the field-angle dependence of ^{93}Nb- and ^{27}Al-NMR signals for single crystal $PrNb_2Al_{20}$. Together with previously reported NQR parameters and band calculation, we succeeded to deconvolute the NMR signals from the Nb, Al 48f, and one of Al 96g sites. The experimental results are well reproduced by using the numerically obtained EFG tensors, thus confirms the availability of the band calculation. Unfortunately, the strong overlapping of the NMR lines prevents us to obtain the precise NMR and NQR parameters for Al 16c site in the present stage. Nevertheless, these results help us to investigate the importance of multipole degrees of freedom for the anomalous properties of $PrNb_2Al_{20}$.

Acknowledgments
This work has been supported by a Grant-in-Aid for Scientific Research on Innovative Areas "J-Physics" (No. 15H05885) and Grants-in-Aid for Scientific Research (Nos. 26400359, 15H05745, and 15H03689) of the Ministry of Education, Culture, Sports, Science and Technology, Japan.

References

[1] Higashinaka R, Nakama A, Ando M, Watanabe M, Aoki Y and Sato H 2011 *J. Phys. Soc. Jpn.* **80** SA048
[2] Sakai A and Nakatsuji S, 2011 *J. Phys. Soc. Jpn.* **80** 063701
[3] Onimaru T, Matsumoto K T, Inoue Y F, Umeo K, Sakakibara T, Karaki Y, Kubota M and Takabatake T 2011 *Phys. Rev. Lett.* **106** 177001
[4] Sakai A, Kuga K and Nakatsuji S 2012 *J. Phys. Soc. Jpn.* **81** 083702
[5] Tsujimoto M, Matsumoto Y, Tomita T, Sakai A and Nakatsuji S 2014 *Phys. Rev. Lett.* **113** 267001
[6] Sakai O, Shiina R, Shiba H and Thalmeier P 1997 *J. Phys. Soc. Jpn.* **66** 3005
[7] Tanida H, Takagi S, Suzuki H S, Satoh I and Komatsubara T 2006 *J. Phys. Soc. Jpn.* **75** 074721
[8] Kiss A and Kuramoto Y 2006 *J. Phys. Soc. Jpn.* **75** 103704
[9] Kikuchi J, Takigawa M, Sugawara H and Sato H 2007 *J. Phys. Soc. Jpn.* **76** 043705
[10] Kubo T, Kotegawa H, Tou H, Higashinaka R, Nakama A, Aoki Y and Sato H 2015 *J. Phys. Soc. Jpn.* **84** 074701
[11] Harris R K, Becker E D, Cabral De Menezes S M, Granger P, Hoffman R E and Zilm K W 2008 *Mag. Res. Chem* **46** 582
[12] Pyykko P 2008 *Mol Phys* **106** 1965
[13] Kubo T, Kotegawa H, Tou H, Higashinaka R, Nakama A, Aoki Y and Sato H 2014 *JPS Conf. Proc.* **3** 012031
[14] Kubo T, Kotegawa H, Tou H, Higashinaka R, Nakama A, Aoki Y and Sato H 2015 *J. Phys.: Conf. Ser.* **592** 012093
[15] Kubo T, Kotegawa H, Tou H, Higashinaka R, Nakama A, Aoki Y and Sato H 2016 *J. Phys.: Conf. Ser.* **683** 012015
[16] Tou H, Doi M, Sera M, Yogi M, Kitaoka Y, Kotegawa H, Zheng G.-q, Harima H, Sugawara H and Sato H 2005 *Physica B* **359** 892
[17] Tou H, Tsugawa N, Sera M, Harima H, Haga Y and Ōnuki Y 2007*J. Phys. Soc. Jpn.* **76** 024705; 2007 *J. Phys. Soc. Jpn.* **76** 058001
[18] Tokunaga Y, Sakai H, Kambe S, Sakai A, Nakatsuji S and Harima H 2013 *Phys. Rev. B* **88** 085124
[19] Taniguchi T, Yoshida M, Takeda H, Takigawa M, Tsujimoto M, Sakai A, Matsumoto Y and Nakatsuji S 2016 *J. Phys.: Conf. Ser.* **683** 012016
[20] Harima H 2006 *Physica B* **378–380** 246
[21] Momma K and Izumi F 2011 *J. Appl. Cryst.* **44** 1272

Evolution of a magnetic order in the quasi-kagome lattice system $CeRh_{1-x}Pd_xSn$ ($x \leq 0.75$)

C L Yang[1], K Umeo[2] and T Takabatake[1,3]

[1] Graduate School of Advanced Sciences of Matter, Hiroshima University, Higashi-Hiroshima 739-8530, Japan
[2] Natural Science Center for Basic Research and Development, Hiroshima University, Higashi-Hiroshima 739-8526, Japan
[3] Institute for Advanced Materials Research, Hiroshima University, Higashi-Hiroshima 739-8530, Japan

takaba@hiroshima-u.ac.jp

Abstract. The equiatomic compound CeRhSn with a quasi-kagome lattice of Ce atoms displays a non-Fermi liquid behavior at low temperatures. We have studied how the ground state changes with the substitution of Pd for Rh in $CeRh_{1-x}Pd_xSn$ ($0 \leq x \leq 0.75$) by measuring the specific heat C, magnetic susceptibility χ, and magnetization M. With increasing x, the lattice parameters a and c increase by 1.6%, the paramagnetic Curie temperature in $\chi(T)$ changes from −112 K to 16 K, and $M(B = 5\text{ T})$ at $T = 1.8$ K increases from $0.05\mu_B$/Ce to $1.4\mu_B$/Ce. A long-range magnetic order manifests in sharp peaks in C at 0.7, 1.0 and 1.5 K for $x = 0.4$, 0.5 and 0.65, respectively. An antiferromagnetic order for $x = 0.75$ is indicated by the observations of peaks in both $\chi(T)$ and $C(T)$ at around 3.0 K and metamagnetic behavior in $M(B)$ at $B = 2$ T. Our results point out that the doping of $4d$ electrons in the non-Fermi liquid system CeRhSn destroys the coherence of the quasi-kagome lattice and weakens the hybridization between the $4f$ state and conduction bands, leading to the evolution of antiferromagnetic order.

1. Introduction

Quantum phase transitions in the $4f$ electron systems have fascinated physicists for decades because they involve competition and/or coexistence of magnetic order, non-Fermi liquid (NFL) state, and unconventional superconductivity [1,2]. Conventional quantum criticality at zero temperature can be reached by tuning the coupling between the $4f$ electron and conduction electrons governing the competition between RKKY-interaction and Kondo effect, well known as the Doniach phase diagram [3]. The ordinary ways to tune the balance of the two interactions are through chemical composition, pressure, and magnetic field, leading from a magnetically ordered state to a paramagnetic one [4]. Recently, a kagome lattice of $4f$ elements with antiferromagnetic nearest neighbor interaction due to the triangular-like geometry is reported to induce an unconventional quantum critical point [5-12]. In this system the competition occurs between a spin liquid state and a Kondo singlet state, which is caused by the interplay among the RKKY-interaction, Kondo effect, and magnetic frustration [13,14]. Hence, a long-range order can be induced by either destruction of symmetry of frustrated structure or breakdown of Kondo screening behavior.

CePdAl has a quasi-kagome lattice of Ce ions in the hexagonal ZrNiAl-type structure [5,15,16]. Within a single quasi-kagome layer, only two thirds of Ce moments participating in the antiferromagnetic (AFM) long-range order occurring at T_N = 2.7 K, while the others remain paramagnetic [6]. By doping with Ni for Pd, the AFM order in $CePd_{1-x}Ni_xAl$ is completely suppressed at x = 0.144. The frustrated Ce moments give rise to a two dimensional quantum critical point (QCP) [9]. The ratio $|\theta_p|/T_N$, where θ_p is the paramagnetic Curie temperature in the magnetic susceptibility χ, increases with x as the Kondo effect surpasses the RKKY-interaction. Another example with QCP is the isostructural alloy system $CeRh_{1-x}Pd_xIn$, in which a mixed valence state changes to an AFM heavy fermion state [12]. Then, an AFM order appears for x = 0.8 at T_N = 0.65 K which rises to 1.7 K for x = 1. Thereby, the increase in the 4d-electron number deepens the 4f level from the Fermi level suppressing the hybridization between 4f states of Ce and 4d states.

In the isostructural compound CeRhSn, layers composed of Ce and Rh1 atoms alternate along the c axis with layers composed of Rh2 and Sn atoms as shown in insert of Figure 1 [11]. Nearest neighbors for the Ce atom are four Rh2 atoms and the second nearest neighbor is one Rh1 atom. This structural aspect suggested that the Ce 4f state is strongly hybridized with the 4d band derived mainly from the Rh2 atoms, leading to the valence-fluctuating state with a high Kondo temperature $T_K \sim$ 200 K. Below 7 K, however, NFL behavior appear: $\chi(T)$ shows a power-law behavior, $\rho(T)$ decreases with T^n ($n \leq 1.5$), and the specific heat divided by temperature C/T shows an upturn and saturation to a large value of 200 mJ/K^{2}mol. These evident NFL behaviors are thought to be manifestation of the frustration of Ce moments which hinders the long-range order and leads to a QCP. Furthermore, a zero-field QCP in CeRhSn has been indicated by the measurements of C, uniaxial thermal expansion, and magnetic Grüneisen parameter [17]. The anisotropy of thermal expansion displaying the critical behavior only along the a axis is in support of the fact that the quantum criticality is driven by geometrical frustration in the quasi-kagome lattice.

Recently, it has been reported that the alloy $CeRh_{1-x}Pd_xSn$ maintains the ZrNiAl-type structure up to x = 0.8 [18]. In the whole range, no magnetic order was found down to 2.5 K in the measurements of $\chi(T)$. We expected that the destruction of the local symmetry of the frustrated structure should induce a magnetic order. Therefore, we have studied the magnetic properties down to 0.4 K in this work.

2. Experiment

Polycrystalline samples with initial compositions of $CeRh_{1-X}Pd_XSn$ (X = 0, 0.2, 0.3, 0.4, 0.5, 0.7, 0.8) were prepared by arc-melting appropriate amounts of constituent elements Ce (99.9%), Rh (99.9%), Pd (99.99%) and Sn (99.999%) under a pure argon atmosphere. In order to improve the homogeneity, these samples were turned over after melting and remelted for several times, and were subsequently annealed at 900°C for 6 days in evacuated quartz ampoules. Total weight loss after preparation was not above 1%. The samples were characterized by powder X-ray diffraction and wavelength dispersive electron-probe microanalysis (EPMA). Lattice parameters of the ZrNiAl-type structure were calculated by least-square refinements of the XRD patterns. The metallographic examination showed that the polycrystalline samples are composed of grains preferentially oriented along the c axis, which is perpendicular to the bottom of the ingot. For the measurements of $\chi(T)$ and magnetization $M(B)$, the magnetic field was applied along the preferentially oriented c axis of the samples which were cut perpendicularly to the bottom. The measurements were performed in the temperature range from 1.8 to 300 K with a Quantum Design MPMS. The specific heat was measured on 10 mg samples from 0.4 to 20 K by the relaxation method in a Quantum Design PPMS.

3. Results and discussion

3.1. Chemical and structural properties

The results of EPMA are consistent with the nominal composition for $X \leq 0.5$. However, with increasing X, the quantity of impure phase $Ce(Rh_{1-X}Pd_X)_2Sn_2$ increased. The real compositions of Pd, x, in $CeRh_{1-x}Pd_xSn$ phase were found to be 0.65 and 0.75 for X = 0.7 and 0.8, respectively. The hexagonal lattice parameters a and c of CeRhSn agree with the reported values [18]. As shown in Figure 1, the a parameter linearly increases as x increases from 0 to 0.75, while the x dependence of the c parameter deviates upward from the Vegard's law, in consistent with the previous report [18]. This deviation hints a change in the 4f state from the valence fluctuating state to the localized state, which is consistent with the results of magnetic susceptibility and magnetization as will be shown in the next subsections. The expansion along the c axis enlarges the distance between Ce and Rh2 out of basal plane, which may decrease the overlap of wave functions between the 4f electron and the 4d electron. Nevertheless, local extension in the basal plane may break the symmetry of the quasi-kagome lattice of Ce atoms.

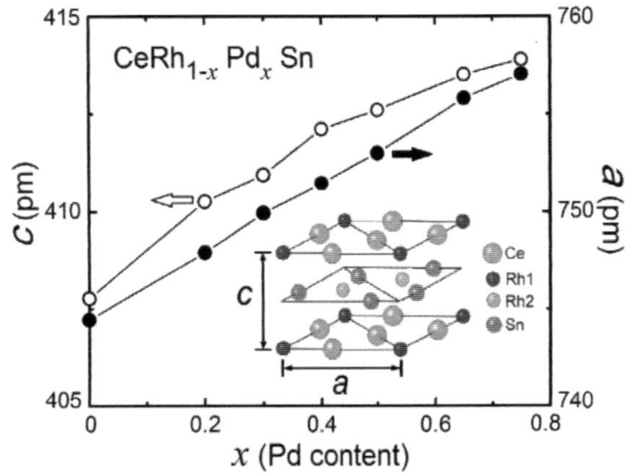

Figure 1. Hexagonal lattice parameters of $CeRh_{1-x}Pd_xSn$ as a function of x. The insert shows the unit cell of CeRhSn [11].

3.2. *Magnetic properties*

Figure 2 shows the inverse susceptibility as a function of temperature for $CeRh_{1-x}Pd_xSn$ ($x \leq 0.75$) measured in a magnetic field of 1 T. The solid lines at $T > 100$ K represent the Curie-Weiss fit with $\chi = C/(T - \theta_p) + \chi_0$, where the last term represents the temperature independent contribution. For $x = 0$, the large negative value of paramagnetic Curie temperature $\theta_p = -112$ K and effective magnetic moment

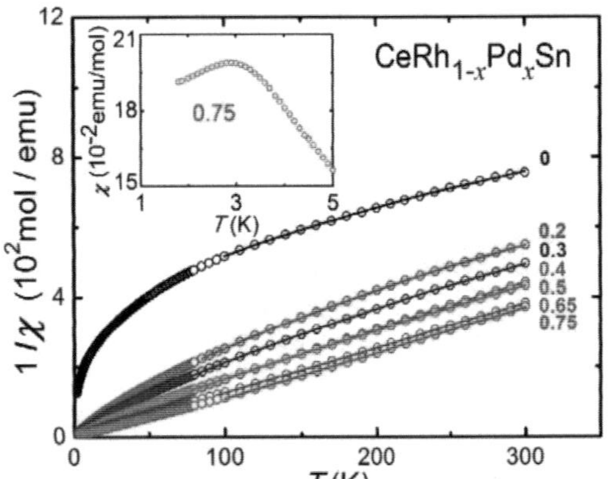

Figure 2. Temperature dependence of the inverse susceptibility of $CeRh_{1-x}Pd_xSn$. Solid lines are Curie-Weiss fits to the data at temperatures above 100 K.

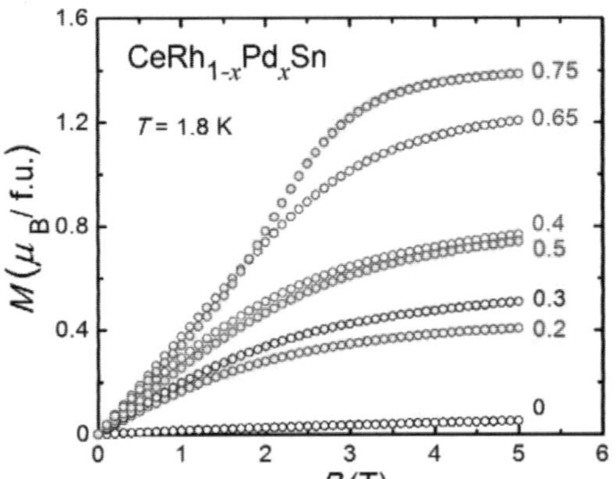

Figure 3. Magnetization of $CeRh_{1-x}Pd_xSn$ as a function of the magnetic field at 1.8 K.

$\mu_{eff} = 1.47\mu_B$ are characteristics for a valence fluctuation system. With the increase in x to 0.2, the value of $|\theta_p|$ decreases to 13 K. This decreasing trend in $|\theta_p|$ suggests that the hybridization is weakened as $|\theta_p|/2$ is a measure of T_K [19]. The sign of θ_p becomes positive for $x \geq 0.65$, which can be attributed either to the crystal field effect on nearly trivalent Ce ions or possible ferromagnetic interaction due to the increase in the Fermi wave vector. For $x \geq 0.65$, μ_{eff} increases to 2.50μ_B/Ce, which value is close to 2.54μ_B expected for a free Ce^{+3} ion. As shown in the inset of Figure 2, the data of $\chi(T)$ for $x = 0.75$ exhibits an obvious maximum at around 3 K, suggesting an antiferromagnetic order.

The magnetization curves $M(B)$ measured in fields up to 5 T at $T = 1.8$ K are shown in Figure 3. For $x = 0$, $M(B)$ increases almost linearly, yielding a value of 0.05μ_B/Ce at 5 T. This value is 67% of that reported on the single crystal for B // c, confirming the preferred orientation of grains along the c axis in the polycrystal. For $x \geq 0.2$, $M(B)$ gradually increases and shows a tendency towards saturation. Especially, $M(B)$ for $x = 0.75$ shows a spin-flop behaviour at $B = 2$ T and reaches a saturated value of 1.4 μ_B/Ce at 5 T. Under the hexagonal symmetry, the $J = 5/2$ multiplet of the $4f$ state splits into three doublets. The magnitude of 1.4 μ_B/Ce is 67% of the value 2.14 μ_B/Ce for a ground state doublet with $J_Z = \pm 5/2$. This fact indicates that the hybridization between $4f$ states of Ce and $4d$ states of Rh2 is weakened because the $4d$ electron doping lowers the energy level of $4f$ state from the Fermi level.

3.3. Specific heat

The data of specific heat C between 0.4 K and 20 K are plotted as C/T vs log$T$ in Figure 4. For $x = 0$, C/T linearly increases on cooling below 7 K and saturates to a value $\gamma = 290$ mJ/K^{2}mol, in agreement with the behaviour reported for a single crystal [11]. For $x = 0.2$ and 0.3, pronounced upturns suggest the approach to a QCP. For x $\geqslant$ 0.4, the C/T data exhibit peaks, whose temperatures are taken as T_N. The T_N increases from 0.7 K for $x = 0.4$ to 1.5 K for $x = 0.65$, indicating the enhancement of AF

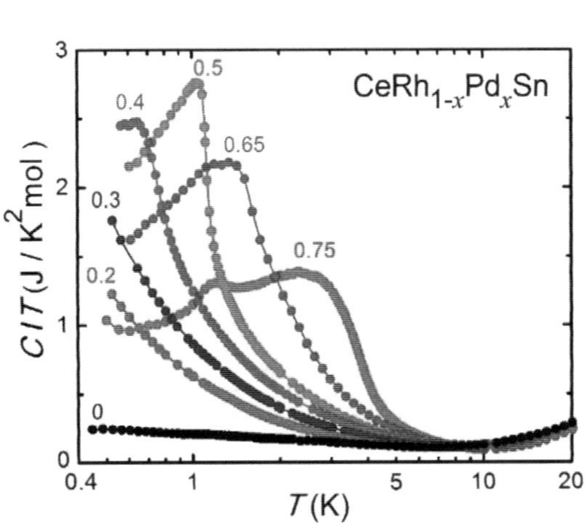

Figure 4. Specific heat of $CeRh_{1-x}Pd_xSn$ plotted as C/T vs log$T$.

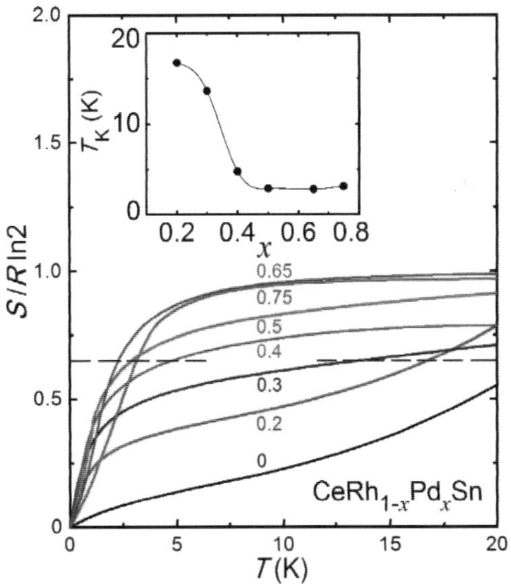

Figure 5. Magnetic entropy of $CeRh_{1-x}Pd_xSn$ as a function of temperature. The dotted line represents the value of 0.65Rln2. The insert shows the x dependence of Kondo temperature T_K.

interaction and/or growth of magnetic moments. For $x = 0.75$, two maxima are found at 1.2 K and 3 K. The latter one is identified as T_N because it agrees with the maximum temperature in $\chi(T)$. It is noted that the peaks in the C/T data broaden with increasing x from 0.4 to 0.75. The magnetic structure in this system should be further studied by the neutron diffraction experiment.

In order to obtain the magnetic entropy $S(T)$, specific data of $CeRh_{1-x}Pd_xSn$ for $0 \leq x \leq 0.3$ and $0.4 \leq x \leq 0.75$ were extrapolated, respectively, with the equations $C/T = \gamma + \beta T^2 + CT^2 \ln T$ [20] and $C/T = \gamma + \beta T^2 + \delta T^2 e^{-\Delta/k_B T}$ [21]. The data of isostructural LaRhSn were used as the phonon contribution. The temperature dependences of $S(T)$ are plotted in Figure 5. The $S(T)$ curves for $x = 0.65$ and 0.75 are saturated to $R\ln 2$ at 20 K, confirming the doublet ground state under the crystal field. The T_K was estimated by the relationship $S(T_K) = 0.65R\ln 2$, which was given for a Kondo impurity [22,23]. The values of T_K as a function of x are shown in insert. The T_K decreases rapidly from 16 K for $x = 0.2$ to 5 K for $x = 0.4$, indicating the strong suppression of Kondo effect.

4. Conclusions

We have reported a set of measurements on X-ray diffraction, magnetic susceptibility, magnetization, and specific heat for $CeRh_{1-x}Pd_xSn$ series in which the Ce atoms form a quasi-kagome lattice. With increasing x up to 0.75, which is the limit for the ZrNiAl-type structure, lattice parameters a and c increase by 1.6%. The $4f$ state of Ce ion changes from a valence fluctuating state to a stable trivalent state with a doublet ground state. Importantly, the specific data show apparent maximum in C/T for $x \geq 0.4$. The spin-flop behaviour in the magnetization curve and the peak in $\chi(T)$ for $x = 0.75$ supports the AFM ordered state. The AFM order may be induced by the combination of two mechanisms. On one hand, the increasing c axis parameter and additional $4d$ electrons weaken the hybridization between $4f$ orbitals of Ce and $4d$ orbitals of Rh2. As a result, the magnetic moment is increased and thus the RKKY interaction is enhanced. On the other hand, local extension in the basal plane near the Pd atoms may break the symmetry of the quasi-kagome lattice of Ce atoms, then the frustrated Ce moments have the chance to be coupled antiparallel. In any case, we have found that the NFL state of CeRhSn is easily transformed to an AFM ordered state. For the study of quantum phase transition, the system $CeRh_{1-x}Pd_xSn$ deserves to be further studied by neutron scattering and μSR experiments.

Acknowledgements
This work was partly supported by KAKENHI Grant Numbers Nos. 26400363 and 16H01076.

References

[1] Si Q, Pixley J H, Nica E, Yamamoto S J, Goswami P, Yu R and Kirchner S 2014 *J. Phys. Soc. Jpn.* **83** 0061005, and references therein
[2] Gegenwart P, Si Q and Steglich F 2008 *Nat. Phys.* **4** 186
[3] Doniach S 1977 *Physica B* **91** 231
[4] Löhneysen H v, Rosch A, Vojta M and Wölfle P 2007 *Rev. Mod. Phys.* **79** 1015
[5] Kitazawa H, Matsushita A, Matsumoto T and Suzuki T 1994 *Physica B* **199-200** 28
[6] Dönni A, Ehlers G, Maletta H, Fischer P, Kitazawa and Zolliker M 1996 *J. Phys.: Condens. Matter* **8** 11213
[7] Goto T, Hane S, Umeo K, Takabatake T and Isikawa Y 2002 *J. Phys. Chem. Solids* **63** 1159
[8] Oyamada A, Maegawa S, Nishiyama M, Kitazawa H and Isikawa Y 2008 *Phys. Rev. B* **77** 064432
[9] Fritsch V, Bagrets N, Goll G, Kittler W, Wolf M J, Grube K, Huang C L and Löhneysen H v 2014 *Phys. Rev. B* **89** 054416
[10] Fritsch V *et al.* 2015 *Eur. Phys. J. Special Topics* **224** 997
[11] Kim M S *et al.* 2003 *Phys. Rev. B* **68** 054416
[12] Brück E, Nakotte H, Bakker K, de Boer F R, de Châtel P F, Li J Y, Kuang J P and Yang F M 1993 *J. Alloys Comp.* **200** 79
[13] Vojta M 2008 *Phys. Rev. B* **78** 125109.

[14] Coleman P and Nevidomskyy A H 2010 *J. Low Temp. Phys.* **161** 182
[15] Woitschach S, Stockert O, Koza M M, Fritsch V, Löhneysen H v and Steglich F 2013 *Phys. Status Solidi B* **250** 468
[16] Fritsch V, Huang C L, Bagrets N, Grube K, Schumann S and Löhneysen H v 2013 *Phys. Status Solidi B* **250** 506
[17] Tokiwa Y, Stingl C, Kim M S, Takabatake T and Gegenwart P 2015 *Sci. Adv.* **1** e1500001
[18] Niehaus O, Abdala P M and Pöttgen R 2015 *Z. Naturforsch.* **70** 253
[19] Gignoux D and Gomez-Sal J C 1984 *Phys. Rev. B* **30** 3967
[20] Brodsky M B 1978 *Rep. Prog. Phys.* **41** 1547
[21] Cooper B R 1962 *Proc. Phys. Soc.* **80** 1225
[22] Desgranges H U and Schotte K D 1982 *Phys. Lett.* **91** 240
[23] Bredl C D, Steglich F and Schotte K D 1978 *Z. Phys. B* **29** 327

Optimizing the Preparation of Monoclinic and Pyrochlore $Mn_2Sb_2O_7$

Darren C Peets[1], †, **Hasung Sim**[1,2], **and Maxim Avdeev**[3]

[1] Center for Correlated Electron Systems, Institute for Basic Science (IBS), Seoul 08826, Korea
[2] Department of Physics and Astronomy, Seoul National University, Seoul 08826, Korea
[3] Australian Nuclear Science and Technology Organisation, Lucas Heights, NSW 2234, Australia
†Current address: Advanced Materials Laboratory, Fudan University, Shanghai 200438, China

E-mail: dpeets@fudan.edu.cn

Abstract. Two polymorphs of $Mn_2Sb_2O_7$ have been reported, in the pyrochlore structure and in a chiral, monoclinic structure. In both cases, magnetization data have shown prominent features attributable to a ferrimagnetic impurity. Accessing the intrinsic physical properties of both materials thus requires first making them phase pure. We describe our efforts to optimize the preparation of both polymorphs and how we concluded this was necessary, in hopes that future work will be aided by detailed knowledge of this process.

1. Introduction

Strong frustration, in which magnetic interactions compete with each other, impedes spin systems from selecting a unique global ground state at low temperature, leading to a wide variety of physics in which quantum mechanical effects and details of the spins' interactions can be crucial. $Mn_2Sb_2O_7$ can be prepared in the pyrochlore structure (pyr-$Mn_2Sb_2O_7$), shown in Fig. 1(c) and (d), by low-temperature routes[1–4]. This is the most-frustrated known three-dimensional arrangement of magnetic ions, and the quenched orbital moments in pyr-$Mn_2Sb_2O_7$ make it especially interesting. However, preparation by standard solid-state synthesis results in the chiral monoclinic structure m-$Mn_2Sb_2O_7$[5–8] as shown in Fig. 1(a) and (b), which is more closely related to the trigonal Weberites. The Mn sublattice in this structure is built up of Mn_4 armchair motifs. Both polymorphs of $Mn_2Sb_2O_7$ are likely to be of considerable interest, but both are readily contaminated by traces of a ferrimagnetic impurity which can obscure the intrinsic physics.

Previous investigation into the magnetic behaviour of the monoclinic polymorph indicated a paramagnetic Curie-Weiss temperature T_{CW} around −45 to −50 K and a paramagnetic moment corresponding to high-spin d^5 Mn^{2+},[9] with a bulk magnetic transition around 13 K. Clear history dependence reported below ∼55 K was attributed to an apparent abrupt onset of short-range correlations[9]. Our work (reported elsewhere) concluded, among other results, that the low-temperature transition was intrinsic while the reported history dependence was spurious[8]. Meanwhile, pyr-$Mn_2Sb_2O_7$ had been reported to form a spin glass around 43 K[2, 3], several times *higher* than the bulk magnetic transition reported in the less-frustrated monoclinic polymorph. Our work on the pyrochlore material concluded that this putative spin-glass transition is not

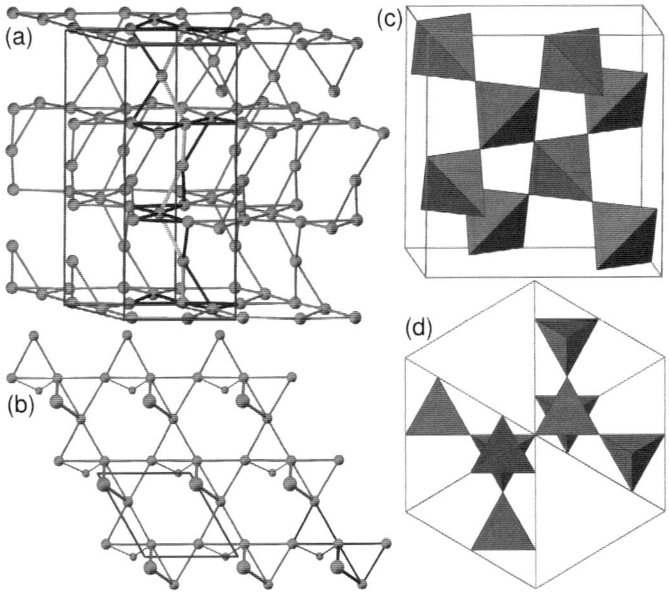

Figure 1. Mn sublattices in m-$Mn_2Sb_2O_7$ and pyr-$Mn_2Sb_2O_7$. (a) m-$Mn_2Sb_2O_7$ is composed of Mn_4 armchair units — one layer is shown in (b) — adding atoms above and below a kagome plane. The armchairs form helices in the three-dimensional structure, and the shortest bonds in our refinement form a helix along the c axis (ligher-coloured bonds). (c) The pyrochlore structure is composed of corner-sharing Mn_4 tetrahedra, with kagome planes perpendicular to all ⟨111⟩ directions (d).

intrinsic, and that the spins actually freeze at a much lower temperature[4]. In both $Mn_2Sb_2O_7$ materials, it would appear that the previous work was limited by phase purity. Since extracting the intrinsic behaviour evidently relies crucially on phase purity, and since we anticipate that other groups will find these materials interesting, this proceedings paper concentrates on how the preparation of these two polymorphs can be optimized.

2. Experimental

Powder samples of m-$Mn_2Sb_2O_7$ were prepared in Al_2O_3 crucibles in air, from intimately mixed MnO_2 (Alfa Aesar, 99.997%) and Sb_2O_3 (Alfa Aesar, 99.999%). Mixed powders were calcined typically 2–3 times with intermediate grindings at temperatures between 1050 and 1150°C, typically for 24h per temperature; the mass was monitored for loss of volatile component oxides, and powder diffraction patterns (Rigaku Miniflex II, Cu$K\alpha$ source) were used for initial verification of phase purity. As will be discussed, magnetometry was more sensitive, and in some cases the powder was reground and recalcined as high as 1175°C after magnetic measurements indicated an incomplete reaction.

Powder samples of pyr-$Mn_2Sb_2O_7$ were prepared following the recipe of Brisse[1], with precursor "antimonic acid" first prepared from $SbCl_5$ (Alfa Aesar, 99.997%) and deionized ice water as described in Ref. [10]. The antimony precursor, once dried, was ground with $Mn(Ac)_2{\cdot}4H_2O$ (Aldrich, 99.99%) under ethanol. This starts the reaction, which releases acetic acid and water. The reaction was allowed to continue in an open glass vial at ~50°C for several days until the material finally dried, at which point it was reground under ethanol. It was then reacted for 12h at each of a series of temperatures from 200 to 550°C in Al_2O_3 crucibles in air, typically in steps of 50°C, with intermediate grinding. Powder diffraction and magnetic measurements were again used to verify phase purity and in this case also Mn site occupancy.

Magnetization measurements in fields up to 5 T were performed in a Quantum Design MPMS-XL SQUID magnetometer in RSO mode. Powder samples were packed inside gelatin capsules, which were closed with Kapton tape and loaded into plastic straws. The contribution from the empty sample holder was below the level of the noise. X-ray powder diffraction patterns were collected at key points using a Bruker D8 Discover diffractometer, and the progress of the reaction was monitored using a Rigaku Miniflex II benchtop powder diffractometer, each with a Cu$K\alpha$ source. Powder neutron diffraction was performed at various temperatures at the ECHIDNA diffractometer at the OPAL research reactor at ANSTO, Australia, from 6.5 to 163.95° in steps of 0.05°, with neutron wavelengths of 1.6620 and 2.4395 Å. Refinements will be reported elsewhere[4, 8].

3. Preparation of m-$Mn_2Sb_2O_7$

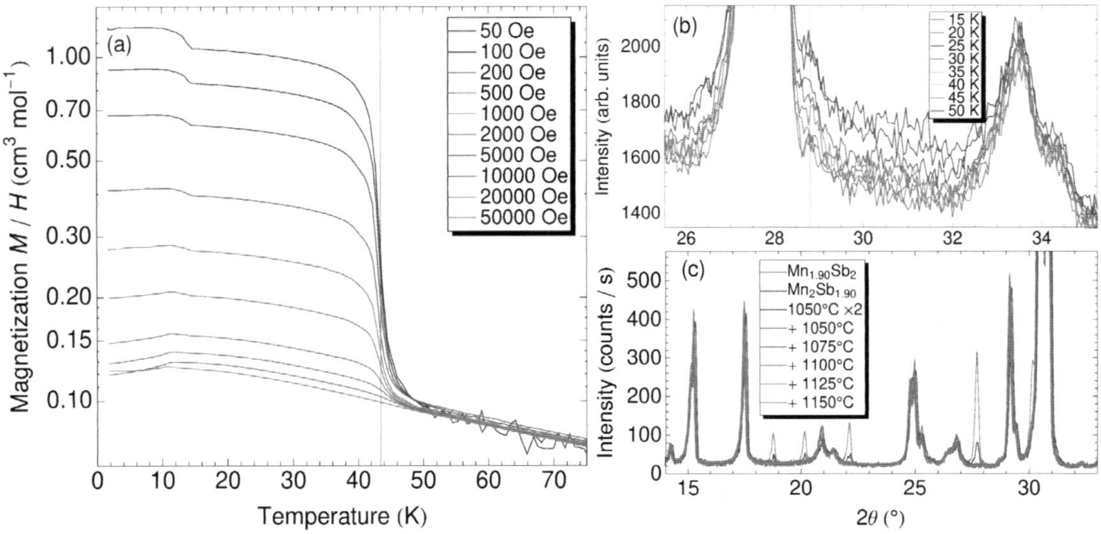

Figure 2. a) Temperature-dependent field-cooled magnetization in an early sample that was deemed phase-pure by x-ray diffraction, showing a significant ferromagnetic contribution below ~43 K. b) A tiny shoulder near 29° in the neutron data, corresponding to the strongest magnetic peak in Mn_3O_4, is suppressed around 40 K. c) Impurity peaks were visible for several-percent nonstoichiometry, but extra peaks due to a reduction in symmetry to $P2$ were not affected.

X-ray diffraction found m-$Mn_2Sb_2O_7$ to be phase pure for up to ~3 % initial cation nonstoichiometry, usually after the second calcine. As shown in Fig. 2a, magnetization measurements at this stage indicated a strong ferro- or ferrimagnetic transition around 43 K. Note that this figure uses a logarithmic vertical scale so that other features remain visible. Neutron diffraction on this material similarly found no obvious impurity peaks, with magnetic Bragg peaks appearing below the ~13 K transition. There was no evident signature of the dominant magnetic transition around 43 K, which was quite troubling. However, close inspection of the neutron diffraction pattern turned up a tiny hump at the angle corresponding to the strongest magnetic reflection of Mn_3O_4 (around 28.8° in Fig. 2b), which disappeared above 40 K. Mn_3O_4 is known to have a broad ferrimagnetic transition around 45 K[11, 12]. It was now clear that the most prominent magnetic transition was actually from trace Mn_3O_4 impurities that were below x-ray diffraction's detection limit, and that this transition was sufficiently strong, in low fields, to completely swamp the signal from the intrinsic transitions at lower temperature. Magnetization measurements in low field were thus identified as exquisitely sensitive to the presence of this impurity, offering an efficient and effective way to optimize the sample stoichiometry.

In optimizing the preparation conditions for this monoclinic polymorph, several approaches were tried, and it is worth commenting on a few of them. Figure 2c shows diffraction data on samples with ~5% cation nonstoichiometry, as well as the effect of additional sintering. Peaks associated with Mn impurities were present in samples prepared with a several-percent excess of Mn (~5% is shown), and in a sample reacted twice at 1050°C. This latter sample was subjected to a third calcine at 1050°C, then a series of additional firings at progressively higher temperatures. After the third calcine, no traces of impurities were visible in the x-ray pattern, while further steps (below 1250°C) caused no damage to the sample. Magnetization data corresponding to these tests, shown in Fig. 3a, indicate a clear reduction in the Mn_3O_4 component with successive firings. So additional calcination steps provided one solution. Given the temperature dependence, this is more likely due to allowing excess volatile oxide components to evaporate, than to allowing an unfinished reaction to reach completion.

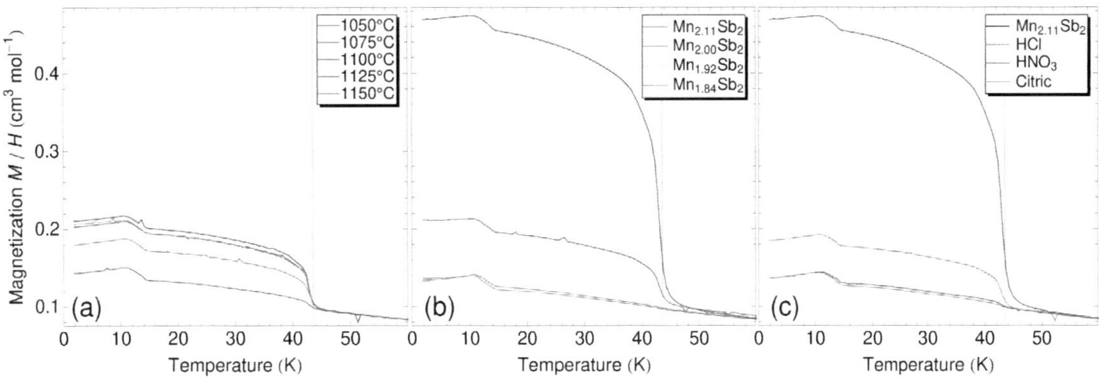

Figure 3. Approaches for reducing Mn_3O_4 contamination, as characterized by field-cooled magnetization in a field of 50 mT: (a) Additional calcining steps at higher temperatures suppress the Mn_3O_4 component. (b) Effect of varying the cation ratio: A small Mn deficiency suppresses the impurity contribution. (c) Effect of various acids, and their Mn-rich starting point.

Tests of the stoichiometry indicated that even a small excess of Sb was sufficient to nearly eliminate the magnetic contribution from Mn_3O_4 after the second calcine (Fig. 3b). This presumably either compensates for loss of volatile oxide components or for minor deviations from ideal oxygen stoichiometry in the starting materials. Impurity phases due to excess antimony oxide would not be expected to contribute to the physical properties, since Sb^{3+} and Sb^{5+} are both full-shell, and their oxides are non-magnetic insulators.

Additionally, Mn-rich m-$Mn_2Sb_2O_7$ samples were treated with several dilute (~10%) acids, to test the acids' ability to dissolve out unreacted Mn oxides. Fig. 3c shows that while HCl, HNO_3 and citric acid all worked, best results were obtained using citric acid. The citric acid was prepared from solid granules and was more concentrated.

While additional calcines, a ~2% Sb excess, and leaching with acid were each able to greatly suppress the magnetic contribution from Mn_3O_4, the combination of any two of these approaches was able to fully eliminate this impurity component. We settled on a slight Sb excess coupled with a third sintering, with verification by magnetization measurements. Our final magnetization data for publication were collected on samples subjected additionally to a citric acid treatment, as a precaution.

4. Mn-Deficient Pyrochlore

Pyrochlore $Mn_2Sb_2O_7$ is most likely formed by substituting Mn^{2+} for water (or the $2H^+$ on water) in pyrochlore 'antimonic acid' $Sb_2O_5 \cdot nH_2O$ ($n \sim 1.5-2$). If this is performed at sufficiently low temperature, the Sb pyrochlore matrix is retained, and the ultimate product is phase-pure pyr-$Mn_2Sb_2O_7$. As shown in Fig. 4a and highlighted in its inset, as this preparation process progresses there is a gradual decrease in the lattice parameter and an increase in the

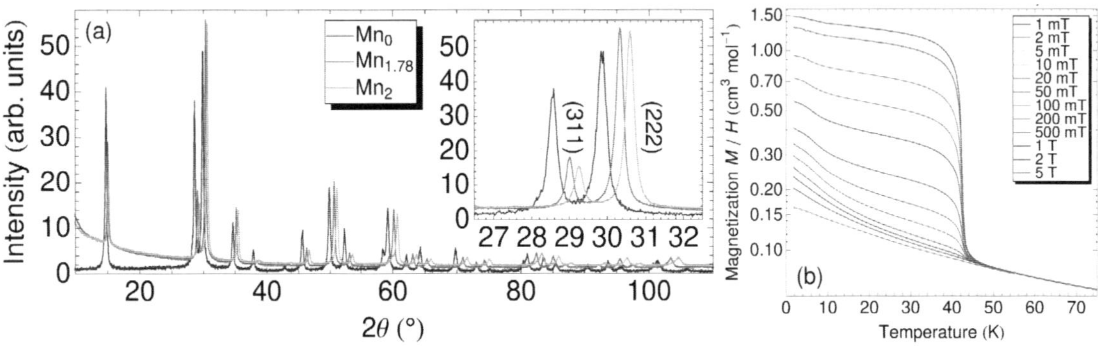

Figure 4. (a) X-ray diffraction patterns on samples with different Mn contents. The inset shows the (311) and (222) peaks. (b) Magnetization measurements clearly indicate the presence of unreacted Mn_3O_4 in a sample with a small Mn excess.

ratio between the strong (311) and (222) Bragg peaks in x-ray diffraction. The process is continuous, and phase separation is not observed. The peak ratio and lattice parameter thus provide means for optimizing the preparation conditions and monitoring completion of the reaction. This compound is particularly prone to excess Mn due to the variable hydration level n affecting the Sb precursor's molecular weight, and due to the nature of the reaction as a gradual substitution. As in the other polymorph, the clear transition in the magnetization around 43 K arising from ferrimagnetic Mn_3O_4 (see Fig. 4b, again using a logarithmic scale) provides an extremely sensitive test for excess or unreacted Mn. In Fig. 4a, the sample identified as Mn_0 is the antimonic acid precursor; Mn_2 is the fully-reacted, stoichiometric powder, which refined to $Mn_{2.00(5)}Sb_2O_7$ based on x-ray data; and $Mn_{1.78}$ refined to $Mn_{1.78(5)}Sb_2O_7$ if the Mn site was constrained to be either vacant or occupied by Mn.

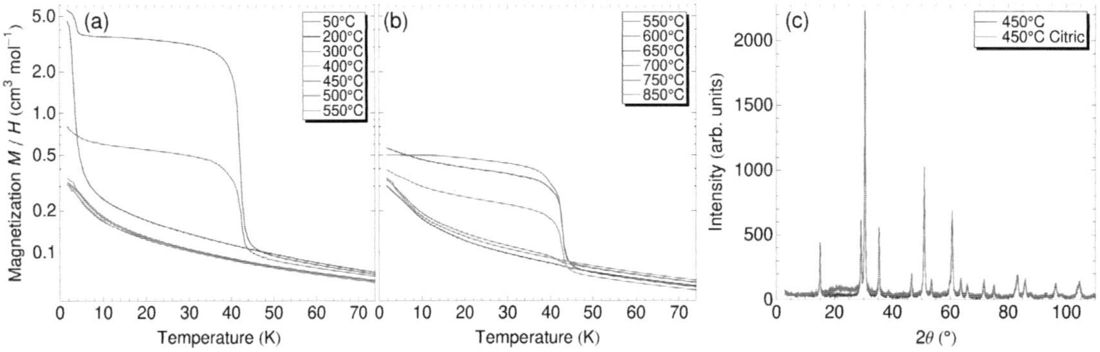

Figure 5. (a) Field-cooled magnetization (H = 10 mT) after successive heat treatments of the same batch, up to 550°C. (b) Increasing the temperature beyond 550°C eventually leads to decomposition. (c) After calcining at 450°C, a portion of this batch was treated with citric acid, producing a diffuse hump in the diffraction between 15 and 40° and an increase to the (311) peak's relative intensity, suggesting damage to the sample and deintercalation of Mn.

Of the approaches identified for eliminating Mn_3O_4 from the monoclinic polymorph, control over the stoichiometry is extremely challenging due to the variable molecular weight of the Sb precursor, but necessary. Heating the sample to higher temperatures causes it to decompose. Acid was found to damage the material somewhat, most likely by leaching Mn atoms back out of the pyrochlore matrix, reversing the process that put them there. This was manifested as an increase in diffuse x-ray scattering and an increase in the relative intensity of the (311) Bragg peak, as shown in Fig. 5c. However, repeated calcines remain an option, and were ultimately able to produce stoichiometric, Mn_3O_4-free samples. Optimal results were obtained for a final calcining temperature of 550°C (see Fig. 5a), with the first hints of decomposition appearing in the magnetization for temperatures of 600–650°C (Fig. 5b) and in the x-ray diffraction pattern by 800°C. By 700°C, there was already an increased diffuse component in the diffraction pattern. The reaction is essentially complete by 400°C, but the magnetization in low fields can continue to show hints of Mn_3O_4 for calcines up to 450°C. Ultimately, for the pyrochlore polymorph, we reacted the material for 12 h at each of a sequence of temperatures up to 550°C.

5. Conclusion

For both $Mn_2Sb_2O_7$ polymorphs, having phase-pure samples allows us to study the intrinsic magnetic behaviour without interference from other sources of magnetism. The optimized process is described above, and our results on the resulting samples will be reported elsewhere[4, 8]. The bottom line here is that when preparing complex oxide materials for fundamental physics research, it is necessary to be careful, to test for reproducibility, and to be cognizant of what the most likely impurity phases will do if present. We hope that by making a thorough description of our methods and optimization available, this report will aid the community in reproducing our work and more fully investigating these interesting materials.

Acknowledgments

This work was supported by the Institute for Basic Science (IBS) in Korea (IBS-R009-G1). We acknowledge the support of the Bragg Institute, Australian Nuclear Science and Technology Organisation, in providing neutron research facilities used in this work.

References

[1] F. Brisse, D. J. Stewart, V. Seidl, and O. Knop, Can. J. Chem. **50**, 3648 (1972).
[2] H. D. Zhou, C. R. Weibe, A. Harter, N. S. Dalal, and J. S. Gardner, J. Phys.: Condens. Matter **20**, 325201 (2008).
[3] H. D. Zhou, C. R. Wiebe, J. A. Janik, B. Vogt, A. Harter, N. S. Dalal, and J. S. Gardner, J. Solid State Chem. **183**, 890 (2010).
[4] D. C. Peets, H. Sim, M. Avdeev, and J.-G. Park, Phys. Rev. B **94**, 174431 (2016).
[5] H. G. Scott, J. Solid State Chem. **66**, 171 (1987).
[6] H. G. Scott, Z. Kristallogr. **190**, 41 (1990).
[7] L. Chelazzi, T. B. Ballaran, G. O. Lepore, L. Bindi, and P. Bonazzi, Solid State Sci. **21**, 85 (2013).
[8] D. C. Peets, H. Sim, S. Choi, M. Avdeev, S. Lee, S. J. Kim, H. Kang, D. Ahn, and J.-G. Park, Phys. Rev. B **95**, 014424 (2017).
[9] J. N. Reimers, J. E. Greedan, C. V. Stager, M. Bjorgvinnsen, and M. A. Subramanian, Phys. Rev. B **43**, 5692 (1991).
[10] D. J. Stewart and O. Knop, Can. J. Chem. **48**, 1323 (1970).
[11] K. Dwight and N. Menyuk, Phys. Rev. **119**, 1470 (1960).
[12] G. B. Jensen and O. V. Nielsen, J. Phys. C: Solid State Phys. **7**, 409 (1974).

Re-entrant spin glass behaviour and large magnetocaloric effect in Er_2PtSi_3

D. X. Li [1], Y. Homma [1], A. Nakamura [1], F. Honda [1], T. Yamamura [2], D. Aoki [1]

[1] Institute for Materials Research, Tohoku University, Oarai, Ibaraki, 311-1313 Japan
[2] Institute for Materials Research, Tohoku University, Sendai, 980-8577 Japan

E-mail: dxli@imr.tohoku.ac.jp

Abstract. We present the magnetic properties and magnetocaloric effect of ternary intermetallic compound Er_2PtSi_3 with the orthorhombic Ba_2LiSi_3-type structure. Dc magnetization, ac susceptibilit, magnetic relaxation, electrical resistivity and specific heat measurements indicate that this system undergoes two magnetic phase transitions from a paramagnetic state sequentially to an antifwerromagnetic state at T_N=5.4 K and then to a random spin freezing state at T_f=2.4 K. In addition, the large magnetocaloric effect is also observed for the Er_2PtSi_3 alloy. Near its Néel temperature T_N, the maximum magnetic entropy change, relative cooling power and maximum adiabatic temperature change are estimated to be 16.1 $Jkg^{-1}K^{-1}$, 290 Jkg^{-1} and 7.7 K, respectively, for a field change of 5 T. These results suggest that Er_2PtSi_3 should be classified as a re-entrant spin-glass system and can be considered as a potential low-temperature magnetic refrigeration material.

1. Introduction

Ternary silicides with the composition R_2PtSi_3 (R=rare earth or uranium) constitute an interesting series of the 2:1:3 nonmagnetic atom disorder (NMAD) intermetallic compounds. They are reported to crystallize in an AlB_2-derived hexagonal structure, and a doubling of the unit-cell parameter c occurs for R=Gd and Y, but not for R=La, Ce, Pr and Nd [1]. Magnetic measurements reveal that compounds R_2PtSi_3 with R = Ce, Nd, and Gd show ferromagnetic (FM) phase transition at Curie temperature T_C=5, 18, and 24 K, respectively [1], while complex magnetic properties are observed for the Pr, Tb and Dy compounds [1-3]. Moreover, our study suggests that more or less Pt and Si atoms in R_2PtSi_3 randomly distribute on the Pt-Si superstructure, and strong or weak spin-glass (SG)-like effects exhibit in these compounds [2]. In addition, 5f electronic compound U_2PtSi_3 crystallizing in the AlB_2-type NMAD structure, shows the simple SG behavior without long-range magnetic order [4]. In order to fully observe and understand the magnetic properties of this series of compounds, we have prepared a polycrystalline Er_2PtSi_3, another member of the R_2PtSi_3 family, and undertaken a systemic measurement on its magnetic and magnetocaloric properties. In this paper, we present the detailed experimental results including ac susceptibility $\chi_{ac}(T)$, dc magnetization $M(H)$, magnetic relaxation $M(t)$, electrical resistivity $\rho(T)$ and specific heat $C(T)$ on a well-annealed Er_2PtSi_3 sample.

2. Experimental

Polycrystalline sample of Er_2PtSi_3 was prepared by arc melting high-purity raw metals (Er: 3N; Pt: 4N; Si: 6N) in an arc furnace under a purified argon atmosphere. The button was flipped and remelted four times to ensure homogeneity, and then wrapped into tantalum foil and annealed in evacuated silica tube at 900 °C for 23 days. The quality of the obtained sample was checked by x-ray powder diffraction at room temperature with CuK_α radiation, and the result indicates that the annealed sample is almost single-phased with negligible impurity phases. The observed diffraction pattern can be indexed based on the orthorhombic Ba_2LiSi_3-type structure model (space group *Fddd*), an ordered superstructure derivative of the hexagonal AlB_2-type. There are five different crystallographic positions in this structure, two of them are occupied by Er atoms, and mixed positions of Pt and Si atoms are distributed on the other three non-equivalent sites as recently determined for another 2:1:3 system, Yb_2AuSi_3 [5].

The dc magnetization, ac susceptibility and magnetic relaxation were measured between 2 and 300 K in magnetic fields up to 7 T using a SQUID magnetometer. The thermal-relaxation technique was employed for specific heat measurement in the temperature range between 2 and 30 K by using a physical properties measurement system (PPMS, Quantum Design). Electrical resistivity measurements were performed between 0.5 and 285 K using a standard four-terminal DC method.

3. Results and discussion

Figure 1 shows the low temperature dc susceptibility χ (=*M*/*H*) of Er_2PtSi_3 between 2 and 7 K measured in magnetic fields of *H*=10 and 100 Oe under zero field-cooled (ZFC) and field-cooled (FC) conditions. It is clear from this figure that both $\chi_{ZFC}(T)$ and $\chi_{FC}(T)$ curves show an evident peak at T_N=5.4 K and no difference between them around T_N indicating an antiferromagnetic (AFM) phase transition in Er_2PtSi_3 at T_N. As the temperature is further decreased to ~4 K, the χ value increases again, while

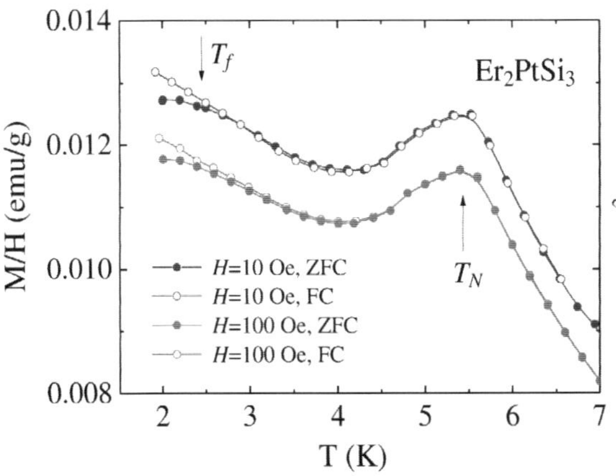

Figure 1. Temperature dependence of the dc susceptibility *(M/H)* of Er_2PtSi_3 measured in magnetic fields of 10 and 100 Oe under FC (○) and ZFC (•) conditions.

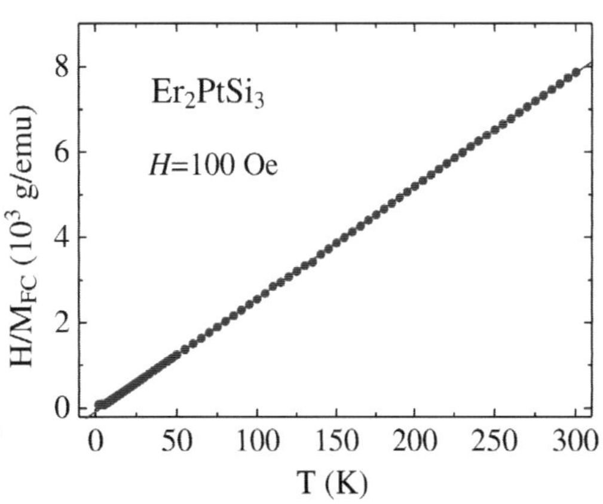

Figure 2. Temperature dependence of the reciprocal dc susceptibility *(H/M$_{FC}$)* of Er_2PtSi_3 measured in a field of 100 Oe under FC condition.

$\chi_{ZFC}(T)$ and $\chi_{FC}(T)$ curves clearly separate from each other below T_f~2.4 K. Such an irreversible magnetic behaviour suggests the occurrence of some kind of magnetic phase transition, that is, the system re-enters a SG-like metastable magnetic state at the temperature T_f (below T_N). Above about 20 K, $\chi(T)$ follows a Curie-Weiss behaviour $\chi(T) = \mu_{eff}^2/8(T-\theta_P)$. The least-squares fitting (solid line in Fig. 2) yields a small paramagnetic Curie temperature of θ_p=3.3 K, and an effective magnetic moment of μ_{eff}=9.6 μ_B/Er, close to the value of 9.58 μ_B expected for free Er^{+3} ion in the *J*=15/2 Hund's rule ground state indicating that the 4f electrons are almost localized within the Er atoms.

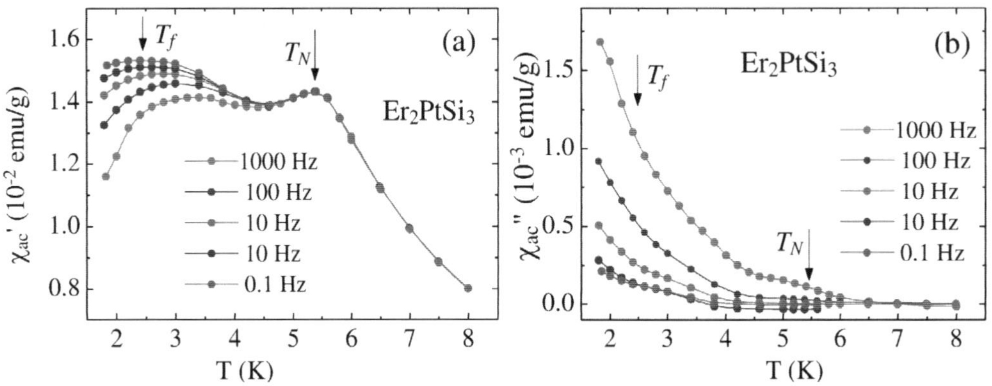

Figure 3. Temperature dependences of real component (a) and imaginary component (b) of the ac susceptibility of Er_2PtSi_3 measured at various frequencies with an excitation field of H_{ac}=1 Oe.

In order to confirm the low-temperature SG effect, we performed an ac susceptibility measurement on the Er_2PtSi_3 sample at frequency range 0.1 Hz $\leq \omega/2\pi \leq$ 1000 Hz. Figure 3 shows the in-phase component $\chi_{ac}'(T,\omega)$ and out-of-phase component $\chi_{ac}''(T,\omega)$ of the ac susceptibility between 2 and 8 K. It is clear from this figure that χ_{ac}' exhibits an evident peak at T_N=5.4 K with the amplitude and position independent of the frequency. This is the typical feature of long-range order magnetic phase transition. Moreover, further reducing the temperature, $\chi_{ac}'(T,\omega)$ shows another characteristic maximum near T_f=2.4 K with the amplitude and position strongly depending on the frequency ω of the applied ac magnetic field. As ω increases, T_f in $\chi_{ac}'(T,\omega)$ shifts to higher temperatures. For example, at $\omega/2\pi$=0.1 Hz, the peak in $\chi_{ac}'(T,\omega)$ appears at 2.4 K, which shifts to 3.3 K at $\omega/2\pi$=1000 Hz. This is a typical behaviour characteristic of SG materials [5]. Moreover, the initial frequency shift of T_f, which is usually defined as $\delta T_f = \Delta T_f/(T_f \Delta\log\omega)$ and is used to distinguish a canonical SG from a SG-like material [5], is 0.08 for Er_2PtSi_3 comparable to the typical values for spin glasses (from a few thousandths to a few hundredths). These results can be considered as important evidence for the re-entrant SG state in Er_2PtSi_3 below T_f.

On the other hand, it is interesting to note that the out-of-phase component $\chi_{ac}''(T,\omega)$, which has rather small magnitude (roughly one order of magnitude smaller than χ_{ac}'), shows a significant increase and strong frequency dependence at the temperatures near T_f suggesting the influence of SG effect. However, there is no distinct peak in $\chi_{ac}''(T,\omega)$ curve around T_N~5.4 K, where $\chi_{ac}'(T,\omega)$ as well as dc susceptibility shows a significant peak. This is an exemplary feature characterizing the AFM phase transition. In this paper, the temperature T_N=5.4 K and T_f=2.4 K observed in $\chi_{ac}'(T,\omega)$ at ω=0.1 Hz is defined as Néel temperature and spin freezing temperature, respectively, for Er_2PtSi_3.

It is well known that remanence and long-time magnetic relaxation effect, relating to the activated processes between the metastable magnetic states, are also the characteristic features of spin glasses [5, 6]. These phenomena have also been observed for the Er_2PtSi_3 sample at the temperatures below T_f. Figure 4 illustrates the field variation of the magnetization $M(H)$ measured at 2 K (<T_f) up to 5.5 T for Er_2PtSi_3 after cooling the sample in zero field from room temperature. The inset shows the low field part in an expanded scale. As increasing H, $M(H)$ increases relatively quickly up to ~1 T, then rises slowly and reaches a value of 5.7 μ_B per Er atom at 5.5 T. This value is much smaller than 9 μ_B/Er, the expected saturation value for Er_2PtSi_3. When H is returned from 5.5 T, weak hysteresis effect appears at about 0.2 T, and a small remanent magnetization of about 0.02 μ_B per Er atom is detected at zero field (inset of Fig. 4). From a careful measurement of the hysteresis loop we determine the coercive field H_C of Er_2PtSi_3 to be 23 Oe at 2 K.

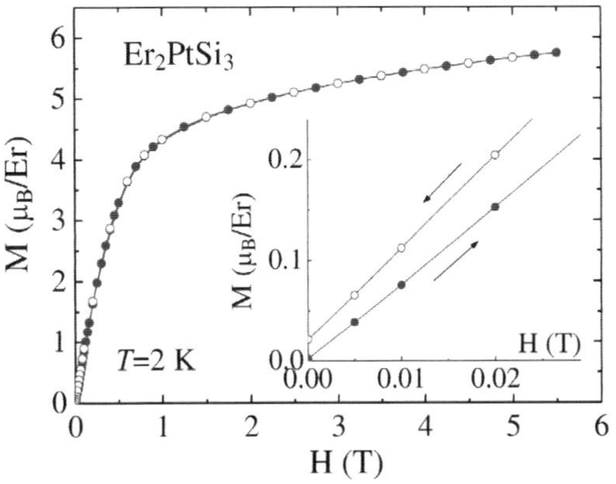

Figure 4. Magnetic field dependence of magnetization of Er_2PtSi_3 measured at 2 K. The inset displays the low field part in an expanded scale.

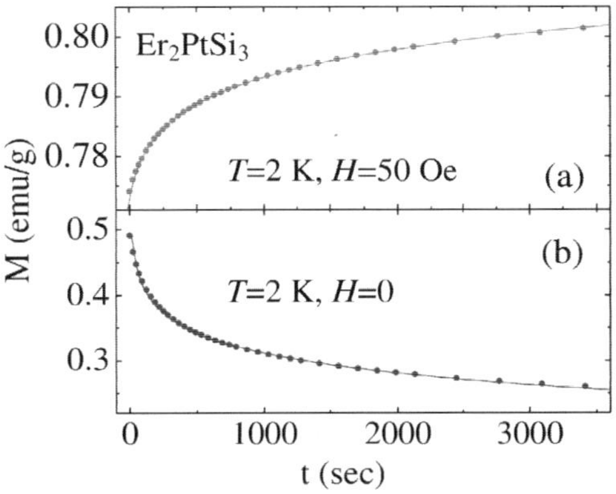

Figure 5. Decay of magnetization of Er_2PtSi_3 measured at 2 K in 50 Oe (a) and in zero field (b).

To measure magnetic relaxation effect in a field, the sample was first cooled in zero field from 50 K to 2 K, then a magnetic field of 50 Oe (much larger than H_C=23 Oe) was applied and the recording started immediately just as the field stabilized (t=0) [Fig. 5(a)]. Furthermore, to measure the relaxation behaviour in zero field, we first cooled the sample in zero field from 50 K to 2 K, then, a field of 5000 Oe was applied for about 5 min before switching it off (at t=0) [Fig. 5(b)]. As seen from Fig. 5, in both cases the decay of $M(t)$ is remarkably slow. After waiting for one hour both nonsaturation magnetization at 50 Oe and nonzero remanence in zero field are still far from saturation. Such a long-time magnetic relaxation feature in Er_2PtSi_3 could be contributed to the SG effect. Using a logarithmic function, $M(t)=M_0+S\ln(t+t_0)$, the obtained relaxation behaviour can be fitted very well over the full time range studied. The best fitting results obtained by using the least-squares method are shown by the solid lines in Fig. 5, which yields following values of fitting parameters for Er_2PtSi_3: M_0=0.7446 emu/g and S=0.0069 emu/g for H=50 Oe [inset of Fig. 5(a)], and M_0=0.6182 emu/g and S= –0.0444 emu/g for H=0 [inset of Fig. 5(b)]. Note that similar logarithmic function had also been used to describe the relaxation behaviour for many other SG systems such as for $Gd_{2-x}Y_xPdSi_3$ (x=0, 0.4, 1.0 and 1.6) [7]. Where M_0 and S, called as initial zero-field magnetization and magnetic viscosity, respectively, are the fitting parameters depending on the temperature. The parameter t_0 depends on the measuring conditions and has only limited physical relevance.

The electric resistivity and specific heat measurements give further evidence for the formation of re-entrant SG state in Er_2PtSi_3. It is clear from Fig. 6 that the temperature dependence of electrical resistivity, $\rho(T)$, shows an upturn as the temperature is lowered near T_N. After passing through a maximum value at about 4 K, $\rho(T)$ decreases smoothly down to 0.5 K. This behaviour is often observed in rare-earth compounds and explained as the decrease of carrier density due to the formation of magnetic Brillouin-zone boundary gaps [8]. Note that no anomaly (peak or rapid decrease) can be detected around T_f (=2.4 K). At high temperatures, $\rho(T)$ shows metallic conductivity and downward bending between 50 and 250 K, which may be resulted from crystal field interaction and/or s-d interband scattering of conduction electrons [9].

The experimental results of temperature dependence specific heat $C(T)$ of Er_2PtSi_3 measured in fields of 0 and 5 T are illustrated in Fig. 7. In zero field, a large peak is observed at 5 K just below T_N, due to the AFM ordering. It is interesting to note that, at the spin freezing temperature T_f where ac susceptibility shows an evident frequency dependent maximum, no anomaly can be detected in the $C(T)$ curve. Experimentally, it is not easy to determine the magnetic contribution C_m to the total specific heat accurately, since the total specific heat consists of vibrational, electronic and magnetic parts. However,

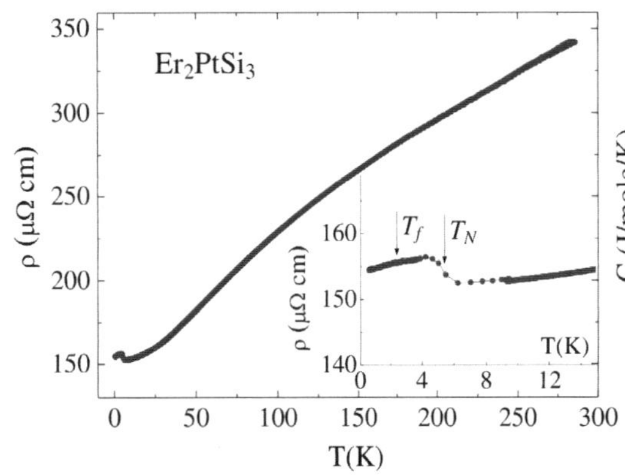

Figure 6. Temperature dependence of electrical resistivity of Er_2PtSi_3 between 0.5 and 285 K. The inset shows the low temperature data.

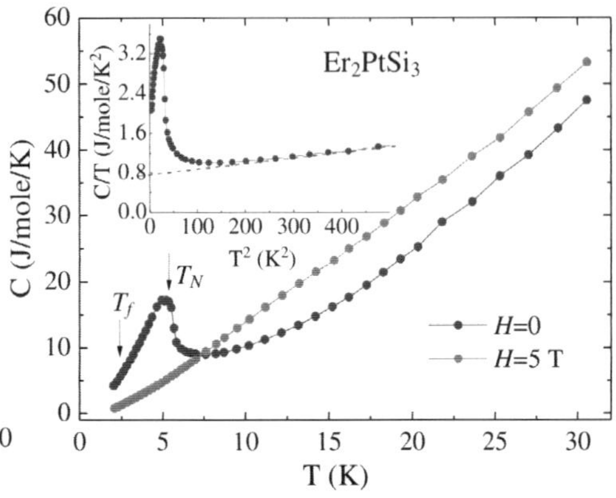

Figure 7. Temperature dependence of specific heat of Er_2PtSi_3 measured in H=0 and 5 T between 2 and 30 K. The inset shows the data plotted as C/T vs. T^2.

at low temperature, the vibrational part is smaller and electronic contribution is proportional to T. Thus the absence of dramatic change in $C(T)$ at T_f means the magnetic part $C_m(T)$ do not behaves anomaly at this temperature. This result excludes the possibility of a long-range magnetic order phase transition at T_f in Er_2PtSi_3. The anomalous magnetic properties around T_f in ac and dc susceptibilities can only be originated from the formation of SG state. Using the data between 12 and 22 K, the γ value (specific heat coefficient of T-linear term) estimated from the C/T vs. T^2 plots to be 400 mJ(mole-Er)$^{-1}$K^{-1} much larger than that of a normal metal. Random distribution of Pt and Si atoms may be responsible for this larger γ value as usually observed in NMAD compounds [10]. Note that the larger specific heat peak observed in zero field is almost completely smoothed out in a field of 5 T. These data are also used to calculate the magnetic entropy change as described below.

It is is very significant to compare the experimental results mentioned above to those observed for other 2:1:3 NMAD intermetallic compound. Among them, long-range magnetic ordering occurs in some systems with strong magnetic exchange interactions such as in ferromagnets Nd_2PtSi_3 [11] and Nd_2RhSi_3 [12] and in antiferromagnets Tb_2RhSi_3 [12] and Tb_2AgIn_3 [13]. In contrast, in some systems with weak magnetic coupling SG state may dominate the magnetic properties such as in Ce_2AgIn_3 [14], Ce_2CuSi_3 [12] and some U_2MSi_3 compounds [15]. For Er_2PtSi_3, we found the coexistence of long-range magnetic ordering and SG state. Such a reentrant SG behaviour has also been observed in some other 2:1:3 compounds such as in Tb_2CuIn_3 [16], Nd_2CuSi_3 [17], Tb_2PdSi_3 and Dy_2PdSi_3 [18]. It seems that in these reentrant SG systems a part of spins could exist as random individual spins or finite-size granules with net magnetic moments (magnetic clusters) due to the NMAD structure and don't participate in long-range magnetic ordering. In the present case, Er atoms in Er_2PtSi_3 occupy two different crystallographic positions, and mixed positions of Pt and Si atoms are distributed on three non-equivalent sites. This characteristic crystal structure may result in the formation of randomly distributed individual spins or magnetic clusters, which could weakly interact with each other at low temperatures mediated by the conduction electrons and frustrated magnetic moments could be formed. As the temperature is decreased crossing the critical point T_f, magnetic moments of these individual spins or clusters would be frozen in disorderly directions forming metastable SG state similar to what happen in amorphous or diluted metallic SG materials. In fact, magnetic clusters have been confirmed to form in similar 2:1:3 NMAD compounds U_2PdGa_3 and U_2PtGa with dimension 120~150 Å based on the neutron powder diffraction measurements [19]. For Er_2PtSi_3, however, since there is no detailed experimental data on the magnetic structure, we cannot give convincing information about the amount of individual spins and the

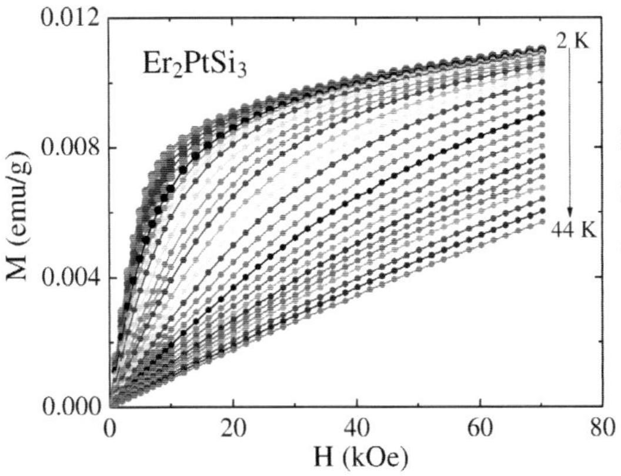

Figure 8. Magnetization of Er_2PtSi_3 as a function of magnetic field measured at different constant temperatures.

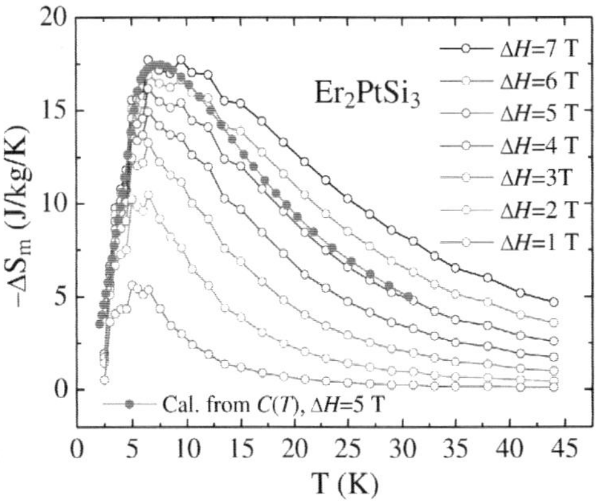

Figure 9. Magnetic entropy change of Er_2PtSi_3 as a function of temperature determined from the magnetization and specific heat measurements.

dimension of magnetic cluster based on the present work. In this sense, further experimental works including neutron powder diffraction measurement on Er_2PtSi_3 are necessary.

On the other hand, in order to obtain more comprehensive information on the magnetic properties of Er_2PtSi_3, magnetocaloric effect (MCE) of this compound is also studied by measuring $M(H)$ isotherms up to 7 T at the temperatures around T_N with increments of 1.5-3.0 K. The measurements are shown in Fig. 8. In general, MCE is characterized by magnetic entropy change ΔS_m caused by a field change. Using the Maxwell relation, $\Delta S_m(T,H) = \int_0^H [\partial M(T,H)/\partial T]_H \, dH$, magnetic entropy change ΔS_m can be calculated from the $M(H)$ data. In the present case, the calculated results of ΔS_m are displayed in Fig. 9 for Er_2PtSi_3. Magnetic hysteresis effect was ignored in this calculation, since almost no magnetic hysteresis can be observed in the $M(H)$ curves at the temperatures near T_N. The negative values of ΔS_m (positive MCE) are observed even in small magnetic field changes. It is clear that at temperature just above T_N, ΔS_m shows a maximum value, ΔS_m^{max}, which increases with the magnetic field change (ΔH) and reaches a value of $-\Delta S_m^{max}$=16.1 Jkg^{-1}K^{-1} for ΔH=5 T.

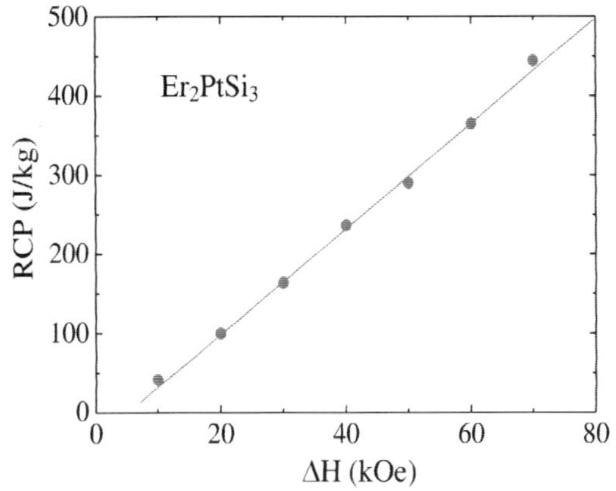

Figure 10. Magnetic field dependence of the relative cooling power of Er_2PtSi_3.

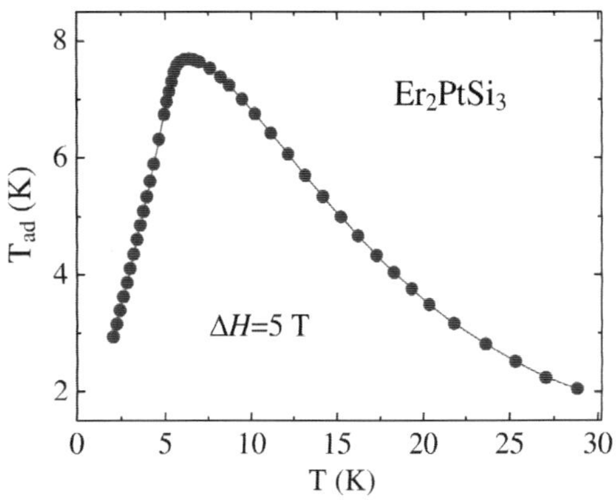

Figure 11. Adiabatic temperature change of Er_2PtSi_3 calculated from the specific heat data for ΔH=5 T.

In order to verify the accuracy of the magnetic entropy change estimated based on the Maxwell relation, ΔS_m of Er_2PtSi_3 was also calculated based on the specific heat data measured in zero and 5 T (see Fig. 7) using the expression $\Delta S_m=\Delta S=S(H,T)-S(0,T)$, where $S(H,T) = \int_0^T [C(H,T)/T]dT$. The calculating results are also displayed in Fig. 9 together with the data calculated by the magnetic method. It is found that the calorimetric and magnetic methods yield the similar results in the temperature range measured, even though the ΔS_m value estimated from $C(H,T)$ is slightly larger than that estimated from $M(H,T)$.

It is well known that relative cooling power (RCP) and adiabatic temperature change (ΔT_{ad}) are also important parameters for evaluating the usefulness of a magnetic refrigerant material, and are usually defined as $RCP = |\Delta S_m^{max}| \times \delta T_{FWHM}$ [20, 21] and $\Delta T_{ad}(\Delta H,T) = [T(S)_H - T(S)_0]_S$ [22], respectively. Where δT_{FWHM} is the full width at half maximum of the $-\Delta S_m$ vs T plot. Figure 10 shows the calculated results for Er_2PtSi_3. The RCP value increases linearly with increasing ΔH and reaches a value of RCP=290 Jkg^{-1} for ΔH=5 T, while ΔT_{ad} increases with T, reaches a maximum at the temperature slightly higher than T_N, and then decreases with further lowering the temperature. The maximum ΔT_{ad} is observed at ~6 K with a larger value of ΔT_{ad}^{max}= 7.7 K for ΔH = 5 T.

In conclusion, we have investigated the magnetic properties and MCE of polycrystalline Er_2PtSi_3 by both magnetic and calorimetric techniques. X-ray diffraction result suggests that the annealed sample crystallizes in the orthorhombic Ba_2LiSi_3-type structure (space group *Fddd*), an ordered superstructure derivative of the hexagonal AlB_2-type. With decreasing the temperature, this system undergoes a magnetic phase transition from PM to AFM state at T_N=5.4 K, and then reenters the SG state at T_f=2.4 K. The clear peak near T_N and lack of anomaly around T_f in both $C(T)$ and $\rho(T)$ curves, the frequency independent peak at T_N and strongly frequency related maximum at T_f observed in $\chi'_{ac}(T,\omega)$ (up-shift of T_f with increasing ω), the long-time magnetic relaxation effect and the irreversible magnetism observed below T_f can be considered as the evidence for the re-entrant SG behaviour. Similar to other NMAD SG systems, random distribution of Pt and Si atoms on the three different crystallographic positions may be the main origin of the low temperature SG state in Er_2PtSi_3. On the other hand, both magnetic and calorimetric measurements reveal a large MCE for the Er_2PtSi_3 compound with the maximum magnetic entropy change $-\Delta S_m^{max}$=16.1 $Jkg^{-1}K^{-1}$, the relative cooling power RCP=290 Jkg^{-1} and the maximum adiabatic temperature change ΔT_{ad}^{max}= 7.7 K for ΔH = 5 T at the temperature near T_N=5.4 K. These data are comparable with those shown in the literatures for some other rare earth based refrigerant materials with similar working temperatures, such as $ErRu_2Si_2$ [23] and $ErNi_5$ [24]. Thus, re-entrant SG compound Er_2PtSi_3 could also be considered as a promising candidate for use as a magnetic refrigerant at low temperatures.

Acknowledgments
A part of the experimental work was performed in the Laboratory of Low Temperature Materials Science, IMR, Tohoku University. The authors gratefully thank Prof. T. Nojima, Dr. S. Nakamura and other members of this laboratory for their kindly assistance. This work was supported by JSPS KAKENHI Grant Numbers 26400347.

References

[1] Majumdar S, Sampathkumaran E V, Brando M, Hemberger J and Loidl A 2001 *J. Magn. Magn. Mater.* **236** 99.
[2] Li D X, Nimori S, Homma Y and Shiokawa Y 2002 Suppl *J. Phys. Soc. Jpn.* **71** 211.
[3] Li D X, Yamamura T, Homma Y, Yubuta K, Shikama T and Aoki D 2013 *J Korean Phys. Soc.* **62** 2233.
[4] Li D X, Shiokawa Y, Haga Y, Yamamoto E and Onuki Y 2002 *J. Phys. Soc. Jpn.* **71** 418.
[5] Mydosh J A 1993 *Spin Glass: An Experimental Introduction* (Taylar & Francis, London).
[6] Binder K and Young A P 1986 *Rev. Mod. Phys.* **58** 801.

[7] Majumdar S, Sampathkumaran E V, Eckert D, Handstein A, Mueller K H, Saha S R, Sugawara H and Sato H 1999 *J Phys.: Condensed Matter.* **11** L329 and references therein.
[8] Sakurai J, Nakatani S, Adam A and Fujiwara H 1992 *J. Magn. Magn. Mater.* **108** 143; Krimmel A, Hemberger J, Nicklas M, Knenel G. Trinkl W, Brando M, Fritsch V and Loidl A 1999 *Phys. Rev. B* **59** R6604.
[9] Kaczorowski D, Czopnik A, Suski W, Nikiforov V N and Gribanov A V 2001 *Solid State Commun.* **117** 373.
[10] Gschneidner K A, Tang J, Dhar S K and Goldman A 1990 *Physica B* **163** 507.
[11] Li D X, Nimori S, Shiokawa Y, Haga Y, Yamamoto E and Onuki Y 2001 *Solid State Commun.* **120** 227.
[12] Szytula A, Leciejewicz J and Maletka K 1993 *J. Magn. Magn. Mater.* **118** 302.
[13] Semitelou J P , Siouris J, Yakinthos J K, Schäfer W and Schmitt D 1999 *J. Alloys Compd.* **283** 12.
[14] Nishioka T, Tabata Y, Taniguchi T and Miyako Y 2000 *J. Phys. Soc. Jpn.* **69** 1012.
[15] Li D X, Kimura A, Haga Y, Nimori S and Shikama T 2011 *J. Phys.: Condensed Matter.* **23** 076003 and references therein.
[16] Siouris I M, Semitelou I P, Yakinthos J K, Schäfer W and Arons R 2001 *J. Alloys Compd.* **314** 1.
[17] Tien C and Luo L 1998 *Solid State Communs.* **107** 295.
[18] Li D X, Nimori S, Shiokawa Y, Haga Y, Yamamoto E and Onuki Y 2003 *Phys. Rev. B* **68** 012413.
[19] Tran V H, Steglich F and André G 2002 *Phys. Rev. B* **65** 134401.
[20] Gschneidner Jr K A, Pecharsky V K and Tsokol A O 2005 *Rep. Prog. Phys.* **68** 1479.
[21] Gschneidner Jr K A and Pecharsky V K 2000 *Annu. Rev. Mater. Sci.* **30** 387.
[22] Pecharsky V K and Gschneidner Jr K A 1999 *J. Appl. Phys.* **86** 65.
[23] Samanta T, Das I and Banerjee S 2007 *Appl. Phys. Lett.* **91** 152506.
[24] Ranke P J von, Mota M A, Grangeia D F, Magnus G Carvalho A, Gandra F C G, Coelho A A, Caldas A, Oliveira N A de and Gama S 1998 *Phys. Rev. B* **70** 134428.

Geometric Spin Frustration in $Zn_{1-x}Ni_xCr_2O_4$ System

A Khan[1], H Kaneko[2], M O Faruk[1], M K Alam[3], C Kim[3], S Naher[1] and H Suzuki[2*]

[1] Department of Physics, Shahjalal University of Science & Technology, Sylhet-3114, Bangladesh

[2] Department of Physics, Kanazawa University, Kakuma-machi, 920-1192, Japan

[3] Department of Emerging Materials Science, Daegu Gyeongbuk Institute of Science and Technology, Daegu, South Korea

Keywords: Geometric frustration, Quantum fluctuation

Abstract

At room temperature, the cubic spinel $ZnCr_2O_4$ is known as the typical geometrically frustrated spin system. Upon cooling this compound undergoes a crystal structure change driven by removing the frustrated degenerated states, accompanying with the magnetic ordering. In the tetragonal phase the c-axis is larger than the a-axis. Another spinel compound $NiCr_2O_4$ takes the cubic structure above 310 K. Below 310 K $NiCr_2O_4$ is tetragonally distorted ($c/a < 1$) due to cooperative Jahn-Teller ordering. In the previous reports, the geometrical frustrated spin fluctuations still exist in $NiCr_2O_4$. In this report we investigated how the geometrical frustration changes and how the ratio of the crystal axis, c/a changes in $Zn_{1-x}Ni_xCr_2O_4$ system by the low temperature x-ray diffraction measurement and the SQUID magnetometer.

1. Introduction

Geometric frustration is one of the most interesting topics in the study of condensed matter systems. Frustrated interactions often cause extensive degeneracy in the ground state of the system and prevent any ordering down to low temperatures. The ground state of a frustrated system is quite intriguing and can be modified into a novel and interesting state, such as a spin liquid, associated with quantum fluctuations. The degenerated states, however, is removed its degeneracy in general by a broken symmetry, such as a crystal distortion at low temperatures.

*H.Suzuki; ult2002@staff.kanazawa-u.ac.jp

S.-H. Lee et al [1] have found the resonance-like excitation mode in the inelastic neutron scattering spectrum in spinel ferrimagnet $ZnCr_2O_4$. The similar resonance like spectrum modes were also observed in the magnetic and crystal structure ordered phase in $NiCr_2O_4$ and $FeCr_2O_4$ [2]. These modes have been identified to be the spin hexamer and heptamer, defined as antiferromagnetic spin correlation in the molecule units. So even the degenerated states of the geometric frustrated states remove its degeneracy in the broken symmetry, the dynamical fluctuations still remain in the large scale molecular unit. The study of the geometric frustration enters into the new fascinating stage. The spinel compound $ZnCr_2O_4$ is considered as a typical geometric fluctuating spin system with Cr^{3+} ions. On the other hand in $NiCr_2O_4$, Jahn-Teller (*J.T.*) ion Ni^{2+} distorts the crystal structure, resulting in the broken symmetry of the geometrical frustration. Moreover the magnetic moment of Ni^{2+} ion also disturbs the geometric frustration, since the magnetic Ni^{2+}-O-Cr^{3+} interaction usually being stronger than the frustrated interaction between the Cr^{3+} ions. We have already reported the study of frustration induced the crystal structure change and the magnetic order on $ZnCr_2O_4$ and $Zn_{0.5}Ni_{0.5}Cr_2O_4$ [3] and also on $Zn_{1-x}Ni_xCr_2O_4$ for x = 1, 0.8 and 0.6 [4]. With the magnetic measurements by SQUID for $Zn_{1-x}Ni_xCr_2O_4$ for x = 1, 0.8 and 0.6, we will discuss the geometric frustration system of $Zn_{1-x}Ni_xCr_2O_4$ from the new point of view based on dynamical spin fluctuation in the hexamer and heptomer.

2. Experiment

a) Sample preparation

The single crystals of $ZnCr_2O_4$ and $Zn_{0.5}Ni_{0.5}Cr_2O_4$ were grown with a Bi_2O_3-flux in Platinum crucibles of 1.5 liters volume. The cooling rate was 0.5°C/hr. For the powder specimens for the powder x-ray diffraction, the single crystals were grinded. The powder samples $Ni_xZn_{1-x}Cr_2O_4$ with x = 0.6, 0.8 and 1 were prepared from the raw material powders of ZnO, NiO, and Cr_2O_3. The powders were mixed in appropriate proportions in an agate mortar under acetone for 3 hrs. The mixtures were then pre-sintered at 800^0 C for 5 hrs. Resultant mixtures were reground and hand milled for 3 hrs under acetone. Pressing the mixture into pellet and sintering at 1200^0C for 5 hrs. Powder x-ray diffraction (XRD) pattern at room temperature confirmed that the final product has a cubic-type spinel single phase for 0.6 and 0.8 and tetragonal spinel single phase for x = 1. Also XRD indicated no impurities for the samples x = 0.6 and 0.8. But the XRD for the sample $NiCr_2O_4$ indicated impurity phases, this is due to NiO_2 or Cr_2O_3 and Cu.

IOP Conf. Series: Journal of Physics: Conf. Series **807** (2017) 042004 doi:10.1088/1742-6596/807/4/042004

b) Magnetic measurement

Magnetic susceptibility was measured between 1.8 and 300 K by a SQUID magnetometer MPMS system (Quantum Design Co. Ltd.). Magnetization vs magnetic field was also measured up to 7 Tesla at several temperatures.

c) Low temperature x-ray diffraction

Two different types of cooling systems were used in our low temperature x-ray diffraction experiments. Above about 10 K, a ^{4}He circulating cryocooler was used and below about 10 K, a ^{3}He-^{4}He dilution refrigerator (D.R.) was used. X-ray diffraction measurements for powder specimens were performed using the RINT 2500 system, Rigaku Co. From the observed diffraction profile the lattice constant d, the integrated intensity (I.I.) and the full width at half maximum (FWHM) were obtained.

d) Debye-Waller Factor

The temperature dependence of the I.I. can be obtained by the Debye-Waller factor. (see; for example, C. Kittel; Introduction to Solid State Physics, John Wiley & Sons, Inc. Sixth edition, Appendix A.)

When a x-ray beam is scattered by a crystal: let the position of atom nominally at r_j contain a term u(t) = r_j + u(t). We suppose each atom fluctuates independently about its own equilibrium position. Then the scattered intensity of the x-ray beam is

$$I = I_0 \exp(-1/3\langle u^2\rangle G^2), \qquad (1)$$

where I_0 is the scattered intensity from the rigid lattice and G is the reciprocal lattice vector. When we assume a classical harmonic oscillator for a crystal atom, then,

$$M\omega^2\langle u^2\rangle = 3k_BT, \qquad (2)$$

where M is the mass of the atom and ω is the frequency of the harmonic oscillator, then

$$I\,(hkl) = I_0 \exp(-k_BTG^2/M\omega^2). \qquad (3)$$

For quantum oscillators $\langle u^2\rangle$ does not vanish even at T = 0; there is zero point motion. The zero point energy is $3/2\hbar\omega$, so that in the ground state

$$\langle u^2\rangle = 3\,\hbar/2M\omega\,, \qquad (4)$$

by (1),

$$I\,(hkl) = I_0 \exp(-\hbar G^2/2M\omega) \qquad (5)$$

at absolute zero. The exponential factor is the Debye-Waller factor. At higher temperatures, the intensity of the x-ray scattering spectrum depends on the temperature following eq. (3). When the temperature is low enough, the intensity gives the constant value due to the eq. (5). But when the frequency of the lattice fluctuation changes, such as the softening of the lattice, that is, ω becomes low, the intensity is going to decrease following eq. (5).

3. Experimental results and Discussion

Experimental results for $ZnCr_2O_4$ and $Zn_{1-x}Ni_xCr_2O_4$ have been reported by some of the present authors [3]. Magnetic properties for $NiCr_2O_4$ have been also reported by some groups [5,6]. The temperature dependences of the magnetic susceptibility were shown in Fig. 1a for $Zn_{0.2}Ni_{0.8}Cr_2O_4$ and in Fig. 1b for $Zn_{0.4}Ni_{0.6}Cr_2O_4$. The Neel temperature T_N can be obtained as the maximum of the magnetic susceptibility. Though the shape of the maximum is different between these two compounds, but the temperature is nearly same as 30 K. All the values of T_N were shown in Table 1 for each compound. From these figures the inverse of the magnetic susceptibility is plotted against temperature for $Zn_{0.2}Ni_{0.8}Cr_2O_4$ and $Zn_{0.4}Ni_{0.6}Cr_2O_4$. From these figures the asymptotic Curie temperature θ_{CW} are obtained. The θ_{CW} in whole $Zn_{1-x}Ni_xCr_2O_4$ system is listed together with the data for other previously reported compounds. The θ_{CW} gives the strength of the antiferromagnetic interaction. Though the antiferromagnetic interaction is rather strong, that is, $|\theta_{CW}|$ is rather high such as from 388 to 870, the magnetic ordering is low due to the geomagnetic frustration.

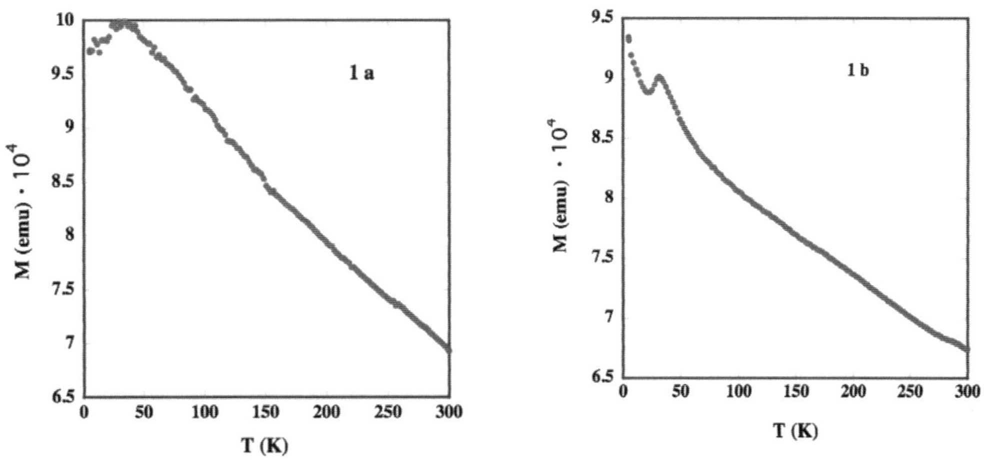

Figure 1. Temperature dependence of the magnetization , a) for $Zn_{0.2}Ni_{0.8}Cr_2O_4$ and b) for $Zn_{0.4}Ni_{0.6}Cr_2O_4$.

The frustration index (F.I.) defined as $|\theta_{CW}/T_N|$ is also listed in Table 1. Though the strongest antiferromagnetic interaction can be observed at around $Zn_{0.4}Ni_{0.6}Cr_2O_4$, the F.I. changes rather smoothly against Ni concentration x as shown in Fig. 2 .

	θ_{CW} (K)	T_N(K)	F.I. ($\lvert\theta_{CW}/T_N\rvert$)
$NiCr_2O_4$	- 470	65	7.2
$Zn_{0.2}Ni_{0.8}Cr_2O_4$	- 510	30	17
$Zn_{0.4}Ni_{0.6}Cr_2O_4$	- 870	30	29
$Zn_{0.5}Ni_{0.5}Cr_2O_4$	-520	16.5	31.5
$ZnCr_2O_4$	-388	12	32.33

Table. 1. List of asymptotic Curie temperature θ_{CW} and Neel temperature T_N. Frustration Index (F.I.) is also calculated and listed.

In $ZnCr_2O_4$, to remove the geometric frustrated degenerated state, the spin-Piels-like crystal distortion occurs at about T_S = 12K accompanying with an antiferromagnetic ordering. In the quantum critical spin-1/2 antiferromagnetic chain, small symmetry breaking perturbations lead to the spin-Peiels transition, that is, a lattice distortion that lowers the energy of the spin system by introducing a gap in the magnetic excitation spectrum. In the case of ZnCr2O4, the spin lattice is not a linear chain but a triangle lattice. But the situation is the same. In the tetragonal phase below Ts, the ratio of the lattice constants c/a – 1 > 0.

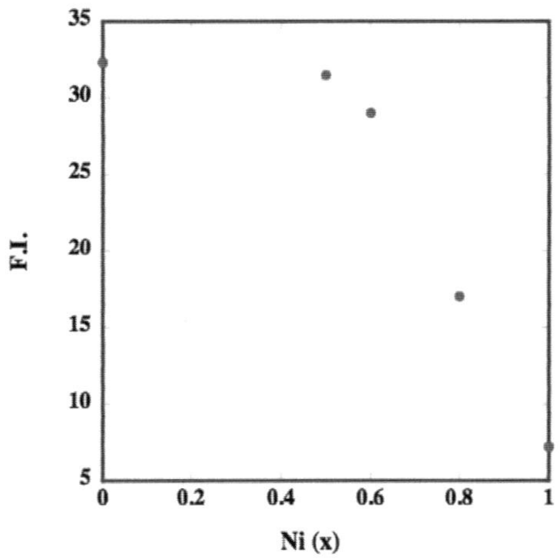

Figure 2. Frustration index against Ni concentration.

On the other hand, in $NiCr_2O_4$ we also observed the ratio c/a - 1 < 0 below Ts = 310 K. So at which composition of $Zn_{1-x}Ni_xCr_2O_4$ the ratio c/a – 1 changes its sign is interesting. In Fig. 3 it is shown that c/a is larger than 1 in $Zn_{0.2}Ni_{0.8}Cr_2O_4$. The crystal structure changes below Ts = 160 K. In $Zn_{0.4}Ni_{0.6}Cr_2O_4$ compound, the temperature dependence of the d value for (440) reflection is shown in Fig.4. Due to the inhomogeneity of the Zn and Ni atoms, the clear split of the reflection spectrum in x-ray diffraction cannot be observed. Only a broad spectrum can be observed. The (440) spectrum in cubic phase should be split into two peaks, (440) and (404), (044). When we assume the equivalent domain structure below tetragonal distortion, the peak of (404), (044) should have twice intensity than the (440) peak. So the temperature dependence of the (440) spectrum in cubic phase can be mostly determined in tetragonal

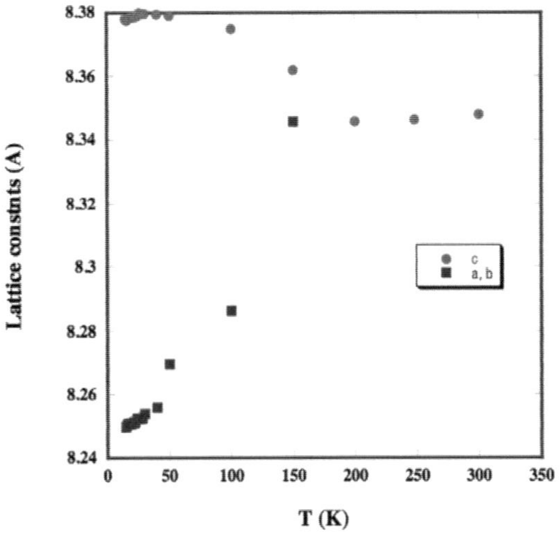

Figure 3. Temperature variation of the lattice constants of $Zn_{0.2}Ni_{0.8}Cr_2O_4$.

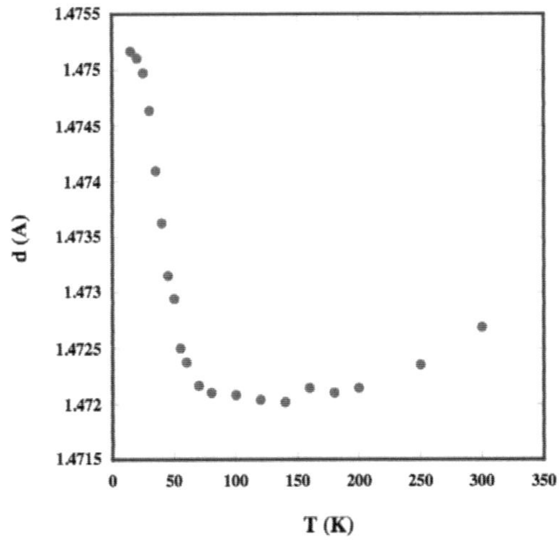

Figure 4. Lattice spacing against temperature for $Zn_{0.4}Ni_{0.6}Cr_2O_4$.

phase by the (404), (044) peak. Fig. 4 tells us lattice constant c is larger than a, b axis. In $Zn_{0.5}Ni_{0.5}Cr_2O_4$, the split of the spectrum below Ts cannot be also observed. But since the temperature dependence of the d value is rather small at around the transition temperature, it seems that $c/a > 1$ and $c/a < 1$ competed to each other. So the value of (c/a -1) changes its sign at around Ni concentration 0.5.

Now we would like to discuss the fluctuation of the lattice from the experimental results of the I.I. of x-ray spectrum. As for the typical result the I.I. in $Zn_{0.2}Ni_{0.8}Cr_2O_4$ compound is shown in Fig. 5. $Zn_{0.2}Ni_{0.8}Cr_2O_4$ shows the crystal distortion at 160 K and the magnetic order around 30 K. The temperature, 160 K is rather high, so the softening due to the crystal structure change should be rather small. In the figure the hump around 150 K can be seen. In $Zn_{0.2}Ni_{0.8}Cr_2O_4$, the structure change is due to the Jahn-Teller effect of Ni^{2+}.

With decreasing temperature magnetic transition accompanying with the crystal distortion occurs at 30K. So the I.I. for decreases rather clearly don to about 30 K. At about 16 K, there is another small anomaly which can correspond to another magnetic transition. In $NiCr_2O_4$, the J.T. distortion occurs at 310 K and the ferrimagnetic transition accompanying with the crystal distortion at 65 K and another magnetic transition around 30 K. Then as shown in Fig. 6 the I.I. first starts from the bottom, then increases following to the temperature dependence of the Debye-Waller factor, eq. (3). Then the I.I. decreases down to

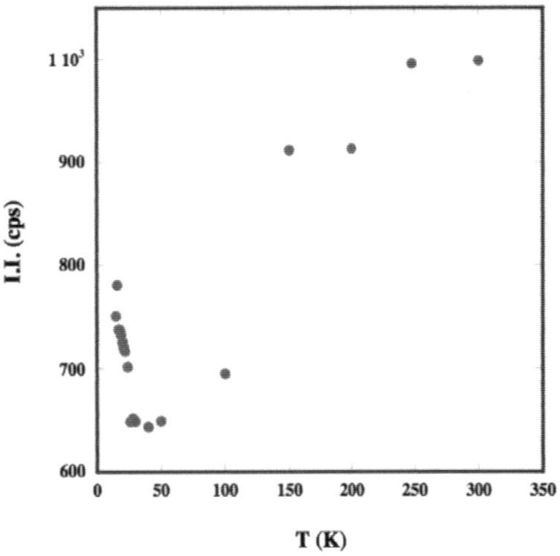

Figure 5. I.I. against T for $Zn_{0.2}Ni_{0.8}Cr_2O_4$.

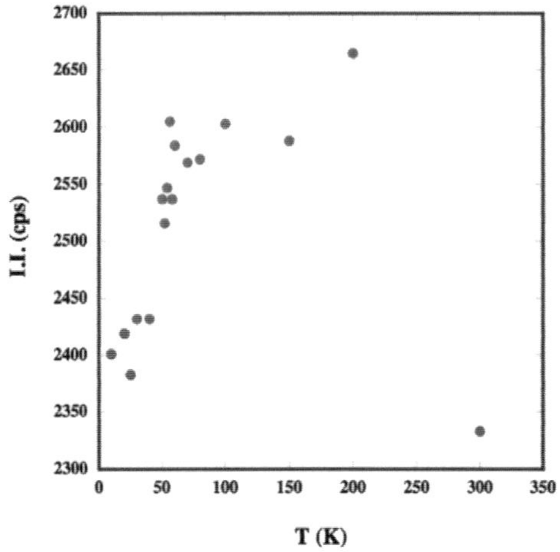

Figure 6. I.I. against T for $NiCr_2O_4$.

about 60 K. In the vicinity of that temperature the I.I. shows the steep peak and then decreases down to about 30 K. In $ZnCr_2O_4$ the geometric frustration induces the crystal structure change accompanying with the magnetic order at about 12 K. The temperature dependence of the I.I. of $ZnCr_2O_4$ compound clearly shows this phase transition as shown in Fig. 7. From these results it can be said that in $Zn_{1-x}Ni_xCr_2O_4$ system the magneto- structure coupling is rather strong. Each magnetic transition corresponds to the anomaly of the I.I. of the x-ray diffraction spectrum.

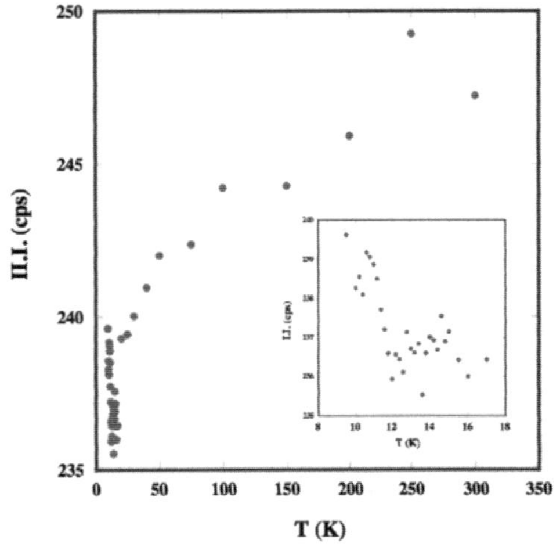

Figure 7. I.I. against T for $ZnCr_2O_4$. The intersect shows low temperature part in the expanded scale.

Neutron inelastic scattering experiments reported the dynamical spin fluctuation modes exist in the paramagnetic and also in the magnetic ordered and the distorted structure phases in the spin molecule hexamer and heptamer. At present experiments we cannot distinguish these mode in our x-ray experiment.

References

[1] Lee S,-H, Broholm C, Kim T H, Ratcliff II W and Cheong S-W, Phys. Rev. Lett. 84 (2000) p.3718.

[2] Tomiyasu K, Hiraka H, Ohyama K and Yamada K, J. Phys. Soc. Jpn. 77 (2008) p.124703-1.

[3] Xue Y, Naher S, Hata F, Kaneko H, Suzuki H and Kino Y, J. Low Temp. Phys. 151 (2008) p.1193.

[4] Khan A, Kaneko H, Suzuki H, Naher S, Ahsan M H, Islam M A, Basith M A, Alam H M and Saha H M. SpringerPlus (2015) 4:468, DOI 10.1186/s40064-015-1224-z.

[5] Klemme S and Miltenburg J C, Phys. Chem. Miner. 29 (2002) p.663.

[6] Tomiyasu K and Kagomiya I, J. Phys. Soc. Jpn. 73 (2004) p.2539.

Magnetic transitions on the pseudo-ternary compounds $Ho_{1-x}Y_xRh_2Si_2$

T Shigeoka[1], T Fujiwara[1], Y Uwatoko[2]

[1]Graduate School of Science and Engineering, Yamaguchi University, Yamaguchi 753-8512, Japan.
[2]Institute for Solid State Physics, University of Tokyo, Kashiwa-shi Chiba 277-8581, Japan.

shigeoka@yamaguchi-u.ac.jp

Abstract. Magnetic measurements on the pseudo-ternary compounds $Ho_{1-x}Y_xRh_2Si_2$, which substitute nonmagnetic Y for Ho, were performed. The lattice parameters a and c are almost constant by the change of Y composition x. The transition temperatures T_{N1}, T_t, and T_{N2} decrease with increasing x. The critical compositions for T_{N1} and T_{N2} are determined to be x_{N1}=0.98 and x_{N2}=0.75, respectively. "The successive component-separated magnetic transitions" appear for the wide x region from x=0 to x=0.75. The partial ordered state, or frustration appears on these compound. The magnetic ordered state persists up to x=0.98 in spite of very dilute Ho-compounds: very weak magnetic interactions. The effective magnetic moments are almost constant from x=0 to 1; μ_{eff}=10.6 $\pm$0.15 μ_B/Ho, and magnetic moment μ=10 μ_B/Ho as well. The $Ho_{0.95}Y_{0.05}Rh_2Si_2$ single crystal shows the quite similar magnetic behaviour to that of $HoRh_2Si_2$. The constant magnetic moment and constant effective magnetic moment irrespective of x are confirmed from the magnetic behaviours of single crystals $Ho_{1-x}Y_xRh_2Si_2$ (x=0.05, 0.6 and 0.8).

1. Introduction

The ternary compound $HoRh_2Si_2$, which crystallizes in the tetragonal $ThCr_2Si_2$-type structure, shows so called "successive component-separated magnetic transitions"; the c- and ab-components of magnetic moments independently order at the different temperatures, T_{N1}=29.1 K and T_{N2}=12.1 K [1]. So the partial ordered state, where only the c-component of magnetic moments orders and the ab-components disorder, appears for $T_{N1}>T> T_{N2}$, This type of transitions often appears in frustration systems. The geometrical frustration, which comes from a competition of multiple interactions due to a geometrical ion arrangement, is responsible for most of frustrations [2, 3]. The geometrical frustration is, however, not like to occur in this compound due to the structure. The origin is very interesting. This compound has an additional transition at T_t=27.3 K [1]. The simple antiferromagnetic structure which is the propagation vector $\boldsymbol{k}$ = (0, 0, 1) with magnetic moments tilted from the c-axis by 28° at low temperatures has been reported from the neutron study [4]. Magnetic studies on the single crystal show complex magnetic behaviors [5, 6]; step-like metamagnetic processes appear along the symmetry directions below T_{N2}. For $T_{N2}<T<T_t$, a two-step metamagnetic one persists along the [001] direction whereas the processes along directions in the basal plane become a paramagnetic like one, suggesting that the ab-components of magnetic moments are disordered, frustrated. The B_{100}-T and B_{110}-T magnetic phase diagrams show the existence of peculiar phase boundary around T_{N2}, suggesting that a quadrupole effect play an important role for the component-separated magnetic transition. The

pseudo-ternary compounds $Ho_{1-x}Gd_xRh_2Si_2$, which substitutes Gd with no quadrupole for Ho, have been studied; the quadrupole effects on the transitions were investigated. The results declare that quadrupole interactions play an important role in this compound system [7]. In order to obtain more information on an origin of the successive component-separated magnetic transitions, magnetic behaviors of pseudo-ternary compounds $Ho_{1-x}Y_xRh_2Si_2$, which substitute non-magnetic Y for Ho, were studied. The ionic radiuses of Y and Ho ions are 1.06 Å and 1.05 Å, respectively; the ionic radiuses of Y and Ho are almost same. So it should be expected that effects of magnetic interactions on the magnetic transitions can be studied without a deformation of the lattice on $Ho_{1-x}Y_xRh_2Si_2$.

2. Experimental

Polycrystalline samples of pseudo-ternary compounds $Ho_{1-x}Y_xRh_2Si_2$ were prepared by arc-melting a stoichiometric mixture of the pure elements (Ho and Y: purity of 3N, Rh: 3N, and Si: 5N) in an argon atmosphere. Subsequently, single crystals (x = 0.05, 0.6 and 0.8) were grown by the Czochralski technique using a tri-arc furnace; the polycrystalline sample was melted on a water-cooled copper hearth in the tri-arc furnace. A crystal was slowly pulled from the melt by a pulling-up rod with a seed crystal. The pulling-up rate was 5 *mm*/*h* [8]. The crystal structure and the single phase nature were confirmed by X-ray powder diffraction. The quality of the single crystals was checked and crystallographic orientations were determined by the back Laue method. The magnetic susceptibility and magnetization were measured using a PPMS and/or MPMS (Quantum Design) in the Institute for Solid State Physic, the University of Tokyo.

3. Results and discussion

3.1. Lattice parameter

The lattice parameters a and c of the tetragonal cell on the $Ho_{1-x}Y_xRh_2Si_2$ were determined by the least square analysis to X-ray diffraction powder patterns for polycrystalline samples. The Y composition x dependences of those are shown in Fig. 1. For $HoRh_2Si_2$ (x = 0) a = 4.026 Å, c = 9.899 Å, and a = 4.033 Å, c = 9.917 Å for YRh_2Si_2 (x = 1), which are agreement with the previous reports [9, 10]. The rate of change in both a and c from x = 0 to 1 is 0.18 %, and is very slight increase; no lattice deformation occurs. It is expected that the crystal field effects in this system don't change by this substitution.

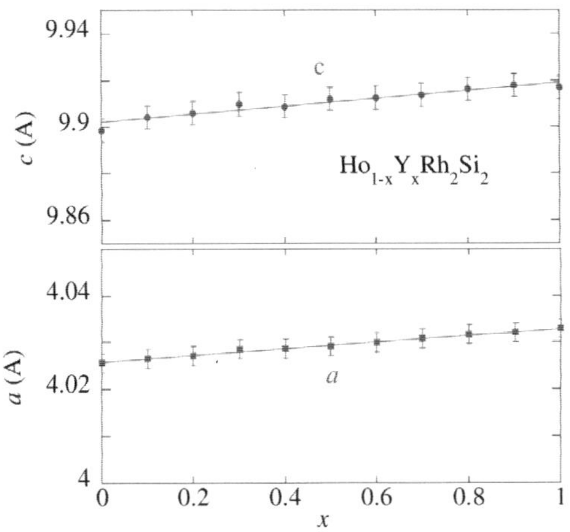

Figure. 1 Y composition x dependence of lattice parameters a and c on the polycrystalline compounds $Ho_{1-x}Y_xRh_2Si_2$.

3.2. Magnetic susceptibility

The temperature dependences of magnetic susceptibility along the main symmetry axes on the single crystals $Ho_{1-x}Y_xRh_2Si_2$ (x = 0.05, 0.6 and 0.8) are shown in Fig. 2, Fig. 3 and Fig. 4, respectively. A magnetic anisotropy between the c-axis and directions in the basal plane is evidenced from the figure which indicates the easy magnetization direction is the [001] direction. In Fig. 2 for $Ho_{0.95}Y_{0.05}Rh_2Si_2$ (x=0.05), the [001] magnetic susceptibility $\chi_{[001]}$ shows a clear maximum at T_{N1}=28.6 K and a small hump at T_{N2}=11.4 K. In contrast, susceptibilities $\chi_{[100]}$ and $\chi_{[110]}$ for directions in the basal plane show a small hump at T_{N1} and a clear maximum at T_{N2}. A small kink at T_t=25.5 K is also observed. This feature is similar to that of $HoRh_2Si_2$, although the transition temperatures become lower than those of $HoRh_2Si_2$ [1]; the successive component-separated magnetic transitions appear. The susceptibilities obey the Curie-Weiss law above 50 K. The χ^{-1}-T curves become linear and are almost parallel with each other at high temperatures. The effective magnetic moment estimated is 10.6 μ_B/Ho on the

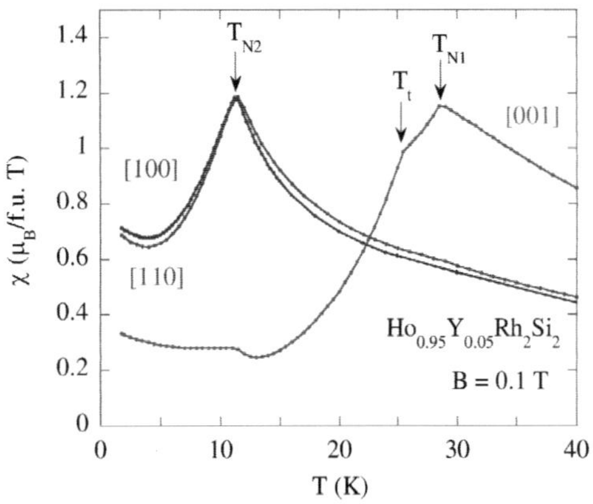

Figure. 2 Magnetic susceptibility along the main symmetry axes of tetragonal cell on the $Ho_{0.95}Y_{0.05}Rh_2Si_2$ (x=0.05) single crystal.

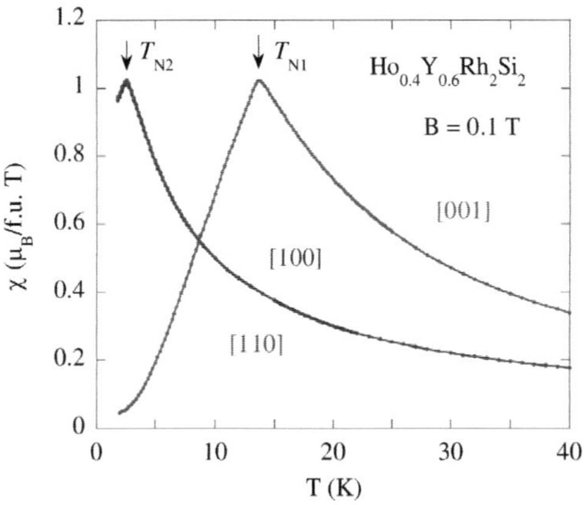

Figure. 3 Magnetic susceptibility along the main symmetry axes of tetragonal cell on the $Ho_{0.4}Y_{0.6}Rh_2Si_2$ (x=0.6) single crystal.

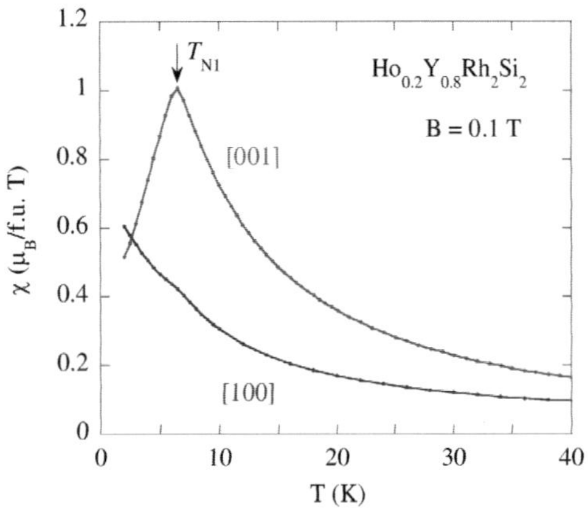

Figure. 4 Magnetic susceptibility along the main symmetry axes of tetragonal cell on the $Ho_{0.2}Y_{0.8}Rh_2Si_2$ (x=0.8) single crystal.

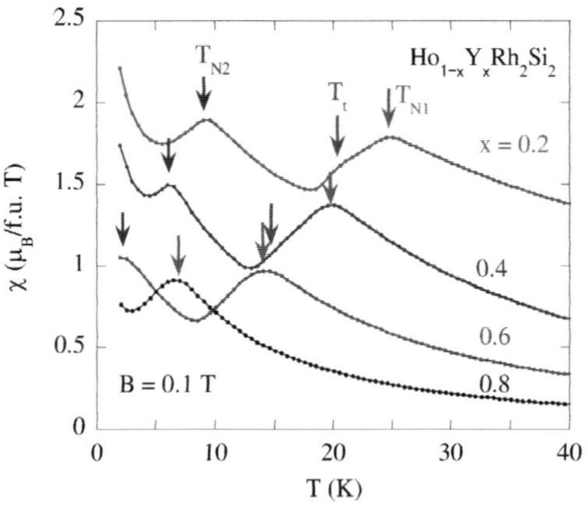

Figure. 5 Temperature dependence of magnetic susceptibility on the polycrystalline compounds $Ho_{1-x}Y_xRh_2Si_2$ (x =0.2, 0.4, 0.6 and 0.8)

compound which is in good agreement with the Ho^{3+} free ion magnetic moment ($g[J(J+1)]^{1/2}$=10.6 μ_B). This shows that 4*f* electrons in this compound are well localized and Rh ion is nonmagnetic. The paramagnetic Curie temperatures along the c-axis and [100], [110] directions in the basal plane are Θ_{001}=18.5 K, Θ_{100}=-5.4 K and Θ_{110}=-7.6 K, respectively, indicating that the dominant interaction along the c-axis and in the basal plane is ferromagnetic and antiferromagnetic, respectively. With increasing *x*, the transitions shift to lower temperatures; for *x*=0.6, T_{N1}=13.6 K and T_{N2}=2.5 K, and for *x*=0.8, T_{N1}=6.4 K and T_{N2}<2 K. The transition of T_t is hardly observed in the figures 3 and 4.

Figure 5 shows temperature dependences of magnetic susceptibility on the polycrystalline compounds $Ho_{1-x}Y_xRh_2Si_2$ (*x* =0.2, 0.4, 0.6 and 0.8). It can be confirmed that the larger *x* make the lower transition temperatures. The magnetic susceptibilities of all the compounds obey the Curie-Weiss law for high temperatures. The effective magnetic moments per Ho atom estimated are almost constant irrespective of change in *x*: μ_{eff}=10.6 ±0.15 μ_B/Ho which is good agreement with the Ho^{3+} theoretical moment. This means that 4*f* electrons in this compound system are well localized and Rh ion is nonmagnetic in this compound system.

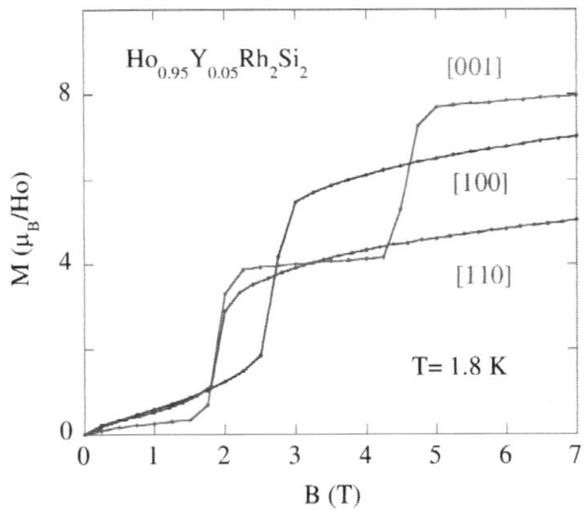

Figure. 6 Magnetization curves at 1.8 K along the main symmetry axes on the $Ho_{0.95}Y_{0.05}Rh_2Si_2$ (*x*=0.05) single crystal.

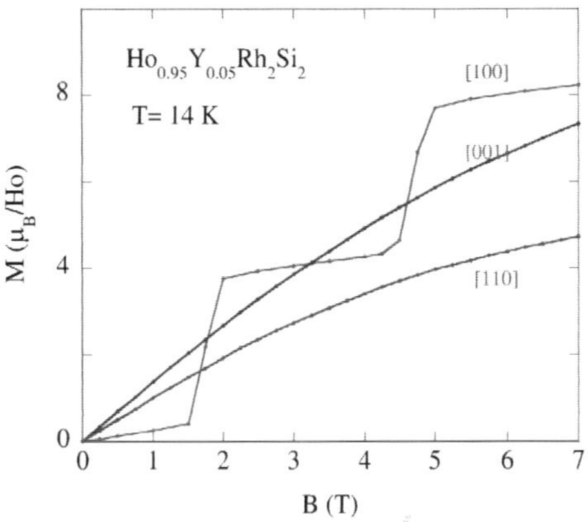

Figure. 7 Magnetization curves at 14 K along the main symmetry axes on the $Ho_{0.95}Y_{0.05}Rh_2Si_2$ (*x*=0.05) single crystal.

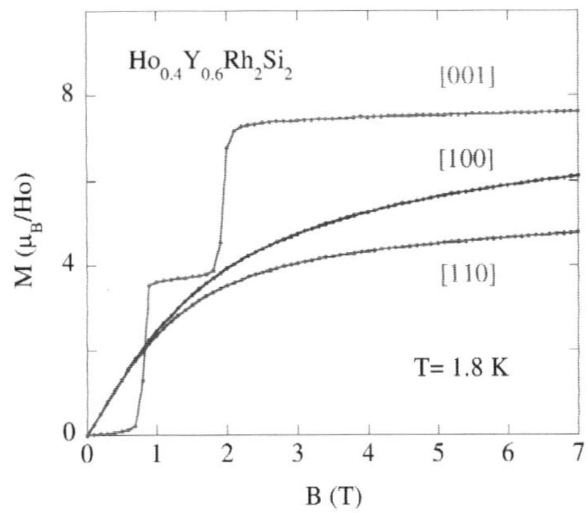

Figure. 8 Magnetization curves at 1.8 K along the main symmetry axes on the $Ho_{0.4}Y_{0.6}Rh_2Si_2$ (*x*=0.6) single crystal.

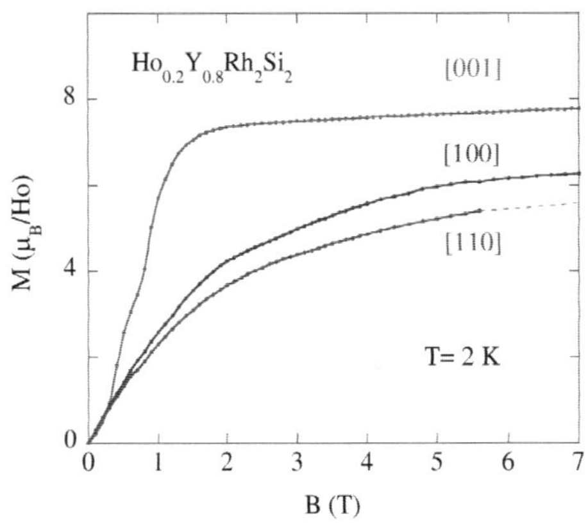

Figure. 9 Magnetization curves at 2 K along the main symmetry axes on the $Ho_{0.2}Y_{0.8}Rh_2Si_2$ (*x*=0.8) single crystal.

3.3. Magnetization process

In order to elucidate more detail magnetic behaviour by a change of x, magnetization processes of some single crystal compounds (x=0.05, 0.6 and 0.8) were studied. Magnetization curves at 1.8 K is shown on the $Ho_{0.95}Y_{0.05}Rh_2Si_2$ single crystal (x=0.05) in Fig.6. Sharp step-like metamagnetic transitions appear along the all symmetry directions. This behaviour is similar to that of $HoRh_2Si_2$ [5], indicating that magnetic moments tilt from the main symmetry axes. The [001] magnetization process, easy axis magnetization one is a two-step metamagnetic one; the magnetization increases rapidly at critical fields B_{c1}=1.9 T and B_{c2}=4.6 T. The magnetization curve is almost linear with a slight slope except around transition fields. The magnetization just above B_{c1} and B_{c2} is M_1=4.0 μ_B/Ho and M_2=8.0 μ_B/Ho, respectively. The M_2 is two times as large as M_1. If a Ho atom in this compound has the theoretical Ho^{3+} free ion moment of gJ=10 μ_B, the tilt angle of magnetic moments from the c-axis can be deduced from M_2 to be 37°. This tilt angle should become 0 ° and the magnetization should saturate 10 μ_B/Ho by higher magnetic fields as the case of $HoRh_2Si_2$ [6]. A clear step-like metamagnetic transition also appears in the basal plan processes. The critical fields are B_{100} =2.6 T and B_{110}=1.9 T in the [100] and [110] processes, respectively. The magnetization just above the transition B_{100} and B_{110} is M_{100}=5.7 μ_B/Ho and M_{110}=3.5 μ_B/Ho, respectively. The higher temperature behaviours shown in Fig. 7 are also quite similar to that of $HoRh_2Si_2$; the two-step metamagnetic process along the [001] direction persists up to T_t whereas the basal plan processes become a paramagnetic one above T_{N2}. This indicates the partial ordered state, where the magnetic moments of *c*-component order while *ab*-components disorder, appears for $T_{N2}<T<T_t$. Figures 8 and 9 show the magnetization curves at 1.8 K and 2 K on the $Ho_{1-x}Y_xRh_2Si_2$ (x=0.6 and 0.8) single crystals, respectively. A two-step metamagnetic transition appears in the [001] process, and paramagnetic like processes appear in the basal plane on the both compounds. These behaviours are similar to that in Fig. 8. They are corresponding to those in the partial ordered state. The critical fields are B_{c1}=0.8 T and B_{c2}=1.9 T for x=0.6, and B_{c1}=0.3 T and B_{c2}=0.8 T for x=0.8. The larger x, weaker magnetic interactions, make lower critical fields, which is reasonable because it should be expected that the antiferromagnetic interactions become smaller with decreasing magnetic Ho. The saturation magnetization (at B=7 T) is 7.7 μ_B/Ho and 7.8 μ_B/Ho for x=0.6 and x=0.8, respectively, which is almost same to the moment of x=0 and x=0.05 around 7 T. It should be speculated the saturation magnetic moment reaches 10 μ_B as same to that of $HoRh_2Si_2$. This means that the Ho magnetic moment keeps the theoretical Ho^{3+} magnetic moment (gJ=10 μ_B) irrespective of x, in the diluted Ho composition compounds as well.

3.4. Composition dependence of transition temperature

The Y composition x dependences of magnetic transition temperatures T_{N1}, T_t, and T_{N2} on the $Ho_{1-x}Y_xRh_2Si_2$ compounds are shown in Fig. 10. The transitions were determined from measurements of χ-T above mentioned. The magnetic ordered state is divided to 3 states by the transition lines, which are

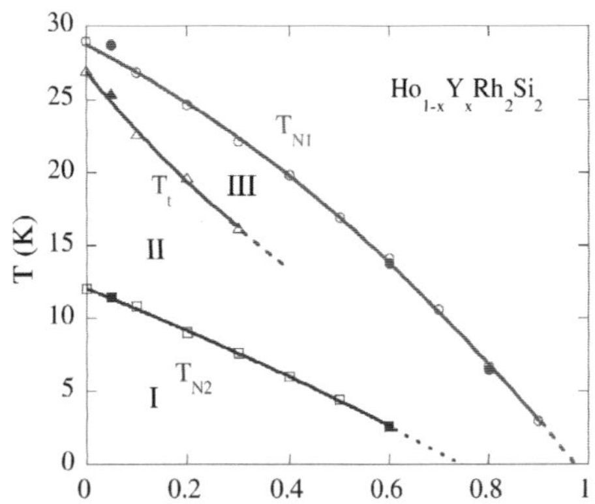

Figure. 10 Y composition x dependences of magnetic transition temperatures T_{N1}, T_t, and T_{N2} on the $Ho_{1-x}Y_xRh_2Si_2$ compounds. Closed and open symbols are transition temperatures determined from single crystal and polycrystalline samples, respectively. The lines are drawn by fitting to quadratic functions.

referred to as phase I, II, and III in the order of increasing temperature as shown in the figure. The phase I has a simple antiferromagnetic structure with the propagation vector $\boldsymbol{k}$ = (0, 0, 1) and magnetic moments tilted from the *c*-axis [4]. The phase II and III are the partial ordered states as reported previously [5]. The transition temperatures T_{N1}, T_t, and T_{N2} decrease with increasing x. The tendency is consistent with the fact that the magnetic interactions become weaker with decreasing Ho composition (increasing x: nonmagnetic Y) in this compound system. The lines in the figure are drawn by fitting to quadratic functions. The dependencies well obey quadratic functions. The curves of T_{N1} and T_{N2} have upward curvatures while one of T_t has a downward curvature. This suggests that the main origin which induces the transitions of T_{N1} and T_{N2} is different from that of T_t. The critical composition, which is a composition where a transition temperature becomes T=0 K, for T_{N1} and T_{N2} is determined to be x_{N1}=0.98 and x_{N2}=0.75, respectively by extrapolating the quadratic functions to T=0. It should be worth noting that the ordered state persists in spite of very dilute Ho compounds: only a few percent Ho-compound orders. The successive component-separated magnetic transitions appear for wide x range up to x_{N2}=0.75. The x-dependency of a magnetic transition should have a downward curvature if it follows the de Gennes factor: $(g\text{-}1)^2J(J+1)$. The x-dependency of T_t is not scaled by de Gennes factor though it has an upward curvature. The RKKY interaction may take an important part for the transition of T_t.

3.5. Summery

Magnetic studies on the pseudo-ternary compounds $Ho_{1-x}Y_xRh_2Si_2$, which substitute non-magnetic Y for Ho, were performed using either single crystal (x= 0.05, 0.6 and 0.8) or polycrystalline (x= 0 ~1 every 0.1) compounds. The lattice parameters *a* and *c* are almost constant by the change of Y composition x. The effects of magnetic interactions on the magnetic transitions can be studied without a change in the crystal field effects on this compound system. The single crystal compound $Ho_{0.95}Y_{0.05}Rh_2Si_2$ (x= 0.05) exhibit quite similar magnetic features to those of $HoRh_2Si_2$. It shows "successive component-separated magnetic transitions" and step-like metamagnetic transitions along all the symmetry directions at low temperatures. The metamagnetic transitions in the basal plane disappear for T>T_{N2} for the phase II: the partial ordered state. The similar behaviours appear on the single crystals $Ho_{1-x}Y_xRh_2Si_2$ (x=0.6 and 0.8) at low temperatures as well. So these compounds is in the phase II at low temperatures. The magnetic moment per Ho atom of the compounds (x=0.95, 0.6, 0.4) is same to that of $HoRh_2Si_2$; a Ho magnetic moment and effective magnetic moment keep constant in the diluted Ho compounds. The partial ordered state, or frustration appears on these compound. "The successive component-separated magnetic transitions" appear in this system as well. The Y composition x dependence of magnetic transitions were determined. The transition temperatures T_{N1}, T_t, and T_{N2} decrease with increasing x. The critical compositions, where the transition temperatures of T_{N1} and T_{N2} become zero, are determined to be x_{N1}=0.98 and x_{N2}=0.75, respectively. It should be noted that the ordered state persists in spite of very dilute Ho region: only a few percent Ho-compound orders magnetically. The effective magnetic moments are almost constant from x=0 to 1; μ_{eff}=10.6 $\pm$0.15 μ_B/Ho which is good agreement with the Ho^{3+} theoretical moment. This means that 4*f* electrons in this compound system are well localized and Rh ion is nonmagnetic. The successive component-separated magnetic transitions appear for wide x range up to x_{N2}=0.75. The x-dependency of T_t indicates the RKKY interaction may take a part of the cause for the transition of T_t. In spite of decreasing magnetic interactions, the partial ordered state is stable. The magnetic interaction is not main cause for the appearance of the partial ordered state, the frustration. Other effects should be investigated.

Acknowledgment

This work was carried out under the Visiting Researcher's Program of the Institute for Solid State Physics, the University of Tokyo. This work was supported by JSPS KAKENHI Grant Number 15H03681.

References

[1] Shigeoka T, Fujiwara T, Munakata K, Matsubayashi K, Uwatoko Y 2011 *J. Phys. Conf. Ser.* **273** 012127.
[2] Achiwa A 1969 *J. Phys. Soc. Jpn.* **27** 561.
[3] Watanuki R, Sato G, Suzuki K, Ishihara M, Yanagisawa T, Nemoto Y, Goto T 2005 *J. Phys. Soc. Jpn.***74** 2169.
[4] Felner I, Nowik I 1984 *J. Phys. Chem. Solids* **45** 419.
[5] Shigeoka T, Fujiwara T, Matsubayashi K, Uwatoko Y 2012 *J. Phys. Conf. Ser.* **391** 012063.
[6] Shigeoka T, Fujiwara T, Kondo A, Kindo K, Uwatoko Y 2014 *J.Phys. Soc. Jpn. Conf. Ser.* **3** 017005.
[7] Shigeoka T, Morita T, Fujiwara T, Matsubayashi K, Uwatoko Y 2015 *Phys. Proc.* **66** 835..
[8] Menovsky A, Snel C.F, Gortenmulder T. J, Tan H. J and Plastra T.T.L 1986 *J. Crystal Growth* **74 231.**
[9] Felner I, Nowik I 1983 *Solid StateCommun.* **47** 831.
[10] Sekizawa K, Takano Y, Takigami H, Takahashi Y 1987 *J. Less-Common Metals* **127** 99.

IOP Conf. Series: Journal of Physics: Conf. Series **807** (2017) 042006 doi:10.1088/1742-6596/807/4/042006

Single crystal growth and magnetic properties of pseudo-kagome lattice RRhPb (R = Nd, Sm and Gd)

Y Matsumoto[1], R Goto[1], Y Haga[2], Z Fisk[2,3], and S Ohara[1]

[1]Department of Engineering Physics, Electronics and Mechanics, Graduate school of Engineering, Nagoya Institute of Technology, Nagoya, Aichi 466-8555, Japan
[2]Advanced Science Research Center, Japan Atomic Energy Agency, Tokai, Ibaraki 319-1195, Japan
[3]Department of Physics and Astronomy, University of California, Irvine, California 92697, USA

E-mail: matsumoto.yuji@nitech.ac.jp

Abstract.
We have synthesized single crystals of RRhPb (R = Nd, Sm and Gd) with hexagonal ZrNiAl-type structure from Pb-flux. The crystal structures were confirmed by powder X-ray diffraction and the compositions were determined by electron-probe micro-analyzer (EPMA). We have measured their magnetic properties of RRhPb. It is found that the RRhPb (R = Nd, Sm and Gd) are antiferromagnets with two successive phase transitions with magnetic ordering occuring at T_{N1} = 3.6 K and T_{N2} = 3.4 K in NdRhPb, T_{N1} = 11.5 K and T_{N2} = 8.3 K in SmRhPb and T_{N1} = 17.6 K and T_{N2} = 15.3 K in GdRhPb.

1. Introduction

Rare earth and actinide intermetallic compounds with the hexagonal ZrNiAl-type structure (RTX : R=rare earth, actinide, T=transition metal, X=Al, Ga, In, Sn and Pb) are one of the extensively studied systems [1, 2]. The R-T(1) layer and T(2)-X layer are stacked along c-axis as shown in Fig. 1. The R ion of the R-T(1) layer coordinates a psudo-kagome lattice possessing geometrical frustration. The typical 120° array of magnetic moments observed in TbNiIn, DyNiIn, YbAuIn, DyAuIn and ErAuIn arises from this frustration [3, 4].

In heavy fermion systems, interesting magnetic order has been found. CePdAl is a heavy fermion antiferromagnet with antiferromagnetic temperature T_N = 2.7 K and specific heat coefficient γ = 250 mJ/mol K^2 [5]. The magnetic structure of CePdAl consists of three inequivalent Ce sites. The magnetic ordering vector q = (1/2, 0, τ) where $\tau \sim 0.35$ weakly depends on temperature. Two-thirds of the Ce ions order while one-third does not order below T_N. The magnetic structure is considered to originate from the geometrical frustration [6].

The search for new compounds with ZrNiAl-type structure is motivated by the possible discovery of new physical phenomena due to the geometrical frustration. Until now, there are few studies on RTPb systems [7, 8, 9, 10] and physical properties have been only reported for CeTPb (T = Pd, Rh and Pt). This prompted the study of single crystal growth and physical properties of RRhPb (R = Nd, Sm and Gd) reported here.

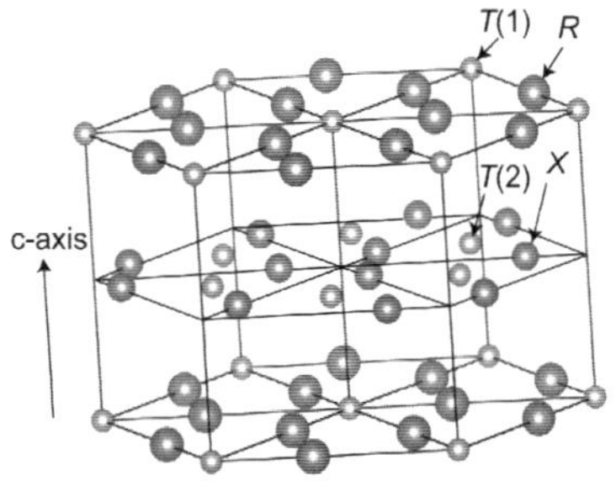

Figure 1. Crystal structure of ZrNiAl-type structure.

2. Experimental

The single crystals of *R*RhPb (*R* = Nd, Sm and Gd) were grown from Pb-flux. 3N (99.9 %) *R* = Nd, Sm and Gd, 4N Rh, and 5N Pb were reacted in with starting composition of 1:1:10. These materials were placed in an alumina crucible and sealed in an evacuated quartz tube. The sealed tubes were heated to 1150 °C, soaked for 12 hours, then cooled down to 700 °C in 90 hours. The excess Pb was spun off in a centrifuge.

The single phase of the hexagonal ZrNiAl-type structure was confirmed by powder X-ray diffraction. The powder was obtained from crashed single crystals. Figure 2 shows the powder X-ray diffraction patterns of the *R*RhPb. We mixed in Si powder as a standard. The X-ray diffraction patterns gave a = 7.706 Å, c = 3.951 Å for NdRhPb, a = 7.695 Å c = 3.876 Å for SmRhPb, and a = 7.697 Å, c = 3.826 Å for GdRhPb.

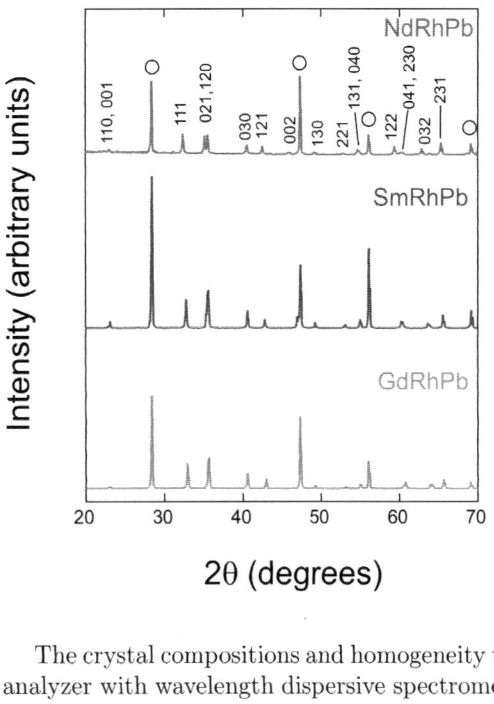

Figure 2. The powder X-ray diffraction patterns of *R*RhPb (*R* = Nd, Sm and Gd) for angle from 20 to 70 degrees. The indexes are for the hexagonal ZrNiAl-type structure. The circles show the peaks from Si powder.

The crystal compositions and homogeneity were determined by using an electron-probe micro-analyzer with wavelength dispersive spectrometers (EPMA-WDS; JEOL-8530). We used NdB_6, SmB_6, $Gd_3Ga_5O_{12}$, Rh and PbF_2 as standard reference materials for EPMA. The chemical compositions were determined to be Nd:Rh:Pb = 1.00:1.02:1.04, Sm:Rh:Pb = 1.00:0.93:0.96 and Gd:Rh:Pb = 1.00:1.05:1.06, which were in good agreement with the ideal 1:1:1 stoichiometry.

The magnetic properties were measured by using a commercial superconducting quantum

IOP Conf. Series: Journal of Physics: Conf. Series **807** (2017) 042006 doi:10.1088/1742-6596/807/4/042006

interference device magnetometer (Quantum Design).

3. Results and discussion

Figure 3 shows the reciprocal magnetic susceptibility $1/\chi(T)$ of NdRhPb as a function of temperature at 0.1 T. Above 70 K, $1/\chi(T)$ can be fit by a Curie-Weiss law. We estimated that the effective moment $\mu_{\rm eff}$ and Weiss temperature $\theta_{\rm p}$ are 3.78 $\mu_{\rm B}$ and -1 K for $H \parallel c$, and 3.83 $\mu_{\rm B}$ and -55 K for $H \parallel a$ respectively, indicating that the effective moment is close to the value expected for Nd^{+3} configuration ($\mu_{\rm eff}$ = 3.62 $\mu_{\rm B}$). $\chi(T)$ has a broad kink around $T_{\rm M1}$ = 3.6 K and a maximum at $T_{\rm M2}$ = 3.4 K for $H \parallel a$. On the other hand, $\chi(T)$ has a maximum $T_{\rm M1}$ = 3.6 K but anomaly not observed at $T_{\rm M2}$ = 3.4 K for $H \parallel c$. The decreasing of $\chi(T)$ below $T_{\rm M1}$ for $H \parallel c$ is about 5 times larger than that for $H \parallel a$, indicating that the NdRhPb is an antiferromagnet with two successive transition and an Ising-like magnetic structure.

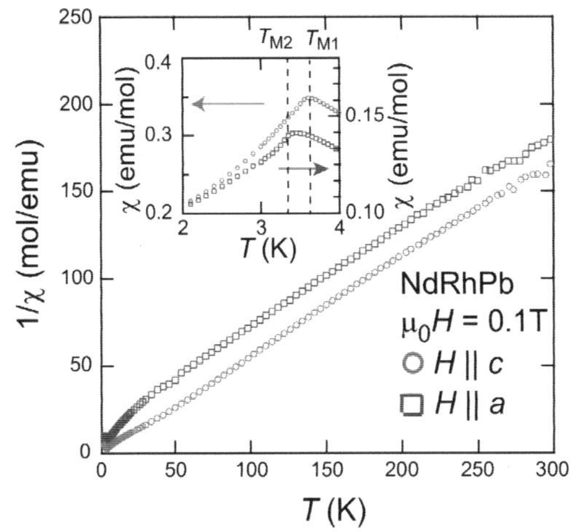

Figure 3. The temperature dependence of reciprocal magnetic susceptibility of NdRhPb at 0.1 T for $H \parallel c$ and a. The inset shows the expanded view of the magnetic susceptibility for $H \parallel c$ (circles, left axis), and for $H \parallel a$ (squares, right axis). The dashed lines indicate the T_{M1} and T_{M2}.

Figure 4 shows the reciprocal magnetic susceptibility of SmRhPb as a function of temperature at 1 T. $\chi(T)$ can not be fit with the susceptibility using a modified Curie-Weiss law $\chi = \chi_0 + C/(T - \theta_{\rm p})$ because the parameters were not fixed. We need to estimate the energy level splitting and the valence of Sm-ion in SmRhPb. The $\chi(T)$ has a maximum at $T_{\rm M1}$ = 11.5 K and a kink at $T_{\rm M2}$ = 8.3 K for $H \parallel c$. For $H \parallel a$, $\chi(T)$ has two anomaly at $T_{\rm M1}$ and $T_{\rm M2}$ and $\chi(T)$ is almost independent to the temperature below $T_{\rm M2}$, indicating that the magnetic hard axis would be around c-axis.

Figure 5 shows $1/\chi(T)$ of GdRhPb measured in 0.1 T. Above 50 K, $1/\chi(T)$ can be fit with a Curie-Weiss law. We estimated that the $\mu_{\rm eff}$ and $\theta_{\rm p}$ are 8.04 $\mu_{\rm B}$ and 9 K for $H \parallel c$, and 8.06 $\mu_{\rm B}$ and 10 K for $H \parallel a$, respectively, indicating that the magnetic anisotropy is weak. The effective moment is close to the value expected for Gd^{+3} ($\mu_{\rm eff}$ = 7.94 $\mu_{\rm B}$). A peak at $T_{\rm M1}$ = 17.6 K and a weak drop at $T_{\rm M2}$ = 15.3 K for $H \parallel c$ and a are observed.

We have measured the magnetic properties of RRhPb (R=Nd, Sm and Gd). We summarized the results in table 1. It is found that the RRhPb (R = Nd, Sm and Gd) are antiferromagnets with two successive phase transitions. The successive transitions with ZrNiAl-type structure have been observed in PrNiAl, NdPdAl, NdNiAl, YbAgGe, and so on [11, 12, 13, 14]. These double transitions appear to be a common phenomena in ZrNiAl-type structure.

This structure has the geometrical frustration. Therefore the suppression of $T_{\rm M}$ is expected, especially strong for SmRhPb because of its magnetic structure. The $T_{\rm M1}$ of RRhPb (R=Nd, Sm and Gd) roughly follows the de-Gennes scaling ($T_{\rm M}$ vs. $(g_{\rm J}\text{-}1)^2\ J(J+1)$, where $g_{\rm J}$ is the

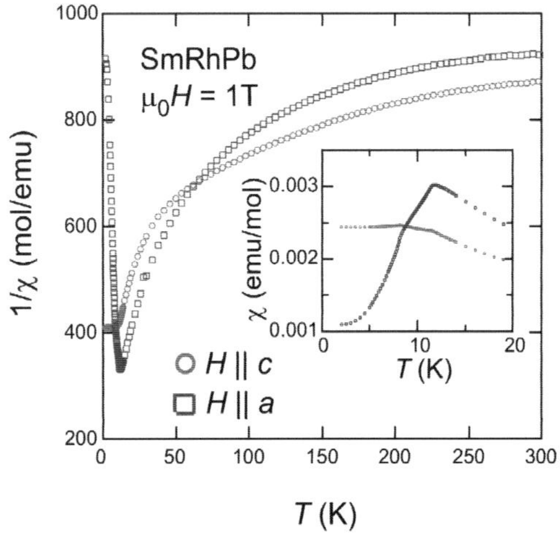

Figure 4. The temperature dependence of the reciprocal magnetic susceptibility of SmRhPb at 1 T for $H \parallel c$ and a. The inset shows the expanded view of the magnetic susceptibility.

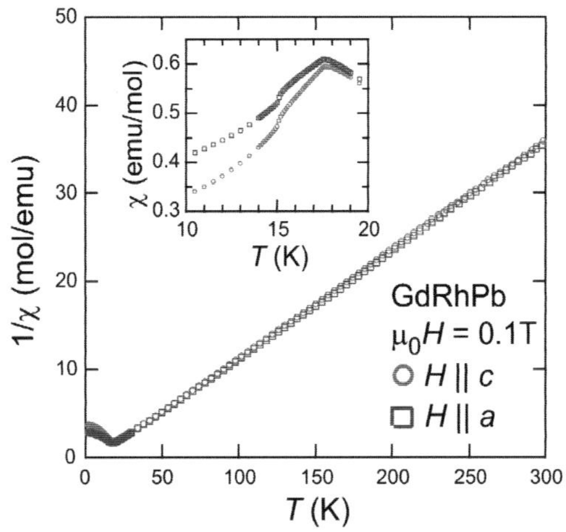

Figure 5. The temperature dependence of reciprocal magnetic susceptibility of GdRhPb at 0.1 T for $H \parallel c$ and a. The inset shows the expanded view of the magnetic susceptibility.

Table 1. T_{M1}, T_{M2}, de-Gennes factor and $|\theta_p|/T_{M1}$ of RRhPb R = Nd, Sm and Gd.

Sample	T_{M1}	T_{M2}	de-Gennes factor	$\lvert\theta_p\rvert/T_{M1}$
NdRhPb	3.6 K	3.4 K	1.84	0.28(H ‖ c), 16.18(H ‖ a)
SmRhPb	11.5 K	8.3 K	4.46	-
GdRhPb	17.6 K	15.3 K	17.6	0.51 (H ‖ c), 0.57 (H ‖ a)

Lande g-factor and J is the total angular momentum), namely the suppression of T_M in SmRhPb is not strong. The $|\theta_p|/T_M$ is less than 1, except for NdRhPb for $H \parallel a$ (probably due to the uniaxial magnetic anisotropy). The effect of geometrical frustration in RRhPb (R = Nd, Sm and Gd) appears to be negligible.

4. Conclusion

We have succeeded in growing the single crystals of RRhPb (R=Nd, Sm and Gd) using Pb-flux. We have confirmed the crystal structures of RRhPb (R=Nd, Sm and Gd) using powder X-ray diffraction. The RRhPb (R = Nd, Sm and Gd) are antiferromagnets with two successive phase transitions. The magnetic transitions occur at T_{N1} = 3.6 K and T_{N2} = 3.4 K in NdRhPb and appear Ising-like. The magnetic transitions take place at T_{N1} = 11.5 K and T_{N2} = 8.3 K in SmRhPb with magnetic hard axis c. GdRhPb exhibits magnetic transitions at T_{N1} = 17.6 K and T_{N2} = 15.3 K. The effect of geometrical frustration in RRhPb (R = Nd, Sm and Gd) appears to be negligible.

Acknowledgments

We thank Ms. A. Iwasaka for technical support of EPMA measurements. We thank Mr. M. Taki and Mr. Y. Minamiguchi for technical support. This work was supported by a Grant-in-Aid for Scientific Research Young Scientists B (No 24740248) and a Grant-in-Aid for Scientific Research C (No 26400333) from the Ministry of Education, Culture, Sports, Science and Technology (MEXT) and Japan Society of the Promotion of Science (JSPS).

References

[1] Gupta S, and Suresh K G, 2015 *J. Alloys Comp.* **618** 562
[2] Sechovsky V, and Havela L, Handbook of magnetic materials (Elsevier, Holland, 1998) Vol. 11, Chap. 1, p. 3
[3] Gondek L, and Szytula A, 2007 *J. Alloys Comp.* **442** 111
[4] Gondek L, Szytula A, Baran S, and Velasco J H, 2004 *J. Magn. Magn. Mater.* **272** e443
[5] Schank C, Jährling F, Luo L, Grauel A, Wassilew C, Borth R, Olesch G, Bredl C D, Geibel C, and Steglich F, 1994 *J. Alloys Compounds* **207-208** 329
[6] Dönni A, Ehlers G, Maletta H, Fischer P, Kitazawa H, and Zolliker M, 1996 *J. Phys.: Condens. Matter* **8** 11213
[7] Melnyk G, Gulay L D, and Tremel W, 2012 *J. Alloys Comp.* **528** 70
[8] Marazza R, Mazzone D, Riani P, and Zanicchi G, 1995 *J. Alloys Comp.* **220** 241
[9] Gulay L D, Kaczorowski D, Szajek A, Pietraszko A, 2008 *J. Phys. and Chem. Solids.* **69** 1934
[10] Hermes W, Payaprol S, and Pöttgen R, 2007 *Z. Naturforsch.* **62b** 901
[11] Javorsky P, Sechovsky V, Arons R R, Burlet P, Ressouche E, Svoboda P, and Lapertoto G, 1996 *J. Magn. Magn. Mater.* **164** 183
[12] Keller L, Dönni A, Kitazawa H, Tang J, Fauth F, and Zolliker M, 1998 *Physica B* **241-243** 660
[13] Bud'ko S L, Morosan E, and Canfield P C, 2004 *Phys. Rev. B* **69** 014415
[14] Umeo K, Yamane K, Muro Y, Katoh K, Niide Y, Ochiai A, Morie T, Sakakibara T, and Takabatake T, 2004 *J. Phys. Soc. Jpn.* **73** 537

Nodal gap structure of the heavy-fermion superconductor URu_2Si_2 revealed by field-angle-dependent specific-heat measurements

Shunichiro Kittaka[1], Yusei Shimizu[1], Toshiro Sakakibara[1], Yoshinori Haga[2], Etsuji Yamamoto[2], Yoshichika Ōnuki[2,3], Yasumasa Tsutsumi[4,5], Takuya Nomoto[6], Hiroaki Ikeda[7], and Kazushige Machida[7]

[1]Institute for Solid State Physics, University of Tokyo, Kashiwa, Chiba 277-8581, Japan
[2]Advanced Science Research Center, Japan Atomic Energy Agency, Tokai, Ibaraki 319-1195, Japan
[3]Department of Physics, University of the Ryukyus, Nishihara, Okinawa 903-0213, Japan
[4]Department of Basic Science, University of Tokyo, Tokyo 153-8902, Japan
[5]Condensed Matter Theory Laboratory, RIKEN, Wako, Saitama 351-0198, Japan
[6]Department of Physics, Kyoto University, Kyoto 606-8502, Japan
[7]Department of Physics, Ritsumeikan University, Kusatsu, Shiga 525-8577, Japan

E-mail: `kittaka@issp.u-tokyo.ac.jp`

Abstract. URu_2Si_2 is a promising candidate for a chiral d-wave superconductor whose gap is composed of a horizontal line node at equator and point nodes at the north and south poles. However, previous measurements of the specific heat and the thermal conductivity did not detect quasiparticle excitations from a horizontal line node in heavy-mass bands. Here, we have provided strong evidence for the presence of a horizontal line node from field-angle-dependent specific-heat measurements by using a high-quality single crystal of URu_2Si_2. We observed the $\sqrt{H}$ behavior of the specific heat at low fields which does not depend on inplane field angle ϕ but shows a shoulder-like anomaly in the polar-angle θ dependence around $\theta \sim 45°$. This feature is more clearly seen at lower temperatures, suggesting the detection of low-energy quasiparticle excitations reflecting the gap structure. From theoretical analyses based on microscopic calculations, we have demonstrated that this anomaly can be explained by the existence of a horizontal line node at $k_z = 0$. Thus, the gap structure of URu_2Si_2 matches well with the E_g chiral $k_z(k_x + ik_y)$ symmetry.

1. Introduction

URu_2Si_2 exhibits unconventional superconductivity below $T_c = 1.4$ K in the mysterious "hidden order" phase. Reflecting the likely exotic pairing mechanism, its superconducting gap appears to be anisotropic; the presence of nodes has been indicated from power-law temperature dependences of thermodynamic quantities such as the specific heat and the nuclear relaxation rate [1, 2, 3]. In addition, invariance of the specific heat [4] as well as the thermal conductivity [5] under a magnetic field rotated within the ab plane supports that the superconducting gap of URu_2Si_2 is rotationally symmetric around the c axis. From these results, E_g chiral d-wave superconductivity described by the gap symmetry of $k_z(k_x + ik_y)$ has been expected to be realized in URu_2Si_2. Indeed, broken time reversal symmetry in the superconducting state has been suggested from recent polar Kerr-effect measurements [6].

In the anticipated chiral d-wave state, the superconducting gap has point nodes at the north and south poles and a horizontal line node at $k_z = 0$. However, the results of specific-heat and thermal-conductivity measurements reported previously [3, 4, 5] cannot detect a horizontal line node in heavy-mass bands. At the time, it had been considered that heavy-mass Fermi surfaces are absent at $k_z = 0$ in the hidden order phase, resulting in disappearance of nodal quasiparticle excitations from the horizontal line node. Nevertheless, it has recently been clarified that heavy-mass bands similar to those in the antiferromagnetic phase [7, 8] exist at $k_z = 0$ also in the hidden order phase [9]. Therefore, the absence of quasiparticle excitations from the horizontal line node is incompatible with the chiral d-wave state.

In order to settle this controversy on the gap structure of URu_2Si_2, we have performed field-angle-dependent specific-heat measurements at low temperatures by using a high-quality single crystal [10]. In this contribution, we report the field-angle ϕ and θ dependence of the $\sqrt{H}$ behavior in the specific heat, evidencing the presence of a horizontal line node in heavy-mass bands at $k_z = 0$. These results ensure that the gap symmetry of URu_2Si_2 is indeed of the $k_z(k_x + ik_y)$ type.

2. Experimental

Single crystals of URu_2Si_2 were grown by the Czochralski pulling method in a tetra-arc furnace [11]. A high-quality single crystal (10.6 mg weight), the same sample in Ref. [10], was used in the present study. The specific heat C was measured by the standard quasi-adiabatic heat-pulse method or the relaxation method. The sample was cooled down to 0.1 K in a dilution refrigerator. Magnetic field was generated by using a vector magnet, up to 5 T in the x direction and 3 T in the z direction. The rotation of a refrigerator by using a stepper motor enables to control the field orientation three dimensionally with a high accuracy of $\sim 0.01°$. The field orientation is described by the azimuthal angle ϕ from the a axis and polar angle θ from the c axis.

3. Results and Discussion

Solid circles in Fig. 1 represent temperature dependence of the specific heat divided by temperature, C/T, of the present sample. For comparison, the C/T data taken from previous reports [4, 11, 12] are also plotted in the same figure. It is evident that the present sample is of the highest quality among them, as demonstrated by a sharp superconducting transition, high T_c, and low value of the residual C/T at low temperatures. Below 0.2 K, C/T shows anomalous increase on cooling. This upturn is too large to be attributed to a nuclear specific-heat contribution. Because the upturn is more prominent in a lower-quality sample [11], it might be caused by inclusion of some impurities.

Figure 2 plots the C/T data at 0.2 K as a function of $\sqrt{H}$ for $H \parallel a$ and $H \parallel c$. In both field directions, $C(H)$ is proportional to $\sqrt{H}$ at low fields, as represented by dashed lines. In the vortex state, energy spectra of quasiparticles are shifted due to the Doppler effect by $\delta E = m_e \boldsymbol{v}_F \cdot \boldsymbol{v}_s$, where m_e is quasiparticle mass, $\boldsymbol{v}_F$ is the Fermi velocity, and $\boldsymbol{v}_s$ is the velocity of supercurrent flowing around vortices. Around nodes, even a small shift of the spectra causes a finite zero-energy density of states (ZEDOS) which increases proportionally with $\sqrt{H}$ [13], in sharp contrast to the H-linear behavior expected Therefore, the observed $\sqrt{H}$ behavior in the low-temperature specific heat, which is a powerful probe to detect ZEDOS, demonstrates the presence of nodes in the superconducting gap.

In general, the field-orientation dependence of this $\sqrt{H}$ behavior reflects the location of gap nodes. When a magnetic field is pointing to a nodal direction, quasiparticle excitations at this node are strongly suppressed because of $\delta E = 0$, i.e., $\boldsymbol{v}_F \perp \boldsymbol{v}_s$. Therefore, the fact that the $\sqrt{H}$

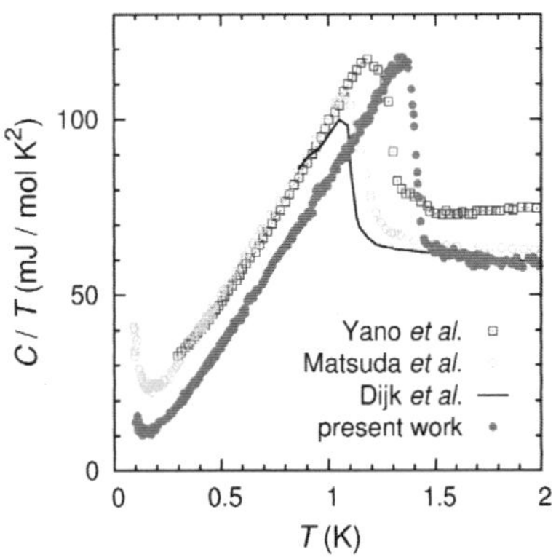

Figure 1. Temperature dependence of the zero-field C/T of the present sample (solid circles), compared with those of other samples in previous reports [4, 11, 12].

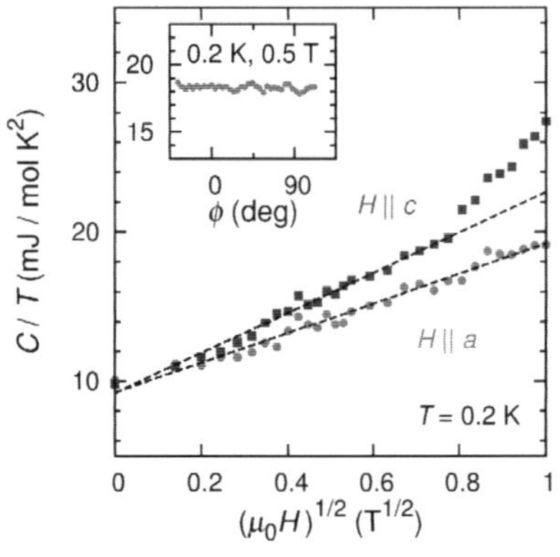

Figure 2. C/T as a function of $\sqrt{H}$ measured at 0.2 K for $H \parallel a$ and $H \parallel c$. Solid lines are fitting results for the low-field data by using a function $a\sqrt{H} + b$. Inset represents the azimuthal field-angle ϕ dependence of C/T at 0.2 K in a magnetic field of 0.5 T rotated within the ab plane.

behavior of $C(H)$ is observed in any field direction excludes a possibility of the gap structure having point nodes at the north and south poles alone [4]. Under a rotating magnetic field across nodal directions, this anisotropic Doppler effect causes an oscillation of the ZEDOS which shows a local minimum for a nodal direction. The inset of Fig. 2 presents the C/T data measured under a rotating field of 0.5 T within the ab plane at 0.2 K, where ϕ denotes the azimuthal field angle measured from the a axis. Any oscillation has not been observed, suggesting that vertical line nodes are not present in the gap.

The absence of vertical line nodes motivates us to search for a horizontal line node. In order to examine its presence and location, we measured the field-angle θ dependence of the specific heat, where θ is a polar angle between the magnetic field and the c axis. Figures 3 and 4 show $C(\theta)/T$ taken at 0.2 and 0.34 K, respectively. As shown in the inset of Fig. 4, $C(\theta)$ measured in the normal state at 1.8 K and 0.2 T is invariant with changing θ. At a low temperature of 0.2 K, a shoulder-like anomaly has been found around $\theta \sim 45°$ in $C(\theta)$ measured in a field of 0.2 T. This anomaly moves to a larger θ with increasing field and becomes obscure by increasing temperature. The latter feature implies that this anomaly indeed comes from low-energy quasiparticle excitations around nodes.

To clarify the origin of the shoulder-like anomaly in $C(\theta)$, we have performed microscopic calculations on the basis of the quasiclassical Eilenberger theory assuming a chiral d-wave gap of the $k_z(k_x + ik_y)$ type and a spherical Fermi surface [14]. In these calculations, the Pauli-paramagnetic effect and the anisotropy of the Fermi velocity have not been considered. Figures 5(a) and 6(a) show the polar-angle θ dependence of the ZEDOS, $N(E = 0)$, at $B = 0.03$ and 0.1, respectively. Here, B is scaled by the Eilenberger unit [15]. Note that Fig. 5(a) reproduces the shoulder-like anomaly at $\theta \sim 45°$. With increasing field, the shoulder-like anomaly moves to higher θ and a clear dip appears at $\theta = 0°$ due to reversal of the $C(\theta)$ anisotropy originating from the slight anisotropy of H_{c2} ($H_{c2}^{\parallel x} \gtrsim H_{c2}^{\parallel z}$) reflecting the gap structure. In the experiment, $C(\theta)$ of URu_2Si_2 does not show the dip at $\theta = 0°$ because the strong Pauli-

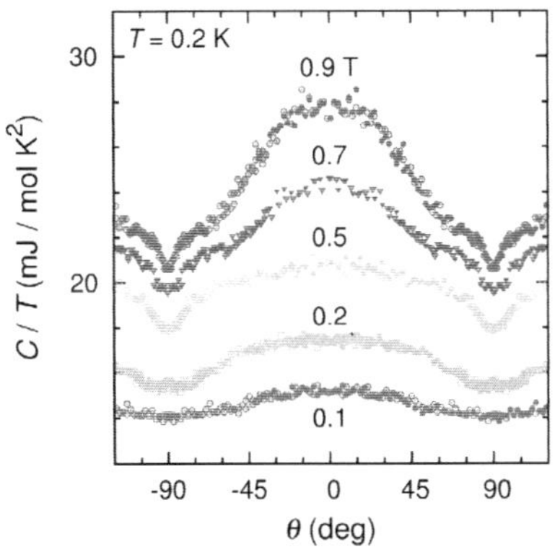

Figure 3. Polar field-angle θ dependence of C/T at 0.2 K in rotating fields within the ac plane. Numbers labelling curves represent the applied magnetic field in tesla.

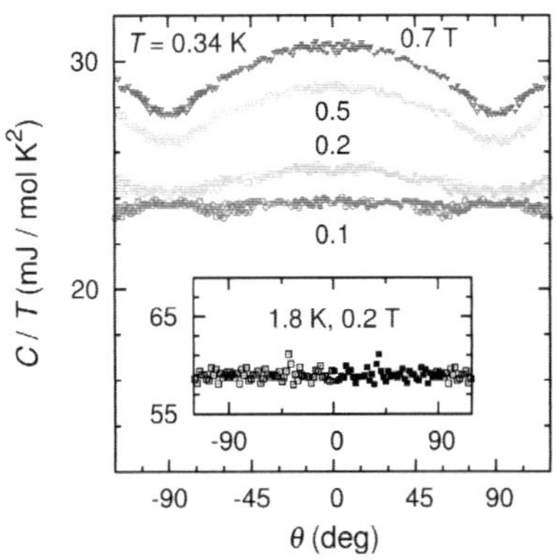

Figure 4. The $C(\theta)/T$ data at 0.34 K in several fields. Inset shows the normal-state $C(\theta)/T$ measured at 1.8 K in 0.2 T.

paramagnetic effect yields a peak at $\theta = 0°$.

In Figs. 5(b) and 6(b), we show the $\boldsymbol{k}$-resolved density of states, $N_{\boldsymbol{k}}(0)$, mapped on the spherical Fermi surface at several θ for $B = 0.03$ and 0.1 rotated within the xz plane, respectively. Here, the integration of $N_{\boldsymbol{k}}(0)$ over the Fermi surface corresponds to $N(E = 0)$. Most of quasiparticle excitations occur around a line node; contribution from point nodes to $N(E = 0)$ is relatively small. Although $N_{\boldsymbol{k}}(0)$ is enhanced in the overall Fermi surface by increasing magnetic field, qualitatively similar features have been seen in both fields; $N_{\boldsymbol{k}}(0)$ is strongly suppressed at the k_x direction when θ becomes larger than 45°, i.e., for $\theta = 75°$ and 90°. This is due to the suppression of the Doppler effect there and yields the shoulder-like anomaly in $C(\theta)$. Because the low-temperature specific heat is dominated by heavy quasiparticles, the shoulder-like anomaly in $C(\theta)$ evidences the presence of a horizontal line node at $k_z = 0$ in heavy-mass bands. This fact further supports that the gap symmetry of URu_2Si_2 belongs to the chiral E_g symmetry of the $k_z(k_x + ik_y)$ type [16].

4. Summary

We report the results of field-angle-dependent specific-heat measurements by using a high quality single crystal of URu_2Si_2. It was found that the specific heat shows the $\sqrt{H}$ behavior in any field direction and the shoulder-like anomaly in its polar-angle θ dependence which is smeared out with increasing temperature. From theoretical analyses based on the microscopic theory, we have demonstrated that these features are strong evidence for the presence of a horizontal line node in the heavy-mass bands. Thus, the present study settles the remaining controversy in URu_2Si_2 and helps establishing that URu_2Si_2 is a chiral d-wave superconductor of the $k_z(k_x + ik_y)$ type.

Acknowledgement

We thank Y. Yanase for useful discussions. A part of the numerical calculations was performed by using the HOKUSAI GreatWave supercomputer system in RIKEN. This work was supported by a Grants-in-Aid for Scientific Research on Innovative Areas "J-Physics" (15H05883) from MEXT, and KAKENHI (15K05158, 15H03682, 25103716, 26400360, 15H05745, 15K17715, 15J05698) from JSPS.

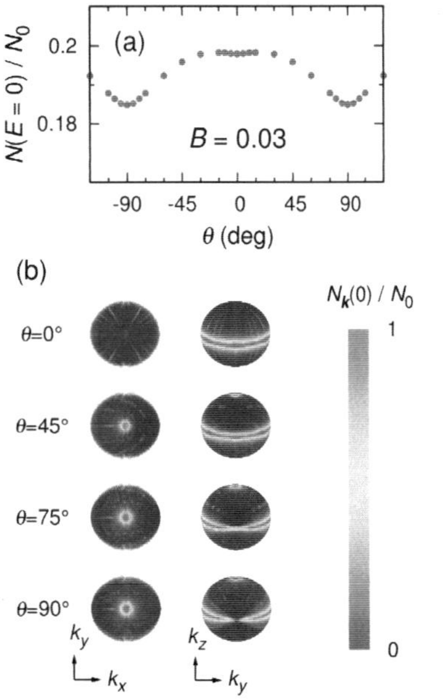

Figure 5. (a) Polar field-angle θ dependence of $N(E = 0)$ at $B = 0.03$ scaled by the normal-state density of states, N_0, obtained by microscopic calculations with the assumption of the gap symmetry of $k_z(k_x + ik_y)$. (b) Angle-resolved density of states $N_{\boldsymbol{k}}(0)$ at $B = 0.03$ for selected θ, mapped on the spherical Fermi surface. Here, the magnetic field is rotated within the xz plane.

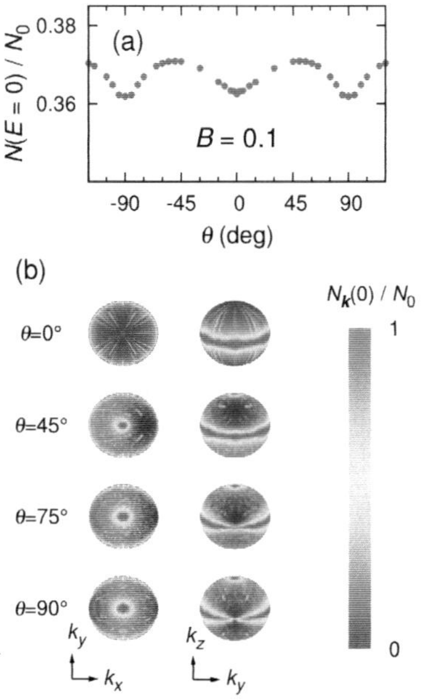

Figure 6. (a) θ dependence of $N(E = 0)/N_0$ at $B = 0.1$. (b) $N_{\boldsymbol{k}}(0)$ for selected θ in a rotating magnetic field of $B = 0.1$, mapped on the spherical Fermi surface.

References

[1] Fisher R A, Kim S, Wu Y, Phillips N E, Mcelfresh M, Torikachili M S and Maple M B 1990 *Physica B* **163** 419
[2] Matsuda K, Kohori Y and Kohara T 1996 *J. Phys. Soc. Jpn.* **65** 679
[3] Kasahara Y, Iwasawa T, Shishido H, Shibauchi T, Behnia K, Haga Y, Matsuda T D, Onuki Y, Sigrist M and Matsuda Y 2007 *Phys. Rev. Lett.* **99** 116402
[4] Yano K, Sakakibara T, Tayama T, Yokoyama M, Amitsuka H, Homma Y, Miranović P, Ichioka M, Tsutsumi Y and Machida K 2008 *Phys. Rev. Lett.* **100** 017004
[5] Kasahara Y, Shishido H, Shibauchi T, Haga Y, Matsuda T D, Onuki Y and Matsuda Y 2009 *New J. Phys.* **11** 055061
[6] Schemm E R, Baumbach R E, Tobash P H, Ronning F, Bauer E D and Kapitulnik A 2015 *Phys. Rev. B* **91** 140506(R)
[7] Ohkuni H, Inada Y, Tokiwa Y, Sakurai K, Settai R, Honma T, Haga Y, Yamamoto E, Yamagami H, Takahashi S and Yanagisawa T 1999 *Phil. Mag. B* **79** 1045
[8] Ikeda H, Suzuki M T, Arita R, Takimoto T, Shibauchi T and Matsuda Y 2012 *Nat. Phys.* **8** 528
[9] Hassinger E, Knebel G, Matsuda T D, Aoki D, Taufour V and Flouquet J 2010 *Phys. Rev. Lett.* **105** 216409
[10] Kittaka S, Shimizu Y, Sakakibara T, Haga Y, Yamamoto E, Onuki Y, Tsutsumi Y, Nomoto T, Ikeda H and Machida K 2016 *J. Phys. Soc. Jpn.* **85** 033704

[11] Matsuda T D, Hassinger E, Aoki D, Taufour V, Knebel G, Tateiwa N, Yamamoto E, Haga Y, Ōnuki Y, Fisk Z and Flouquet J 2011 *J. Phys. Soc. Jpn.* **80** 114710
[12] van Dijk N H, Bourdarot F, Klaasse J C P, Hagmusa I H, Brück E and Menovsky A A 1997 *Phys. Rev. B* **56** 14493
[13] Volovik G E 1993 *JETP Lett.* **58** 469
[14] Tsutsumi Y, Nomoto T, Ikeda H, and Machida K, arXiv:1604.02806.
[15] Miranović P, Ichioka M, Machida K and Nakai N 2005 *J. Phys.: Condens. Matter* **17** 7971
[16] Sigrist M and Ueda K 1991 *Rev. Mod. Phys.* **63** 239

Effect of ruthenium doping on superconductivity in Nb_2PdS_5

Q Chen[1], C Y Shen[1], X H Yang[1], X J Yang[1], Y P Li[1], C M Feng[1,2,3], Z A Xu[1,2,3]

[1]Department of Physics and State Key Laboratory of Silicon Materials, Zhejiang University, Hangzhou 310027, China
[2]Zhejiang California International NanoSystems Institute, Zhejiang University, Hangzhou 310058, China
[3]Collaborative Innovation Centre of Advanced Microstructures, Nanjing 210093, China

E-mail: zhuan@zju.edu.cn

Abstract. We synthesized a series of $Nb_2Pd_{1-x}Ru_xS_5$ polycrystalline samples by solid-state reaction method and investigated systematically the Ru doping effect on superconductivity by transport and magnetic measurements. It is found that superconductivity is enhanced with Ru doping and the superconducting transition temperature (T_c) reaches a maximum of 6.86 K at $x = 0.2$. Ru doping is regarded as the hole-type doping similar to the case of Ir doping. The evolution of $T_c{}^{mid}$ with Ru doping is compared with the case of Ir or Ag doping.

1. Introduction

The recent discovery of superconductivity in the new transition metal-chalcogenide compound Nb_2PdS_5 with $T_c \sim 6$ K has attracted much attention[1]. This quasi-one-dimensional(Q1D) material which is argued to be a multi-band superconductor[1, 2, 3, 4] has remarkably high and anisotropic upper critical field ($H_{c2}^{\|b} >37$ T). Chemical substitution has been investigated in this T_2PdCh_5 (T = Nb or Ta, Ch = S or Se) system and superconductivity was found at 2.5 K and 6 K in Ta_2PdSe_5[3] and Ta_2PdS_5[5], respectively. Partial substitution of Pd by Ni[6] or Ir[7] has slightly enhanced T_c but superconductivity is suppressed in the cases of Pt-for-Pd doping [6] or Ag-for-Pd doping[7], and Se-for-S doping [8]. However, the feature of large H_{c2} relative to T_c which surpasses by far the expected Pauli limiting field ($H_{c2} = 1.84T_c$[9]) is found to be robust against these substitutions. This common feature provide a strong evidence of unconventional superconductivity in this system.

The band structure calculations[1, 4, 10] present that the Fermi surface of Nb_2PdS_5 is composed of multiple sheets and this system is in proximity to a magnetically or charge density wave(CDW) ordered state due to the nesting properties of those Q1D fermi surface sheets. The change in d electron population in the Pd site is expected to flatten the Q1D fermi surface sheets and thus it may enhance the nesting properties and the proposals for unconventional superconducting pairing scenario were suggested[1, 4]. Moreover, it is also suggested that the strong spin-orbit coupling associated with the heavy Pd element could lead to the high H_{c2}[5, 11]. Through controlling SOC by Pt and Ni partial substitutions on the Pd sites, H_{c2}/T_c is found to be tunable, indicating a crucial role of SOC on the enhancement of H_{c2}[6, 12]. These theoretical and experimental studies have verified the importance of the presence of Pd irons with a large Z number to the unconventional properties. Recently, it has been reported that the charge carrier density (or band filling) which can be modulated by hole(electron)-type doping could be a crucial factor to tune superconductivity[7] in this system. So far, the origin of large H_{c2} and exotic superconductivity in T_2PdCh_5 is far from conclusive.

In this paper, we investigate the doping effect on the Pd site by heterovalent transition metal ruthenium. The Ru doping is considered as a hole-type dopant but with weaker SOC compared to the Ir doping, thus it may help on distinguishing the different roles of charge carrier density and SOC upon superconductivity. The evolution of superconductivity is obtained by measuring electrical resistivity and magnetic susceptibility for $Nb_2Pd_{1-x}Ru_xS_5$. T_c reaches a maximum of 6.86 K at $x = 0.2$. The superconducting window is wide, indicating a robust superconductivity in Nb_2PdS_5.

2. Experimental details

We synthesized a series of $Nb_2Pd_{1-x}Ru_xS_5$ polycrystalline samples by usual solid-state reaction method using stoichiometric amounts of powders of Nb (99.99%), Pd (99.99%), Ir, Ru or Ag (99.99%), and S (99.9%). All the starting materials were thoroughly ground and pressed into pellets and then sealed in evacuated quartz tubes. The quartz ampoules were slowly heated to 1073$\sim$ 1123 K and held for 48 h. This procedure was finally repeated again for homogeneity.

Powder x-ray diffraction (XRD) was carried out at room temperature on a PANalytical X-ray diffractometer using Cu Kα radiation, and lattice constants were determined using the program X'Pert HighScore. Electrical resistivity was measured by a standard four-terminal method using Oxford-15 T cryostat with Keithley 2400 source-measure meters and 2182A nanovoltmeters.

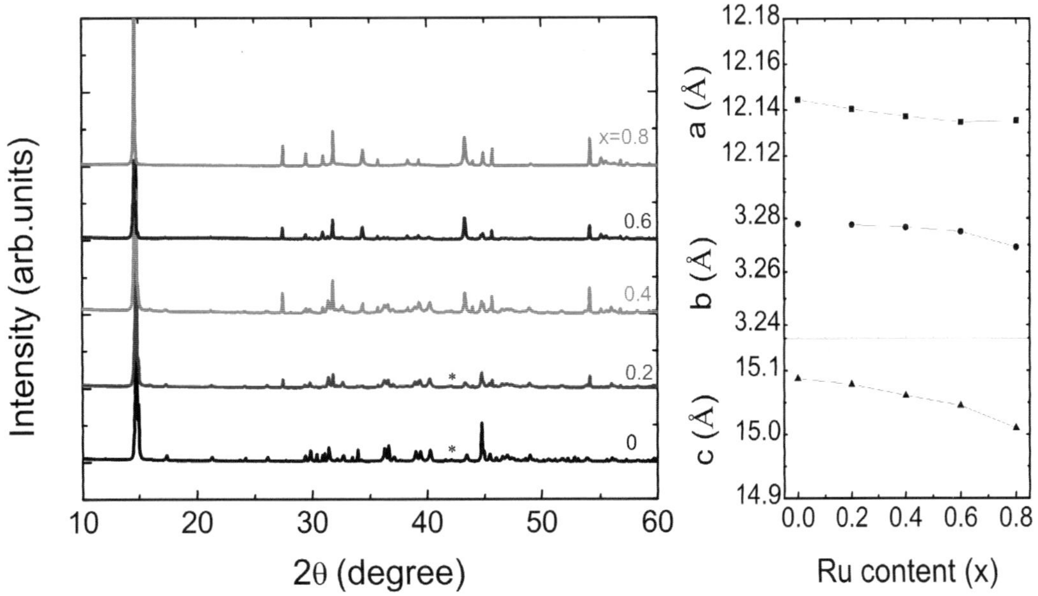

Figure 1. Left panel: room temperature powder X-ray diffraction patterns for $Nb_2Pd_{1-x}Ru_xS_5$ with x= 0, 0.2, 0.4, 0.6 and 1, respectively. Right panel: lattice constants as a function of nominal doping content x.

IOP Conf. Series: Journal of Physics: Conf. Series **807** (2017) 052002 doi:10.1088/1742-6596/807/5/052002

3. Result and discussion

Fig. 1 shows the XRD patterns of $Nb_2Pd_{1-x}Ru_xS_5$ (x = 0, 0.2, 0.4, 0.6, 1). All the main peaks can be well indexed with a monoclinic structure (space group C2/m), except for a few minor peaks due to unknown impurities marked by the asterisks. The variation of lattice parameters was shown in the right panel. It is found that a and b remain almost constant while the c-axis shrinks significantly with increasing Ru content. This observation indicates that Ru atoms are successfully doped into this compound which is consistent with the smaller ion radius of Ru compared to Pd.

The temperature-dependent resistivity is shown in Fig. 2. The resistivity at T = 300 K ranges from 1.55 mΩ·cm to 4.6 mΩ·cm. For the undoped compound (x = 0), the resistivity decreases slowly with decreasing temperature and then shows a sharp drop around $T_c \sim 6$ K. This metallic behavior is consistent with previous reports[1][7]. Upon doping with Ru, a small upturn shows up at low temperature which can be ascribed to Anderson localization[5] or grain boundary effect[13]. The inset of Fig. 2 is the enlarged plot of resistivity versus T, showing the variation of superconducting transition. Superconductivity was firstly enhanced then suppressed with further doping which resembles the case of Ir doping [7].

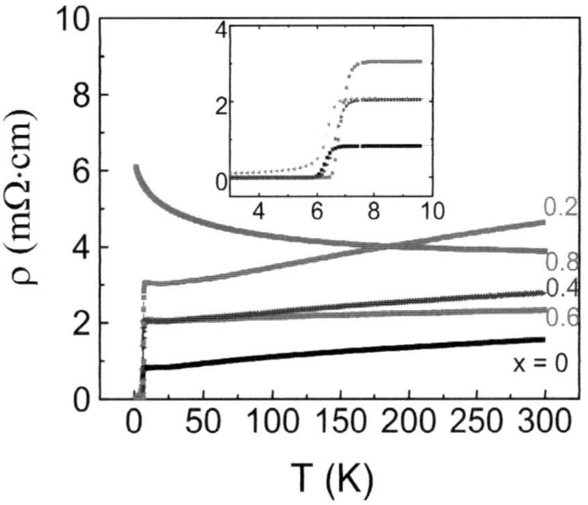

Figure 2. Temperature dependence of the electrical resistivity of $Nb_2Pd_{1-x}Ru_xS_5$. Inset: the enlarged plot of resistivity around the superconducting transition temperatures.

Figure. 3 presents temperature dependence of magnetic susceptibility. The samples with $0 \leq x \leq 0.4$ show strong diamagnetic signal confirming bulk superconductivity. The transition temperature is consistent with the resistivity measurements (shown in fig. 4). For x = 0.8, superconductivity no longer survives.

According to our previous work[7], partial substitution of Pd by Ir which can be considered as hole-type doping could obviously increase T_c for low Ir content, while Ag doping which is regarded as electron-type doping destroys superconductivity quickly. It is interesting to compare the effect of Ru doping with Ir doping both of which are regarded as hole-type dopants. Since the strength of SOC is proportional to Z^4, where Z is the atomic mass number, Ru doping may hardly change SOC associated with Pd site, while Ir or Pt doping could increase SOC in the system [6, 12]. Therefore we may distinguish the effect of SOC from charge carrier density through comparing Ru doping and Ir doping.

Based on the resistivity and magnetic measurements, a phase diagram of T_c vs. doping level is shown in Fig. 4, the data of Ir or Ag doping are plotted for comparison. Circle symbols stand for the characteristic temperature T_c^{mid} where $\rho(T)$ drops to 50% of the normal state value. Ru and Ir doping are shown by hollow symbols while Ag doping is in solid character. Pentagonal symbols represent superconducting transition temperature derived from magnetic

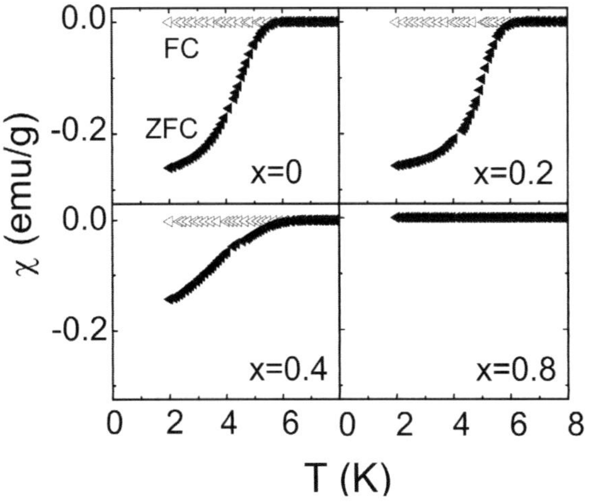

Figure 3. Temperature dependence of magnetic susceptibility measured under H = 10 Oe for $Nb_2Pd_{1-x}R_xS_5$ (x= 0, 0.2, 0.4, and 0.8).

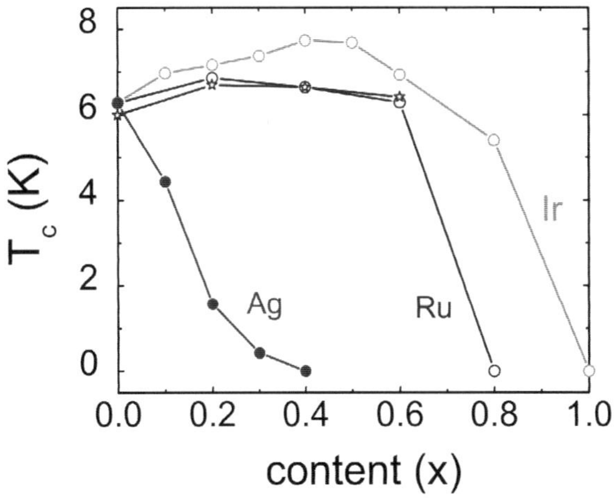

Figure 4. The phase diagram of transition temperature T_c as a function of doping level x. The blue, magenta and wine circle lines stand for T_c^{mid} of Ru, Ir and Ag doped $Nb_2Pd_{1-x}R_xS_5$, respectively. Pentagonal symbols represent T_c extracted from magnetic susceptibility data.

susceptibility. It can be found that superconductivity is enhanced with a maximum ${T_c}^{mid}$ of 6.86 K by Ru (x = 0.2) doping, and 7.73 K by Ir (x = 0.4) doping, respectively. Both Ru and Ir doping are expected to increase the hole-type carrier density, which may drive the system far away from the magnetic order or a possible CDW order and thus favor superconductivity, while the electron-type dopings (i.e., Ag doping) have an opposite effect. In this scenario, we could understand why T_c initially increases for both Ru doping and Ir doping. However, the maximum of T_c is even higher and the superconducting range (up to x of 0.8) is wider for the Ir doping case. We proposed that the enhanced SOC strength due to heavier Ir may account for this difference. This result implies that not only the charge carrier density but also SOC could also play an important role in controlling superconductivity of Nb_2PdS_5. It should also be noted that superconductivity can survive up to very high doping level in the cases of Ru or Ir doping, which demonstrates that T_c is quite robust upon disorder in the hole-type doping, as compared with the iron-based superconductors $LaFe_{1-x}Co_xAsO$[14] and $BaFe_{2-x}Ni_xAs_2$[15]. Careful studies on the single-crystalline $Nb_2Pd_xS_{5-\delta}$ have found that superconductivity occurs in a wide range of Pd ($0.6 < x < 1$) and S ($0 < \delta < 0.61$) contents[16], suggesting again that superconductivity in this system is very robust.

4. Conclusion

In summary, we studied the Ru doping effect on superconductivity in Nb_2PdS_5 and superconductivity with a maximum ${T_c}^{mid}$ of 6.86 K at x =0.2 has been observed. Bulk superconductivity is confirmed by magnetic measurements. An electronic phase diagrams is presented showing the enhancement of T_c by partial substitution of Pd by Ru and the comparison with the Ir or Ag doping is made, which indicates that the charge carrier density as well as SOC has a significant effect on the superconductivity.

Acknowledgments

This work is supported by the National Basic Research Program of China (Grant Nos. 2014CB921203 and 2012CB927404), NSF of China (Contract Nos. U1332209 and 11190023), the Ministry of Education of China (Contract No. 2015KF07), and the Fundamental Research Funds for the Central Universities of China.

References

[1] Zhang Q et al (2013) "Superconductivity with extremely large upper critical fields in $Nb_2Pd_0.81S_5$." *Sci. Rep.* **3**.

[2] Zhang Q R, Rhodes D, Zeng B, Besara T, Siegrist T, Johannes M D, and Balicas L (2013) "Anomalous metallic state and anisotropic multiband superconductivity in $Nb_3Pd_{0.7}Se_7$. "*Phys. Rev.* B **88**(2), 024508.

[3] Zhang J, Dong J K, Xu Y, Pan J, He L P, Zhang L J and Li S Y (2015) "Superconductivity at 2.5 K in the new transition-metal chalcogenide $Ta_2Pd_xSe_5$." *Supercond. Sci. Tech.* **28**(11), 115015.

[4] Singh D J (2013)" Electronic structure and upper critical field of superconducting $Ta_2Pd_xS_5$." *Phys. Rev.* B **88**(17), 174508.

[5] Lu Y, Takayama T, Bangura A F, Katsura Y, Hashizume D and Takagi H (2013) "Superconductivity at 6 K and the violation of Pauli limit in $Ta_2Pd_xS_5$." *J. Phys. Soc. Jpn* **83**(2), 023702.

[6] Zhou N et al (2014) "Controllable spin-orbit coupling and its influence on the upper critical field in the chemically doped quasi-one-dimensional Nb_2PdS_5 superconductor." *Phys. Rev.* B **90**(9), 094520.

[7] Shen C Y, Si B Q, Bai H, Yang X J, Tao Q, Cao G H and Xu Z A (2016) "Pd site doping effect on superconductivity in $Nb_2Pd_{0.81}S_5$." *Europhys. Lett.* **113**(3), 37006.

[8] Niu C Q et al (2013) "Effect of selenium doping on the superconductivity of $Nb_2Pd(S_{1-x}Se_x)_5$." *Phys. Rev.* B **88**(10), 104507.

[9] Clogston A M (1962) "Upper limit for the critical field in hard superconductors." *Phys. Rev. Lett.* **9**(6), 266.

[10] Seunghyun K et al (2013) "Enhanced upper critical fields in a new quasi-one-dimensional superconductor $Nb_2Pd_xSe_5$." *New. J. Phys.* **15**(12), 123031.

[11] Carbotte J P (1990) "Properties of boson-exchange superconductors." *Rev. Mod. Phys.* **62**(4), 1027.

[12] Werthamer N R, Helfand E and Hohenberg P C (1966) "Temperature and purity dependence of the superconducting critical field, H_{c2}. III. Electron spin and spin-orbit effects." *Phys. Rev.*, **147**(1), 295.

[13] Carrington A and Cooper J R (1994) "Influence of grain-boundary scattering on the transport properties of a high-T_c oxide superconductor." *Physica* C: Superconductivity. **219**(1), 119-22.

[14] Wang C et al (2009) "Effects of cobalt doping and phase diagrams of $LFe_{1-x}Co_xAsO$ (L = La and Sm)." *Phys. Rev.* B **79**(5), 054521.

[15] Li L J et al (2009) "Superconductivity induced by Ni doping in $BaFe_2As_2$ single crystals." *New. J. Phys.* **11**(2), 025008.

[16] Yu H, Zuo M, Zhang L, Tan S, Zhang C, and Zhang Y (2013). "Superconducting fiber with transition temperature up to 7.43 K in $Nb_2Pd_{1-x}S_5$ ($0.6 < x < 1$)." *J. Am. Chem. Soc.* **135**(35), 12987-89.

Anomalous Scaling Relations and Pairing Mechanism of the Fe-based Superconductors

Yunkyu Bang[1] **and G. R. Stewart**[2]

[1]Department of Physics, Chonnam National University, Kwangju 500-757, Republic of Korea
[2] Physics Department, University of Florida, Gainesville, FL 32611-8440, USA

E-mail: [1]ykbang@chonnam.ac.kr

E-mail: [2]stewart@phys.ufl.edu

Abstract. The anomalous scaling behavior of the specific heat jump ΔC *vs.* T_c ($\Delta C \sim T_c^{\alpha}, \alpha \approx 3$), first observed by Bud'ko *et al*[1], has been confirmed with several families of the Fe-based superconducting compounds with various dopings. And more recently, Xing *et al*[2] have reported that the condensation energy (CE) *vs.* T_c also follows an anomalous power law ($\Delta E \sim T_c^{3.5}$) with about forty different samples of the Fe-based superconductors. These apparently non-BCS power law relations have strongly hinted at a non-BCS pairing mechanism of the Fe-based superconductors. We have studied a minimal two band BCS model to understand these anomalous scaling behaviors and showed that these non-BCS relations of $\Delta C \sim T_c^3$ as well as $\Delta E \sim T_c^{3.5}$ can be well reproduced by the simple two-band BCS model paired by a dominant interband repulsive interaction ($V_{inter} > V_{intra} > 0$). The origin of these anomalous scaling relations are the kinematic constraint of $\frac{\Delta_h}{\Delta_e} \sim \sqrt{\frac{N_e}{N_h}}$ near T_c of the two band BCS model, as realized in the $S_{\pm}$-wave pairing state. Our results imply that these seemingly non-BCS-like scaling behaviors, on the contrary to the common expectations, are in fact strong experimental evidences that the pairing mechanism of the Fe-based superconductors is genuinely a BCS mechanism, meaning that *the Cooper pairs are formed by the itinerant carriers glued by a pairing interaction.*

1. Introduction

The pairing mechanism of the Fe-based superconductors (IBS) still remain an open question. It is because there is almost no direct experimental probe for identifying a pairing mechanism while there are numerous experimental probes for identifying the pairing symmetry or gap functions. In this regard, the anomalous scaling behavior of the specific heat jump ΔC *vs.* T_c ($\Delta C \sim T_c^3$), first observed by Bud'ko *et al*[1] and confirmed by many researchers[3, 4, 5, 6, 7] has stimulated many people to speculate this non-BCS relation as a strong evidence of the quantum critical point (QCP) driven superconductivity in IBS[8]. And more recently, Xiang *et al*[2] has reported another non-BCS scaling behavior with the condensation energy (CE) ΔE *vs.* T_c ($\Delta E \sim T_c^{3.5}$) with about forty different samples of the IBS, which therefore reenforced the speculation that the QCP is the underlying reason for these anomalous scaling behaviors and its strong connection to the superconducting (SC) pairing mechanism itself.

While we do not know the microscopic pairing mechanism of the IBS, we noted that the multi-band nature itself of the IBS has produced many unusual deviations[9, 10] from a standard single band BCS superconductors. Motivated by this simple idea, we studied a minimal two band BCS model with dominant interband repulsive interaction ($V_{inter} > 0$) to calculate the ΔC *vs.* T_c and the CE ΔE *vs.* T_c. To our surprise, we have found that the simple two band BCS model naturally produces both anomalous scaling relations simultaneously. Our results imply that these seemingly non-BCS-like scaling behaviors observed in the thermodynamic quantities, on the contrary to the common expectations, are in fact strong experimental evidences of a BCS pairing mechanism in the IBS. All strong correlation effects, abundantly observed in the normal state of the IBS, should renormalize the effective mass m^*_{qp} of quasiparticles, density of states $N_{h,e}$, the pairing interactions $V_{inter,intra}(q)$, etc., but when the system enters the SC transition, the pairing mechanism itself seems to be governed by the BCS mechanism.

2. Minimal two Band model for the SH jump ΔC vs. T_c

2.1. Formalism

We consider a minimal two band model for the IBS[9] consisting of one hole band and one electron band; each band represents a thermodynamic average of a group of hole bands around Γ point and a group of electron bands around M point in the Brillouin zone (BZ). We assume that the pairing interactions $V_{ab}(a,b = h,e)$ are all repulsive ($V_{ab} > 0$) but with a dominant interband repulsion as $V_{he,eh} > V_{hh,ee} > 0$. This is an extremely simplified phenomenological model and defined by the following coupled BCS gap equations.

$$\begin{aligned} \Delta_h &= -[V_{hh} N_h \chi]\Delta_h - [V_{he} N_e \chi]\Delta_e, \\ \Delta_e &= -[V_{ee} N_e \chi]\Delta_e - [V_{eh} N_h \chi]\Delta_h, \end{aligned} \tag{1}$$

where $N_{h,e}$ are the DOS of the hole and electron bands, respectively, and the pair susceptibility χ at T_c is defined as $\chi(T_c) = T_c \sum_n \int_{-\Lambda_0}^{\Lambda_0} d\xi \frac{1}{\omega_n^2+\xi^2} \approx \ln\left[\frac{1.14\Lambda_0}{T_c}\right]$, with $\omega_n = \pi T_c(2n+1)$, and Λ_0 is a pairing energy cut-off. This model produces the $s_{\pm}$-gap solution in which the order parameter (OP) $\Delta_{h,e}$ on each band has opposite sign and different magnitudes, in general[9]. In particular, in the limit $V_{intra}/V_{inter} \rightarrow 0$, the above gap equations can be analytically solved and provides the following kinematic relation[9]

$$\frac{\Delta_h}{\Delta_e} \sim \sqrt{\frac{N_e}{N_h}} \quad \text{as } T \rightarrow T_c, \tag{2}$$

and the critical temperature is also obtained as

$$T_c \approx 1.14\Lambda_0 \exp\left[-1/(V_{inter}\sqrt{N_e N_h})\right]. \tag{3}$$

And the total SH jump ΔC should be a summation of the SH jump contribution of each band as follows

$$\Delta C = \sum_{i=h,e} N_i(0)\left(\frac{-d\Delta_i^2}{dT}\right)\Big|_{T_c}. \tag{4}$$

Now in order to calculate the experimental data of ΔC vs. T_c for a Fe-122 compound with a series of doping, we need a modelling of doping. First, we notice that the undoped parent compound such as $BaFe_2As_2$ is a compensated metal, hence has the same number of electrons and holes, i.e. $n_h = n_e$. Therefore it is a reasonable approximation to take $N_h = N_e$ at no doping and then the doping of holes (K, Na, etc.) or electrons (Co, Ni, etc.) is simulated by varying N_h and N_e while keeping $N_e + N_h = N_{tot} = const.$ We have several remarks on this modelling of doping: (1) The assumption $N_{tot} = const.$ is only for convenience. The sensitive parameters of our model are the relative values between N_e and N_h but not the total DOS N_{tot}. (2) We do not literally mean $n_{h(e)} = N_{h(e)}$. The actual change of DOS $N_{h(e)}$ with doping "x" (holes or electrons) should be complicated and also non-linear. What our model assumes is: $N_h \approx N_e$ at no doping and the difference of DOSs $|N_h - N_e|$ increases with doping. This assumption is consistent with the Angle-Resolved-Photo-Emission-Spectroscopy (ARPES) measurements of $(Ba_{1-x}K_x)Fe_2As_2$[11] and $Ba(Fe_{1-x}Co_x)_2As_2$[12] which show the systematic changes of hole (electron) Fermi surface (FS) sizes with dopings. For the rest of this paper, we use the normalized DOSs as $\bar{N}_{h,e} = N_{h,e}/N_{tot}$ and accordingly define the dimensionless coupling constants as $\bar{V}_{intra/inter} = N_{tot} \cdot V_{intra/inter}$.

In the case of $V_{intra} = V_{hh,ee} = 0$, we can obtain the analytic solutions $\Delta_{h,e}(T)$ near T_c as

$$\Delta_h^2(T) \approx \frac{2}{1+N_h/N_e}\Delta_{BCS}^2(T), \qquad (5)$$
$$\Delta_e^2(T) \approx \frac{2}{1+N_e/N_h}\Delta_{BCS}^2(T)$$

with $\Delta_{BCS}^2(T) = \pi^2 \frac{8}{7\zeta(3)} T_c^2(1 - T/T_c)$. Substituting the results of Eq.(5) into Eq.(4) yields

$$\frac{\Delta C}{T_c} \approx 4 \times (3.06)^2 N_{tot} \cdot (\bar{N}_h \bar{N}_e). \qquad (6)$$

In contrast to the one band BCS superconductor, Eq.(6) clearly shows that $\Delta C/T_c$ can have a strong T_c dependence through $\bar{N}_h \bar{N}_e$ even with a constant N_{tot}. With doping in a given Fe-Pnictide/Chalcogenide (FePn/Ch) compound, $\bar{N}_h$ and $\bar{N}_e (= 1 - \bar{N}_h)$ can vary over the range of $\bar{N}_{h,e} = [0, 1]$.

In the next section, we will show numerical results for the cases of $V_{intra} = 0$ and $V_{intra} \neq 0$, and also include the impurity scatterings. We found that the non-pair-breaking impurity scattering plays a crucial role in order to explain the ideal BNC scaling $\Delta C/T_c \propto T_c^2$ in $Ba(Fe_{1-x}TM_x)_2As_2$ (TM=Co,Ni) as well as a strong deviation in $Ba_{1-x}K_xFe_2As_2$[13].

2.2. Numerical results

In Fig.1, we show the numerical results of $\Delta C(\bar{N}_h)$ *vs* $T_c(\bar{N}_h)$ for the case of $\bar{V}_{intra} = 0.0$ for different inter-band potentials $\bar{V}_{inter} = 1.0, 1.5, 2.0, 3.0$ and 5.0, respectively, where $\Delta C(\bar{N}_h)$ and $T_c(\bar{N}_h)$ are implicitly related through $\bar{N}_h \in [0, 1]$. With extensive numerical experiments, we found: (1) $\Delta C/T_c$ becomes $\sim T_c^2$ for the whole region if $\bar{V}_{inter} > 5.0$, however, this strength of pairing potential might be unrealistically strong. (2) Including $V_{intra} \neq 0.0$ does not change the general behavior shown in Fig.1 as long as $\bar{V}_{intra} < \bar{V}_{inter}/2$.

Although it is not completely successful to reproduce the ideal BNC scaling behavior, the results of the generic two band model in Fig.1 is remarkably non-BCS and very encouraging in that it shows that $\Delta C/T_c$ has changed more than one order of magnitude with $N_{tot} = const.$ However, the BNC scaling behavior is still limited in a region near the higher T_c region (where $\Delta N = |\bar{N}_e - \bar{N}_h|$ is small) and we need an extra mechanism to extend the power law scaling behavior at the lower temperature region. We noticed from the experimental data that the size of jump ΔC is more severely reduced by impurity scattering when the jump size is small (low T_c region in Fig.1) compared to when the jump size is large (higher T_c region in Fig.1).

Phenomenologically we can consider two kind of impurity scattering in the two band model: Γ_0 (intra-band scattering) and Γ_π (inter-band scattering)[14]. As we assumed the $s_\pm$-wave state, Γ_π causes strong pair-breaking effect (e.g. suppression of T_c and reduction of $\Delta_{h,e}$), while Γ_0 doesn't affect the superconductivity itself[15]. Besides the standard impurity effects on T_c and $\Delta_{h,e}$[14], the quasiparticle spectra becomes broadened by the impurity scattering. Therefore the SH jump of ΔC of Eq.(4) should be modified with this quasiparticle broadening as follows[16],

$$\Delta C = \sum_{i=h,e} N_i \Big(\frac{-d\Delta_i^2}{dT}\Big)\Big|_{T_c} \int_0^\infty \frac{dx}{2}\Big[\frac{1}{\cosh^2(\frac{x}{2})}\Big]\frac{x^2}{x^2 + (\frac{\Gamma_{tot}}{T_c})^2} \qquad (7)$$

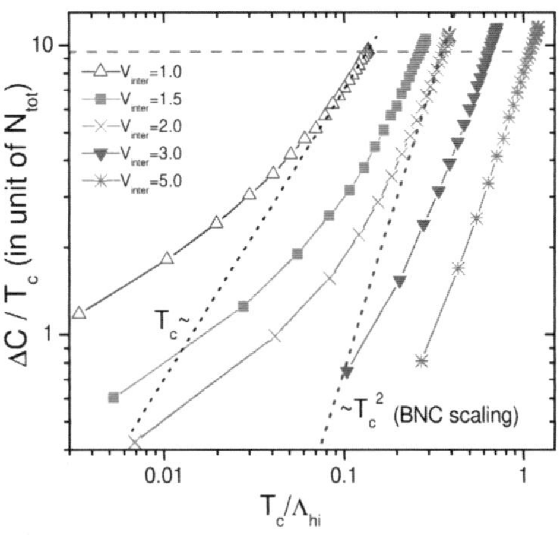

Figure 1. (Color online) Numerical calculations of $\Delta C/T_c$ *vs.* T_c with $\bar{V}_{intra} = 0.0$ but with different $\bar{V}_{inter} = 1.0, 1.5, 2.0, 3.0$ and 5.0, respectively. Horizontal dashed line is the BCS limit of $9.36N_{tot}$ and the dotted lines of $\sim T_c$ and $\sim T_c^2$ (BNC scaling) are guides for the eyes.

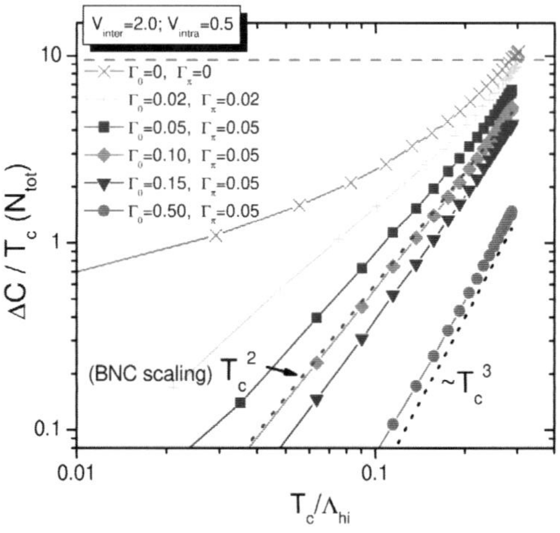

Figure 2. (Color online) Numerical calculations of $\Delta C/T_c$ *vs.* T_c with $\bar{V}_{inter} = 2.0$ and $\bar{V}_{intra} = 0.5$, for different impurity scattering strengths of Γ_0 and Γ_π (in unit of Λ_{hi}). Horizontal dashed line is the BCS limit of $9.36N_{tot}$ and the dotted lines of $\sim T_c^2$ (BNC scaling) and $\sim T_c^3$ (super-strong scaling) are guides for the eyes.

where $x = \omega/T_c$ and the total scattering rate $\Gamma_{tot} = \Gamma_0 + \Gamma_\pi$. Eq.(7) shows that the non-pair-breaking scattering rate Γ_0 entering the thermal average part can strongly reduce ΔC without affecting T_c.

In Fig.2, we show the numerical results of $\Delta C/T_c$ *vs.* T_c with the above impurity scattering included. The calculations were done with moderate strength of the pairing potentials, $\bar{V}_{inter} = 2.0$ and $\bar{V}_{intra} = 0.5$, and with the varied impurity scattering rates Γ_0 and Γ_π. The case without impurity scattering (red " $\times$ " symbols, $\Gamma_0 = \Gamma_\pi = 0$), $\Delta C/T_c$ shows the T_c^2 scaling only for the limited region near the maximum T_c and it quickly becomes flattened and slower than $\sim T_c$. Interestingly, this behavior looks very similar to the experimental data of $Ba_{1-x}K_xFe_2As_2$ [13]. Therefore, we speculate that $Ba_{1-x}K_xFe_2As_2$ system belongs to the clean limit superconductor. This is understandable since K and Ba ions are out of the Fe-As planes and hence the doped K ions would introduce much weaker impurity potentials for the superconductivity. Next, only a small increase of impurity scattering (green " + " symbols, $\Gamma_0 = \Gamma_\pi = 0.02$ in unit of Λ_{hi}) immediately changes $\Delta C/T_c$ *vs.* T_c steeper such as $\Delta C/T_c \sim T_c^{1.5}$ over the whole T_c range. Further increasing the impurity scattering rate the case with $\Gamma_0 = 0.1$ and $\Gamma_\pi = 0.05$ (pink "♦" symbols) displays an ideal BNC scaling $\Delta C/T_c \sim T_c^2$ for the entire range of T_c. For the demonstration purposes, we also calculated the case with unrealistically large impurity scattering rates, $\Gamma_0 = 0.5$ and $\Gamma_\pi = 0.05$ (dark yellow " • " symbols), which displays $\Delta C \sim T_c^4$, a super-strong scaling.

The main message of the numerical calculations in Fig.1 and Fig.2 is: (1) the strong non-BCS relation of $\Delta C \sim T_c^\alpha$ with a power α larger than $\alpha = 1$(BCS limit) is a generic feature of the multiband superconductors, and this is due to a pure kinematic constraint of Eq.(2); (2) however, the system needs some amount of non-pair-breaking impurity scattering (for example,

$\Gamma_0/T_c^0 \approx 1/3$, the pink "♦" symbols in Fig.2) to have the observed BNC scaling. If the impurity scattering strength is weaker, a deviation from the BNC scaling becomes stronger, in particular in the low T_c region. These theoretical features can consistently explain why a majority of the Fe-pnictide systems[1, 3, 4, 5, 6, 7] follow the BNC scaling but some systems[13, 17] show a deviation from it.

3. Condensation Energy ΔE vs. T_c

3.1. Formalism

Using the exactly same model of the previous section, we can calculate the CE ΔE that is the zero temperature property. Therefore we need to solve the coupled gap equation Eq.(1) at $T = 0$ which is rewritten as

$$\begin{aligned} \Delta_h &= -\sum_k V_{hh} b_k^h - \sum_k V_{he} b_k^e \\ \Delta_e &= -\sum_k V_{eh} b_k^h - \sum_k V_{ee} b_k^e \end{aligned} \tag{8}$$

with the Cooper pair amplitudes $b_k^h =< h_{k\uparrow} h_{-k\downarrow} >$ and $b_k^e =< e_{k\uparrow} e_{-k\downarrow} >$, respectively, and at $T = 0$, $b^{h(e)} = \sum_k b_k^{h(e)} = \sum_k \frac{1}{2} \frac{\Delta_{h(e)}}{\sqrt{\xi^2_{h(e)}(k) + \Delta^2_{h(e)}}} = N_{h(e)} \Delta_{h(e)} \log \frac{\Lambda_{hi} + \sqrt{\Lambda_{hi}^2 + \Delta_{h(e)}^2}}{|\Delta_{h(e)}|}$. In general, the CE consists of the kinetic energy loss (increase) and the potential energy gain (decrease) as $\Delta E \equiv < H >_s - < H >_n = \Delta KE + \Delta PE$, which are written as the following general expressions

$$\Delta KE = \sum_{a=h,e} -\frac{1}{2} N_a \Delta_a^2 + N_i \Delta_a^2 \log \frac{\Lambda_{hi} + \sqrt{\Lambda_{hi}^2 + \Delta_a^2}}{|\Delta_a|}, \tag{9}$$

$$\Delta PE = V_{hh} b^{*h} b^h + V_{ee} b^{*e} b^e + V_{he} b^{*h} b^e + V_{eh} b^{*e} b^h. \tag{10}$$

The total ΔE has no further simplification and should be calculated numerically.

3.2. Numerical Results

In Fig.3, we plotted the results of the case of $\bar{V}_{intra} = (\bar{V}_{ee} = \bar{V}_{hh}) = 0$ with varying inter-band potential $\bar{V}_{inter} (= \bar{V}_{he} = \bar{V}_{eh}) = 1.0, 2.0, 3.0, 4.0$, and 5.0, respectively. To our surprise, this simple model immediately produces a strongly non-BCS power law scaling $\Delta E \sim T_c^\beta$ ($\beta \approx 3$) for a wide range of T_c, and ΔE varies over three orders of magnitude. This is exactly the key feature of the experimental data[2].

In Fig.4, we studied the effect of the intra-band interaction $\bar{V}_{intra} = (\bar{V}_{ee} = \bar{V}_{hh}) \neq 0$. We experimented both repulsive ($\bar{V}_{intra} > 0$) and attractive ($\bar{V}_{intra} < 0$) intra-band interactions while fixing the inter-band interaction $\bar{V}_{inter} = 2.0$ at a moderate value. First, adding even very weak attractive ($\bar{V}_{intra} < 0$) intra-band interaction (black squares, pink stars), the scaling relation quickly converges to the BCS limit $\Delta E \sim T_c^2$. This result strongly suggests that the intra-band interaction in the Fe-based superconductors should be repulsive ($\bar{V}_{intra} > 0$) – at least, not attractive. On the other hand, a repulsive intra-band interaction ($\bar{V}_{intra} > 0$) causes the opposite trend although its effect is rather weak, so that we cannot reach the experimental power $\beta \approx 3.5$ with reasonable parameter set.

Finally, in Fig.5, we studied the effect of the pair-breaking impurity scattering (interband impurity scattering) Γ_π. We have also studied the non-pair-breaking impurity scattering (intraband impurity scattering) Γ_0 and found that it has no effect on the scaling which is understandable with Anderson theorem. The results in Fig.5 shows that a very weak impurity scattering rate $\Gamma_\pi/\Lambda_{hi} = 0.05$ is already sufficient to make the scaling as $\Delta E \sim T_c^{3.5}$ as consistent with the experimental data[2].

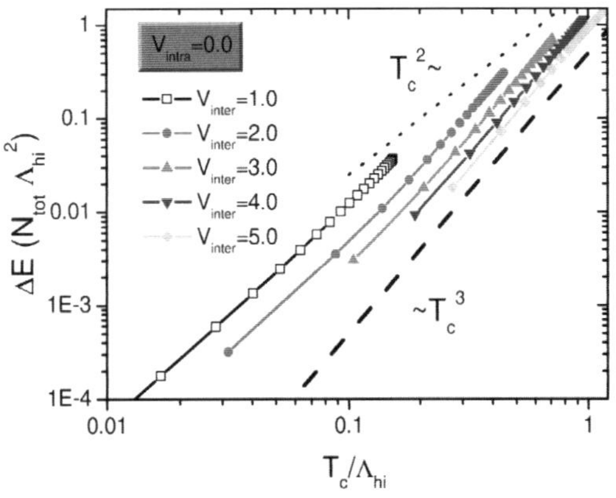

Figure 3. (Color online) Numerical calculations of ΔE *vs.* T_c of the two band model with $\bar{V}_{intra}(= N_{tot}V_{intra}) = 0.0$ for $\bar{V}_{inter}(= N_{tot}V_{inter}) = 1.0, 2.0, 3.0, 4.0$ and 5.0, respectively. The dotted line ($\sim T_c^2$, BCS scaling) and the dashed line ($\sim T_c^3$) are guides for the eyes.

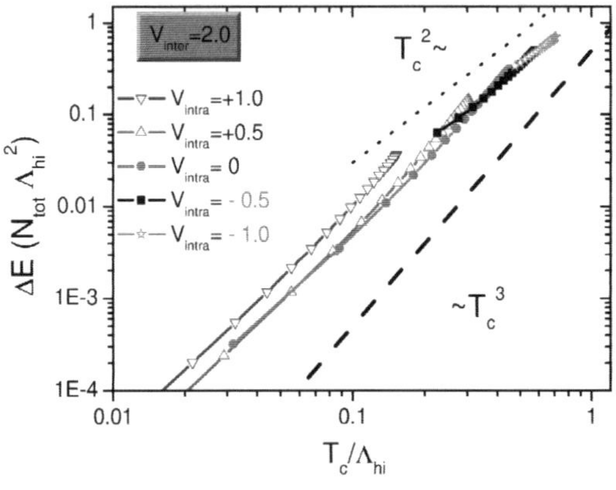

Figure 4. (Color online) Numerical calculations of ΔE *vs.* T_c of the two band model with $\bar{V}_{inter} = 2.0$ for $\bar{V}_{intra} = +1.0, +0.5, 0.0, -0.5$ and -1.0, respectively.

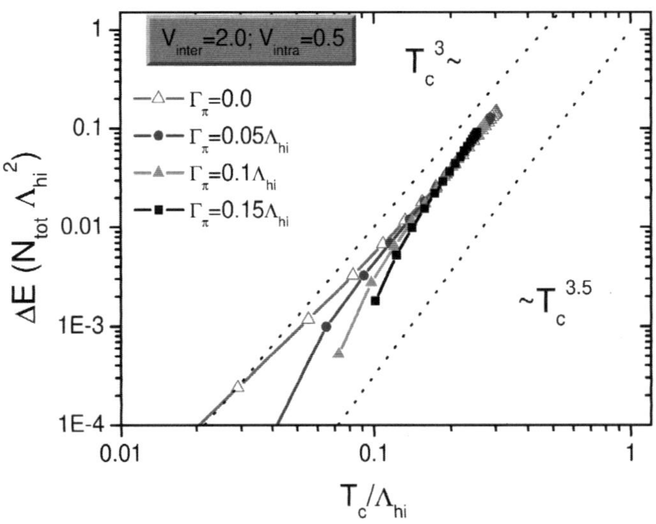

Figure 5. (Color online) Numerical calculations of ΔE *vs.* T_c of the two band model with $\bar{V}_{inter} = 2.0$ and $\bar{V}_{intra} = 0.5$ for the pair-breaking impurity scattering rates $\Gamma_\pi/\Lambda_{hi} = 0.0, 0.05, 0.1$, and 0.15, respectively. The dotted lines of $\sim T_c^3$ and $\sim T_c^{3.5}$ are guides for the eyes.

4. Summary and Conclusions

We have studied the scaling relations of the ΔC *vs.* T_c (BNC scaling) and ΔE *vs.* T_c based on the minimal two band BCS model. Despite the extreme simplicity of the model, our results almost immediately reproduced the two anomalous scaling relations – $\Delta C \sim T_c^3$[1] and $\Delta E \sim T_c^{3.5}$[2] – simultaneously. Adding a small amount of impurity scattering effect, our numerical results can perfectly fit all experimental data. Important implication of our study is that the experimentally observed seemingly non-BCS scaling relations of $\Delta C \sim T_c^3$ and $\Delta E \sim T_c^{3.5}$[2] are in fact strong experimental evidences that the Fe-based superconductors are the BCS superconductors, but with multi-bands, mediated by a dominant repulsive inter-band pairing interaction. Strong correlation effects, abundantly observed in the normal state, should renormalize the effective mass m^*_{qp} of quasiparticles, DOS $N_{h,e}$, the pairing interactions $V_{inter,intra}(\mathbf{q})$, etc., but when the system enters the SC transition, the pairing mechanism itself seems to be governed by the BCS mechanism. Finally, the results of this paper does not have direct relevance to the FeSe monolayer and related systems where only electron FSs exist[18].

4.1. Acknowledgments

YB was supported by Chonnam National University Grant 2014 and Grant No. 2013-R1A1A2-057535 of the National Research Foundation of Korea. GRS was supported by the US Department of Energy, Office of Basic Energy Sciences, contract no. DE-FG02-86ER45268.

References

[1] Budko S L, Ni N, and Canfield P C, 2009 *Phys. Rev. B* **79**, 220516.
[2] Xing J *et al.* 2014 *Phys. Rev. B* **89**, 140503.
[3] Kim J S, Stewart G. R, Kasahara S, Shibauchi T, Terashima T, and Matsuda Y 2011 *J. Phys.: Condens. Matter* **23**, 222201.
[4] Hardy F *et al.* 2010 **Phys. Rev. B 81**, 060501(R) ; Hardy F *et al.* 2010 *Europhys. Lett.* **91**, 47008.
[5] Gofryk K *et al.* 2010 *Phys. Rev. B* **81**, 184518; K. Gofryk K, Vorontsov A B, Vekhter I, Sefat A S, Imai T, Bauer E D, Thompson J D, and Ronning F 2011 *ibid.* **83**, 064513.
[6] Kim J S, Faeth B D, and Stewart G. R 2012 *Phys. Rev. B* **86**, 054509.
[7] Bud'ko S L *et al.* 2014 *Phys. Rev. B* **89**, 014510.
[8] Zaanen J 2009 *Phys. Rev. B* **80**, 212502(R).
[9] Bang Y and Choi H Y 2008 *Phys. Rev. B* **78**, 134523.
[10] Bang Y 2010 *Phys. Rev. Lett.* **104**, 217001.
[11] Sato T, Nakayama K, Sekiba Y, Richard P, Xu Y M, Souma S, Takahashi T, Chen G F, Luo J L, Wang N L, and Ding H 2009 *Phys. Rev. Lett.* **103**, 047002
[12] Sudayama T, Wakisaka Y, Takubo K, Morinaga R, Sato T J, Arita M, Namatame H, Taniguchi M, and Mizokawa T 2010 *Phys. Rev. Lett.* **104**, 177002
[13] Bud'ko S L, Sturza M, Chung D Y, Kanatzidis M G, and Canfield P C 2013 *Phys. Rev. B* **87**, 100509
[14] Bang Y, Choi H Y, and Won H 2009 *Phys. Rev. B* **79**, 054529
[15] Abrikosov A A and Gorkov L P 1961 *Sov. Phys. JETP* **12**, 1243
[16] Skalski S, Betbeder-Matibet O, and Weiss P R 1957 *Phys. Rev.* **136**, A1500.
[17] Grinenko V *et al.* 2014 *Phys. Rev. B* **89**, 060504
[18] Wang Q Y *et al* 2012, *Chin. Phys. Lett.* **29**, 037402.

IOP Conf. Series: Journal of Physics: Conf. Series **807** (2017) 052004 doi:10.1088/1742-6596/807/5/052004

Enhancement of T_c in CeIr$(In_{1-x}Cd_x)_5$ studied by In-NQR

Mitsuharu Yashima[1], Kyohei Tani[1], Kazuhiro Nishimoto[1], Hidekazu Mukuda[1], Yoshio Kitaoka[1], Fuminori Honda[2], Rikio Settai[3], Yoshichika Ōnuki[4]

[1]Graduate School of Engineering Science, Osaka University, Toyonaka, Osaka 560-8531, Japan
[2]Institute for Materials Research, Tohoku University, Oarai, Ibaraki 311-1313, Japan
[3]Department of Physics, Niigata Univ., Ikarashi, Niigata, 950-2181, Japan
[4]Department of Physics and Earth Sciences, Faculty of Science, University of the Ryukyus, Nishihara, Okinawa 903-0213, Japan

E-mail: `mitsuharu@nmr.mp.es.osaka-u.ac.jp`

Abstract. We report on superconducting nature under pressure in CeIr$(In_{1-x}Cd_x)_5$ by In-Nuclear-Quadrupole-Resonance (NQR) studies. In CeIr$(In_{0.925}Cd_{0.075})_5$, the inhomogeneous antiferromagnetic order at $T_N \sim 2.3$ K is induced by Cd-dopants and superconductivity disappears at ambient pressure. However, the measurements of a nuclear-spin-lattice-relaxation rate $1/T_1$ have revealed that the superconductivity suddenly occurs above 2.1 GPa. It is observed that the superconducting gap is enhanced in the Cd-doped sample, indicating that the Cd-doping induces the strong coupling superconductivity leading to the enhancement of T_c in the CeIrIn$_5$ system. Furthermore, we found that the residual density of states at the Fermi level increases with increasing pressure, suggesting that the superconducting nature of CeIrIn$_5$ is quite different from those of CeCoIn$_5$ and CeRhIn$_5$.

1. Introduction

The cerium (Ce)-based heavy-fermion compounds, CeMIn$_5$ (M = Co, Rh, and Ir) provide us with the opportunity to systematically investigate an interplay between antiferromagnetism (AF) and superconductivity (SC) [1, 2, 3, 4]. CeIrIn$_5$ and CeCoIn$_5$ show SC at ambient pressure (P) below $T_c = 0.4$ and 2.3 K, respectively [1, 2, 3, 4]. In CeRhIn$_5$, which is an antiferromagnet with T_N = 3.8 K at ambient P, SC occurs at $T_c \sim 2.2$ K under P over 2 GPa. Since the existence of two superconducting phases was reported in CeRh$_{1-y}$Ir$_y$In$_5$ [5], it has been suggested that the superconducting nature in CeIrIn$_5$ is distinguished from those in CeCoIn$_5$ and CeRhIn$_5$. Figs.1(a) and 1(b) show the respective phase diagrams of SC (denoted as SC1 and SC2) for CeRh$_{1-y}$Ir$_y$In$_5$ and CeIrIn$_5$ under P. SC1 is closely related to antiferromagnetic interactions, as argued extensively in CeCoIn$_5$[6] and CeRhIn$_5$[7], whereas it is reported from the previous NQR measurements that the SC2 in CeIrIn$_5$ occurs in a different situation from SC1 as follows [8]. The maximum of T_c(= 0.9 K) around 3 GPa takes place under the absence of antiferromagnetic correlations that is confirmed from the fact that the $(T_1T)^{-1}$-constant behavior is observed above T_c, where T_1 is a nuclear-spin-lattice-relaxation time. However, the values of $(T_1T)^{-1}$ still remain an order of magnitude larger than those in LaIrIn$_5$, indicating that some fluctuation except for AF spin fluctuations are dominant under high P. Actually, the large electronic specific heat

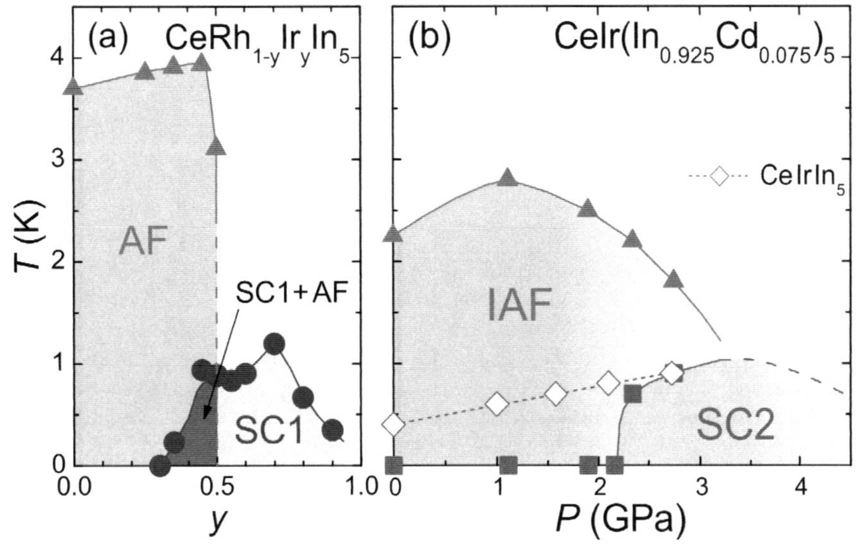

Figure 1. (Color online) (a) The phase diagram for $CeRh_{1-x}Ir_xIn_5$ as a function of Ir concentration. T_N and T_c are referred from Refs. [24, 25]. SC1 is expected to be mediated by magnetic correlations. (b) The $P-T$ phase diagram for $CeIr(In_{0.925}Cd_{0.075})_5$ partly from Ref. [26]. The T_N(solid) and T'_N(open triangles) are determined from the peak in the T dependence of $1/T_1$ and the broadening of NQR spectrum due to the onset of the antiferromagnetic order, respectively. The diamonds indicates T_c in pure $CeIrIn_5$ [8].

coefficient $\gamma \sim 0.38$ J/molK^2 at $P = 1.56$ GPa is observed in $CeIrIn_5$ [9]. Furthermore, from the resistivity measurements, a non-Fermi liquid behavior that the resistivity approximately follows a $T^{1.5}$ law has been observed in the wide P range up to 3.1 GPa [3].

Meanwhile, it was reported that all the anomalous transport properties observed in $CeRh_{0.2}Ir_{0.8}In_5$(SC1) and $CeIrIn_5$(SC2) originate from the AF spin fluctuations irrespective of the superconducting phase to which the system belongs. [10]. The detailed characteristics of AFM spin fluctuations including the dimensionality at ambient P in $CeIrIn_5$ are reported from the NMR measurements [11, 12]. In this context, the reason why the maximum T_c of SC2 in $CeIrIn_5$ is realized far away from an AF quantum critical point (QCP) is not clarified yet. The existence of two superconducting domes(SC1 and SC2) was first reported in $CeCu_2(Si_{1-x}Ge_x)_2$ [13]. It is suggested from extensive experimental and theoretical studies that the occurrence of SC2 in the high P region is mediated by valence fluctuations [14, 15, 16, 17]. In order to clarify the characters of SC2, we report the nuclear-quadrupaole-resonance (NQR) measurement under P in pure and Cd-doped $CeIrIn_5$.

Single crystals of $CeIr(In_{1-x}Cd_x)_5$ were grown by the In-flux method. They were crushed into coarse powder in order to allow RF pulses to easily penetrate the sample for NQR measurements. T_c's for the samples are determined by a sudden decrease in $1/T_1$ below T_c. Hydrostatic P was applied by utilizing a piston-cylinder cell(NiCrAl-BeCu double cylinder) filled with Daphne 7474 as a P-transmitting medium [18]. To calibrate P at low T's, the shift in T_c of Sn metal at P was monitored by resistivity measurements. $CeMIn_5$ consists of alternating layers of CeIn and MIn_4, and there are two sites—In(1) and In(2)—per unit cell. The In(1) and In(2) sites are located in the CeIn and MIn_4 layers, respectively. The measurements of T_1 were mainly performed at the transition of $2\nu_Q$ for the high symmetry In(1) site in $CeMIn_5$. T_1 was measured by the

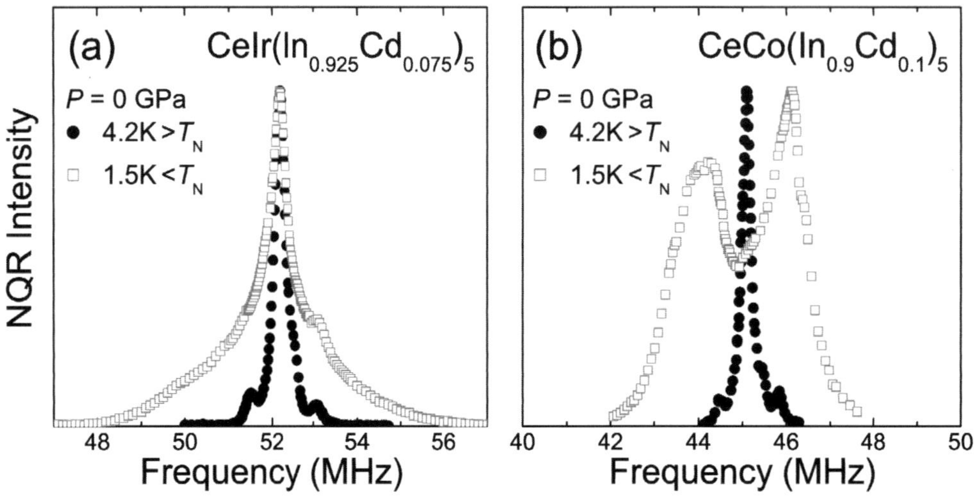

Figure 2. (Color online) (a) $3\nu_Q$ spectra at the In(2) site above and below $T_N \sim 2.3$ K in $CeIr(In_{0.925}Cd_{0.075})_5$. (b) $3\nu_Q$ spectra at the In(2) site above and below $T_N \sim 3$ K in $CeCo(In_{0.9}Cd_{0.1})_5$.

conventional saturation-recovery method. Here, an NQR frequency (ν_Q) is defined by the NQR Hamiltonian, $\mathcal{H}_Q = (h\nu_Q/6)[3I_z{}^2 - I(I+1) + \eta(I_x{}^2 - I_y{}^2)]$, where η is the asymmetric parameter of the electric field gradient. $\eta = 0$ at the In(1) site is expected because of its good symmetry.

2. Experiments and discussion

Figure 1(b) shows the $P-T$ phase diagram in $CeIr(In_{0.925}Cd_{0.075})_5$. The antiferromagnetic order induced by the Cd-doping is observed in the wide P range. Fig. 2(a) shows the NQR spectra at the $3\nu_Q$ of the In(2) site above and below T_N at ambient P in $CeIr(In_{0.925}Cd_{0.075})_5$. The full width at half maximum (σ) of the paramagnetic NQR spectrum at $T = 4.2$ K is less than 350 kHz and considerably sharp, indicating the relatively high-quality of sample in spite of the Cd-doping as impurities. The main peak from ^{115}In nuclei is observed at ~ 52.2 MHz and the small peak at a little lower frequency is derived from the isotope ^{113}In. The origin of another peak at higher frequency than the main peak is not identified. Since it hardly broadens and is distinguishably observed below T_N in spite of no detection of the signal from ^{113}In, it may be associated with some impurity mixed into the sample. It is also possible that another peak at higher frequency than the main peak is derived from impurities induced by the Cd doping. The antiferromagnetic NQR spectrum at $T = 1.5$ K exhibits no clear splitting, but a large broadening at its tail in $CeIr(In_{0.925}Cd_{0.075})_5$. From the previous NQR measurements at the In(2) site in Cd-doped $CeCoIn_5$, as shown in Fig. 2(b), the clear splitting of the NQR spectrum is observed below T_N, indicating that the homogeneous antiferromagnetic order with a uniform magnetic moment M_{AF} over the whole sample is realized [19]. Therefore, the spectral shape below T_N in $CeIr(In_{0.925}Cd_{0.075})_5$ indicates the inhomogeneous antiferromagnetic order (IAF) with a large distribution of M_{AF}. The amplitude of M_{AF} may be locally enhanced near Cd-dopants through the reduction of hybridization between $4f$ and conduction electrons. The difference in the character of the AFM order is probably relevant to the fact that $CeCoIn_5$ is much closer to an AFM QCP than $CeIrIn_5$.

Next, we discuss superconducting characteristics for $CeIr(In_{1-x}Cd_x)_5$. As shown in Fig. 1(b), T_c is enhanced by the application of P in $CeIrIn_5$ in spite of moving the system further away from

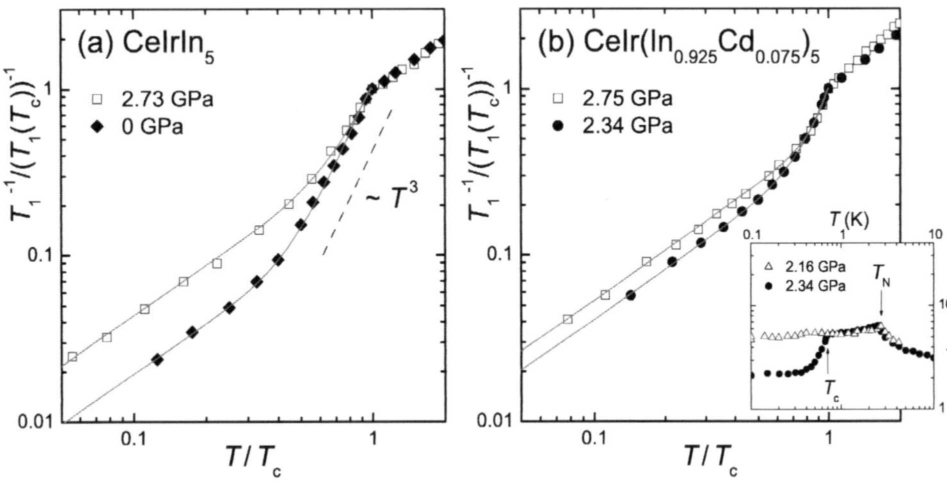

Figure 3. (Color online) (a) The T dependence of $1/T_1$ normalized at T_c at P = 0 and 2.73 GPa in $CeIrIn_5$. (b) The T dependences of $1/T_1$ normalized at T_c at P = 2.34 and 2.75 GPa in $CeIr(In_{0.925}Cd_{0.075})_5$. The inset is the T dependence of $(T_1T)^{-1}$ at P = 2.16 and 2.34 GPa. The solid lines in both figures are the calculated results (see text).

the antiferromagnetic QCP. The IAF phase is observed in wide P region in $CeIr(In_{0.925}Cd_{0.075})_5$. Therefore, SC is suppressed by the onset of the antiferromagnetic order and/or the impurity effect from Cd-dopants. The anomaly in the T dependence of $(T_1T)^{-1}$ at T_N = 2.7 K is derived from the onset of the IAF order at 2.16 GPa in the inset of Fig. 3, and no anomalies are found below T_N, indicating that SC does not occurs up to 2.16 GPa. When P is slightly more increased to 2.34 GPa, the rapid decrease of $(T_1T)^{-1}$ suddenly appears at 0.7 K, suggesting the occurrence of SC. The diamagnetism is also confirmed from the ac-susceptibility measurement using the in-situ NQR coil. Above 2.34 GPa, T_c continues to increase and reaches about 0.9 K, which is the maximum value of T_c around 2.8 GPa in pure $CeIrIn_5$.

The T dependences of $1/T_1$ normalized at T_c are shown for the pure and Cd-doped samples in Figs. 3(a) and (b), respectively. The T dependence of $1/T_1$ below T_c allows us to estimate a superconducting gap ($2\Delta_0/k_BT_c$) and a residual density of states (RDOS) at the Fermi level due to some impurity effect by assuming a certain pairing symmetry. Several models as a pairing symmetry in $CeIrIn_5$ have been suggested from experimental studies, but a $d_{x^2-y^2}$-wave model is recently supported by the thermal-conductivity, magnetic-penetration-depth, and specific-heat measurements [20, 21, 22]. Therefore, we tentatively assume a $d_{x^2-y^2}$-wave model to analyze $1/T_1$ data obtained from the present NQR measurements. The solid lines in Figs. 3 are the calculated results for $CeIrIn_5$ with parameters of ($2\Delta_0/k_BT_c$, RDOS) = (5.2, 0.44) at P = 0 and (5.2, 0.66) at P = 2.73 GPa. $2\Delta_0/k_BT_c$ is almost independent of P in $CeIrIn_5$, consistent with the previous NQR measurements [8]. The value of RDOS in $CeIrIn_5$ is fairly large. RDOS = 0.08 and 0.14 are confirmed at P = 0 in $CeCoIn_5$ [6] and at P = 2.35 GPa in $CeRhIn_5$ [23], respectively. Since the linewidth (σ) of the NQR spectrum at $2\nu_Q$ for the In(1) site in $CeIrIn_5$ is about 50 kHz, it is expected that the sample quality is considerably high. In the case of $CeCoIn_5$ and $CeRhIn_5$, $\sigma \sim$ 20 and 50 kHz, respectively. Considering the existence of the large RDOS in spite of its high purity, it is possible that a superconducting gap is not formed or very small to be easily broken even by weak impurity scattering in one or a few bands of the Fermi surface. Here, we highlight the fact that RDOS in $CeIrIn_5$ increases from 0.44 to 0.66

despite the enhancement of T_c from 0.4 to 0.9 K. When T_c increases, it is simply expected that RDOS is reduced through the formation of a strong-coupling superconductivity. Nevertheless, RDOS is unexpectedly enhanced by the application of P. Such a behavior contrasts with those in $CeCoIn_5$ and $CeRhIn_5$ in which RDOS remains almost unchanged and is reduced as P increases, respectively. The unexpected enhancement of RDOS by the application of P cannot be understood in terms of a simple impurity effect. Furthermore, since $CeIrIn_5$ is probably farther from an AFM QCP than $CeCoIn_5$, the influence of antiferromagnetic spin fluctuations on RDOS should be small in $CeIrIn_5$, as compared with that in $CeCoIn_5$. In fact, antiferromagnetic spin fluctuations are greatly suppressed by the application of P in $CeIrIn_5$ [8]. The cause of the unexpected enhancement by P cannot be explained by the effect of antiferromagnetic spin fluctuation. Analogous SC characteristics are also observed in the Cd-doped sample with the parameters of ($2\Delta_0/k_BT_c$, RDOS) = (6.4, 0.64) and (7.7, 0.73) at P = 2.34 and 2.75 GPa, respectively, indicating that it is intrinsic in the $CeIr(In_{1-x}Cd_x)_5$ system. The value of RDOS in the Cd-doped sample is larger than that in the pure sample due to the impurity effect. It should be noted that $2\Delta_0/k_BT_c$ is enhanced by the Cd-doping in spite of the impurity effect and the induced IAF order. In order to solve the mechanism of SC2, it would be important to understand the role of the Cd-doping for SC2.

3. Conclusion

In conclusion, the present In-NQR measurements have revealed that the nature of SC2 in pure and Cd-doped $CeIrIn_5$ differs from that of SC1 that is closely related to the AFM correlation. In pure $CeIrIn_5$, the unexpected enhancement of RDOS by the application of P is observed in spite of no addition of impurity and the increase in T_c. The similar behavior is also observed in Cd-doped $CeIrIn_5$. The unexpected enhancement of RDOS may be related with the mechanism of SC2. The similar T_c to the maximum value($T_c^{max} \sim 0.9$ K) in pure $CeIrIn_5$ is realized even in Cd-doped $CeIrIn_5$ in spite of the existence of Cd impurities and the AF order. The Cd-doping induces the strong coupling superconductivity in the $CeIrIn_5$ system, leading to the enhancement of T_c for SC2. In order to understand the mechanism of SC2 in more detail, it is inevitable to clarify the cause for the unexpected enhancement of RDOS and T_c in the Cd-doped $CeIrIn_5$ system.

4. Acknowledgments

We thank S. Watanabe and K. Miyake for enlightening discussions and theoretical comments. This work was supported by Grants-in-Aid for Specially Promoted Research (Grant No. 20001004) and for Scientific Research (C) (Grant No. 24540372) from the Ministry of Education, Culture, Sports, Science and Technology (MEXT) of Japan. It was partially supported by the Global COE Program from MEXT.

5. References

[1] C. Petrovic, P. G. Pagliuso, M. F. Hundley, R. Movshovich, J. L. Sarrao, J. D. Thompson, Z. Fisk, and P. Monthoux, J. Phys.: Cond. Mat. **13**, L337 (2001).
[2] H. Hegger, C. Petrovic, E. G. Moshopoulou, M. F. Hundley, J. L. Sarrao, Z. Fisk, and J. D. Thompson, Phys. Rev. Lett. **84**, 4986 (2000).
[3] T. Muramatsu, N. Tateiwa, T. C. Kobayashi, K. Shimizu, K. Amaya, D. Aoki, H. Shishido, Y. Haga, and Y. Ōnuki, J. Phys. Soc. Jpn. **70**, 3362 (2001), and unpublished.
[4] C. Petrovic, R. Movshovich, M. Jaime, P. G. Pagliuso, M. F. Hundley, J. L. Sarrao, Z. Fisk, and J. D. Thompson, Europhys. Lett. **53**, 354 (2001).
[5] M. Nicklas, V. A. Sidorov, H. A. Borges, P. G. Pagliuso, J. L. Sarrao, and J. D. Thompson, Phys. Rev. B **70**, 020505(R) (2004).
[6] M. Yashima, S. Kawasaki, Y. Kawasaki, G.-q. Zheng, Y. Kitaoka, H. Shishido, R. Settai, Y. Haga and Y. Ōnuki, J. Phys. Soc. Jpn. **73**, 2073 (2004).

[7] M. Yashima, H. Mukuda, Y. Kitaoka, H. Shishido, R. Settai, and Y. Ōnuki, Phys. Rev. B **79**, 214528 (2009).
[8] S. Kawasaki, G.-q. Zheng, H. Kan, Y. Kitaoka, H. Shishido, and Y. Ōnuki, Phys. Rev. Lett. **94**, 037007 (2005).
[9] R. Bortha, E. Lengyela, P. G. Pagliusob, J. L. Sarraob, G. Sparna, F. Steglicha, and J. D. Thompson, Physica B **312-313**, 136 (2002).
[10] Y. Nakajima, H. Shishido, H. Nakai, T. Shibauchi, M. Hedo, Y. Uwatoko, T. Matsumoto, R. Settai, Y. Onuki, H. Kontani, and Y. Matsuda, Pys. Rev. B **77**, 214504 (2008).
[11] S. Kambe, Y. Tokunaga, H. Sakai, H. Chudo, Y. Haga, T. D. Matsuda, and R. E. Walstedt, Pys. Rev. B **81**, 140405(R) (2010).
[12] S. Kambe, H. Sakai, Y. Tokunaga, and R. E. Walstedt, Pys. Rev. B **82**, 144503 (2010).
[13] H. Q. Yuan, F. M. Grosche, M. Deppe, C. Geibel, G. Sparn, and F. Steglich, Science **302**, 2104 (2003).
[14] A. T. Holmes, D. Jaccard, and K. Miyake, Phys. Rev. B **69**, 024508 (2004).
[15] K. Fujiwara, Y. Hata, K. Kobayashi, K. Miyoshi, J. Takeuchi, Y. Shimaoka, H. Kotegawa, T. C. Kobayashi, C. Geibel, and F. Steglich, J. Phys. Soc. Jpn. **77**, 123711 (2008).
[16] Y. Onishi. and K. Miyake, J. Phys. Soc. Jpn. **69**, 3955 (2000).
[17] S. Watanabe, M. Imada, and K. Miyake, J. Phys. Soc. Jpn. **75**, 043710 (2006).
[18] K. Murata, K. Yokogawa, H. Yoshino, S. Klotz, P. Munsch, A. Irizawa, M. Nishiyama, K. Iizuka, T. Nanba, T. Okada, Y. Shiraga, and S. Aoyama, Rev. Sci. Instrum. **79**, 085101 (2008).
[19] R.R. Urbano, B.-L. Young, N. J. Curro, J. D. Thompson, L. D. Pham, and Z. Fisk, Phys. Rev. Lett. **99**, 146402 (2007).
[20] Y. Kasahara, T. Iwasawa, Y. Shimizu, H. Shishido, T. Shibauchi, I. Vekhter, and Y. Matsuda, Phys. Rev. Lett. **100**, 207003 (2008).
[21] D. Vandervelde, H. Q. Yuan, Y. Ōnuki, and M. B. Salamon, Phys. Rev. B **79**, 212505 (2009).
[22] S. Kittaka, Y. Aoki, T. Sakakibara, A. Sakai, S. Nakatsuji, Y. Tsutsumi, M. Ichioka, and Kazushige Machida, Phys. Rev. B **85**, 060505(R) (2012).
[23] M. Yashima, H. Mukuda, Y. Kitaoka, H. Shishido, R. Settai, and Y. Ōnuki, Phys. Rev. B **79**, 214528 (2009).
[24] G.-q. Zheng, N. Yamaguchi, H. Kan, Y. Kitaoka, J. L. Sarrao, P. G. Pagliuso, N. O. Moreno, and J. D. Thompson, Phys. Rev. B **70**, 014511 (2004).
[25] S. Kawasaki, M. Yashima, Y. Mugino, H. Mukuda, Y. Kitaoka, H. Shishido,and Y. Ōnuki, Phys. Rev. Lett. **96**, 147001 (2006).
[26] M. Yashima, N. Tagami, S. Taniguchi, T. Unemori, K. Uematsu, H. Mukuda, Y. Kitaoka, Y. Ōta, F. Honda, R. Settai, and Y. Ōnuki, Phys. Rev. Lett. **109**, 117001 (2012).

International Conference on Strongly Correlated Electron Systems (SCES 2016) IOP Publishing
IOP Conf. Series: Journal of Physics: Conf. Series **807** (2017) 052005 doi:10.1088/1742-6596/807/5/052005

Coherence Effects of Caroli–de Gennes–Matricon Modes in a Nodal Topological Superconductor UPt_3

Yasumasa Tsutsumi and Yusuke Kato

Department of Basic Science, The University of Tokyo, Meguro, Tokyo 153-8902, Japan

E-mail: `tsutsumi@vortex.c.u-tokyo.ac.jp`

Abstract. Coherence effects by the impurity scattering of Caroli–de Gennes–Matricon (CdGM) modes in a vortex for the E_{1u} planar state and the E_{1u} chiral state which are possible superconducting states in UPt_3 have been studied. First, we analytically derived the eigenvalue and eigenfunction of the CdGM modes under magnetic elds producing a vortex without impurities. Then, we studied impurity effects on the CdGM modes by introducing the impurity self-energy, which are dominated by the coherence factor depending on the eigenfunction of the CdGM modes. The difference of the impurity effects for the E_{1u} planar and E_{1u} chiral states owing to the magnetic eld is possibly utilized to identify the gap function in UPt_3.

1. Introduction

The heavy fermion superconductor UPt_3 [1, 2] is a highly probable nodal topological superconductor. As a result of numerous experimental and theoretical studies over three decades, the possible gap functions in UPt_3 have been narrowed to the E_{1u} planar state [3, 4], E_{1u} chiral state [5], and E_{2u} chiral state [6]. Since each superconducting gap structure in the momentum space di ers between the E_{1u} states and the E_{2u} state, the distinction between them will be made by further experiments. In the E_{1u} states, however, the E_{1u} planar state, described by the d-vector $\boldsymbol{d}(\boldsymbol{k}) \propto (\boldsymbol{x}k_y + \boldsymbol{y}k_x)(5k_z^2 - k^2)$, and the E_{1u} chiral state, $\boldsymbol{d}(\boldsymbol{k}) \propto \boldsymbol{z}(k_x + ik_y)(5k_z^2 - k^2)$, have the same gap structure. Moreover, the results of distinguishable measurements between them have been in con ict with each other. On one hand, the suppression of the upper critical eld along the c-axis like the Pauli limiting [2] is compatible with the E_{1u} chiral state, but, on the other, the anisotropy of the Knight shift [7] corresponds to the E_{1u} planar state. The presence or absence of the time reversal symmetry in the superconducting state also has not been clear [8, 9, 10]. Then, we hope other experimental methods in order to distinguish between the E_{1u} planar state and the E_{1u} chiral state.

In this paper, we have investigated impurity e ects of the Caroli–de Gennes–Matricon (CdGM) modes [11] in a quantized vortex for the E_{1u} planar state and the E_{1u} chiral state including in uence of the magnetic eld which produces a vortex. The di erence of the impurity e ects for them may be utilized to identify the gap function in UPt_3. First, in Sec. 2, we analytically derive the eigenvalue and eigenfunction of the CdGM modes under weak magnetic elds producing a quantized vortex along the c-axis for the E_{1u} planar and E_{1u} chiral states. The impurity e ects of the CdGM modes are clari ed in Sec. 3 through the analytically derived coherence factor and the numerically calculated density of states (DOS). In the nal section, we summarize the results and discuss how to utilize the impurity e ects for a distinction between the E_{1u} planar state and the E_{1u} chiral state.

2. Caroli–de Gennes–Matricon modes

First, we derive CdGM modes without impurities from the Bogoliubov–de Gennes (BdG) equation. The BdG equation for an inhomogeneous order parameter is described by [12]

$$\int d\boldsymbol{r}_2 \begin{pmatrix} \hat{\epsilon}(\boldsymbol{r}_1, \boldsymbol{r}_2) & \hat{\Delta}(\boldsymbol{r}_1, \boldsymbol{r}_2) \\ \hat{\Delta}^\dagger(\boldsymbol{r}_2, \boldsymbol{r}_1) & -\hat{\epsilon}^{\mathrm{T}}(\boldsymbol{r}_2, \boldsymbol{r}_1) \end{pmatrix} \vec{u}_\nu(\boldsymbol{r}_2) = E_\nu \vec{u}_\nu(\boldsymbol{r}_1). \tag{1}$$

In the normal-state Hamiltonian $\hat{\epsilon}(\boldsymbol{r}_1, \boldsymbol{r}_2)$, the vector potential can be neglected when considering the CdGM modes in extreme type II superconductors with a large Ginzburg–Landau parameter $\kappa \gg 1$ [11]. Then, the magnetic eld $\boldsymbol{H}$ producing a vortex along the z-axis is included in the normal-state Hamiltonian only as the Zeeman term $\mu_{\mathrm{B}} \boldsymbol{H} \cdot \hat{\boldsymbol{\sigma}}$:

$$\hat{\epsilon}(\boldsymbol{r}_1, \boldsymbol{r}_2) = \ (\boldsymbol{r}_1 - \boldsymbol{r}_2) \left\{ \frac{\hbar^2}{2m} \left[-\partial^2 - \frac{1}{\ }\partial \ - \frac{1}{\ ^2}\partial_\phi^2 - \partial_z^2 - k_{\mathrm{F}}^2 \right] \hat{\sigma}_0 + \mu_{\mathrm{B}} \boldsymbol{H} \cdot \hat{\boldsymbol{\sigma}} \right\}, \tag{2}$$

where m is the particle mass, k_{F} is the Fermi wave number, μ_{B} is the Bohr magneton, $\hat{\sigma}_0$ is the 2×2 unit matrix, $\hat{\boldsymbol{\sigma}}$ is the Pauli matrix, and $(\partial \ , \partial_\phi, \partial_z)$ are di erential operators in cylindrical coordinates. The pair potential is

$$\hat{\Delta}(\boldsymbol{r}_1, \boldsymbol{r}_2) = \int \frac{d\boldsymbol{k}}{(2\pi)^3} \hat{\Delta}(\boldsymbol{r}, \boldsymbol{k}) e^{i\boldsymbol{k}\cdot\boldsymbol{r}'}, \tag{3}$$

with $\boldsymbol{r} = (\boldsymbol{r}_1 + \boldsymbol{r}_2)/2$ and $\boldsymbol{r}' = \boldsymbol{r}_1 - \boldsymbol{r}_2$, and the wave function is

$$\vec{u}_\nu(\boldsymbol{r}) = \begin{pmatrix} u_\nu^\uparrow(\boldsymbol{r}) \\ u_\nu^\downarrow(\boldsymbol{r}) \\ v_\nu^\uparrow(\boldsymbol{r}) \\ v_\nu^\downarrow(\boldsymbol{r}) \end{pmatrix}. \tag{4}$$

2.1. E_{1u} planar state

For the E_{1u} planar state [3, 4], the gap function is described by

$$\hat{\Delta}(\boldsymbol{r}, \boldsymbol{k}) \equiv i\boldsymbol{d}(\boldsymbol{r}, \boldsymbol{k}) \cdot \hat{\boldsymbol{\sigma}}\hat{\sigma}_y = i\Delta(\boldsymbol{r})(\boldsymbol{x}k_y + \boldsymbol{y}k_x)(5k_z^2 - k_{\mathrm{F}}^2)/k_{\mathrm{F}}^3 \cdot \hat{\boldsymbol{\sigma}}\hat{\sigma}_y. \tag{5}$$

Note that the two spin components of the gap function have the same spatial dependence as long as the eld parallel to the z-axis [4]. By substituting the gap function into the BdG equation (1) and following the procedure in Ref. [13], the BdG equation can be separated into the equations for the up-spin state and for the down-spin state:

$$\begin{pmatrix} \epsilon(\boldsymbol{r}) + \mu_{\mathrm{B}} H & \frac{i}{k_{\mathrm{F}}}\left[\Delta(\boldsymbol{r})\mathcal{D}(\boldsymbol{r}) + \frac{1}{2}\mathcal{D}(\boldsymbol{r})\Delta(\boldsymbol{r})\right] \\ \frac{i}{k_{\mathrm{F}}}\left[\Delta^*(\boldsymbol{r})\mathcal{D}^*(\boldsymbol{r}) + \frac{1}{2}\mathcal{D}^*(\boldsymbol{r})\Delta^*(\boldsymbol{r})\right] & -\epsilon(\boldsymbol{r}) - \mu_{\mathrm{B}} H \end{pmatrix} \begin{pmatrix} u_\nu^\uparrow(\boldsymbol{r}) \\ v_\nu^\uparrow(\boldsymbol{r}) \end{pmatrix} = E_\nu^\uparrow \begin{pmatrix} u_\nu^\uparrow(\boldsymbol{r}) \\ v_\nu^\uparrow(\boldsymbol{r}) \end{pmatrix},$$
$$\begin{pmatrix} \epsilon(\boldsymbol{r}) - \mu_{\mathrm{B}} H & -\frac{i}{k_{\mathrm{F}}}\left[\Delta(\boldsymbol{r})\mathcal{D}^*(\boldsymbol{r}) + \frac{1}{2}\mathcal{D}^*(\boldsymbol{r})\Delta(\boldsymbol{r})\right] \\ -\frac{i}{k_{\mathrm{F}}}\left[\Delta^*(\boldsymbol{r})\mathcal{D}(\boldsymbol{r}) + \frac{1}{2}\mathcal{D}(\boldsymbol{r})\Delta^*(\boldsymbol{r})\right] & -\epsilon(\boldsymbol{r}) + \mu_{\mathrm{B}} H \end{pmatrix} \begin{pmatrix} u_\nu^\downarrow(\boldsymbol{r}) \\ v_\nu^\downarrow(\boldsymbol{r}) \end{pmatrix} = E_\nu^\downarrow \begin{pmatrix} u_\nu^\downarrow(\boldsymbol{r}) \\ v_\nu^\downarrow(\boldsymbol{r}) \end{pmatrix}, \tag{6}$$

where

$$\epsilon(\boldsymbol{r}) \equiv \frac{\hbar^2}{2m} \left[-\partial^2 - \frac{1}{\ }\partial \ - \frac{1}{\ ^2}\partial_\phi^2 - \partial_z^2 - k_{\mathrm{F}}^2 \right], \tag{7}$$

and

$$\mathcal{D}(\boldsymbol{r}) \equiv ie^{i\phi}\left(\partial \;\; + i\frac{1}{\;\;}\partial_\phi\right)\frac{5\partial_z^2 + k_{\rm F}^2}{k_{\rm F}^2}. \tag{8}$$

Thus, since the magnetic eld just gives the Zeeman shift of the Fermi energy, we can neglect the contribution from low magnetic elds.

Here, we consider the singly quantized vortex state. In low elds, another spin component of the gap function, $\boldsymbol{d} \parallel \boldsymbol{z}$, compensates the vortex core and prohibits the zero energy CdGM modes [4, 14]. Since we mainly focus on the topologically protected zero energy CdGM modes, we discuss the CdGM modes under the substantial eld along the z-axis to suppress $\boldsymbol{d} \parallel \boldsymbol{z}$. The condition gives the axisymmetric order parameter $\Delta(\boldsymbol{r}) = \Delta(\;)e^{i\phi}$ and the spin-degenerate eigenvalue of the CdGM modes [15]

$$E_\nu = -l\omega_q, \tag{9}$$

where

$$\omega_q \equiv |5\cos^2\alpha - 1|\frac{\int_0^\infty \frac{|\Delta(\;')|}{k_{\rm F}\;'} e^{\;\;2\chi_q(\;')}d\;'}{\int_0^\infty e^{\;\;2\chi_q(\;')}d\;'}, \tag{10}$$

with

$$\chi_q(\;) = \frac{|5\cos^2\alpha - 1|}{\hbar v_{\rm F}}\int_0 |\Delta(\;')|d\;', \tag{11}$$

by using the Fermi velocity $v_{\rm F}$. The quantum number $\nu = (l, q)$ consists of the angular momentum l and the wave number along the vortex line $q \equiv k_{\rm F}\cos\alpha$. Thus, the eigenvalue of the CdGM modes is discretized by $\omega_q \sim |\Delta(\infty)|^2/E_{\rm F}$ for each l, where $E_{\rm F}$ is the Fermi energy. The eigenfunctions of the CdGM modes for the up- and down-spin states are given by [15]

$$\begin{pmatrix} u_\nu^\uparrow(\boldsymbol{r}) \\ v_\nu^\downarrow(\boldsymbol{r}) \end{pmatrix} = \mathcal{N}_\nu^\uparrow \begin{pmatrix} J_{l+1}(k_q\;)e^{i\phi} \\ s_q J_{l\;\;1}(k_q\;)e^{\;\;i\phi} \end{pmatrix} e^{\;\;\chi_q(\;)}e^{il\phi}e^{iqz}, \quad \begin{pmatrix} u_\nu^\downarrow(\boldsymbol{r}) \\ v_\nu^\uparrow(\boldsymbol{r}) \end{pmatrix} = \mathcal{N}_\nu^\downarrow \begin{pmatrix} J_l(k_q\;) \\ -s_q J_l(k_q\;) \end{pmatrix} e^{\;\;\chi_q(\;)}e^{il\phi}e^{iqz}, \tag{12}$$

with the Bessel function J_l, $k_q \equiv k_{\rm F}\sin\alpha$, $s_q = \mathrm{sgn}(5\cos^2\alpha - 1)$, and the normalization factor $\mathcal{N}_\nu^{\uparrow(\downarrow)}$ for the up- (down-)spin state.

2.2. E_{1u} chiral state

For the E_{1u} chiral state [5], the gap function is described by

$$\hat{\Delta}(\boldsymbol{r}, \boldsymbol{k}) = i\Delta(\boldsymbol{r})\boldsymbol{z}(k_x + ik_y)(5k_z^2 - k_{\rm F}^2)/k_{\rm F}^3 \cdot \hat{\boldsymbol{\sigma}}\hat{\sigma}_y. \tag{13}$$

Here, we neglect the opposite chiral state proportional to $k_x - ik_y$. The neglected component does not show qualitative e ects on the CdGM modes as long as we consider uniaxially symmetric simple bands. By substituting Eq. (13) into the BdG equation (1), since the BdG Hamiltonian is block-diagonal, the BdG equation is separated into two equations as

$$\begin{pmatrix} \epsilon(\boldsymbol{r}) + \mu_{\rm B}H & \frac{1}{k_{\rm F}}\left[\Delta(\boldsymbol{r})\mathcal{D}(\boldsymbol{r}) + \frac{1}{2}\mathcal{D}(\boldsymbol{r})\Delta(\boldsymbol{r})\right] \\ -\frac{1}{k_{\rm F}}\left[\Delta^*(\boldsymbol{r})\mathcal{D}^*(\boldsymbol{r}) + \frac{1}{2}\mathcal{D}^*(\boldsymbol{r})\Delta^*(\boldsymbol{r})\right] & -\epsilon(\boldsymbol{r}) + \mu_{\rm B}H \end{pmatrix}\begin{pmatrix} u_\nu^\uparrow(\boldsymbol{r}) \\ v_\nu^\downarrow(\boldsymbol{r}) \end{pmatrix} = E_\nu^+ \begin{pmatrix} u_\nu^\uparrow(\boldsymbol{r}) \\ v_\nu^\downarrow(\boldsymbol{r}) \end{pmatrix},$$

$$\begin{pmatrix} \epsilon(\boldsymbol{r}) - \mu_{\rm B}H & \frac{1}{k_{\rm F}}\left[\Delta(\boldsymbol{r})\mathcal{D}(\boldsymbol{r}) + \frac{1}{2}\mathcal{D}(\boldsymbol{r})\Delta(\boldsymbol{r})\right] \\ -\frac{1}{k_{\rm F}}\left[\Delta^*(\boldsymbol{r})\mathcal{D}^*(\boldsymbol{r}) + \frac{1}{2}\mathcal{D}^*(\boldsymbol{r})\Delta^*(\boldsymbol{r})\right] & -\epsilon(\boldsymbol{r}) - \mu_{\rm B}H \end{pmatrix}\begin{pmatrix} u_\nu^\downarrow(\boldsymbol{r}) \\ v_\nu^\uparrow(\boldsymbol{r}) \end{pmatrix} = E_\nu \begin{pmatrix} u_\nu^\downarrow(\boldsymbol{r}) \\ v_\nu^\uparrow(\boldsymbol{r}) \end{pmatrix}. \tag{14}$$

The magnetic eld shifts the eigenvalue in Eq. (6) as $E_\nu^\pm = E_\nu^\uparrow \pm \mu_B H$.

Again we consider the singly quantized axisymmetric vortex state. As long as neglecting the opposite chiral component induced near the vortex core, the order parameter is described by the axisymmetric form, $\Delta(\boldsymbol{r}) = i\Delta(\ )e^{i\kappa\phi}$, under the eld along the z-axis. Note that the opposite chiral component is possible to break the axisymmetry [16]. The eigenfunction for the E_{1u} chiral state depends on the vorticity $\kappa = \pm 1$ owing to the chirality of the gap function. By the same mannar for the E_{1u} planar state, the eigenvalue of the CdGM modes is given by

$$E_\nu^\pm = -\kappa l \omega_q \pm \mu_B H. \tag{15}$$

The energy shift is signi cant even for small magnetic elds $\mu_B H \quad |\Delta(\infty)|$ because $\omega_q \sim |\Delta(\infty)|^2/E_F \quad |\Delta(\infty)|$. The eigenfunctions of the CdGM modes are given by

$$\begin{aligned} \begin{pmatrix} u_\nu^\uparrow(\boldsymbol{r}) \\ v_\nu^\downarrow(\boldsymbol{r}) \end{pmatrix} &= \mathcal{N}_\nu^+ \begin{pmatrix} J_{l+\frac{\kappa+1}{2}}(k_q^+ \ )e^{i\frac{\kappa+1}{2}\phi} \\ \kappa s_q J_{l \ \frac{\kappa+1}{2}}(k_q \ )e^{\ i\frac{\kappa+1}{2}\phi} \end{pmatrix} e^{\ \chi_q(\ )} e^{il\phi} e^{iqz}, \\ \begin{pmatrix} u_\nu^\downarrow(\boldsymbol{r}) \\ v_\nu^\uparrow(\boldsymbol{r}) \end{pmatrix} &= \mathcal{N}_\nu \begin{pmatrix} J_{l+\frac{\kappa+1}{2}}(k_q \ )e^{i\frac{\kappa+1}{2}\phi} \\ \kappa s_q J_{l \ \frac{\kappa+1}{2}}(k_q^+ \ )e^{\ i\frac{\kappa+1}{2}\phi} \end{pmatrix} e^{\ \chi_q(\ )} e^{il\phi} e^{iqz}, \end{aligned} \tag{16}$$

with the normalization factor $\mathcal{N}_\nu^\pm$ and

$$k_q^\pm \equiv k_F \sin\alpha \pm \frac{\mu_B H}{\hbar v_F \sin\alpha}. \tag{17}$$

The order of di erence between k_q^+ and k_q is small as $k_F(\mu_B H/E_F) \quad k_F$.

3. Impurity effects on Caroli–de Gennes–Matricon modes

Next, we consider impurity e ects on the CdGM modes. The Dyson equation obeyed by the Matsubara Green's function with impurity self-energy $\widehat{\Sigma}_{\rm imp}(\boldsymbol{r}_1, \boldsymbol{r}_2, \omega_n)$ is described by

$$\widehat{G}(\boldsymbol{r}, \boldsymbol{r}', \omega_n) = \widehat{G}^{(0)}(\boldsymbol{r}, \boldsymbol{r}', \omega_n) + \int d\boldsymbol{r}_1 \int d\boldsymbol{r}_2 \widehat{G}^{(0)}(\boldsymbol{r}, \boldsymbol{r}_1, \omega_n)\widehat{\Sigma}_{\rm imp}(\boldsymbol{r}_1, \boldsymbol{r}_2, \omega_n)\widehat{G}(\boldsymbol{r}_2, \boldsymbol{r}', \omega_n). \tag{18}$$

For nonmagnetic impurities, the impurity self-energy is given by

$$\widehat{\Sigma}_{\rm imp}(\boldsymbol{r}_1, \boldsymbol{r}_2, \omega_n) = \frac{\Gamma_{\rm n}}{\pi N_F} F(\boldsymbol{r}_1 - \boldsymbol{r}_2)\widehat{G}(\boldsymbol{r}_1, \boldsymbol{r}_2, \omega_n) \equiv \gamma F(\boldsymbol{r}_1 - \boldsymbol{r}_2)\widehat{G}(\boldsymbol{r}_1, \boldsymbol{r}_2, \omega_n), \tag{19}$$

where N_F is the DOS per spin in the normal state at the Fermi energy, $\Gamma_{\rm n}$ is the impurity scattering rate in the normal state, and $F(\boldsymbol{r}_1 - \boldsymbol{r}_2)$ describes the spatial dependence of the squared impurity potential. Here, $F(\boldsymbol{r}_1 - \boldsymbol{r}_2) = \ (\boldsymbol{r}_1 - \boldsymbol{r}_2)$ gives the impurity self-energy in the self-consistent Born approximation [17, 18, 19]. Instead, we introduce the spatial dependence of the impurity potential in the z-direction as $F(\boldsymbol{r}_1 - \boldsymbol{r}_2) = \ (\boldsymbol{\rho}_1 - \boldsymbol{\rho}_2)f(z_1 - z_2)$, where $\boldsymbol{\rho}$ indicates the two-dimensional coordinates perpendicular to the vortex line. We will consider the columnar defects with uniform $f(z_1 - z_2)$ which give the remarkable impurity e ects on the CdGM modes in nodal topological superconductors [15]. Note that the columnar defects can be arti cially spread by ion irradiation [20, 21]. The Green's function without impurities is derived from the BdG wave function in Eq. (4) as

$$\widehat{G}^{(0)}(\boldsymbol{r}, \boldsymbol{r}', \omega_n) = \sum_\nu \frac{\widehat{\tau}_z \vec{u}_\nu(\boldsymbol{r})\vec{u}_\nu^\dagger(\boldsymbol{r}')}{E_\nu - i\omega_n}, \tag{20}$$

where $\widehat{\boldsymbol{\tau}}$ is the Pauli matrix in the Nambu space. By deriving $\widehat{G}(\boldsymbol{r}, \boldsymbol{r}, \omega_n)$ from Eq. (18), we can obtain the DOS as

$$N(\omega) = \int d\boldsymbol{r} \mathrm{Im} \left[\frac{1}{\pi} \mathrm{Tr} \widehat{\tau}_z \, \widehat{G}(\boldsymbol{r}, \boldsymbol{r}, \omega_n) \Big|_{i\omega_n \to \omega + i0^+} \right]. \tag{21}$$

Here, we consider the DOS of the CdGM modes by the approximation of neglecting the contributions from the continuum state to the impurity self-energy of the CdGM modes. Within the approximation, we should consider two states of the CdGM modes for both the E_{1u} planar and E_{1u} chiral states. In this situation, the DOS is obtained by [18]

$$N(\omega) = \sum_{\nu,s} \mathrm{Im} \left[-\frac{1}{\pi} \frac{\left[(\bar{1} - \bar{\sigma}_\nu(\omega))^{-1} \right]_{s,s}}{\omega - E_\nu^s + i0^+} \right] \equiv \sum_{\nu,s} N_\nu^s(\omega), \tag{22}$$

with the quantum number $\nu = (l, q)$. The matrix $\bar{\sigma}_\nu(\omega)$ is constituted by [18]

$$[\bar{\sigma}_\nu(\omega)]_{s_1,s_2} \equiv \sigma_\nu^{s_1,s_2}(\omega) = \frac{\gamma}{(2\pi)^2} \sum_{l',s'_1,s'_2} \int dq' \tilde{f}(q - q') \frac{M_{\nu,\nu'}^{s_1,s_2,s'_1,s'_2} \left[(\bar{1} - \bar{\sigma}_{\nu'}(\omega))^{-1} \right]_{s'_1,s'_2}}{(\omega - E_{\nu'}^{s'_1} + i0^+)(\omega - E_\nu^{s_2} + i0^+)}, \tag{23}$$

where $\tilde{f}(q)$ is the Fourier transform of $f(z)$ and $M_{\nu,\nu'}^{s_1,s_2,s'_1,s'_2}$ is the coherence factor de ned by

$$M_{\nu,\nu'}^{s_1,s_2,s'_1,s'_2} \equiv \int d \;\; \left[\vec{u}_\nu^{s_1 \dagger}(\;) \widehat{\tau}_z \vec{u}_{\nu'}^{s'_1}(\;) \right] \left[\vec{u}_{\nu'}^{s'_2 \dagger}(\;) \widehat{\tau}_z \vec{u}_\nu^{s_2}(\;) \right]. \tag{24}$$

Here, $\vec{u}_\nu^s(\;)$ is the radial part of the eigenfunction in Eq. (12) or Eq. (16). The E_{1u} planar and E_{1u} chiral states are distinguished by di erent $N(\omega)$ owing to di erent eigenvalues E_ν^s and di erent eigenfunctions $\vec{u}_\nu^s(\;)$ giving self-energy $\bar{\sigma}_\nu(\omega)$. For the E_{1u} planar and E_{1u} chiral states, since the eigenfunction of the CdGM modes in each state is constituted by two components in the di erent space, the coherence factor vanishes except when $s_1 = s'_1$ and $s_2 = s'_2$. This restriction gives a solution $\sigma_\nu^{s_1,s_2} = 0$ for $s_1 \neq s_2$ in Eq. (23). For simplicity, we rede ne σ and M as follows:

$$\sigma_\nu^s(\omega) \equiv -(\omega - E_\nu^s + i0^+) \sigma_\nu^{s,s}(\omega), \tag{25}$$

$$M_{\nu,\nu'}^s \equiv M_{\nu,\nu'}^{s,s,s,s}. \tag{26}$$

The DOS is rewritten with this notation by [15]

$$N_\nu^s(\omega) = \mathrm{Im} \left[-\frac{1}{\pi} \frac{1}{\omega - E_\nu^s + \sigma_\nu^s(\omega) + i0^+} \right], \tag{27}$$

with

$$\sigma_\nu^s(\omega) = -\frac{\gamma}{(2\pi)^2} \sum_{l'} \int dq' \tilde{f}(q - q') \frac{M_{\nu,\nu'}^s}{\omega - E_{\nu'}^s + \sigma_{\nu'}^s(\omega) + i0^+}. \tag{28}$$

The modi ed self-energy $\sigma_\nu^s(\omega)$ re ects the eigenfunction of the CdGM modes given by Eq. (12) or Eq. (16) through the coherence factor

$$M_{\nu,\nu'}^s = \int d \;\; \left| m_{\nu,\nu'}^s(\;) \right|^2 e^{\;\; 2\chi_q(\;) \;\; 2\chi_{q'}(\;)}, \tag{29}$$

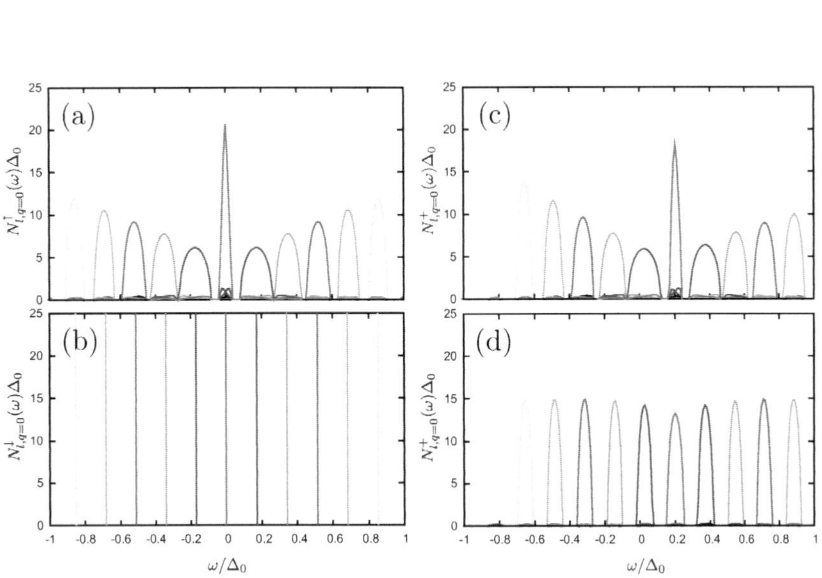

Figure 1. DOS of CdGM modes for $-5 \leq l \leq 5$. (a) $N^{\uparrow}_{l,q=0}$ with $E^{\uparrow}_{l,q=0} = -l\omega_0$ in the E_{1u} planar state, (b) $N^{\downarrow}_{l,q=0}(\omega) = \ (\omega + l\omega_0)$ in the E_{1u} planar state, (c) $N^{+}_{l,q=0}$ for $\kappa = +1$ with $E^{+}_{l,q=0} = -l\omega_0 + \mu_B H$ in the E_{1u} chiral state, and (d) $N^{+}_{l,q=0}$ for $\kappa = -1$ with $E^{+}_{l,q=0} = +l\omega_0 + \mu_B H$ in the E_{1u} chiral state. Note that the DOS $N_{l,q=0}(\omega)$ for another spin state in the E_{1u} chiral state is given by $N^{+}{}_{l,q=0}(-\omega)$. Color code indicates the absolute value of l: $l = 0$ (red), $|l| = 1$ (blue), $|l| = 2$ (green), $|l| = 3$ (magenta), $|l| = 4$ (cyan), and $|l| = 5$ (yellow).

where for the E_{1u} planar state,

$$m^{\uparrow}_{\nu,\nu'}(\ ) = \mathcal{N}^{\uparrow}_{\nu}\mathcal{N}^{\uparrow}_{\nu'}[J_{l+1}(k_q\ )J_{l'+1}(k_{q'}\ ) - s_q s_{q'} J_{l\ 1}(k_q\ )J_{l'\ 1}(k_{q'}\ )],$$
$$m^{\downarrow}_{\nu,\nu'}(\ ) = \mathcal{N}^{\downarrow}_{\nu}\mathcal{N}^{\downarrow}_{\nu'}[J_{l}(k_q\ )J_{l'}(k_{q'}\ ) - s_q s_{q'} J_{l}(k_q\ )J_{l'}(k_{q'}\ )], \tag{30}$$

and for the E_{1u} chiral state,

$$m^{\pm}_{\nu,\nu'}(\ ) = \mathcal{N}^{\pm}_{\nu}\mathcal{N}^{\pm}_{\nu'}[J_{l+1}(k^{\pm}_q\ )J_{l'+1}(k^{\pm}_{q'}\ ) - s_q s_{q'} J_{l\ 1}(k^{\mp}_q\ )J_{l'\ 1}(k^{\mp}_{q'}\ )], \tag{31}$$

with $\kappa = +1$ and

$$m^{\pm}_{\nu,\nu'}(\ ) = \mathcal{N}^{\pm}_{\nu}\mathcal{N}^{\pm}_{\nu'}[J_{l}(k^{\pm}_q\ )J_{l'}(k^{\pm}_{q'}\ ) - s_q s_{q'} J_{l}(k^{\mp}_q\ )J_{l'}(k^{\mp}_{q'}\ )], \tag{32}$$

with $\kappa = -1$. The di erence of the coherence factor between the E_{1u} planar and E_{1u} chiral states is signi cant for columnar defects giving $s_q s_{q'} = 1$.

In Fig. 1, we show the DOS of each CdGM mode $N^{s}_{l,q=0}$ for $-5 \leq l \leq 5$. Here, we consider simple columnar defects without k_z-transfer by putting $\tilde{f}(q-q') = k_F\ (q-q')$ in Eq. (28), where $s_q s_{q'}$ of the coherence factor in Eqs. (30)-(32) is given by unity. In the calculation of the modi ed self-energy in Eq. (28), we sum the CdGM modes with $|l'| \leq 10$. We take $\Delta(\ ) = \Delta_0 \tanh(\ /\xi)$ with $k_F\xi = 5$, which xes $\chi_q(\ )$ in the coherence factor and ω_q in the eigenvalue of the CdGM modes.

The DOS $N^{\uparrow}_{l,q=0}$ for the E_{1u} planar state with $\Gamma_n = 0.02\pi\Delta_0$ is shown in Fig. 1(a). For the up-spin CdGM modes, only the zero-energy mode has a sharp spectrum of the DOS owing to $M^{\uparrow}_{(0,0),(0,0)} = 0$. Note that the slight width of the zero-energy DOS is due to the contribution from nite angular momentum modes. The down-spin CdGM modes have the in nite DOS $N^{\downarrow}_{l,q=0}(\omega) = \ (\omega + l\omega_0)$ as schematically shown in Fig. 1(b) owing to the coherence factor $M^{\downarrow}_{(l,0),(l',0)} = 0$ for any l and l'.

In Figs. 1(c) and 1(d), the DOS of CdGM modes $N^{+}_{l,q=0}$ for the E_{1u} chiral state is shown. The chirality of the gap function is parallel to the vorticity ($\kappa = +1$) in Fig. 1(c) and that is antiparallel to the vorticity ($\kappa = -1$) in Fig. 1(d). Here, we set magnetic eld to $\mu_B H = 0.2\Delta_0$. In the limit $H \to 0$, $N^{\pm}_{l,q=0}$ should correspond to $N^{\uparrow}_{l,q=0}$ in Fig. 1(a) for $\kappa = +1$ and $N^{\downarrow}_{l,q=0}(\omega) = \ (\omega + l\omega_0)$ in Fig. 1(b) for $\kappa = -1$, where $N_{l,q=0}(\omega)$ is given by $N^{+}{}_{l,q=0}(-\omega)$.

A main di erence between Figs. 1(a) and 1(c) with the same parameter $\Gamma_n = 0.02\pi\Delta_0$ is the energy shift of the $l = 0$ mode with the sharp DOS. The height of the DOS for the $l = 0$ mode is slightly di erent by the di erence between the coherence factors $m^{\uparrow}_{\nu,\nu'}(\ )$ in Eq. (30) and $m^{+}_{\nu,\nu'}(\ )$ in Eq. (31); however, the width of the DOS is mainly provided by the contribution from nite angular momentum modes, similarly to $N^{\uparrow}_{0,0}$.

The DOS $N^{+}_{l,q=0}(\omega)$ for $\kappa = -1$ in Fig. 1(d), which is calculated with $\Gamma_n = 0.2\pi\Delta_0$, is qualitatively di erent from the in nite DOS $N^{\downarrow}_{l,q=0}(\omega) = \ (\omega + l\omega_0)$ in Fig. 1(b). Since $m^{+}_{(l,0),(l',0)}(\ )$ in Eq. (32) is small but nite, the DOS $N^{+}_{l,q=0}(\omega)$ has a nite width in each CdGM mode increased by the magnetic eld $\mu_B H$ and the impurity scattering rate Γ_n. On the other hand, since $m^{\downarrow}_{(l,0),(l',0)}(\ )$ in Eq. (30) exactly vanishes for any l and l', the DOS $N^{\downarrow}_{l,q}$ is given by the delta function independent of $\mu_B H$ and Γ_n within the approximation of neglecting the contributions from the continuum state to the CdGM modes.

The observable DOS is the sum of that from the two spin states, $N^{\uparrow}_{l,q}$ and $N^{\downarrow}_{l,q}$ for the E_{1u} planar state and $N^{+}_{l,q}$ and $N_{l,q}$ for the E_{1u} chiral state. Then, the sharp zero energy DOS from the $l = 0$ CdGM mode, which does not shift by the magnetic eld, in the E_{1u} chiral state is highly contrasted with the shifted or broadened DOS in the E_{1u} chiral state.

4. Summary and Discussion

In summary, we have investigated impurity e ects on the CdGM modes by nonmagnetic columnar defects for the E_{1u} planar state and the E_{1u} chiral state which are possible superconducting states in UPt_3. First, we have derived the eigenvalue and eigenfunction of the CdGM modes under magnetic elds producing a vortex without impurities. For the E_{1u} planar state, in uence of the magnetic eld upon the CdGM modes can be neglected. On the other hand, the magnetic eld causes an energy shift of the CdGM modes and slight deformation of the eigenfunction of them for the E_{1u} chiral state. There are two types of impurity e ects in the E_{1u} chiral state for the "parallel state" where the chirality of the gap function is parallel to the vorticity and for the "antiparallel state" where the chirality is antiparallel to the vorticity. In the parallel state, the CdGM modes have the sharp DOS at the energy proportional to the amplitude of the magnetic eld, namely $\omega = \pm\mu_B H$. In contrast, the sharp DOS of the zero-energy up-spin CdGM modes in the E_{1u} planar state never shifts by the magnetic eld, which is guaranteed by topological invariance [15]. In the antiparallel state, all the shifted CdGM modes have the DOS with a nite width depending on $\mu_B H$ and Γ_n, which is contrasted with the in nite DOS of the down-spin CdGM modes in the E_{1u} planar state.

Through the utilization of the di erence of the impurity e ects on the CdGM modes, the E_{1u} planar state and the E_{1u} chiral state are possibly distinguished by some experiments. The coherence factor, which dominates the impurity e ects, can be directly observed by inelastic neutron scattering experiments or quasiparticle interference imaging. Concrete estimations of the observable physical quantities remain in a future work.

Acknowledgments
We thank T. Hanaguri and Y. Masaki for helpful discussion. Y.T. acknowledges nancial support from the Japan Society for the Promotion of Science (JSPS). This work was supported by JSPS KAKENHI Grant Numbers 15K17715 and 15J05698.

References
[1] Stewart G R, Fisk Z, Willis J O and Smith J L 1984 *Phys. Rev. Lett.* **52** 679
[2] Joynt R and Taillefer L 2002 *Rev. Mod. Phys.* **74** 235
[3] Machida Y, Itoh A, So Y, Izawa K, Haga Y, Yamamoto E, Kimura N, Onuki Y, Tsutsumi Y and Machida K 2012 *Phys. Rev. Lett.* **108** 157002
[4] Tsutsumi Y, Machida K, Ohmi T and Ozaki M 2012 *J. Phys. Soc. Jpn.* **81** 074717
[5] Izawa K, Machida Y, Itoh A, So Y, Ota K, Haga Y, Yamamoto E, Kimura N, Onuki Y, Tsutsumi Y and Machida K 2014 *J. Phys. Soc. Jpn.* **83** 061013
[6] Choi C H and Sauls J A 1991 *Phys. Rev. Lett.* **66** 484
[7] Tou H, Kitaoka Y, Ishida K, Asayama K, Kimura N, Onuki Y, Yamamoto E, Haga Y and Maezawa K 1998 *Phys. Rev. Lett.* **80** 3129
[8] Luke G M, Keren A, Le L P, Wu W D, Uemura Y J, Bonn D A, Taillefer L and Garrett J D 1993 *Phys. Rev. Lett.* **71** 1466
[9] de Réotier P D, Huxley A, Yaouanc A, Flouquet J, Bonville P, Imbert P, Pari P, Gubbens P C M and Mulders A M 1995 *Phys. Lett. A* **205** 239
[10] Schemm E R, Gannon W J, Wishne C M, Halperin W P and Kapitulnik A 2014 *Science* **345** 190
[11] Caroli C, de Gennes P and Matricon J 1964 *Phys. Lett.* **9** 307
[12] Kawakami T, Mizushima T and Machida K 2011 *J. Phys. Soc. Jpn.* **80** 044603
[13] Matsumoto M and Heeb R 2001 *Phys. Rev. B* **65** 014504
[14] Tsutsumi Y, Ishikawa M, Kawakami T, Mizushima T, Sato M, Ichioka M and Machida K 2013 *J. Phys. Soc. Jpn.* **82** 113707
[15] Tsutsumi Y and Kato Y 2016 *J. Phys. Soc. Jpn.* **85** 053704
[16] Tokuyasu T A, Hess D W and Sauls J A 1990 *Phys. Rev. B* **41** 8891
[17] Kopnin N B and Kravtsov V E 1976 *JETP Lett.* **23** 578
[18] Masaki Y and Kato Y 2016 *J. Phys. Soc. Jpn.* **85** 014705
[19] Masaki Y and Kato Y 2015 *J. Phys. Soc. Jpn.* **84** 094701
[20] Civale L, Marwick A D, Worthington T K, Kirk M A, Thompson J R, Krusin-Elbaum L, Sun Y, Clem J R and Holtzberg F 1991 *Phys. Rev. Lett.* **67** 648
[21] Toulemonde M, Bouffard S and Studer F 1994 *Nucl. Instrum. Methods Phys. Res., Sect. B* **91** 108

Superconductivity and Antiferromagnetic Spin Fluctuations in $LaFe(As_{1-x}P_x)(O_{1-y}F_y)$ probed by ^{31}P-NMR

T. Shiota[a], H. Mukuda[a], M. Uekubo[b], F. Engetsu[a], M. Yashima[a], Y. Kitaoka[a], K. T. Lai[b], H. Usui[b], K. Kuroki[b], S. Miyasaka[b], and S. Tajima[b]

[a]Graduate School of Engineering Science, Osaka University, Osaka 560-8531, Japan;
[b]Graduate School of Science, Osaka University, Osaka 560-0043, Japan

E-mail: mukuda@mp.es.osaka-u.ac.jp

Abstract. ^{31}P-NMR study on $LaFe(As_{1-x}P_x)(O_{0.86}F_{0.14})$ unraveled that antiferromagnetic spin fluctuations (AFMSFs) emerge significantly at x=0.4 where the T_c is markedly enhanced, indicating that the AFMSFs are one of the important factors for raising T_c. From extensive comparison over wide compositions for $0 \leq x \leq 1$ and $0 \leq y \leq 0.14$, we revealed that there are two different types in temperature evolution of the AFMSFs: One is enhanced particularly at low energies that evolves only at low temperatures, which mainly derives from the two orbitals of $d_{xz/yz}$. The other is distributed broadly at finite energies that appears up to high temperatures, which derives from three orbitals of d_{xy} and $d_{xz/yz}$. The highest T_c(=27 K) state in the present compositions appears at (x,y)=(0.4, 0.1) where two characteristics of AFMSFs merge, suggesting the contribution of the AFMSFs over wide energies to the onset of SC. The nonmonotonic variation of T_c in $LaFe(As_{1-x}P_x)(O_{1-y}F_y)$ is attributed to the AFMSFs from degenarated multiple-3d-orbitals on iron-pnictide superconductors.

1. Introduction

Superconducting (SC) transition temperature (T_c) of a layered iron(Fe)-pnictide $LaFeAsO_{1-y}F_y$ exhibits a broad maximum depending on F^- content, with the highest T_c of ~26 K at y ~0.1[1, 2]. Further substitution of either H^- or F^- for O^{2-} uncovered the presence of the second SC dome with higher T_c than that of the first one [3, 4]. It has been known that the T_c of Fe-pnictides enhances when a $FeAs_4$ block forms a nearly regular tetrahedral structure [5] or the height of pnictogen ($h_{\rm Pn}$) from the Fe plane is close to ~1.38 Å [6]. However, in $LaFe(As_{1-x}P_x)(O_{1-y}F_y)$, the T_c values do not vary monotonously irrespective of monotonous variation of thier lattice parameters with x[7, 8, 9, 10]. Previous ^{31}P-NMR studies on these compounds have revealed that T_c reaches the maximum, as a result of the marked enhancement of AFM spin fluctuations (AFMSFs) at low energies[11, 12]. This AFMSFs come from the depression of re-emergent AFM2 phase in $LaFe(As_{1-x}P_x)O$, which exhibits at $0.4 \leq x \leq 0.7$ of y=0 that intervenes between two superconductivity (SC) phases[12, 13]. These results provide clear evidence that AFMSFs at low energies are one of the key elements for enhancing T_c in Fe-pnictide SCs even though the lattice parameters deviate from their optimum values. In this paper, we report ^{31}P-NMR study on

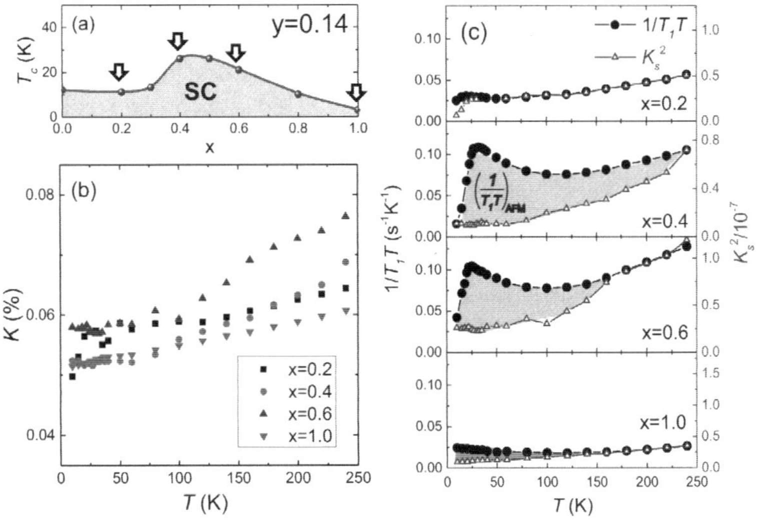

Figure 1. (Color online) (a) T_c vs x for $LaFe(As_{1-x}P_x)(O_{0.86}F_{0.14})$ [10]. T dependences of (b) Knight shift and (c) $(1/T_1T)$ and $K_s^2(T)$ for $0.2< x <1.0$. As for x=0.4 and 0.6, $(1/T_1T)$ and $K_s(T)$ show the different behavior at low temperatures due to the development of AFMSFs with finite wave vectors.

$LaFe(As_{1-x}P_x)(O_{0.86}F_{0.14})$ to unravel the general relationship between superconductivity and the AFMSFs in the La-based 1111 series.

2. Experimental Procedures

Polycrystalline samples of $LaFe(As_{1-x}P_x)(O_{0.86}F_{0.14})$ for $0.2\leq x \leq 1$ were synthesized by the solid-state reaction method [10, 8]. T_cs are determined from an onset of SC diamagnetism as shown in Fig. 1(a)[10]. ^{31}P-NMR(I=1/2) measurement was performed on coarse powder samples of $LaFe(As_{1-x}P_x)(O_{0.86}F_{0.14})$ with nominal contents x=0.2, 0.4, 0.6, and 1.0, as shown by arrows in Fig. 1(a).

3. Results and discussion

Figures 1(b) and 1(c) show the T dependences of $K(T)$ and $(1/T_1T)$ for $0.2< x <1.0$, respectively. In general, the K incluides the temperature(T)-dependent spin shift $K_s(T)$ and the T-independent chemical shift $K_{\rm chem}$. The $K_{\rm chem}$ ~0.04% was assumed to be the same with those of previous studies in the series[11, 12, 14, 15]. The T dependence of $K_s^2(T)$ is shown in Fig. 1(c), together with that of $(1/T_1T)$. In x=0.2 and 1.0, the T dependence of $(1/T_1T)$ follows that of $K_s^2(T)$ for wide T region, which corresponds to Korringa's relation in normal metals. On the other hand, the $(1/T_1T)$s at x=0.4 and 0.6 increase as T lowers, although $K_s(T)$s decrease. This contrasted behavior between $(1/T_1T)$ and $K_s(T)$ demonstrates the development of AFMSFs with finite wave vectors as T lowers. In order to deduce the development of AFMSFs following the previous studies [11, 12, 14], we assume that $(1/T_1T)$ is decomposed as, $(1/T_1T) = (1/T_1T)_{\rm AFM} + (1/T_1T)_0$, where the first term represents the contribution of AFMSFs with finite wave vectors Q presumably around $(0,\pi)$ and $(\pi,0)$ and the second one does the q-independent one in proportion to $N(E_{\rm F})^2$ i.e. K_s^2. Figure 2(a) shows a contour plot of $(1/T_1T)_{\rm AFM}$ for y=0.14. This plot shows that the AFMSFs develop upon cooling

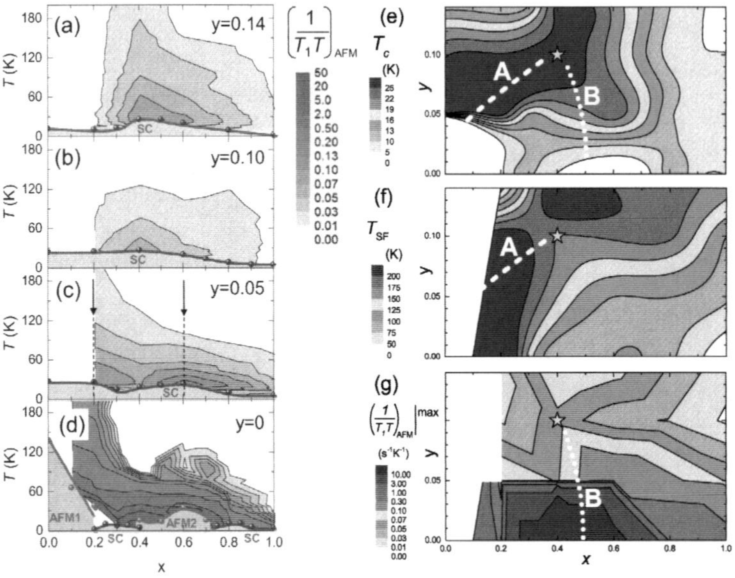

Figure 2. (Color online) Contour plots of $(1/T_1T)_{\mathrm{AFM}}$ for (a) y=0.14, along with (b) y=0.05, (c) 0.10[11], and (d) 0[12]. Arrows in (c) for y=0.05 denote two typical types of T evolution of $(1/T_1T)_{\mathrm{AFM}}$: The AFMSFs at x=0.2 gradually develop upon cooling from higher T, whereas the AFMSFs at x=0.6 develops rapidly only at low T, although the value of $(1/T_1T)_{\mathrm{AFM}}$ at low T is smaller in former case than in the latter case. Contour plots of (e) T_c [1, 2, 7, 8, 9, 10], (f) T_{SF}, and (g) $(1/T_1T)_{\mathrm{AFM}}|^{\max}$. Here, we define T_{SF} being the temperature below which the AFMSFs start to develop, and $(1/T_1T)_{\mathrm{AFM}}|^{\max}$ being the maximum value of $(1/T_1T)_{\mathrm{AFM}}$. The respective curves A and B correspond to the branches that T_{SF} and $(1/T_1T)_{\mathrm{AFM}}|^{\max}$ are relatively large. The star at (x, y)=(0.4, 0.1) denotes the highest T_c(=27 K) point.

for x=0.4 and 0.6 exhibiting relatively high T_c, while they are drastically suppressed for x=0.2 and 1.0 exhibiting very low T_c. The contour plots for y=0, 0.05, and 0.1 previous obtained are shown in Figs 2(b)-2(d), providing a clear evidence that the AFMSFs play a significant role in raising T_c in wide compositions of x and y.

However, we note that there are the two types in the T-evolution of the AFMSFs upon cooling. For example, in the two samples of y=0.05 as shown by arrows in Fig. 2(c), the AFMSFs at x=0.6 largely develop only at low temperatures (T <100 K), whereas those at x=0.2 gradually develop upon cooling from high temperatures (T <250 K), although a value of $(1/T_1T)_{\mathrm{AFM}}$ at T_c is larger at x=0.6 than at x=0.2. In order to gain further insight into these features of the AFMSFs, we define T_{SF} being the temperature below which the $(1/T_1T)_{\mathrm{AFM}}$ start to develop, and $(1/T_1T)_{\mathrm{AFM}}|^{\max}$ being the maximum value of $(1/T_1T)_{\mathrm{AFM}}$. Thus, the $(1/T_1T)_{\mathrm{AFM}}|^{\max}$ probes a low energy limit of dynamical spin susceptibility $\chi''(Q,\omega)$, representing how large the spectral weight of AFMSFs are at low energies. The T_{SF} indicates the temperature that $\chi''(Q,\omega)$ significantly starts to increase upon cooling, roughly pointing to a characteristic energy of AFMSFs. Figures 2(e) , 2(f), and 2(g) show the contour plots of T_c[1, 2, 7, 8, 9, 10], T_{SF}, and $(1/T_1T)_{\mathrm{AFM}}|^{\max}$, respectively. The curves A and B in these figures represent the relatively high T_c regions. The curve A is along the high values of T_{SF} in Fig. 2(f), in other words, roughly along the high values of the characteristic energy of AFMSFs. It indicates that the development of AFMSFs below T_{SF} is mostly responsible for raising T_c. On the one hand, the curve B are along the large values of $(1/T_1T)_{\rm AFM}|^{\rm max}$, as shown in Fig. 2(g), demonstrating that the development of the AFMSFs at low energies is also responsible for raising T_c. These two different types of T-evolution of AFMSFs are theoretically reproduced by multiple AFMSFs derived from different orbitals: The $d_{xz/yz}$-derived AFMSFs around the AFM2 at (x, y)=(0.6, 0) are largely enhanced at low energies, whereas the $d_{xz/yz}$+d_{xy} derived AFMSFs around the AFM1 at (x, y)=(0, 0) are characteristic at finite energies rather than at low energies[17, 16]. It is notable that the highest T_c=27 K is denoted at (x, y)=(0.4, 0.1) by a star in Fig. 2(e) around which the curves A and B merge, suggesting the contribution of the AFMSFs over wide energies to the onset of SC.

4. Summary

^{31}P-NMR studies in $LaFe(As_{1-x}P_x)(O_{1-y}F_y)$ revealed that the T_c is markedly enhanced when the AFMSFs emerge significantly, indicating that the AFMSFs are one of the important factors for enhancing T_c, even though the lattice parameters deviate from their optimum values for $FePn_4$ regular tetrahedron. Systematic comparison of the NMR results over a series of $LaFe(As_{1-x}P_x)(O_{1-y}F_y)$ ($0\leq x \leq 1$ and $0\leq y \leq 0.14$) has unraveled that there are two types in the T-evolution of the AFMSF: One is the AFMSFs that develops rapidly only at low temperature and reach the relatively high value of $(1/T_1T)_{\rm AFM}$ at T_c, by contrast, the other is the AFMSFs that develops gradually from high temperature upon cooling whereas the maximum value of $(1/T_1T)_{\rm AFM}$ is lower than that for the former. According to the theory, the former arises from the collapse of AFM2 order at (x,y)=(0.6, 0), which derives from the good nesting of Fermi surfaces dominated mostly by $d_{xz,yz}$ orbitals[17, 16], by contrast, the latter arises from the AFM1 order that appears at (x,y)=(0, 0). The non-monotonic variation of T_c in $LaFe(As_{1-x}P_x)(O_{1-y}F_y)$ is attributed to the AFMSFs derived from the multiple-orbital nature inherent to Fe-pnictide SCs.

References

[1] Y. Kamihara *et al.*, J. Am. Chem. Soc. **130**, 3296 (2008).
[2] H. Luetkens *et al.*, Nat. Mater. **8**, 305 (2009).
[3] S. Iimura *et al.*, Nat. Commun. **3**, 943 (2012).
[4] J. Yang *et al.*, Chin. Phys. Lett. **32**, 107401 (2015).
[5] C. H. Lee *et al.*, J. Phys. Soc. Jpn. **77**, 083704 (2008).
[6] Y. Mizuguchi *et al.*, Supercond. Sci. Technol. **23**, 054013 (2010).
[7] S. Saijo *et al.*, Physica C **470**, S298 (2010).
[8] K. T. Lai *et al.*, Phys. Rev. B **90**, 064504 (2014).
[9] C. Wang *et al.*, Europhys. Lett. **86**, 47002 (2009).
[10] M. Uekubo *et al.*, submitted.
[11] H. Mukuda *et al.*, Phys. Rev. B **89**, 064511 (2014).
[12] H. Mukuda *et al.*, J. Phys. Soc. Jpn. **83**, 083702 (2014).
[13] S. Kitagawa *et al.*, J. Phys. Soc. Jpn. **83**, 023707 (2014).
[14] M. Miyamoto *et al.*, Phys. Rev. B **92**, 125154 (2015).
[15] T. Shiota *et al.*, J. Phys. Soc. Jpn. **85**, 053706 (2016).
[16] H. Arai *et al.*, Phys. Rev. B **91**, 134511 (2015).
[17] H. Usui *et al.*, Sci. Rep. **5**, 11399 (2015).

IOP Conf. Series: Journal of Physics: Conf. Series **807** (2017) 052007 doi:10.1088/1742-6596/807/5/052007

Ultra-low temperature ac susceptibility of the heavy-fermion superconductor $YbRh_2Si_2$

L Steinke[1], E Schuberth[1,2], S Lausberg[1], M Tippmann[2], A Steppke[1], C Krellner[1,3], C Geibel[1], F Steglich[1,4,5] and M Brando[1]

[1] Max Planck Institute for Chemical Physics of Solids, 01187 Dresden, Germany
[2] Walther Meissner Institut für Tieftemperaturforschung der Bayerischen Akademie der Wissenschaften, 85748 Garching, Germany
[3] Physics Institute, University of Frankfurt, 60438 Frankfurt, Germany
[4] Center for Correlated Matter, Zhejiang University, Hangzhou, Zhejiang 310058, China
[7] Institute of Physics, Chinese Academy of Sciences, Beijing 100190, China

E-mail: brando@cpfs.mpg.de

Abstract. Recently, we reported on the discovery of superconductivity in $YbRh_2Si_2$ at the very low temperature $T_c = 2\,\mathrm{mK}$. We present here complementary measurements of the temperature and field dependence of the ac susceptibility $\chi_{ac}(T)$ of this heavy-fermion system for temperatures down to 1 mK and magnetic fields in the basal plane up to 75 mT. The in-phase response $\chi'_{ac}(T)$ shows a steep drop at T_c, but it does not become negative, possibly because of the presence of the earth field that was not compensated. The out-of-phase response $\chi''_{ac}(T)$ shows an increase right below T_c indicating that the transition is first order. We also observe two kinks in the field dependence of $\chi_{ac}(B)$ at $B_1 \approx 50\,\mathrm{mT}$ and $B_2 \approx 60\,\mathrm{mT}$, which imply a more complex $T-B$ phase diagram for $YbRh_2Si_2$ with $B \perp c$. These features possibly indicate two second-order phase transitions or domain reorientation at B_1 and a phase transition from the antiferromagnetic to the paramagnetic state at B_2.

1. Introduction

$YbRh_2Si_2$ is a canonical quantum critical heavy-fermion (HF) compound [1, 2]. It is characterized by a Kondo temperature $T_K = 25\,\mathrm{K}$ and antiferromagnetic (AF) order below $T_{AF} = 70\,\mathrm{mK}$ with Yb-derived ordered moments of the order of $10^{-3}\mu_B$ [3, 4, 5]. The AF phase can be continuously suppressed by a magnetic field applied within the basal tetragonal ("easy magnetic") plane. The critical field in the basal plane was estimated to be $B_N \approx 50 - 60\,\mathrm{mT}$ [4, 6]. At B_N isothermal magnetotransport [7, 8] and thermodynamic measurements [4, 6] have provided evidence that an unconventional quantum critical point (QCP) exists, but no superconductivity was found down to 20 mK, which is typically observed in several Ce-based HF compounds [9, 10, 11, 12, 13, 14].

Recently, we reported on the discovery of superconductivity in this system at the very low temperature of 2 mK [16]. $YbRh_2Si_2$ is only the second Yb-based HF superconductor, following β-$YbAlB_4$ with $T_c = 80\,\mathrm{mK}$ [17]. A T_c of 2 mK is much smaller than the typical T_c of Ce-based HF superconductors ($\lesssim 2.5\,\mathrm{K}$). Such a ratio is also found for the magnetic ordering temperatures, e.g., $T_{AF} = 70\,\mathrm{mK}$ for $YbRh_2Si_2$ compared to $T_{AF} \lesssim 4\,\mathrm{K}$ in $CeRhIn_5$; this can be attributed to the much smaller radius of the $4f$ electrons in Yb^{3+} compared to that in Ce^{3+} [18]. In

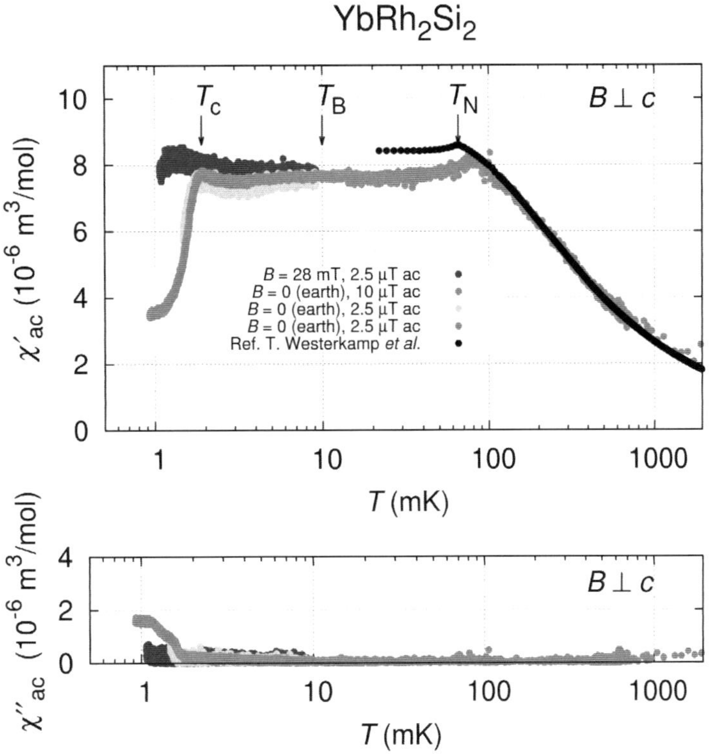

Figure 1. Temperature dependence of the ac susceptibility of $YbRh_2Si_2$ measured in earth magnetic field and earth magnetic field plus 28 mT. With this setup it was possible to measure temperature sweeps from 1 mK up to about 1000 mK and scale the data with those taken in a standard Kelvinox 400 dilution refrigerator (Oxford Instruments) down to 20 mK (black points) [15]. We observe a kink when entering the AFM phase at $T_{AF} \approx 80$ mK, a weak signature at T_B, i.e. at a crossover into the regime with A-phase fluctuations, and a strong drop of $\chi'_{ac}(T)$ at the superconducting transition temperature $T_c = 1.9$ mK (onset). Here, $\chi''_{ac}(T)$ shows a clear increase, suggesting that the transition is of the first order.

addition, we could identify a rich $T - B$ phase diagram for $YbRh_2Si_2$ (with $B \perp c$), by means of magnetisation, susceptibility and heat capacity measurements. This phase diagram has a new A + SC phase below $T_A \gtrsim 2$ mK which includes superconductivity. In addition, there is evidence for strong AF (A-phase) fluctuations below $T_B \simeq 10$ mK. While the AF phase is of purely electronic nature, the new A phase involves a coupling between the magnetic moments due to the Yb-derived $4f$ electrons and the Yb nuclear moments. The size of the critical field B_A at which $T_A \to 0$ (30 – 60 mT) indicates that the A phase must be dominantly of nuclear origin. This led us to conclude that the superconductivity develops from quantum critical fluctuations associated with the disappearance of the AF order induced by the nuclear spin order.

Here, we present new complementary field-dependent measurements of the ac susceptibility $\chi_{ac}(B)$ taken at constant temperatures (down to 11 mK) and in fields as high as 75 mT. We have discovered new features in $\chi'_{ac}(B)$ in form of kinks at $B_1 \approx 50$ mT and $B_2 \approx 60$ mT, which point to a more complex $T - B$ phase diagram for $YbRh_2Si_2$ with $B \perp c$.

2. Experiemntal setup

We have carried out our experiments in the ultra-low-temperature adiabatic demagnetisation refrigerator (ADR) at the Walther Meissner Institute in Garching. It consists of a 0.9 mole $PrNi_5$ nuclear demagnetisation stage with a final temperature of 0.4 mK. The temperature of the nuclear stage was determined by a simulated warm-up curve, which was verified to be in excellent agreement with temperatures obtained by pulsed nuclear magnetic resonance on ^{63}Cu and ^{195}Pt nuclei. This method results in a temperature accuracy of 2 – 3 % even down to 0.8 mK (Ref. [19]). Temperatures above 10 mK were measured with a resistive thermometer.

A 5.44 mg $YbRh_2Si_2$ single crystal of very high quality (RRR $\approx$ 50) was used in our experiments. This thin $YbRh_2Si_2$ platelet was clamped on a 5N silver wire attached to a 5N silver rod which by itself was screwed to the nuclear stage. The crystal was placed in the pick-up coil of the susceptometer with its flat surface (the basal tetragonal plane) aligned parallel to the coil axis. Since the electrical conductivity of the samples is high (of the order of 5000 S) we have no indication that, after thermalisation, its temperature deviated from that of the nuclear stage by more than a few tens of μK, even at the lowest temperatures. Unfortunately, heating the sample to temperatures above 600 mK resulted in a thermal drift of the signal due to a warm-up of the mixing chamber of the dilution refrigerator, making measurements unreliable at higher temperatures.

The ac susceptibility $\chi_{ac}(T)$ was measured by a conventional mutual inductance setup thermally coupled to the nuclear stage and mounted in the centre of a superconducting 8 T magnet. The left and right pick-up coils consist of 18 (1694 windings) and 19 (1786 windings) layers, respectively, of 5 μm copper wire, with a total resistance of 1176 kΩ at 300 K. For the excitation coil we used 3 layers (761 windings) of 0.1 mm superconducting NbTi wire in CuNi matrix which had a resistance of 1.439 kΩ at 300 K. The susceptometer was balanced systematically at 300, 77 and 4.2 K. The in-phase $\chi'_{ac}(T)$ and out-of-phase $\chi''_{ac}(T)$ responses were measured with a two-channel lock-in amplifier. This allowed frequencies up to a few hundred Hz. Here the smallest field was the earth field: According to the international geomagnetic reference field (IGRF), the vertical component of the earth field at the location of the laboratory is -0.0433 mT. In Ref. [16] we presented a measurement of $\chi'_{ac}(T)$ obtained by modulating a rf SQUID system at which the earth field could be compensated with an extremely small dc field between 2 and 5 nT. This allowed us to see negative values of $\chi'_{ac}(T)$ below $T \approx 2$ mK which are not observed in the measurements presented here.

3. Results

Fig. 1 shows the temperature dependence of the in-phase $\chi'_{ac}(T)$ and out-of-phase $\chi''_{ac}(T)$ responses of the ac susceptibility measured in earth magnetic field and at 28 mT (with $B \perp c$). The measurements were taken with increasing temperature. The conventional mutual inductance setup allowed us to measure T-sweeps from 1 mK up to about 1 K and to scale the data with those taken in a standard Kelvinox 400 (Oxford Instruments) down to 20 mK (black points in Fig. 1) [15]. We used an excitation field of 2.5 μT and a frequency of 117 Hz. With higher excitation fields it was not possible to cool the sample to below 10 mK (see, e.g., magenta points in Fig. 1).

The results observed here are very similar to those seen in other and better crystals investigated in Ref. [16]: In particular, we observe i) a clear kink in $\chi'_{ac}(T)$ at $T_{AF} \approx 80$ mK when entering the AF phase, ii) a weak signature at T_B, i.e. at the crossover into the regime with A-phase fluctuations and iii) a strong drop of $\chi'_{ac}(T)$ at the superconducting transition temperature $T_c = 1.9$ mK. Although the drop in $\chi'_{ac}(T)$ is quite large we do not reach negative values for $T \to 0$, but the susceptibility saturates at about 3.5×10^{-6} m^3/mol. With this setup it was not possible to screen the earth magnetic field, and this might be the reason why the superconducting transition is not manifested by negative values. We tried to use a μ-metal

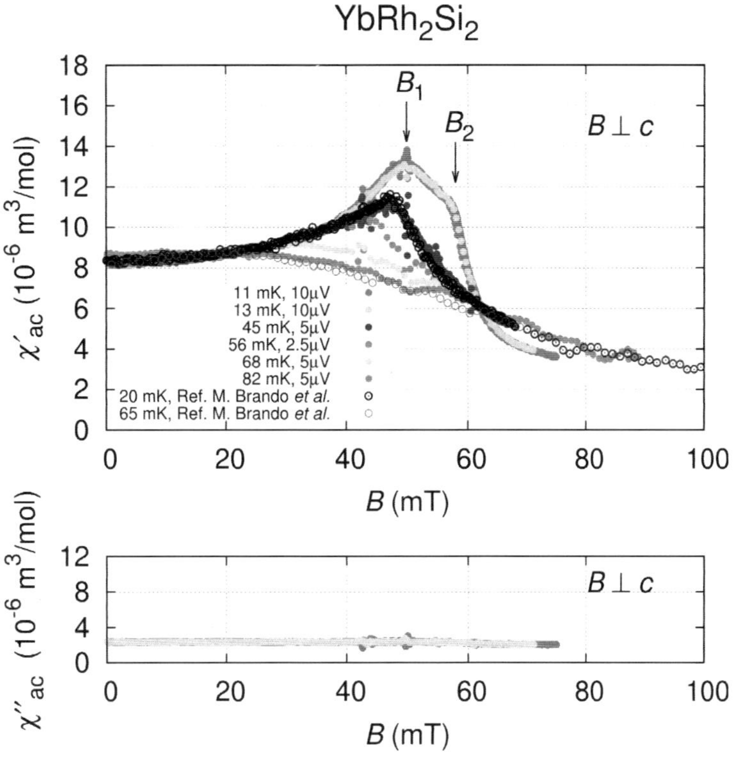

Figure 2. Isothermal field sweeps of the ac susceptibility $\chi'_{ac}(B)$ of $YbRh_2Si_2$ taken at selected temperatures. The data are plotted with those taken in a standard Kelvinox 400 dilution refrigerator (Oxford Instruments) at 20 and 65 mK (black empty points) on a sample of similar quality [20]. The fields B_1 and B_2 mark the positions of kinks in $\chi'_{ac}(B)$ (observed only at 11 and 13 mK) which possibly indicate phase transitions of the second order since $\chi''_{ac}(T)$ is not showing any feature at those fields. However, domain orientation might also be possible at B_1. At higher temperatures $T < T_{AF}$ we observe a single kink at B_1 which decreases with increasing T as previously observed.

cylinder to screen the susceptometer, but its large size resulted in too high a thermal load and did not allow us to cool the low-T stage below 2 mK. We estimate T_c from the onset of the transition which appears broadened compared to that observed at $B = 0$ and $T_c = 2$ mK in Ref. [16] and T_c is only slightly lower. This might be due to the presence of a larger field than earth field. Variations of sample quality are unlikely to play a role since all crystals investigated in Ref. [16] and in this work show a very similar T_c despite of their different residual resistivity ratio. As discussed in Ref. [16], it is possible that this insensitivity of T_c on disorder is most likely due to the short superconducting coherence length, which corresponds to the large effective mass of the charge carriers, i.e., several hundred times the bare electron mass.

At $T_c = 1.9$ mK the imaginary part of the susceptibility shows a clear increase strongly suggesting that the superconducting transition measured in the earth magnetic field is of first order. This resembles the case of A/S-type $CeCu_2Si_2$ [21], implying the absence of microscopic coexistence between electronic AF order and superconductivity.

In Fig. 2 we present isothermal field sweeps of the ac susceptibility $\chi'_{ac}(B)$ of $YbRh_2Si_2$ taken

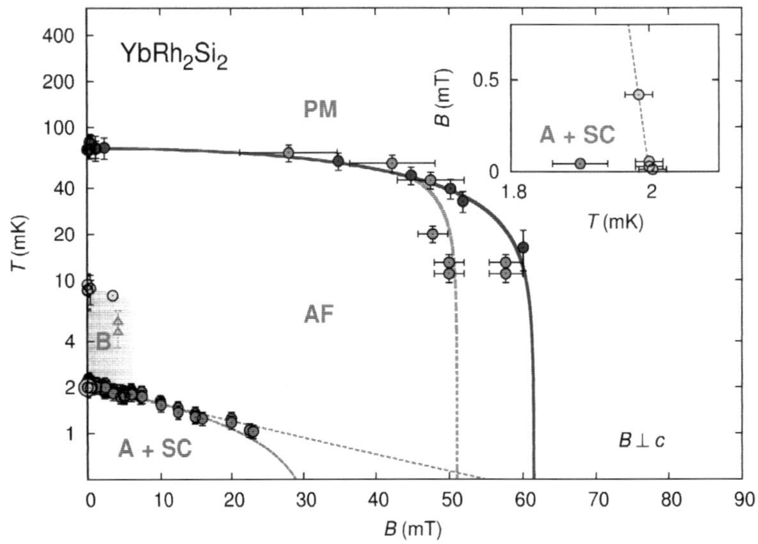

Figure 3. Extended phase diagram of $YbRh_2Si_2$ [16]. We added new points and a new phase boundary line (magenta) inferred from the measurements in this work. PM and AF indicate the paramagnetic and antiferromagnetic states, respectively. The hatched light-blue area indicates the onset of A-phase fluctuations (phase B) which give rise to a reduction of the staggered magnetisation and the beginning of shielding due to superconducting fluctuations. The A + SC phase represents the concurring (dominantly) nuclear AF order and superconductivity. Only at $B = 0$ almost full shielding is observed. The two red dashed lines mark the range within which the A-phase boundary line may end. The green circle indicates the superconducting transition temperature seen in the ac susceptibility at $B = 0$ in Ref. [16] while the yellow circles (partially covered by the green point) result from the shielding signals in the zero-field-cooled dc magnetisation. In the inset, these shielding transitions are shown separately on an enlarged scale. The superconducting transition temperature (onset) observed with earth field in our set-up is $T_c \approx 1.9\,\mathrm{mK}$, slightly lower than that measured at $B = 0$.

at selected temperatures. The data are plotted with those taken in a standard Kelvinox 400 dilution refrigerator (Oxford Instruments) at 20 and 65 mK (black empty points) on a sample of similar quality [20]. Remarkably, the measurement obtained in the Kelvinox 400 at 20 mK overlaps quite well with that obtained in our ADR at 45 mK and shows a single peak at $B_1 \approx 50\,\mathrm{mT}$. This was interpreted as the critical field B_N [20]: In fact, B_1 decreases with increasing T. However, at the lowest temperature of 11 and 13 mK we see two clear features (kinks) in $\chi'_{ac}(B)$ at the fields $B_1 \approx 50\,\mathrm{mT}$ and $B_2 \approx 60\,\mathrm{mT}$. They possibly indicate phase transitions of the second order since $\chi''_{ac}(T)$ is not showing any feature at those fields.

4. Discussion

The features observed in the field dependent ac susceptibility suggest the presence of an additional phase line in the high-field side of the AF phase. This is indicated by the magenta coloured points at B_1 that we added to the $T - B$ phase diagram of Ref. [16] (Fig. 3). The points at B_2 are surely related to the transition from the AF phase into the field-polarised state. Although phase diagrams can be extremely complex, a comparison with the parent compound $YbCo_2Si_2$ could help to elucidate the nature of this new line [22]. In the case of $YbCo_2Si_2$ the

Yb moments are fully localized and the magnetic anisotropy is about 10 times smaller than in $YbRh_2Si_2$, which makes the analysis of the magnetic order much more easier. The magnetic phase diagrams of $YbCo_2Si_2$ with magnetic field along the [100], [110] and [001] directions have been intensively investigated [23] and they are all different. Most importantly, in $YbCo_2Si_2$ there are some lines in the phase diagrams within the basal plane ($B \parallel [100]$ and [110]) which correspond to a depopulation of unfavoured AF domains, without modifying the propagation vector. On the other hand, phase lines which indicate a change of the ordering wave vector were observed to be first order. In our data we definitely see the transition at B_1 to be of the second order, and this might be just due to domain reorentation. Since we did not check the exact orientation of our samples - crystal platelets grow preferentially with one side parallel to the [100] or [110] direction - and because of a possible strong in-plane anisotropy in $YbRh_2Si_2$ as in $YbCo_2Si_2$, the phase diagrams with $B \parallel [100]$ might differ from that with $B \parallel [110]$, with slightly different critical fields of 50 and 60 mT. This could explain why measurements performed in our ADR at 13 mK are apparently different from those measured in the Kelvinox 400 at 20 mK.

5. Acknowledgments

We are indebted to K. Andres, P. Coleman, P. Gegenwart, S. Paschen, Q. Si, S. Wirth and R. Yu for useful discussions and R. Gross for his support of the project at the WMI. Part of the work at Dresden was supported by the DFG Research Unit 960 "Quantum Phase Transitions".

References

[1] Gegenwart P, Si Q and Steglich F 2008 *Nature Phys.* **4** 186–197
[2] Si Q and Steglich F 2010 *Science* **329** 1161–1166
[3] Trovarelli O, Geibel C, Mederle S, Langhammer C, Grosche F M, Gegenwart P, Lang M, Sparn G and Steglich F 2000 *Phys. Rev. Lett.* **85** 626–629
[4] Custers J, Gegenwart P, Wilhelm H, Neumaier K, Tokiwa Y, Trovarelli O, Geibel C, Steglich F, Pépin C and Coleman P 2003 *Nature* **424** 524–527
[5] Ishida K, MacLaughlin D E, Young B L, Okamoto K, Kawasaki Y, Kitaoka Y, Nieuwenhuys G J, Heffner R H, Bernal O O, Higemoto W, Koda A, Kadono R, Trovarelli O, Geibel C and Steglich F 2003 *Phys. Rev. B* **68**(18) 184401
[6] Gegenwart P, Westerkamp T, Krellner C, Tokiwa Y, Paschen S, Geibel C, Steglich F, Abrahams E and Si Q 2007 *Science* **315** 969
[7] Paschen S, Lühmann T, Wirth S, Gegenwart P, Trovarelli O, Geibel C, Steglich F, Coleman P and Si Q 2004 *Nature* **432** 881–885
[8] Friedemann S, Oeschler N, Wirth S, Krellner C, Geibel C, Steglich F, Paschen S, Kirchner S and Si Q 2010 *Proc. Natl. Acad. Sci. USA* **107** 14547–14551
[9] Mathur N D, Grosche F M, Julian S R, Walker I R, Freye D M, Haselwimmer R K W and Lonzarich G G 1998 *Nature* **394** 39–43
[10] Stockert O, Arndt J, Faulhaber E, Geibel C, Jeevan H S, Kirchner S, Loewenhaupt M, Schmalzl K, Schmidt W, Si Q and Steglich F 2011 *Nature Phys.* **7** 119–124
[11] Petrovic C, Pagliuso P G, Hundley M F, Movshovich R, Sarrao J L, Thompson J D, Fisk Z and Monthoux P 2001 *J. Phys.: Condens. Matter* **13** L337
[12] Shishido H, Settai R, Harima H and Ōnuki Y 2005 *J. Phys. Soc. Jpn.* **74** 1103–1106
[13] Park T, Ronning F, Yuan H Q, Salamon M B, Movshovich R, Sarrao J L and Thompson J D 2006 *Nature* **440** 65
[14] Knebel G, Aoki D, Brison J P and Flouquet J 2008 *J. Phys. Soc. Jpn.* **77** 114704
[15] Westerkamp T, Gegenwart P, Krellner C, Geibel C and Steglich F 2008 *Physica B* **403** 1236–1238
[16] Schuberth E, Tippmann M, Steinke L, Lausberg S, Steppke A, Brando M, Krellner C, Geibel C, Yu R, Si Q and Steglich F 2016 *Science* **351** 485–488
[17] Nakatsuji S, Kuga K, Machida Y, Tayama T, Sakakibara T, Karaki Y, Ishimoto H, Yonezawa S, Maeno Y, Pearson E, Lonzarich G, Balicas L, Lee H and Fisk Z 2008 *Nature Phys.* **4** 603
[18] Freeman A, Min B and Norman M 1987 *High Energy Spectroscopy* (*Handbook on the Physics and Chemistry of Rare Earths* vol 10) (Elsevier) pp 165 – 229
[19] Schuberth E 1984 *Rev. Sci. Instr.* **55** 1486–1488

[20] Brando M, Pedrero L, Westerkamp T, Krellner C, Gegenwart P, Geibel C and Steglich F 2013 *physica status solidi (b)* **250** 485–490

[21] Feyerherm R, Amato A, Geibel C, Gygax F N, Hellmann P, Heffner R H, MacLaughlin D E, Müller-Reisener R, Nieuwenhuys G J, Schenck A and Steglich F 1997 *Phys. Rev. B* **56** 699–710

[22] Klingner C, Krellner C, Brando M, Geibel C and Steglich F 2011 *New Journal of Physics* **13** 083024

[23] Pedrero L, Klingner C, Krellner C, Brando M, Geibel C and Steglich F 2011 *Physical Review B* **84** 224401

Resonant x-ray scattering explorations of charge order and broken symmetries in underdoped cuprates

The spontaneous self-arrangement of electrons into periodically modulated patterns, a phenomenon commonly termed as charge order or charge-density-wave, has recently resurfaced as a prominent, universal ingredient for the physics of high-temperature superconductors. Its antagonist coexistence with superconductivity, together with its possible connection to a quantum critical point beyond optimal doping, are symptomatic of a very fundamental role played by this symmetry-broken collective electronic state. Resonant x-ray scattering (RXS) has rapidly become the technique of choice for the study of charge order in momentum space, owing to its ability to directly identify a breaking of translational symmetry in the electronic density. Here we will review our RXS work on underdoped Bi2201 and electron-doped NCCO to detect charge-density-waves even in presence of short-ranged order, exploring a realm previously accessible using real-space probes. In addition, we will discuss how the information available from the full two-dimensional momentum space can be used to demonstrate the presence of charge stripes in YBCO. Finally, we will review our explorations of polarization-dependent scattering intensities to uncover the local symmetry in the charge distribution around the Cu atoms, which was found to be predominantly of a *d*-wave bond-order type.

Keywords: charge order; superconductivity; cuprates; resonant x-ray scattering

Introduction

Recently, the study of charge order and its competition with superconductivity has resurged as a prominent topic of research for the physics of cuprate high-temperature superconductors [1]. An outpouring of experimental results in the last 2-3 years enabled a rapid and unprecedented advancement in our characterization of charge order across several cuprate families. Resonant soft x-ray scattering has represented a key experimental probe of charge order in these materials, leveraging the peculiar nature of the resonant light-matter interaction process and its unique sensitivity to reveal the universality of charge order in cuprates, as well as its orbital symmetry and local structure.

Revamped by the recent discovery of charge order in YBCO [2–5], the presence of this instability has now been confirmed in all cuprate families – Bi2201 [6], Bi2212 [7], Hg1201 [8], and including electron-doped cuprates [9], a situation which is summarized in Fig. 1(a,b). At last, charge order can be qualified as a universal, defining instability of the doped copper-oxygen planes. Most importantly, resonant x-ray scattering, owing to its peculiar sensitivity to the electronic degrees of freedom, has enabled further investigations of the microscopic structure of charge order in the cuprates [10–13]. These studies revealed different manifestations of the charge order symmetry across different cuprate families, thus exposing a variegated but essential phenomenology for the physics of copper-based high-temperature superconductors.

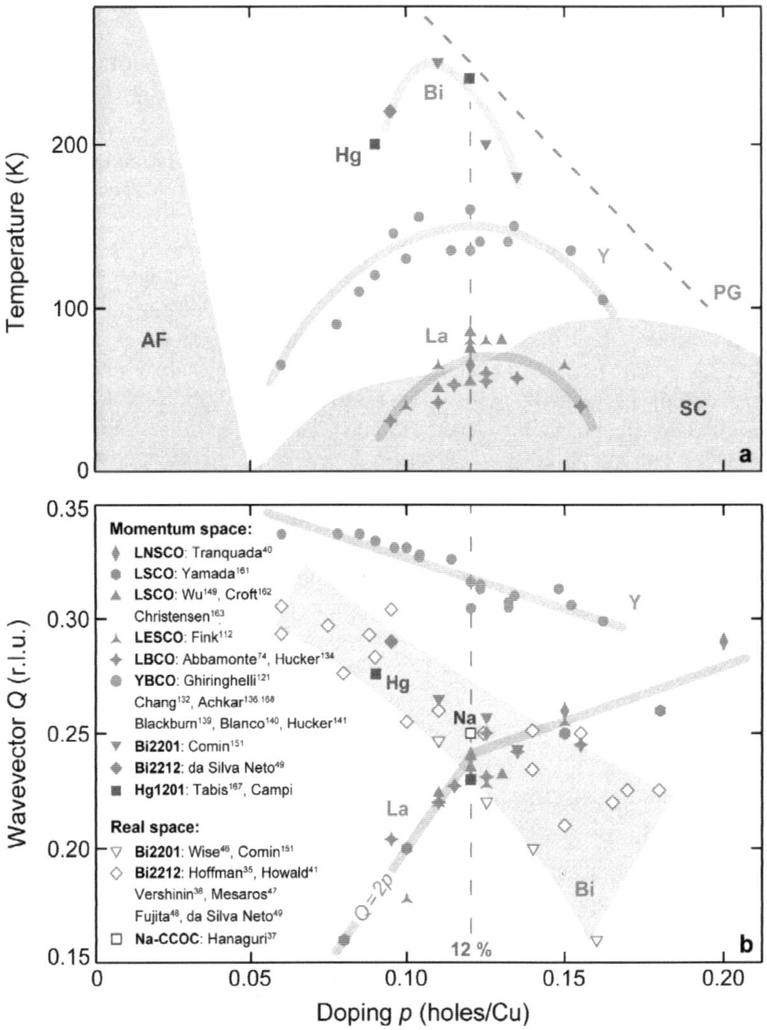

Figure 1. (a,b) Charge order onset temperature (a) and wavevector (b) vs. doping chart for hole-doped (b) cuprates. The shaded areas in (a) denoting the antiferromagnetic (AF), superconducting (SC), and pseudogap (PG) regions define the phase diagram of YBCO. Shaded lines and areas in (b) are guides-to-the-eye to highlight different regimes in the doping-dispersion of the wavevector. From [1].

Methods

Resonant x-ray scattering is a photon-in, photon-out diffraction technique, where the excitation is performed at element-specific x-ray resonances. With this method, it is possible to look for the fingerprints of ordered structures in reciprocal space, while enhancing the sensitivity to selected species. In the context of charge order, RXS has established itself as a prime experimental probe to detect weak, static oscillations of the charge density with unprecedented sensitivity, and has been applied to several families of charge-ordered materials, especially transition metal oxides [14–16]. The enhanced sensitivity of RXS to charge and spin is directly inherited from the nature of the resonant scattering process, which involves a high-energy intermediate state with a core-hole and a valence electron. Such process corresponds to a second-order transition in linear perturbation theory from the nonrelativistic minimal coupling Hamiltonian, and in general its cross section is proportional to the quantum-mechanical probability $w_{i\to f}$ to start from a many-body ground state $|i\rangle$ with zero energy and momentum and leave the system in a final state $|f\rangle$ with finite energy E and momentum $\mathbf{Q}$:

$$w_{i\to f} \propto \left| \sum_{\mathbf{R}} \sum_m \frac{\langle f|\boldsymbol{\varepsilon}' \cdot \mathbf{p}|m\rangle \cdot \langle m|\boldsymbol{\varepsilon} \cdot \mathbf{p}|i\rangle}{E_i - E_n + \hbar\omega_{in} - i\Gamma_m} e^{i\mathbf{Q}\cdot\mathbf{R}} \right|^2 \times \delta(\hbar\omega_{in} - \hbar\omega_{out} - E) \qquad (1)$$

where $\mathbf{R}$ denotes the set of atomic positions, $\mathbf{p}$ is the momentum operator, $\boldsymbol{\varepsilon}(\boldsymbol{\varepsilon}')$ represents the polarizations of incoming (outgoing) photons with energy $\hbar\omega_{in}$ ($\hbar\omega_{out}$) and momentum $\hbar\mathbf{k}_{in}$ ($\hbar\mathbf{k}_{out}$), $|m\rangle$ is a set of intermediate states with a core hole and an excited valence electron above the Fermi energy. Note that energy and momentum conservation laws applied to the scattering process impose that $\hbar\omega_{in} = \hbar\omega_{out} + E$ and $\hbar\mathbf{k}_{in} = \hbar\mathbf{k}_{out} + \mathbf{Q}$; the case of elastic scattering, which is the subject of this article, corresponds to $E = 0$. For a more comprehensive review of the foundations of RXS theory, see also [17–21].
In the case of cuprates, the maximum sensitivity is attained at the oxygen and copper soft x-ray resonances (at 530 eV and 930 eV, respectively), where the RXS intermediate involves an excited electron in a Cu-*3d* or O-*2p* orbital within the CuO_2 planes. The experimental results presented hereafter have all been measured at the Cu-L_3 resonance.

Results

Charge order in Bi2201 and confirmation of real space – reciprocal space duality

Our first study of charge order aimed at elucidating the connection between scattering probes, which are bulk-sensitive and operating in reciprocal space, and microscopy techniques, which are surface-sensitive and access the real space domain. Our first RXS study was performed on underdoped single crystals of the single-layered Bi-based cuprate compound $Bi_2Sr_{2-x}La_xCuO_{6+d}$ (Bi2201), with La content x of 0.8, 0.6, and 0.5, corresponding to hole doping p of 0.11, 0.13, and 0.14, respectively [6]. In particular, by comparing deep-penetrating scattering probes to surface-based scanning tunnelling microscopy (STM), we sought to disclose the connection between signatures of charge order in the bulk of the crystal and at its surface, where static modulations of the local density of states had been reported since the early 2000's [22]. Figure 2(a) shows RXS scans of the momentum axis parallel to the Cu-O bond direction, taken at the Cu-L_3 resonance in the most underdoped sample (p~0.11), where the emergence of a diffuse peak at low temperature can be clearly discerned, despite the weak scattered intensity atop the large fluorescent background. This broad feature signals the presence of a short-ranged electronic density wave in the CuO_2 planes at low temperature. Notably, the scattering signal vanishes when the photon energy is tuned away from the resonance [Fig. 2(b)], as the scattering process becomes insensitive to the states carrying the modulating charge. Figure 2(d) shows the STM tunnelling conductance map on the very same samples, with corresponding Fourier transform in Fig. 2(c). The equivalence between bulk and surface probes becomes self-evident, thus demonstrating the connection and unity between these two complementary techniques in this field of research.

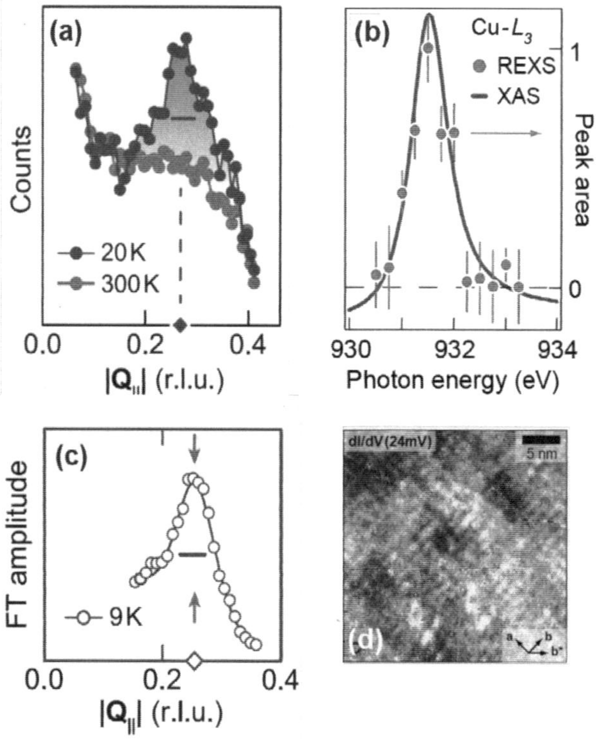

Figure 2. (a,b) Charge-density-wave peak in cuprate compound $Bi_2Sr_{2-x}La_xCuO_{6+d}$ as seen in reciprocal space at low temperatures (20 K) using resonant scattering (a); and corresponding intensity profile across the copper resonance (b). (c,d) Charge-density oscillations in the same samples from STM conductance map (d) and corresponding Fourier transform (c), showing the equivalence with the momentum space representation from scattering (a). From [6].

The other side of the phase diagram: Charge order in electron-doped NCCO

As mentioned in the introduction, charge order has been rapidly found in all cuprate families over a few years of intensive quest, however the electron-doped side of the phase diagram has remained long unexplored. The scientific question that arises in this context is a fundamental one for this phenomenology: what is the driving force behind charge order, and is it affected by the nature of the electronic states near the Fermi energy? To address this important point, we investigated charge order in electron-doped compounds, whose electronic states are predominantly of Cu-*3d* character, as opposed to the O-*2p* orbitals which form the valence band in hole-doped compounds [9]. Figure 3(a) and 3(b) show a series of low-temperature RXS scans at different photon energies for a 0.14 and 0.15 superconducting electron-doped $Nd_{2-x}Ce_xCuO_4$ (NCCO) samples, respectively. In both cases, a broad charge order peak becomes apparent when the photon energy is tuned right at the Cu-L_3 resonance, and disappears completely less than 1 eV away from it. This finding demonstrates the persistence of charge order

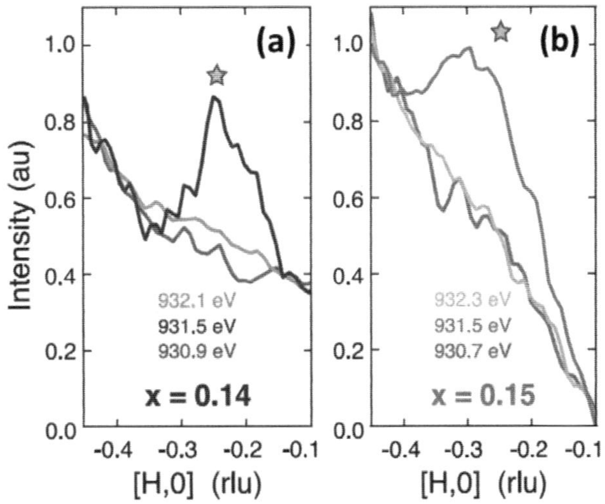

Figure 3. (a,b) RXS scans from a 0.14 (a) and a 0.15 electron-doped $Nd_{2-x}Ce_xCuO_4$ sample, showing the sudden emergence of a charge order peak only in the close vicinity of the Cu-L_3 resonance around 931.5 eV. From [9].

across the extended phase diagram and confirms the centrality of this instability for the low-energy electronic states in the copper-oxygen planes in cuprate superconductors.

The microscopic texture of charge order: stripes and d-wave form factor in YBCO

While being a powerful technique in revealing the momentum-space signatures of weak electronic modulations, RXS has been used to mostly extract the wavevector, strength, and correlation length (inverse peak width) of charge order in cuprates. The resonant process and its cross-section however provide additional information on the local symmetries (at the Cu site, in the case of the Cu resonance) that are imprinted on the form factor tensor which reflects the quantum-mechanical amplitudes for the x-ray transition from core to valence. This information cannot be readily retrieved from a single RXS measurement, however a partial reconstruction of the form factor entries by applying a procedure well-known for magnetic scattering, named azimuthal scan. The schematics of this configuration are shown in Fig. 4(a), and essentially require measuring momentum scans for different angular orientations of the crystallographic axis around an axis (the azimuthal axis) collinear with the probed wavevector **Q**. The corresponding momentum scans vs. azimuthal angle are shown in Fig. 4(b). Using this method on charge-ordered, underdoped $YBa_2Cu_3O_{6+y}$ (YBCO) crystals, a modulation of the scattered intensity as a function of the azimuthal angle could be observed [Fig. 4(f)], whose shape was best reproduced by using a general model for the form factor assuming a local orbital symmetry of the kind depicted in Fig. 4(g) (red pattern), also known as *d-wave bond order* [11].

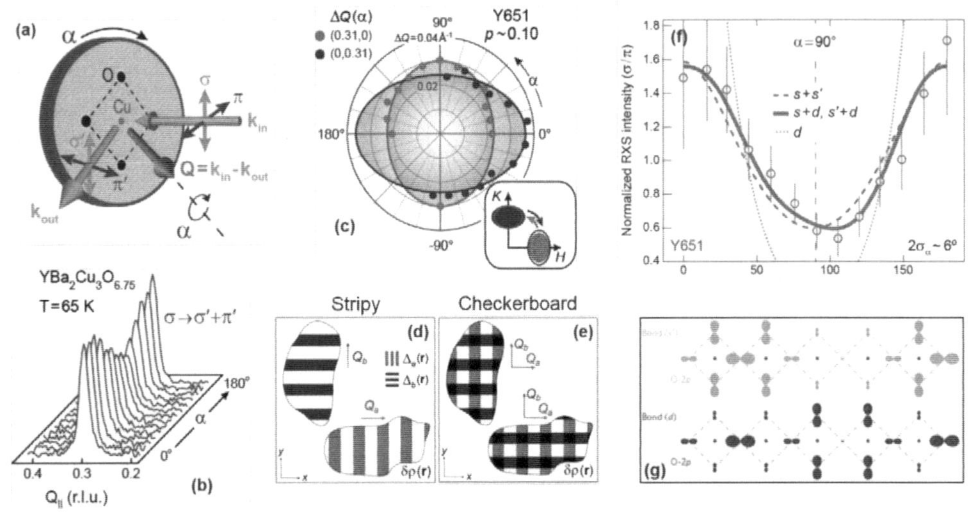

Figure 4. (a) Special scattering geometry used for the tomographic reconstruction of the charge-density-wave peak in reciprocal space [11]. (b) Corresponding momentum scans of charge order peak vs azimuthal angle (α) in YBa2Cu3O6.75. (c) Reconstruction of peak shape from the azimuthal angle dependence of the momentum broadening. (d,e) The resulting peak shape reveals asymmetries that are only compatible with a stripe scenario (b) and not with a checkerboard pattern (c) [10]. (f,g) Azimuthal dependence of RXS intensity (f) and corresponding fit to a multicomponent form factor demonstrating a *d*-wave type orbital pattern of O-2p orbitals (g). From [11].

Lastly, we have adopted a similar approach, again in YBCO but this time focusing on the momentum structure of the RXS scans (as opposed to the scattered intensity) in order to perform a circular reconstruction of the two-dimensional profile of the charge order peak [Fig. 4(c)], which revealed inner anisotropies incompatible with preservation of fourfold rotational symmetry [10]. The latter, which imposes certain constraints on the shape of the charge order peak, is associated with a rotationally-invariant checkerboard state [Fig. 4(e)]. The violation of fourfold symmetry in the RXS two-dimensional peak profiles thus provided support for locally unidirectional modulations of the electronic density, much alike charge stripes [Fig. 4(d)].

Conclusion

To summarize, the phenomenology of charge order in cuprates has been explored to an unprecedented level in recent years, leading to tremendous advancements of our understanding and description of the related phenomenology, on both theoretical and experimental grounds. Recent works have further clarified the connection between charge order and other coexisting and competing phases, such as antiferromagnetism and superconductivity, but more work lies ahead to fully uncover the fundamental role of this instability and the ultimate extent of its relevance for the physics of cuprates and of unconventional superconductors in general.

Acknowledgments
The authors are indebted to P. Abbamonte, A. J. Achkar, D. A. Bonn, L. Braicovich, A. Cavalleri, L. Chauviere, J. C. Davis, H. Eisaki, I. Elfimov, J. Fink,

A. Frano, J. Geck, G. Ghiringhelli, C. Giannetti, R. L. Greene, M. Greven, W. N. Hardy, D. G. Hawthorn, F. He, Y. He, J. E. Hoffman, Y. Jiang, M.-H. Julien, B. Keimer, M. Le Tacon, P. A. Lee, W. S. Lee, R. Liang, C. Mazzoli, M. R. Norman, S. Sachdev, G. A. Sawatzky, E. Schierle, A. Soumyanarayanan, R. Sutarto, L. Taillefer, J. M. Tranquada, E. Weschke, M. M. Yee, and Y. Yoshida, for contributions to the studies presented here and for valuable discussions.
This work was supported by the Max Planck-UBC Centre for Quantum Materials, the Killam, A. P. Sloan, A. von Humboldt, and NSERC's Steacie Memorial Fellowships (A.D.), the Canada Research Chairs Program (A.D.), NSERC, CFI and CIFAR Quantum Materials.

References

[1] R. Comin and A. Damascelli, *Resonant X-Ray Scattering Studies of Charge Order in Cuprates*, Annu. Rev. Condens. Matter Phys. 7 (2016), pp. 369–405.
[2] T. Wu, H. Mayaffre, S. Krämer, M. Horvatić, C. Berthier, W.N. Hardy et al., *Magnetic-field-induced charge-stripe order in the high-temperature superconductor YBa2Cu3Oy*, Nature 477 (2011), pp. 191–194.
[3] G. Ghiringhelli, M. Le Tacon, M. Minola, S. Blanco-Canosa, C. Mazzoli, N.B. Brookes et al., *Long-Range Incommensurate Charge Fluctuations in (Y,Nd)Ba2Cu3O6+x*, Science 337 (2012), pp. 821–825.
[4] J. Chang, E. Blackburn, A.T. Holmes, N.B. Christensen, J. Larsen, J. Mesot et al., *Direct observation of competition between superconductivity and charge density wave order in YBa2Cu3O6.67*, Nat. Phys. 8 (2012), pp. 871–876.
[5] A.J. Achkar, R. Sutarto, X. Mao, F. He, A. Frano, S. Blanco-Canosa et al., *Distinct Charge Orders in the Planes and Chains of Ortho-III-Ordered YBa2Cu3O6+δ Superconductors Identified by Resonant Elastic X-ray Scattering*, Phys. Rev. Lett. 109 (2012), .
[6] R. Comin, A. Frano, M.M. Yee, Y. Yoshida, H. Eisaki, E. Schierle et al., *Charge Order Driven by Fermi-Arc Instability in Bi2Sr2−xLaxCuO6+δ*, Science 343 (2014), pp. 390–392.
[7] E.H. da Silva Neto, P. Aynajian, A. Frano, R. Comin, E. Schierle, E. Weschke et al., *Ubiquitous Interplay between Charge Ordering and High-Temperature Superconductivity in Cuprates*, Science 343 (2014), pp. 393–396.
[8] W. Tabis, Y. Li, M.L. Tacon, L. Braicovich, A. Kreyssig, M. Minola et al., *Charge order and its connection with Fermi-liquid charge transport in a pristine high-Tc cuprate*, Nat. Commun. 5 (2014), .
[9] E.H. da S. Neto, R. Comin, F. He, R. Sutarto, Y. Jiang, R.L. Greene et al., *Charge ordering in the electron-doped superconductor Nd(2–x)Ce(x)CuO4*, Science 347 (2015), pp. 282–285.
[10] R. Comin, R. Sutarto, E.H. da S. Neto, L. Chauviere, R. Liang, W.N. Hardy et al., *Broken translational and rotational symmetry via charge stripe order in underdoped YBa2Cu3O6+y*, Science 347 (2015), pp. 1335–1339.
[11] R. Comin, R. Sutarto, F. He, E.H. da Silva Neto, L. Chauviere, A. Fraño et al., *Symmetry of charge order in cuprates*, Nat. Mater. 14 (2015), pp. 796–800.
[12] A.J. Achkar, M. Zwiebler, C. McMahon, F. He, R. Sutarto, I. Djianto et al., *Nematicity in stripe-ordered cuprates probed via resonant x-ray scattering*, Science 351 (2016), pp. 576–578.

[13] A.J. Achkar, F. He, R. Sutarto, C. McMahon, M. Zwiebler, M. Hücker et al., *Orbital symmetry of charge-density-wave order in La1.875Ba0.125CuO4 and YBa2Cu3O6.67*, Nat. Mater. advance online publication (2016), .

[14] K.J. Thomas, J.P. Hill, S. Grenier, Y.-J. Kim, P. Abbamonte, L. Venema et al., *Soft X-Ray Resonant Diffraction Study of Magnetic and Orbital Correlations in a Manganite Near Half Doping*, Phys Rev Lett 92 (2004), pp. 237204.

[15] P. Abbamonte, A. Rusydi, S. Smadici, G.D. Gu, G.A. Sawatzky and D.L. Feng, *Spatially modulated "Mottness" in La(2-x)Ba(x)CuO4*, Nat Phys 1 (2005), pp. 155–158.

[16] C. Schüßler-Langeheine, J. Schlappa, A. Tanaka, Z. Hu, C.F. Chang, E. Schierle et al., *Spectroscopy of Stripe Order in La1.8Sr0.2NiO4 Using Resonant Soft X-Ray Diffraction*, Phys. Rev. Lett. 95 (2005), pp. 156402.

[17] M. Blume, *Magnetic scattering of x rays (invited)*, J. Appl. Phys. 57 (1985), pp. 3615–3618.

[18] G. Materlik and C.J. Sparks, *Resonant Anomalous X-Ray Scattering: Theory and Applications*, North-Holland, 1994.

[19] P. Carra and B.T. Thole, *Anisotropic x-ray anomalous diffraction and forbidden reflections*, Rev. Mod. Phys. 66 (1994), pp. 1509–1515.

[20] S. Di Matteo, *Resonant x-ray diffraction: multipole interpretation*, J. Phys. Appl. Phys. 45 (2012), pp. 163001.

[21] J. Fink, E. Schierle, E. Weschke and J. Geck, *Resonant elastic soft x-ray scattering*, Rep. Prog. Phys. 76 (2013), pp. 56502.

[22] K. Fujita, M. Hamidian, I. Firmo, S. Mukhopadhyay, C.K. Kim, H. Eisaki et al., *Spectroscopic Imaging STM: Atomic-Scale Visualization of Electronic Structure and Symmetry in Underdoped Cuprates*, in *Strongly Correlated Systems*, A. Avella and F. Mancini, eds., Springer Berlin Heidelberg, 2015, pp. 73–109.

IOP Conf. Series: Journal of Physics: Conf. Series **807** (2017) 052009 doi:10.1088/1742-6596/807/5/052009

Ce-substitution effects on the spin excitation spectra in $Pr_{1.4-x}La_{0.6}Ce_xCuO_{4+\delta}$

S Asano[1], K Tsutsumi[1], K Sato[1], and M Fujita[2]

[1]Department of Physics, Tohoku University, Sendai, Japan, [2]Institute for Materials Research, Tohoku University, Sendai, Japan

E-mail: shun.asano@imr.tohoku.ac.jp

Abstract. We performed inelastic neutron scattering measurements on as-grown single crystals of electron-doped $Pr_{1.4-x}La_{0.6}Ce_xCuO_{4+\delta}$ with x = 0, 0.08 and 0.18. We succeeded in observing the spin excitation spectra up to 300meV in the all samples. In both parent and electron-doped samples, a magnetic signal was observed at the magnetic zone center (0.5, 0.5) in the low-energy region below 40 meV. In the high-energy region above 100 meV, a double-peak structure reflecting the spin-wave excitation was detected in the constant-energy spectrum for the parent sample, while a single broad peak centered at (0.5, 0.5) was observed for the Ce-doped samples. These results suggest that the dispersion of spin excitation spectra become steeper upon doping. Furthermore, we newly found a drastic reduction of the magnetic intensity with Ce-substitution.

1. Introduction

The high-transition-temperature superconductivity in cuprate oxide emerges by the carrier-doping in antiferromagnetically ordered Mott insulators. To understand the mechanisms of the superconductivity, clarifying the electron-hole symmetry in physical properties is quite important. In the hole-doped system, the effect of carrier-doping on spin excitation has been extensively studied by neutron scattering measurements. An hourglass-shaped excitation, which consists of an inwardly dispersive low energy excitation and a spin wave like high energy excitation, is commonly observed in the superconducting phase of $La_{2-x}Sr_xCuO_4$ and $YBa_2Cu_3O_{6+\delta}$ [1, 2]. Appearance of the hourglass excitation in the superconductivity phase suggests the connection between characteristic spin correlations and the superconductivity. In contrast to the study for hole-doped system, systematic experiments of spin excitation in electron-doped systems is limited due to the difficulties in preparing the high quality crystal. Although high-energy inelastic neutron scattering measurements have revealed a steeply standing magnetic dispersion along the energy direction and a weak dynamic spin susceptibility for the superconducting $Pr_{1-x}LaCe_xCuO_4$ [3, 4], the origin of such characteristic feature is still unclear. It has been widely accepted that both the Ce substitution and the reduction annealing are essential for the emergence of superconductivity in the electron-doped cuprate. Therefore, the individual effect on the magnetism should be clarified. Our previous neutron and X-ray studies reported that the Ce-substitution effect is characterized by an elongation of spin excitation spectrum along the energy direction, which is quite different from the negligible doping effect on the high-energy spin excitation in the hole-doped system [5]. However, the data analysis was not done for the neutron scattering measurements. In this paper, we present the results of analysis, which shows the reduction of magnetic intensity as well as the elongation of magnetic excitation by Ce-substitution.

IOP Conf. Series: Journal of Physics: Conf. Series **807** (2017) 052009 doi:10.1088/1742-6596/807/5/052009

2. Experimental Details

The single crystals of $Pr_{1.4-x}La_{0.6}Ce_xCuO_{4+\delta}$ (x = 0, 0.08 and 0.18) were grown by a traveling-solvent floating-zone method. The set of single crystals are co-aligned with the mass of 41.3g (x = 0), 42.6g (x = 0.08) and 28.1g (x = 0.18), respectively. Inelastic neutron scattering measurements were performed with Fermi chopper spectrometer 4SEASONS at Materials and Life Science experimental facility in J-PARC. The incident neutron energies of 357 meV, 88 meV and 39 meV were selected with the multi-E_i method [6]. These experimental conditions are identical to those for previously reported measurements [7]. The scattering intensity was converted into the absolute value of $S(\boldsymbol{Q},\hbar\omega)$ using the vanadium standard. The differential cross section for the inelastic neutron scattering can be written as

$$\frac{d^2\sigma}{d\Omega dE} = \frac{2(\gamma r_e)^2}{\pi g^2 \mu_B^2}\frac{k_f}{k_i} S(\boldsymbol{Q},\hbar\omega)|F(\boldsymbol{Q})|^2 , \tag{1}$$

where $F(\boldsymbol{Q})$ is the magnetic form factor for Cu $3d_{x^2-y^2}$orbital. $(\gamma r_e)^2 = 0.2905$ barn, $g = 2$, and μ_B is Bohr magneton. k_i and k_f are incident and scattered neutron wave vectors, respectively. $S(\boldsymbol{Q},\hbar\omega)$ is a Fourier transform of the spin-spin correlation function, and is connected with the dynamic spin susceptibility $\chi''(\boldsymbol{Q},\hbar\omega)$ via the fluctuation-dissipation theorem.

$$S(\boldsymbol{Q},\hbar\omega) = \chi''(\boldsymbol{Q},\hbar\omega)\frac{1}{1-e^{-\frac{\hbar\omega}{k_B T}}} \tag{2}$$

In this paper, we label the momentum transfer in the tetragonal notation.

3. Results and Discussion

Figure 1 represents the constant energy spectra of the spin excitations at ω = (a) 40 meV, (b) 160 meV, (c) 200 meV and (d) 240 meV for x = 0, 0.08 and 0.18. The excitations were sliced along [100] direction through the antiferromagnetic zone center. In the figure, the spectra were shifted along the vertical direction for the clear visualization. To evaluate peak-position, peak-width and the

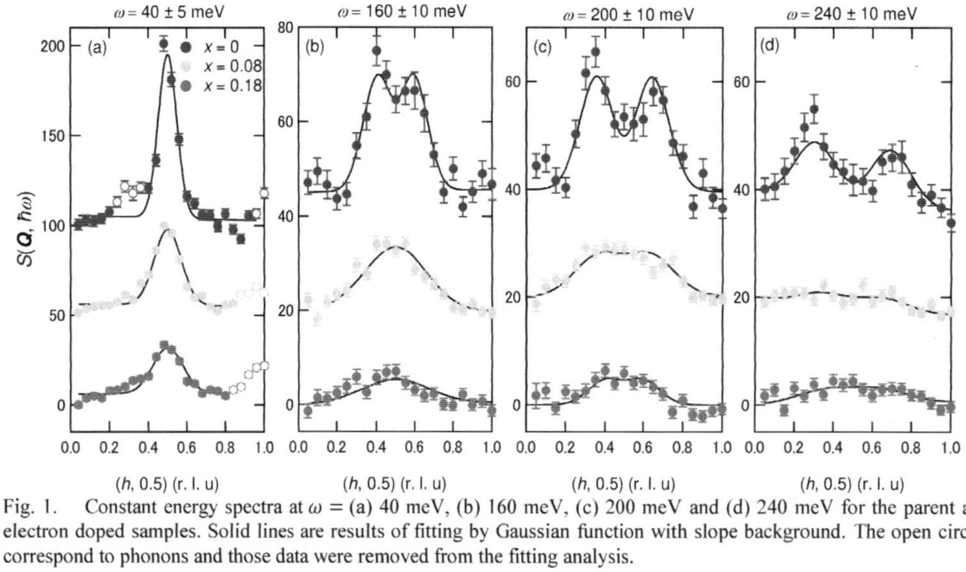

Fig. 1. Constant energy spectra at ω = (a) 40 meV, (b) 160 meV, (c) 200 meV and (d) 240 meV for the parent and electron doped samples. Solid lines are results of fitting by Gaussian function with slope background. The open circles correspond to phonons and those data were removed from the fitting analysis.

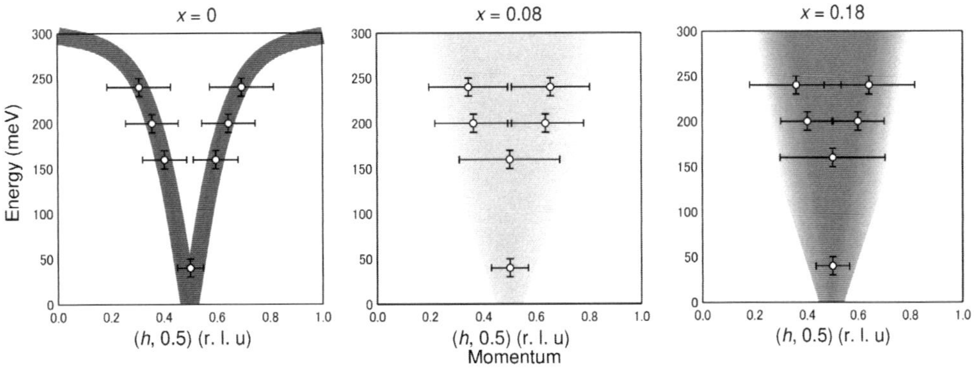

Fig. 2. Schematic pictures for the doping evolution of spin excitation spectra in as-grown $Pr_{1.6-x}La_{0.6}Ce_xCuO_{4+\delta}$. Open circles are the peak-positions evaluated from the fitting analysis. Vertical and horizontal bars represent energy width for the intensity-integration and the peak-width in half-width at half-minimum.

intensity, we carried out fitting analysis. With assuming the slop background, the spectra above 160 meV in the $x = 0$ sample and 200 meV in $x = 0.08$ and 0.18 sample were fitted by two Gaussian function and other spectra were fitted by using single Gaussian function. At $\omega = 40$ meV, as seen in Fig. 1(a), a magnetic peak was observed at the antiferromagnetic zone center (0.5, 0.5) in all samples. Although the peak broadens with Ce-doping, their integrated intensity in the momentum space is comparable (see Fig. 3).

In the high-energy region at 160 meV, 200 meV and 240 meV, a double-peak structure was observed in the $x = 0$ sample, and the distance between the two peaks becomes larger at higher energy transfer, reflecting the outwardly dispersive spin wave excitation as seen in the La_2CuO_4. In contrast, a broad single peak centered at (0.5, 0.5) was observed for the Ce-substituted samples at 160 meV. Even at higher-energy of 240 meV, no well-defined double-peak structure was detected. The intensity is rather gathered at the zone center, indicating that the excitation spectrum becomes steeper by Ce-substitution. Schematic picture for the evolution of spectra induced by Ce-substitution is shown in Fig. 2. This doping evolution is quite different from that in the hole-doped $La_{2-x}Sr_xCuO_4$, which shows weak doping dependence of high energy dispersion with remaining the considerable intensity [5]. Therefore, present results combined with those for $La_{2-x}Sr_xCuO_4$ indicates the doping asymmetry of spin excitation spectrum against the type of carrier. The elongation of the spectra toward the higher energy region was more clearly confirmed in the previous our resonant inelastic X-ray scattering study. [7] We newly found the drastic reduction of magnetic intensity in the high-energy region.

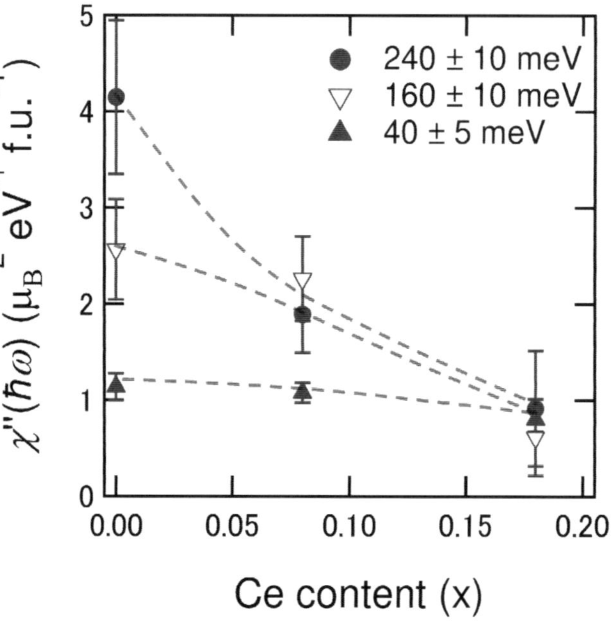

Fig. 3. Ce-substitution dependence of local spin susceptibility $\chi''(\hbar\omega)$.

In Fig. 3, Ce-substitution dependences of the local spin susceptibility $\chi''(\hbar\omega)$ obtained from the integration of intensity in the antiferromagnetic Brillouin zone is summarized. The $\chi''(\hbar\omega)$ at 40 meV exhibits the weak Ce-concentration dependence, while that at 240 meV firmly decreases upon Ce-substitution and the value for x = 0.18 is less than half of that for x = 0. Therefore, the redistribution of intensity by Ce-substitution shows the energy-dependence. This behaviour is similar to the suppression of high energy inelastic neutron intensity reported for the superconducting $YBa_2Cu_3O_{6.5}$. [8] Stock and co-workers claimed that the intensity reduction in $YBa_2Cu_3O_{6.5}$ is originated from the opening of pseudogap. As we mentioned above, the doping evolution of high-energy dispersion is different between the hole-doped and electron-doped systems. The boundary energy for PLCCO with x = 0.18 exceeds 600 meV, which is approximately double of that for x = 0. Therefore, the intensity reduction in PLCCO is mainly attributed to the elongation of excitation spectrum, and the origin of intensity reduction between PLCCO and YBCO would be different.

4. Summary

To study the Ce-substitution effects on the spin excitation, we performed inelastic neutron scattering measurements on the as-grown single crystals of $Pr_{1.4-x}La_{0.6}Ce_xCuO_{4+\delta}$ (x = 0, 0.08 and 0.18). We confirmed that the change of spectrum by Ce-substitution is characterized by the intensity reduction remarkably seen in the high energy region and the elongation of spectrum along the energy direction. These results are different from those in the hole-doped $La_{2-x}Sr_xCuO_4$, indicating the doping asymmetry of spin excitation spectrum against the type of carrier.

Acknowledgement
We thank K. Ishii and T. Tohyama for the fruitful discussion. The experiment at J-PARC was carried out under Project No. 2012A0104 and supported by R. Kajimoto and K. Ikeuchi. MF is supported by Grant-in-Aid for Scientific Research (B) (23340093) and (A) (16H02125).

References

[1] S. M. Hayden, H. A. Mook, Pengcheng Dai, T. G. Perring, and F. Doğan, Nature. **429**, 531 (2004).
[2] J. M. Tranquada, H. Woo, T. G. Perring, H. Goka, G. D. Gu, G. Xu, M. Fujita, and K. Yamada, Nature. **429**, 534 (2004).
[3] M. Fujita, M. Matsuda, B. Fåk, C. D. Frost, and K. Yamada, J. Phys. Soc. Jpn. **75**, 093704 (2006).
[4] S. D. Wilson, S. Li, H. Woo, Pengcheng Dai, H. A. Mook, C. D. Frost, S. Komiya, and Y. Ando, Phys. Rev. Lett. **96**, 157001 (2006).
[5] M. P. M. Dean, G. Dellea, R. S. Springell, F. Yakhou-Harris, K. Kummer, N. B. Brookes, X. Liu, Y-J. Sun, J. Strle, T. Schmitt, L. Braicovich, G. Ghiringhelli, I. Božović, and J. P. Hill, Nature Materials. **12**, 1019 (2013).
[6] M. Nakamura, R. Kajimoto, Y. Inamura, F. Mizuno, M. Fujita, T. Yokoo, and M. Arai, J. Phys. Soc. Jpn. **78**, 093002 (2009).
[7] K. Ishii, M. Fujita, T. Sasaki, M. Minola, G. Dellea, C. Mazzoli, K. Kummer, G. Ghiringhelli, L. Braicovich, T. Tohyama, K. Tsutsumi, K. Sato, R. Kajimoto, K. Ikeuchi, K. Yamada, M. Yoshida, M. Kurooka, and J. Mizuki, Nature. Commun. **5**, 3714 (2014)
[8] C. Stock, R. A. Cowley, W. J. L. Buyers, C. D. Frost, J. W. Taylor, D. Peets, R. Liang,D. Bonn, and W. N. Hardy, Phys. Rev. B. **82**, 174505 (2010).

Vortex states in a superconductor under a helical magnetic field

Saoto Fukui[1], Masaru Kato[1] and Yoshihiko Togawa[2]

[1] Department of Mathematical Sciences, Osaka Prefecture University, 1-1, Gakuencho, Nakaku, Sakai, Osaka 599-8531, Japan

[2] Department of Physics and Electronics, Osaka Prefecture University, 1-1, Gakuencho, Nakaku, Sakai, Osaka 599-8531, Japan

E-mail: `st110035@edu.osakafu-u.ac.jp`

Abstract. We have studied vortex states in a chiral helimagnet / superconductor bilayer system numerically. We consider a two-dimensional superconductor subsystem. An effect of a chiral helimagnet on a superconductor is taken as a magnetic field, which oscillates spatially. Solving the Ginzburg-Landau equations, we obtain various vortex states. Comparing free energies, we find the most stable vortex states under only the oscillating magnetic field H_{CHM} and under the H_{CHM} and the homogeneous applied magnetic field H_{appl}.

1. Introduction

Recently, a chiral helimagnet (CHM) attracts attention in the field of magnetism [1]. This magnetic structure consists of spins that form a helical rotation along a helical axis in Fig.1(a). In the CHM, there are two interactions between nearest neighbor spins; a ferromagnetic exchange interaction and the Dzyaloshinsky-Moriya (DM) interaction[2, 3]. The ferromagnetic interaction causes nearest neighbor spins to be parallel. On the other hand, the DM interaction causes nearest neighbor spins to be perpendicular to each other. The competition between two interactions leads to the small finite angle between nearest neighbor spins. Then, all spins rotates along the helical axis. Under an applied magnetic field, this helical magnetic structure in the CHM changes into a chiral soliton lattice (CSL), which consists of ferromagnetic domains periodically partitioned by 360° domain walls in Fig.1(b). We can regard domain walls as solitons. In the experiment, the magnetic structure of the CHM and the CSL are observed recently[1].

We focus on the effect of the chiral helimagnet on a superconductor(SC). In general, there appear vortices in type-II SCs under a homogeneous magnetic field. These vortices are affectied by the strength of the magnetic field and external current. Also, vortices are affected by a ferromagnet(FM). The FM / SC hybrid structure enhances critical current because of a pinning of vortices[4]. We expect that the CHM also affects vortices strongly. The helical magnetic structure of the CHM causes a helical magnetic field outside the CHM. So, we investigate vortex states in the SC under the helical magnetic field. For this purpose, we solve the Ginzburg-Landau equations numerically[5, 6, 7], and obtain stable vortex states.

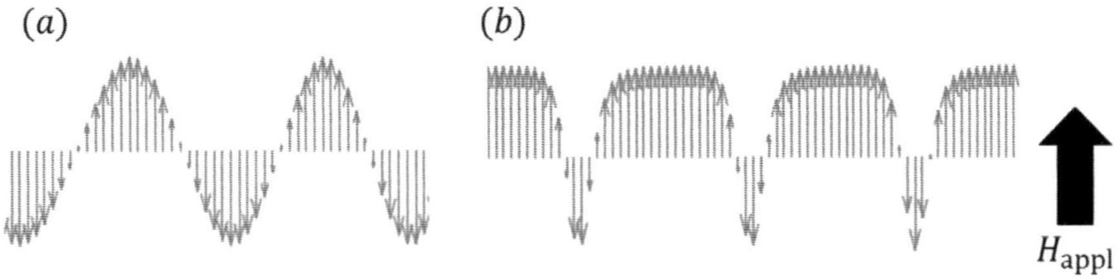

Figure 1. Spin configurations of (a) a chiral helimagnetic structure, and (b) a chiral soliton lattice under an applied magnetic field $H_{\rm appl}$.

2. Methods

We consider a CHM / SC bilayer system in Fig.2. First, we neglect effects of the SC on the CHM. On the other hand, the effect of the CHM on SC is taken as an external magnetic field $\boldsymbol{H}_{\rm CHM}$. We assume the distribution of $\boldsymbol{H}_{\rm CHM}$ is same as the spin configuration of the CHM. So, $\boldsymbol{H}_{\rm CHM}$ is helical along the helical axis. We consider a two-dimentional SC system. Then, only the perpendicular component of $\boldsymbol{H}_{\rm CHM}[(\boldsymbol{H}_{\rm CHM})_z]$ can be taken into account. Since $(\boldsymbol{H}_{\rm CHM})_z$ oscillates spatially, this situation is same as the two-dimenstional SC system under the oscillating magnetic field $(\boldsymbol{H}_{\rm CHM})_z$. $\boldsymbol{H}_{\rm CHM}$ is obtained from a Hamiltonian for the CHM[8];

$$\mathcal{H} = -J\sum_n \boldsymbol{S}_n \cdot \boldsymbol{S}_{n+1} + \boldsymbol{D}\cdot\sum_n \boldsymbol{S}_n \times \boldsymbol{S}_{n+1} + 2\mu_B \boldsymbol{H}_{\rm appl}\sum_n \boldsymbol{S}_n^z, \tag{1}$$

where S_n is a n-th spin, μ_B is the Borh magneton, $\boldsymbol{H}_{\rm appl} = (0, 0, H_{\rm appl})$ is the homogeneous applied magnetic field, J is the coefficient of the exchange interaction, and $\boldsymbol{D} = (D, 0, 0)$ is a DM vector that is parallel to the helical axis. The spin is expressed in a polar coordinate;

$$\boldsymbol{S}_n = S(\sin\theta_n \cos\phi,\ \sin\theta_n \sin\phi, \cos\theta_n). \tag{2}$$

In stable magnetic structures, $\phi = \pi/2$. From the Hamiltonian [Eq.(1)] in the continuum limit, we obtain distribution of angle $\theta(x)$ of spins;

$$\theta(x) = 2\sin^{-1}\left[\mathrm{sn}\left(\frac{\sqrt{H^*}}{k}x \mid k\right)\right] + \pi, \tag{3}$$

where $\mathrm{sn}(u \mid k)$ is the Jacobi's elliptic function and k is a modulus of the elliptic function. k is determined by,

$$\frac{\pi\alpha}{4\sqrt{H^*}} = \frac{E(k)}{k}. \tag{4}$$

$E(k)$ is the complete elliptic integral of the second kind and $\alpha = \tan^{-1}(D/J)$. H^* in Eqs.(3) and (4) is a normalized applied magnetic field,

$$H^* = \frac{2\mu_B H_{\rm appl}}{a^2 S^2\sqrt{J^2 + D^2}}, \tag{5}$$

where a is a lattice constant and we assume $a \sim \xi_0$ (ξ_0 is a coherence length at zero temperature). Finally, from Eqs.(2) and (3), $(\boldsymbol{H}_{\rm ext})_z$ is given by,

$$(\boldsymbol{H}_{\rm ext})_z(x) = H_0\cos\theta(x) + H_{\rm appl}. \tag{6}$$

The first term is a magnetic field from the CHM $(\boldsymbol{H}_{\rm CHM})_z$ and the second term is the homogeneous applied magnetic field. The period of the helical rotation L' is determined by,

$$L' = \frac{2kK(k)}{\sqrt{H^*}}, \tag{7}$$

where $K(k)$ is the complete elliptic integral of the first kind. Under this magnetic field $(\boldsymbol{H}_{\rm ext})_z$, we solve the Ginzburg-Landau equations;

$$\alpha|\psi|^2 + \beta|\psi|^2\psi + \frac{1}{2m^*}\left(\frac{\bar{h}}{i}\nabla - \frac{e^*}{c}\boldsymbol{A}\right)^2\psi = 0, \tag{8}$$

$$\mathrm{curl}(\mathrm{curl}\ \boldsymbol{A} - \boldsymbol{H}_{\rm ext}) = \frac{4\pi}{c}\boldsymbol{J} = \frac{4\pi}{c}\left\{\frac{e^*\bar{h}}{2m^*i}(\psi^*\nabla\psi - \psi\nabla\psi^*) - \frac{e^{*2}}{m^*c}\psi^*\psi\boldsymbol{A}\right\}, \tag{9}$$

where $\alpha = \alpha_0(T - T_c)$, T is a temperature, T_c is a critical temperature, and α_0, β are coefficients. ψ is an superconuducting order parameter, e^* is an effective charge, m^* is an effective mass, $\boldsymbol{A}$ is a magnetic vector potential, and $\boldsymbol{J}$ is a supercurrent density. When we solve these equations using the finite element method, we give the initial values of the order parameter ψ randomly and obtain several stable solutions by iterations[9].

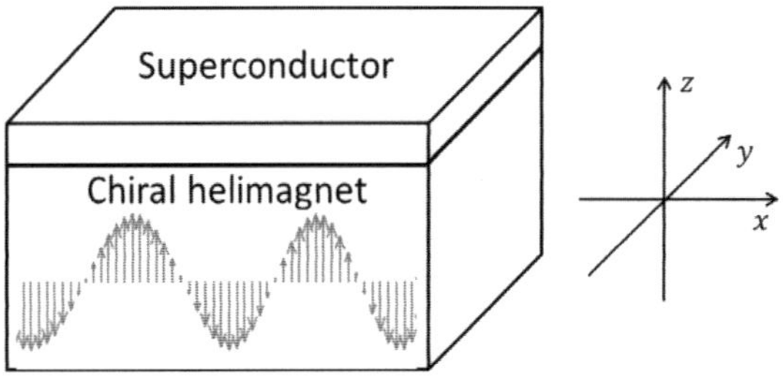

Figure 2. Chiral halimegnet (CHM) / Superconductor (SC) bilayer system.

3. Results and Discussions

We show vortex configurations in the SC under $(\boldsymbol{H}_{\rm ext})_z$ in Eq.(6). We take the Ginzburg-Landau parameter $\kappa = \lambda_0/\xi_0 = 10$ (λ_0 is a penetration length at zero temperature), the temperature $T = 0.3T_c$. The ratio between the ferromagnetic exchange interaction and the DM interaction is taken as $D/J = 0.16$, which is an experimental data of the CHM, $\mathrm{Cr_{1/3}NbS_2}$[10]. The system sizes are $7.0L'\xi_0 \times 20\xi_0$, where L' is the helical period in Eq.(7). For $H_{\rm appl}/(\Phi_0/\xi_0^2) = 0.00$ ($\Phi_0 = h/2e$ is a quantum flux), L'/ξ_0 is approximately 39.2699.

We solve the Ginzburg-Landau equations and obtain a vortex configuration for $H_0/(\Phi_0/\xi_0^2) = 0.020$, $H_{\rm appl}/(\Phi_0/\xi_0^2) = 0.000$, which is shown in Fig.3. From the distributions of phases (b) and magnetic field (c), we find that there are two kind of vortices, which have different directions of quantum fluxes. We call these vortices up- and down-vortices, respectively. In Fig.3, we find that up- and down-vortices appear alternately. Considering a following interaction between vortices and the external magnetic field, this behavior can be explained;

$$E_{VF} = -\frac{1}{4\pi}\Phi_0 \cdot \boldsymbol{H}_{\rm ext}. \tag{10}$$

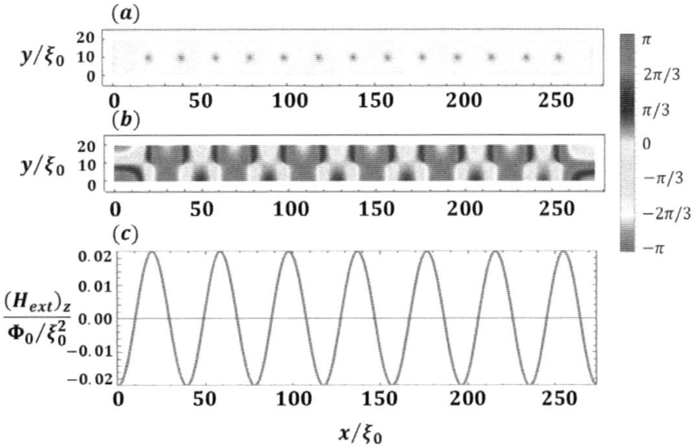

Figure 3. Distributions of (a) order parameter (vortex configuration), (b) phase, and (c) magnetic field for $H_0/(\Phi_0/\xi_0^2) = 0.020$, $H_{\rm appl}/(\Phi_0/\xi_0^2) = 0.000$.

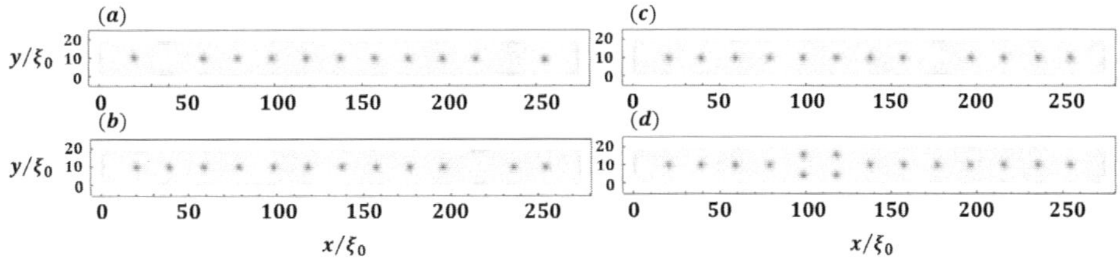

Figure 4. Vortex configurations for $H_0/(\Phi_0/\xi_0^2) = 0.020$, $H_{\rm appl}/(\Phi_0/\xi_0^2) = 0.000$.

Table 1. Free energies $\mathcal{F}$ for vortex configurations in Figs.3 and 4.

Vortex state	Free energy $\mathcal{F}$
Fig.3	4166.04062646254
Fig.4(a)	4139.90157232138
Fig.4(b)	4152.76358564515
Fig.4(c)	4152.79599929134
Fig.4(d)	4197.60960397317

When we solve the GL equations with other initial states, other vortex configurations are obtained and stable configurations are shown in Fig.4 (a) - (d). In the experiment, these vortex configurations may be found due to experimental circumstances. But, we determine the most stable vortex configuration theoretically. Therefore, we calculate the Ginzburg-Landau free energy $\mathcal{F}(\psi, \boldsymbol{A})$;

$$\mathcal{F}(\psi, \boldsymbol{A}) = \int_\Omega \left[\frac{1}{\xi(T)^2} \frac{1}{2} (|\psi|^2 - 1)^2 + |(i\nabla - \tilde{\boldsymbol{A}})\psi^2| \right] d\Omega + \kappa^2 \xi(T)^2 \int_\Omega \left| \text{curl } \boldsymbol{A} - \frac{2\pi}{\Phi_0} \right|^2 d\Omega, \quad (11)$$

where $\xi(T)$ is a coherence length and $\tilde{\boldsymbol{A}} = (2\pi/\Phi_0)\boldsymbol{A}$. These free energies are listed in Table 1. From the Table 1, vortex state in Fig.4(a) is the most stable of these states. In Fig.4(a), there are the least vortices in these states. In Fig.4(b) and (c), the numbers of vortices are same, but vortex configurations are different. Then, free energies are different slightly. The free energy in Fig.4(d) is the largest of these states. In Fig.4(d), there are the most vortices and two vortices form pairs in the fifth- and sixth- rows from the left side.

Therefore, the most stable vortex state is Fig.4(a) for $H_0/(\Phi_0/\xi_0^2) = 0.020$ and $H_{\mathrm{appl}}/(\Phi_0/\xi_0^2) = 0.000$. If the $(\boldsymbol{H}_{\mathrm{ext}})_z$ changes, the most stable state may change from the state in Fig.4(a). When the $(\boldsymbol{H}_{\mathrm{ext}})_z$ becomes larger, the most stable state may change to states in Fig.4(b) and (c). Further $(\boldsymbol{H}_{\mathrm{ext}})_z$ becomes larger and then the vortex state in Fig.3 may become the most stable.

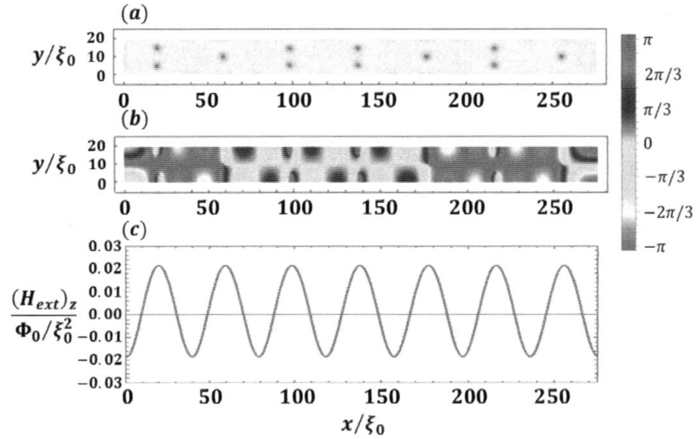

Figure 5. Distributions of (a) order parameter (vortex configuration), (b) phase, and (c) magnetic field for $H_0/(\Phi_0/\xi_0^2) = 0.020$, $H_{\mathrm{appl}}/(\Phi_0/\xi_0^2) = 0.0015$.

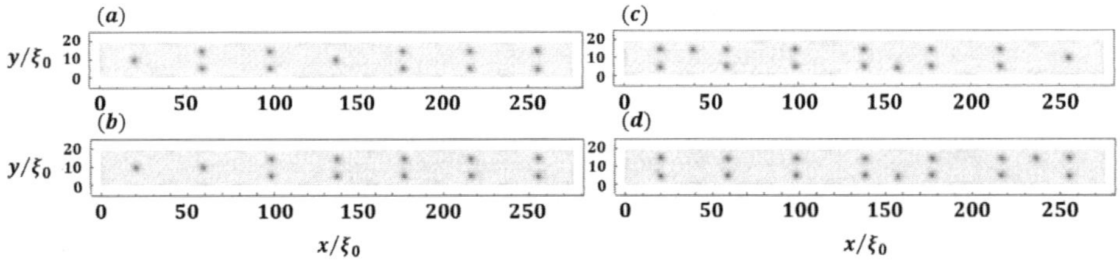

Figure 6. Vortex configurations for $H_0/(\Phi_0/\xi_0^2) = 0.020$, $H_{\mathrm{appl}}/(\Phi_0/\xi_0^2) = 0.0015$.

Next, we show effects of the CHM under the homogeneous applied magnetic field on vortex states. $H_0/(\Phi_0/\xi_0^2)$ is fixed to 0.020. The vortex configuration for $H_{\mathrm{appl}} = 0.000$ has been already shown in Figs.3 and 4. We obtain vortex configurations for $H_{\mathrm{appl}} = 0.0015$ in Figs.5 and 6. These vortex configurations result from the calculation with different initial values of the order parameter. Under the applied magnetic field ($H_{\mathrm{appl}} > 0$), the magnitude of the magnetic field in the negative magnetic field region becomes smaller. Then, the number of down-vortices decreases in Fig.6(c) and (d), or down-vortices disappear in Fig.5, 6(a) and (b). On the other hand, the magnitude of the magnetic field in the positive magnetic field region becomes larger and the number of up-vortices increases.

Table 2. Free energies $\mathcal{F}$ for vortex configurations in Figs.5 and 6.

Vortex state	Free energy $\mathcal{F}$
Fig.5	4243.23216334796
Fig.6(a)	4261.39881390694
Fig.6(b)	4261.40355039939
Fig.6(c)	4307.83875655273
Fig.6(d)	4324.99159993312

Again, we compare free energies of vortex states in Figs.5 and 6. These free energies are listed in Table 2. From the Table 2, free energies for Figs.6(c) and (d) are larger. In these vortex configurations, down-vortices remain in the second- and sixth-rows from the left side (c) and in the second- and fifth-rows from the right side (d), respectively. These down-vortices lead to the increase of free energies. On the other hand, free energies in Figs.5, 6(a) and (b) are smaller, where the down-vortices disappear. Then, we focus on the number and location of up-vortices. In Figs.6(a) and (b), the numbers of up-vortices are same, but vortex configurations are different. In particular, locations of the single vortex rows are different between these two vortex configurations. The distance between two single vortex rows in Fig.6(a) is longer than that in Fig.6(b). Then, the free energy in Fig.6(a) is slightly smaller than that in Fig.6(b). Finally, the most stable vortex state is Fig.5, where there are the least up-vortices. Similarly, the most stable state also may change from Fig.5, if the homogeneous applied magnetic field changes. From the Figs.3-6, the most stable state is controlled by the magnetic field $(\boldsymbol{H}_{\rm ext})_z$.

4. Conclusions

We have obtained vortex states in the CHM / SC bilayer systems numerically. Solving the GL equations, we found that up- and down-vortices appear in the SC. Moreover, using various initial values of the order parameter, we obtained various stable vortex configurations. Experimentally, many patterns of vortex configurations may be obtained. We compared free energies of these configurations and found the most stable state theoretically. Then, we found that the most stable state depends on the magnetic field from the chiral helimagnet and the homogeneous applied magnetic field.

Acknowledgments

This work was supported by JPSJ KAKENHI Grant Number 26400367.

References

[1] Togawa Y, Koyama T, Takayanagi K, Mori S, Kousaka Y, Akimitsu J, Nishihara S, Inoue K, Ovschinnikov A S, and Kishine J 2012 *Phys. Rev. Lett.* **108** 107202
[2] Dzyaloshinskii I E 1958 *J. Phys. Chem. Solids* **4** 241
[3] Moriya T 1960 *Phys. Rev.* **120** 91
[4] Lyuksyutov I F and Pokrovsky V L 2005 *Adv. Phys.* **54** 67
[5] Fukui S, Kato M, and Togawa Y 2015 *Physics Procedia* **65** 85
[6] Fukui S, Kato M, and Togawa Y 2016 *arXiv*:1605.05467
[7] Fukui S, Kato M, and Togawa Y 2016 *arXiv*:1605.05637
[8] Kishine J, Inoue K, and Yoshida Y 2005 *Prog. Theor. Phys. Sppl.* **159** 82
[9] Kato M, Ishida T, Koyama T, and Machida M 2013 *Superconductors-Materials, Properties and Applications, INTECH* 319-342
[10] Benjamin C, Alexander B, Nirmal G, David M, Minbhyea L 2014 *Appl. Phys. Lett.* **105** 072405

A note on the upper critical field of Sr_2RuO_4 under strain

Aline Ramires[1] **and Manfred Sigrist**[2]

[1] Institute for Theoretical Studies, ETH Zürich, CH-8092, Zürich, Switzerland
[2] Institute for Theoretical Physics, ETH Zürich, CH-8093, Zürich, Switzerland

E-mail: `aline.ramires@eth-its.ethz.ch`

Abstract. In the light of the recently discussed mechanism for suppression of superconductivity in multi-orbital systems called *inter-orbital effect*, here we extend our analysis of the upper critical field in Sr_2RuO_4 now under strain. We show that the presence of the standard orbital effect and the new mechanism can consistently account for the qualitative changes observed in the upper critical field for the strained system, in particular, combining the overall enhancement of the critical field with the reduction in the anisotropy. The proposed picture holds for a triplet superconducting state, showing that a singlet state is not the only possibility to account for the observations. We suggest further experiments in order to clarify our understanding about the superconducting phase of Sr_2RuO_4.

1. Introduction

Sr_2RuO_4 has been extensively explored experimentally since it is one of the few putative chiral triplet superconductors [1]. The actual character of the order parameter and the origin of the pairing mechanism in this material are still under debate. The lack of agreement in the community steams mainly from the fact that this is a multi-band system, and different bands may favor distinct pairing mechanisms and consequently order parameters. One mechanism suggests the dominance of the γ-band due to its proximity to a van-Hove singularity [2, 3, 4, 5, 6], while another focuses on the enhancement of spin and charge fluctuations due to nesting [7, 8, 9] in the almost 1D α- and β-bands. The dominance of one over the other is not very clear on a theoretical level and might be dependent on the calculation scheme chosen. Recent studies, based on the renormalization group technique, considering all three bands on the same footing suggests that actually all have gaps of similar magnitude, and in such a case it is difficult to reduce the problem to a subset of the bands [10].

Symmetry aspects have always been useful, giving us robust arguments for the understanding of phases of matter and their stability. In particular for superconducting phases, Anderson's theorems [11, 12] give a qualitative understanding on the stability of different superconducting states in presence of key symmetry breaking fields. For example, singlet superconductivity is suppressed by the application of an external magnetic field, while triplet superconductivity is not robust in presence of certain inversion symmetry breaking fields. Recently [13, 14], the authors showed that these criteria cannot be directly transferred to multi-orbital system, and introduce the concept of *superconducting fitness* to address the question of stability of superconductivity in arbitrarily complex multi-orbital systems. With this tool in hand we can evaluate the stability of different gap structures in presence of inter-orbital effects and external symmetry breaking fields, leading to robust arguments to be contrasted with experiments.

The concept of superconducting fitness, $F(\mathbf{k})$ below, generalizes the idea of symmetry breaking encoded in Anderson's theorems. It does so by capturing the breaking of the degeneracy of the states to be paired, which can be understood as a band splitting. The concept is based on the modified commutator:

$$F(\mathbf{k})(i\sigma_2) = H_0(\mathbf{k})\hat{\Delta}(\mathbf{k}) - \hat{\Delta}(\mathbf{k})H_0^*(-\mathbf{k}), \quad (1)$$

where $H_0(\mathbf{k})$ is the non-interacting Hamiltonian and $\hat{\Delta}(\mathbf{k})$ is the normalized gap matrix, both in an orbital basis [13]. $F(\mathbf{k})$ gives a direct measure of the incompatibility between an arbitrary pairing state and the normal state Hamiltonian. If $F(\mathbf{k}) = 0$ the gap matrix and the electronic structure are compatible, leading to a robust BCS-like instability towards a superconducting state, whereas if $F(\mathbf{k})$ is finite, there is some degree of incompatibility. This statement can be made explicit if we perturb the system by introducing $\delta H(\mathbf{k})$ to the non-interacting Hamiltonian and evaluate the superconducting fitness. In a single band picture, the instability of the superconducting phase under a perturbation can be directly seen as a reduction of the critical temperature following:

$$T_c \sim T_c^0\left(1 - \frac{7\xi(3)}{64\pi^2}\frac{\langle Tr|F(\mathbf{k})|^2\rangle_{FS}}{(T_c^0)^2}\right). \quad (2)$$

Here T_c and T_{c0} are the critical temperatures in presence and absence of the perturbation, respectively, and $\langle ...\rangle_{FS}$ denotes the average over the Fermi surface. Details on the derivation of this relation can be found in reference [13].

Applying this concept to Sr_2RuO_4 we find that the most compatible triplet superconducting (SC) state has the d-vector oriented along the z-axis, as a consequence of the interplay of inter-orbital hopping and spin-orbit coupling [13]. Concerning magnetic field effects, we were able to identify a new mechanism for the suppression of superconductivity in multi-orbital systems referred to as *inter-orbital effect* (IEO) [13], in order to discriminate it from the standard orbital depairing mechanism. The IEO, together with the standard orbital depairing, is a good candidate to explain the unusual behaviour of the upper critical field in this material.

In the next section we introduce the 3-band model for Sr_2RuO_4 and review our previous results applying the concept of superconducting fitness to this system under magnetic field. We highlight some of the interesting experimental facts observed in strained Sr_2RuO_4 [15] in Sec. 3 and extend our analysis for this new condition. We conclude and discuss further interesting directions for experiments in Sec. 4.

2. Unstrained Sr_2RuO_4

Sr_2RuO_4 is modelled by a multi-orbital Hamiltonian including the three low-lying $|yz\rangle$, $|xz\rangle$ and $|xy\rangle$ orbitals (here we label these as $|1\rangle$, $|2\rangle$ and $|3\rangle$, respectively). Based on the symmetries of the orbitals and the underlying lattice, one can construct a tight-binding model with hopping up to next-nearest neighbours and SOC. Introducing the six-dimensional Nambu basis:

$$\Psi^\dagger{}_{\mathbf{k}} = (c^\dagger{}_{1\mathbf{k}\uparrow}, c^\dagger{}_{1\mathbf{k}\downarrow}, c^\dagger{}_{2\mathbf{k}\uparrow}, c^\dagger{}_{2\mathbf{k}\downarrow}, c^\dagger{}_{3\mathbf{k}\uparrow}, c^\dagger{}_{3\mathbf{k}\downarrow}), \quad (3)$$

the matrix form of the Hamiltonian is:

$$H_0 = \sum_{\mathbf{k}} \Psi^\dagger{}_{\mathbf{k}} H_{SRO}(\mathbf{k})\Psi_{\mathbf{k}}, \quad (4)$$

with

$$H_{SRO}(\mathbf{k}) = \begin{pmatrix} \epsilon_{1\mathbf{k}}\sigma_0 & t_{\mathbf{k}}\sigma_0 + i\eta\sigma_3 & -i\eta\sigma_2 \\ t_{\mathbf{k}}\sigma_0 - i\eta\sigma_3 & \epsilon_{2\mathbf{k}}\sigma_0 & i\eta\sigma_1 \\ i\eta\sigma_2 & -i\eta\sigma_1 & \epsilon_{3\mathbf{k}}\sigma_0 \end{pmatrix}, \tag{5}$$

where $\epsilon_{n\mathbf{k}}$ is the dispersion for each of the orbitals originated from intra-orbital hopping, $t_{\mathbf{k}}$ concerns inter-orbital hopping and η is the magnitude of the SOC. Here σ_0 is the 2x2 identity matrix and σ_i are Pauli matrices ($i = 1, 2, 3$). The most important feature of the Hamiltonian above is its structure, which is determined by the character of the low-lying orbitals and the lattice symmetry. By symmetry, inter-orbital hopping is allowed only between orbitals $|1\rangle$ and $|2\rangle$, and the presence of SOC introduces mixing of all orbitals.

The gap matrix can also be written in the six-dimensional basis introduced in Eq. 3. Considering only intra-orbital pairing it can be written as:

$$\Delta(\mathbf{k}) = \begin{pmatrix} \Delta_1(\mathbf{k}) & 0 & 0 \\ 0 & \Delta_2(\mathbf{k}) & 0 \\ 0 & 0 & \Delta_3(\mathbf{k}) \end{pmatrix}, \tag{6}$$

where

$$\Delta_a(\mathbf{k}) = \phi_a \left(\mathbf{d}^a(\mathbf{k}) \cdot \boldsymbol{\sigma}\right)(i\sigma_2), \tag{7}$$

for a triplet state, where the index $a = 1, 2, 3$ refers to different orbitals, $\mathbf{d}^a(\mathbf{k}) = (d_x^a(\mathbf{k}), d_y^a(\mathbf{k}), d_z^a(\mathbf{k}))$ is a three-component vector and an odd function of momentum which parametrizes the order parameter and $\boldsymbol{\sigma} = (\sigma_1, \sigma_2, \sigma_3)$; or

$$\Delta_a(\mathbf{k}) = \phi_a d_0^a(\mathbf{k})(i\sigma_2), \tag{8}$$

for a singlet superconducting state, where the scalar function $d_0^a(\mathbf{k})$ is even in $\mathbf{k}$. Here ϕ_a carries information about the magnitude and phase of the gap in orbital a and $d_0^a(\mathbf{k})$ and $\mathbf{d}^a(\mathbf{k})$ are normalized basis functions which encode the momentum dependence of the gap.

2.1. Magnetic field effects

The presence of external magnetic fields leads to time-reversal symmetry breaking and possible suppression of the SC state. Apart from the standard orbital depairing which is not included here, the effects of magnetic fields can appear in two different ways: by direct coupling to the spin degree of freedom as a Zeeman term or by coupling to the orbital angular momentum. For the specific case of Sr_2RuO_4, we can write the explicit matrix form for these effects in the multi-orbital basis introduced in Eq. 3. We start with the Zeeman term:

$$\delta H_Z = \begin{pmatrix} -\mathbf{h}\cdot\boldsymbol{\sigma} & 0 & 0 \\ 0 & -\mathbf{h}\cdot\boldsymbol{\sigma} & 0 \\ 0 & 0 & -\mathbf{h}\cdot\boldsymbol{\sigma} \end{pmatrix}, \tag{9}$$

where $\mathbf{h} = (h_x, h_y, h_z)$ is the external magnetic field. Note that this perturbation is independent of the orbital character. Focusing only on the field dependent part of the superconducting fitness, we find that the coupling of magnetic field to the spin degree of freedom leads to the same suppression of the critical temperature as in the single band case. Singlet SC is always suppressed in the presence of magnetic field, while triplet SC is suppressed if the magnetic field has a component in the same direction as the d-vector.

On the other hand, considering the coupling of the magnetic field to the orbital degree of freedom the perturbation has the form:

$$\delta H_{Orb} = \begin{pmatrix} 0 & ih_{oz}\sigma_0 & -ih_{oy}\sigma_0 \\ -ih_{oz}\sigma_0 & 0 & ih_{ox}\sigma_0 \\ ih_{oy}\sigma_0 & -ih_{ox}\sigma_0 & 0 \end{pmatrix}. \tag{10}$$

Here the orbital content is important to determine the structure of the matrix δH_{Orb} and ultimately the effects of magnetic field on the superconducting state. Computing the superconducting fitness in order to verify the stability of different gap structures in the presence of orbital effects, we find, omitting the momentum dependence:

$$\begin{aligned} F_{Orb} &= h_{ox}(\hat{\phi}^2 d_j^2 + \hat{\phi}^3 d_j^3)M_{7j} \\ &- h_{oy}(\hat{\phi}^1 d_j^1 + \hat{\phi}^3 d_j^3)M_{5j} \\ &+ h_{oz}(\hat{\phi}^1 d_j^1 + \hat{\phi}^2 d_j^2)M_{2j}, \end{aligned} \tag{11}$$

where $M_{ij} = \lambda_i \otimes \sigma_j$, with λ_i Gell-Mann matrices and σ_i Pauli matrices. Here $j = 0$ for singlet SC and $j = \{x, y, z\}$ for triplet SC. Note that for any gap structure among the different orbitals it is impossible to have all terms in F_{Orb} equal to zero. The magnetic field is potentially detrimental to all SC states irrespective of its direction. So we conclude that for multi-orbital systems the susceptibility of the critical temperature in presence of magnetic fields is not trivial and goes beyond the standard discussion of limiting effects based on symmetry arguments. We call this new effect *inter-orbital effect* (IOE) in order to distinguish it from the usual orbital depairing mechanism.

Concerning experiments, there are a two features that are still not completely understood in the upper critical field of Sr_2RuO_4:

- Anisotropy of the upper critical field: the angular dependence of the upper critical field does not follow the prediction of a pure anisotropic orbital effect described by an effective mass model (EMM) as expected for an ordinary layered type-II superconductor [16, 17, 18];
- Transition becomes first order below $0.8K$ for fields within 2° from the Ru-O plane, as recently observed by magnetocaloric effect [17] and specific heat measurements in ultra pure samples [18].

The presence of a first order transition might suggest that the Pauli paramagnetic limiting effect [19] is the dominant mechanism by which superconductivity is being suppressed. This would be inconsistent with NMR [20, 21, 22] and neutron scattering [23] experiments which show no observable change in the spin susceptibility through the superconducting transition for in-plane fields. An explanation for this unusual behaviour, compatible with all experimental results, has not been established so far, and experimentalists have been suggesting that it is due to a new pair breaking mechanism [16, 18]. In our recent work we suggest that the mechanism could be related to IOE, which can potentially account for *both* the anisotropy and the change in the character of the transition, as we sumamrize in the following.

In type-II SC the orbital effect is the dominant pair breaking mechanism and superconductivity is suppressed in a second order phase transition [24]. In anisotropic materials this effect is captured by the EMM [17, 25]:

$$h_{c2}^{EMM}(\theta) = \frac{h_{c2}(90^o)}{\sqrt{\sin^2\theta + \cos^2\theta/\Gamma^2}}, \tag{12}$$

where $\Gamma = \sqrt{M/m}$ is the anisotropy parameter, with M the out of plane and m the in-plane effective masses and θ is the angle the magnetic field relative to the conducting plane. As can

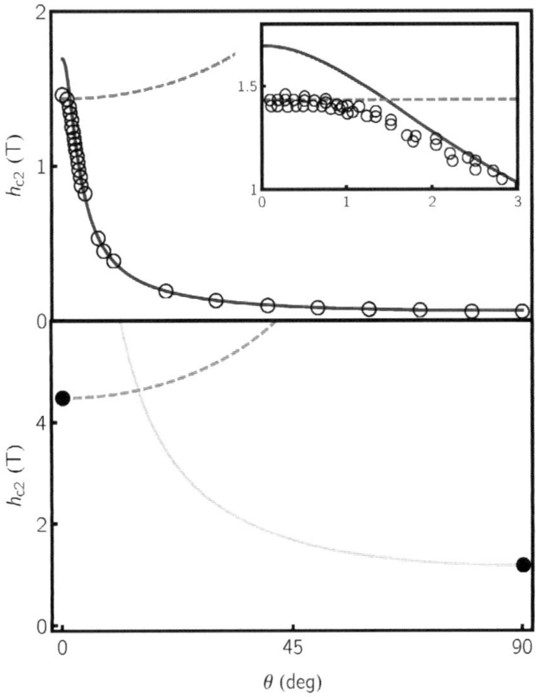

Figure 1. Angular dependence of the upper critical field. Top: Unstrained Sr_2RuO_4. The blue (full) line is the best fit for angles $\theta > 2^o$ with the EMM of Eq. 12 with parameters $\Gamma = 25$ and $h_{c2}(90^o) = 0.07T$. The red (dashed) line refers to the IOE plotted phenomenologically as $h_{c2}^{IOE} = \frac{h_{c2}(0^o)}{\cos\theta}$, with $h_{c2}(0^o) = 1.44T$. The open circles are experimental data from Yonezawa et al. [17]. The inset shows the small angle region. Bottom: Strained Sr_2RuO_4. The green (full) line refers to the orbital effect described by the effective mass model EMM with parameters $\Gamma = 25$ and $h_{c2}(90^o) = 1.2T$. The orange (dashed) line refers to the IOE with $h_{c2}(0^o) = 4.5T$. The full circles are experimental data from Steppke et al. [15].

be seen in Fig. 2.1 top, the EMM (blue, full line) describes the angular dependence of the upper critical field very well for angles larger than 2°, but for angles within 2° of the plane there are clear deviations from the EMM [16, 17, 18]. The extra suppression of the SC state in this region indicates that there should be a second mechanism which is most effective for in-plane fields. The standard paramagnetic limiting effect cannot account for this extra suppression of the SC state for in-plane fields if the d-vector is along the z-direction.

Considering now the effects of the coupling of the magnetic field to the orbital angular momentum described by the IOE, we would expect a suppression of the critical temperature following Eq. 2 leading to a second order phase transition. In analogy to the standard paramagnetic limiting effect, the transition due to IOE is of first order and the upper critical field is in fact determined by comparing the changes in the magnetic and condensation energies across the superconducting transition. The condensation energy can be estimated as:

$$E_C = -\frac{N(0)|\Delta|^2}{2}, \tag{13}$$

where $N(0)$ is the density of states at the Fermi level and $|\Delta|$ the magnitude of the superconducting gap. In the IOE, the magnetic energy comes from the polarizarion of the orbitals within a given band, so here we use $\Delta\chi_{Orb}$ as the change in the orbital susceptibility from the normal to the SC state:

$$E_M = -\frac{\Delta\chi_{Orb} h_x^2}{2}, \tag{14}$$

which leads to the critical field

$$h_{c2}^{IOE}(\theta) \approx \sqrt{\frac{N(0)}{\Delta\chi_{Orb}} \frac{T_c^0}{\cos\theta}}. \tag{15}$$

Here we would like to emphasize that the unusual angular dependence of the upper critical field as shown in Fig. 2.1 can be understood if both mechanisms, standard orbital and inter-orbital, are present. Moreover, we have seen that the standard EMM would lead to a second order transition, while depairing following from the field-induced inter-orbital coupling favours a discontinuous transition much analogous to the expectations for paramagnetic limiting. Therefore we believe that inter-orbital coupling for in-plane fields provides a strong candidate to explain the unusual suppression of superconductivity seen in experiments on Sr_2RuO_4.

3. Magnetic field effects in Sr_2RuO_4 under strain

We now turn to the analysis of the upper critical field for strained Sr_2RuO_4. Under uniaxial stress Sr_2RuO_4 can have its critical temperature enhanced up to $3.4K$ [15] (from the original $1.5K$ for the unstrained material [1]). Compared to the unstrained sample, there are a few quantitative changes on the upper critical field of the strained sample which we highlight below:

- Larger critical fields: compared to the unstrained material, the critical field along the z-axis is enhanced by a factor of ~ 20, while the in-plane critical field is enhanced by a factor of ~ 3 [15, 16, 17, 18]. Given the fact that the critical temperature was enhanced by a factor of ~ 2.3, an enhancement of the upper critical field is expected, but the rather different factors for different orientations is unusual;
- Less anisotropy: the ratio of the upper critical field (in-plane/z-axis) for unstrained Sr_2RuO_4 is around 25 [16, 17, 18], while for the strained sample it is approximately 4 [15]. It is difficult to understand this dramatic reduction of the anisotropy purely within orbital effects by a reduction in the mass anisotropy. The strong 2D character for the strained sample is observed by experiments based on the very different initial slope of h_{c2} versus temperature near T_c for fields along the plane and the z-axis [15]. It is also corroborated by recent DFT calculations for this system under strain [15];
- The transition turns to first order at higher temperature: for unstrained SRO the transition becomes first order for in plane fields only below $0.8K$ [17, 18]. For the strained case the transition is first order below $1.8K$ [15]. This indicates that the mechanism leading to the first order transition takes over at larger energy scales.

The presence of both the standard orbital effect and the IOE can still be in agreement with the findings of the upper critical field under strain. So far we do not have access to the entire angular dependence of the upper critical field, but we can focus on the two extreme points from Steppke et al. [15]. Assuming that the dominant mechanism for suppression of superconductivity with fields along the z-axis is the standard orbital effect which is captured by the EMM, and that under strain the effective mass ratio is not strongly modified (so we still assume $\Gamma \sim 25$ as for the unstrained sample), we can plot the angular dependence in Fig. 2.1 bottom, green full line. Note that the prediction for the in-plane upper critical field overestimates the experimental value by factor of approximately 7, much larger than in the unstrained case discussed above.

Assuming orbital effects are dominant for fields along the z-axis, in order to have agreement with the value of the upper critical field along the z-axis we choose the value $h_{c2}(90^o) = 1.2T$ at $T = 0K$ for the plot, much larger the the value for the unstrained sample. Within Ginzburg-Landau theory $h_{c2} = \phi_0/2\pi\xi^2$, so an enhancement of the z-axis upper critical field by a factor of 20 indicates a reduction in the coherence length by a factor of approximately 4.5. From BCS theory the coherence length can be written as $\xi \approx \hbar v_F/\pi T_c$, so the reduction of the coherence length is consistent with the enhancement of the critical temperature. The critical temperature increases only by a factor of 2.3. But given that this estimate is rather crude (it does not take into account changes in the Fermi velocity, for example), we believe this argument is consistent with the fact that we still have the suppression of superconductivity by the standard orbital effect for fields along the z-axis.

The presence of a first order transition for in-plane fields in the strained sample again suggests the presence of Pauli paramagnetic limiting effects, and naively of a singlet superconducting state. In the previous section we review our recent work which points out that the IEO can also account for such a phenomenology with a triplet superconducting state [13]. We believe this picture to be more appealing since it is in agreement with other experiments that indicate the presence of a triplet state for the unstrained sample. An extension of this idea to the strained case would be natural if there is no phase transition under strain.

Concerning the in-plane upper critical field, from BCS theory, the standard Pauli limiting effect predicts $h_{c2}/T_c \approx 1.25$ [24, 26]. For magnetic field along the z-axis experiments show that $h_{c2}/T_c \approx 1$ for the unstrained sample and $h_{c2}/T_c \approx 1.3$ for the strained sample, indicating that a Pauli limiting-like effect (with h_{c2} scaling linearly with T_c) is consistent with the observations. We plot in Fig. 2.1 bottom, orange dotted line, the respective IEO. We find that this mechanism overcomes the standard orbital effect for much larger angles (at $\theta \approx 20^o$). This reflects that fact that a smaller magnitude of in-plane magnetic field is necessary for suppression of superconductivity. From another perspective it indicates that this mechanism taking over at higher temperatures, being in accordance with the observed first order transition over an extended temperature range for the strained sample. Also, the fact that this mechanism dominates the behaviour of h_{c2} for small angles dramatically reduces the anisotropy in comparison to the expected value due to the mass anisotropy. In conclusion, the IEO is again able to consistently put together two puzzling features of the upper critical field.

4. Conclusion

Here we would like to emphasize that one can understand the phenomenology of the upper critical field in Sr_2RuO_4 assuming a triplet superconducting state. This is possible based the recently discussed mechanism for the suppression of superconductivity in multi-orbital systems called inter-orbital effect. The presence of a first order phase transition under a magnetic field suggests that the Pauli paramagnetic limiting effect is at play, and a naive analysis would lead us to think that a singlet superconducting state is the only possibility which is compatible with the observations. We have argued that the IEO can also lead to first order phase transitions under external magnetic fields by the polarization of the orbital content within each band, in analogy to the spin polarization in the Pauli paramagnetic limiting mechanism. Furthermore, a quantitative analysis of the changes in the values of the upper critical fields and critical temperatures leads to a consistent picture for the unstrained and uniaxially strained samples. In particular, the two extreme angles are dominated by different mechanisms such that the upper critical field scales differently with the critical temperature.

Our discussion does not rule out the possibility of a phase transition from triplet to singlet superconductor as a function of strain. In that case a discontinuity or an inflection point in h_{c2} as a function of strain should be observed experimentally. We believe that experiments along this line would be of great interest for a better understanding of the superconducting phase of Sr_2RuO_4. It would also be important to follow the complete angular dependence of the anisotropy in order to observe at which angles the deviations of the EMM appear in the strained samples. If such an experiment is realized one could also determine by how much the anisotropy parameter Γ changes in comparison to the unstrained sample, so we can understand if the reduction of anisotropy is due to the change in the effective masses or due to the IOE taking over at larger angles, as we discuss above.

Acknowledgements

We thank M. H. Fischer, K. Ishida and C. W. Hicks for stimulating discussions. This work was supported by Dr. Max Rössler, the Walter Haefner Foundation and the ETH Zurich Foundation (AR) and by the Swiss National Science Foundation (MS).

References

[1] Mackenzie A P and Maeno Y 2003 *Rev. Mod. Phys.* **75** 657-712
[2] Nomura T and Yamada K 2000 *J. Phys. Soc. Jpn.* **69** 3678-3688
[3] Nomura T and Yamada K 2002 *J. Phys. Soc. Jpn.* **71** 404-407
[4] Yanase Y and Ogata M 2003 *J. Phys. Soc. Jpn.* **72** 673-687
[5] Nomura T 2005 *J. Phys. Soc. Jpn.* **74** 1818-1829
[6] Wang Q H, Platt C, Yang Y, Honerkamp C, Zhang F C, Hanke W, Rice T M and Thomale R 2013 *Eur. Phys. Lett.* **104** 17013
[7] Raghu S, Kapitulnik A and Kivelson S A 2010 *Phys. Rev. Lett.* **105** 136401
[8] Chung S B, Raghu S, Kapitulnik A and Kivelson S A 2012 *Phys. Rev.* B **86** 064525
[9] Firmo I A, Lederer S, Lupien C, Mackenzie A P, Davis J C and Kivelson S A 2013 *Phys. Rev.* B **88** 134521
[10] Scaffidi T, Romers J C and Simon S H 2014 *Phys. Rev.* B **89** 220510
[11] Anderson P W 1959 *J. Phys. Chem. Solids* **11** 26-30
[12] Anderson P W 1984 *Phys. Rev.* B **30** 4000-4002
[13] Ramires A and Sigrist M 2016 *Prepint* arXiv:1605.03827
[14] Fischer M H *New J. of Phys.* **15** 073006
[15] Steppke A et al. 2016 *Preprint* arXiv:1604.06669
[16] Deguchi K, Tanatar M A, Mao Z, Ishiguro T and Maeno Y 2002 *J. Phys. Soc. Jpn.* **71** 2839-2842
[17] Yonezawa S, Kajikawa T and Maeno Y 2013 *Phys. Rev. Lett.* **110** 077003
[18] Yonezawa S, Kajikawa T and Maeno Y 2014 *J. Phys. Soc. Jpn.* **83** 083706
[19] Amano Y, Ishihara M, Ichioka M, Nakai N and Machida K 2015 *Phys. Rev.* B **91** 144513
[20] Ishida K, Mukuda H, Kitaoka Y, Asayama K, Mao Z Q, Mori Y and Maeno Y 1998 *Nature* **396** 658-660
[21] Ishida K, Mukuda H, Kitaoka Y, Mao Z Q, Fukazawa H and Maeno Y 2001 *Phys. Rev.* B **63** 060507
[22] Murakawa H, Ishida K, Kitagawa K, Mao Z Q and Maeno Y 2004 *Phys. Rev. Lett.* **93**, 167004
[23] Duffy J A, Hayden S M, Maeno Y, Mao Z, Kulda J and McIntyre G J 2000 *Phys. Rev. Lett.* **85** 5412-5415
[24] Tinkham M 1996 *Introduction to Superconductivity* (New York: McGraw-Hill)
[25] Morris R C, Coleman R V and Bhandari R 1972 *Phys. Rev.* B **5** 895-901
[26] Sigrist M 2008 *AIP Conf. Proc.* **1162** 55-97

Single Crystal Growth and Superconductivity in La_7Ni_3 without Inversion Symmetry in the Crystal Structure

Ai Nakamura[a], Fuminori Honda[a], Yoshiya Homma[a], Dexin Li[a], Kengo Nishimura[b], Masashi Kakihana[b], Masato Hedo[c], Takao Nakama[c], Yoshichika Onuki[c] and Dai Aoki[a,d]

[a]Institute for Materials Research, Tohoku University, Oarai, Ibaraki 311-1313, Japan
[b]Graduate School of Engineering and Science, University of the Ryukyus, Nishihara, Okinawa 903-0213, Japan
[c]Faculty of Science, University of the Ryukyus, Nishihara, Okinawa 903-0213, Japan
[d]INAC/PHELIQS, CEA-Grenoble, Grenoble 38054, France

E-mail: nakamuraai@imr.tohoku.ac.jp

Abstract. We succeeded in growing single crystals of La_7Ni_3 with the non-centrosymmetric hexagonal structure by the Bridgman method, and measured the electrical resistivity and speci c heat under magnetic elds. The superconducting transition temperature T_{sc} is 2.4 K. The electronic speci c heat coefficient γ is 44 mJ/(K^2·f.u.-mol). Correspondingly, the upper critical eld $H_{c2}(0)$ in superconductivity is large, $H_{c2} = 27$ kOe. The electronic speci c heat shows the jump at $T_{sc} = 2.4$ K, indicating the bulk superconductivity with $\Delta C_e/\gamma T_c = 1.2$, which is close to the weak coupling BCS value of 1.43. Although the temperature dependence of speci c heat and some parameters of superconductivity are well explained by the conventional BCS model, the temperature dependence of H_{c2} is not consistent with the simple BCS model.

1. Introduction

Superconductivity is a fundamental phenomenon in metallic compounds, which is observed in the s- and p-electron systems as well as the strongly correlated d- and f-electron systems. The pairing mechanism is based on phonons in the BCS-type superconductors, while it is due to magnetic spin uctuations in the strongly correlated electron systems. The typical d-electron systems are high-T_c cuprates [1]. Heavy-fermion superconductors such as $CeCoIn_5$ and UPt_3 belong to the f-electron systems [2]. In these strongly correlated electron systems, the superconducting pairing favors the d-wave spin singlet and/or p- (or f-) wave spin triplet states.

Even if superconductivity belongs to the BCS-type, precise measurements revealed unconventional properties, for example, in $CeRu_2$. The electronic speci c heat coefficient C_e/T below T_{sc} does not increase linearly as a function of applied magnetic eld H but indicates a nearly $\sqrt{H}$-dependence at 0.5 K ($\sim 0.08T_c$) [3]. From the eld-angle-resolved speci c heat measurement, it was de nitely clari ed the gap minimum [4].

Similarly, multigap/multiband superconductivity is another concept of BCS-type superconductivity, which was precisely studied in MgB_2, for example. Namely, two di erent Fermi surfaces based on the σ and π bands possess two di erent gaps [5, 6].

Recently, superconductivity in the non-centrosymmetric compounds attracts our attention [7, 8]. A lack of inversion symmetry or non-centrosymmetry in the crystal structure brings about interesting phenomena in superconductivity and magnetism. The antisymmetric spin-orbit interaction lifts not only the degeneracy of conduction electrons but also the spin. The Fermi surface thus splits due to the antisymmetric spin-orbit interaction. The magnetic interaction is based on the so-called Dzyaloshinskii-Moriya interaction. Unique magnetism so-called skyrmion is observed in MnSi [9]. Theoretically, superconductivity is admixture of the spin singlet and spin triplet pairings. The unique upper critical eld $H_{c2}(0)$ is observed, for example, in $CeIrSi_3$ with the non-centrosymmetric tetragonal structure; $H_{c2}(0) \simeq 450$ kOe for $H \parallel [001]$ and 95 kOe for $H \parallel [110]$ are obtained [8]. Pauli paramagnetic suppression of H_{c2} is not realized only for $H \parallel [001]$ because all the spins of conduction electrons are perpendicular to the magnetic eld direction of $H \parallel [001]$.

Considering these points, we studied the superconducting properties of La_7Ni_3 with the Th_7Fe_3-type non-centrosymmetric hexagonal structure (space group: $P6_3mc$, #186, C^4_{6v}). We grew single crystals by the Bridgman method and measured the electrical resistivity and speci c heat under magnetic elds.

2. Experimental

We succeeded in growing single crystals of La_7Ni_3 by the Bridgman method. Both compounds are congruent in the binary phase alloy diagram. First, we melted the starting materials of 3N-La and 5N-Ni in an arc furnace. Note that the melting point is 530 °C in La_7Ni_3. The ingot was crashed into small pieces, which were inserted into a yttria crucible. The crucible was encapsulated in a quartz ampoule, which was heated to 650 °C and slowly cooled to 480 °C, taking 6 days, and nally cooled to room temperature. The quartz ampoule was placed in an electrical furnace with the temperature gradient ($\Delta T \sim 10$°C.) Figure 1 shows the obtained ingot of La_7Ni_3, cleaved in the c-plane in the hexagonal structure.

The electrical resistivity was measured by the four-probe AC method and the speci c heat by the thermal relaxation method at temperatures down to 0.4 K and at elds up to 40 kOe.

3. Experimental Results and Analyses

We measured the electrical resistivity and speci c heat for La_7Ni_3, as shown in Figs. 2 and 3, respectively. The temperature dependence of resistivity shows the convex curvature at high temperature, which is not common in La-based compounds. This is probably due to the $3d$ electrons of Ni. The resistivity follows the T^2-dependence below 11 K. The coefficient A and the residual resistivity $_0$ are 0.016 $\mu\Omega \cdot \mathrm{cm/K^2}$ and 71.6 $\mu\Omega \cdot \mathrm{cm}$, respectively. At low temperature, the superconductivity is observed at $T_{sc} = 2.4$ K, de ned by the midpoint of the resistivity drop.

The superconductivity is also observed in the speci c heat measurements, indicating the bulk superconductivity, as shown in Fig. 3 the transition temperature is $T_{sc} = 2.4$ K, and

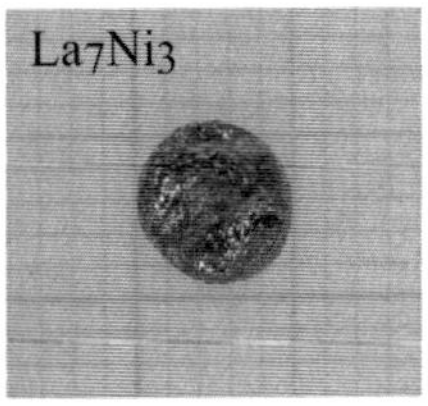

Figure 1. Single crystal ingot of La_7Ni_3.

IOP Conf. Series: Journal of Physics: Conf. Series **807** (2017) 052012 doi:10.1088/1742-6596/807/5/052012

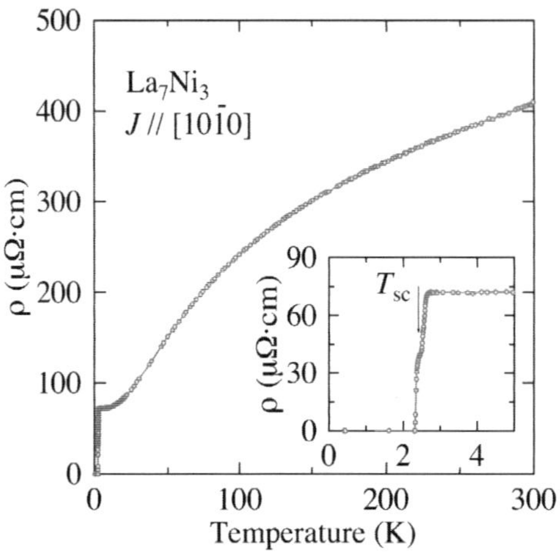

Figure 2. Temperature dependences of electrical resistivity in La_7Ni_3.

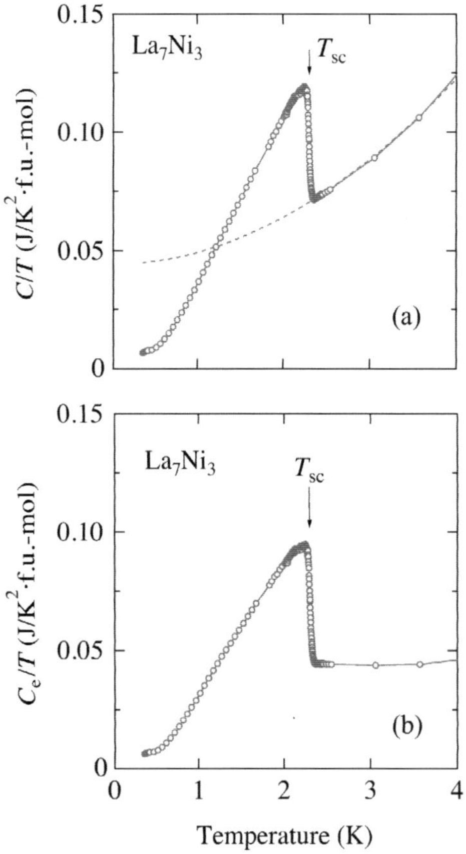

Figure 3. (a) Temperature dependence of the speci c heat in La_7Ni_3. The broken line is the result of tting above T_{sc} by $C/T = \gamma + \beta T^2$. (b) Temperature dependence of the electronic speci c heat in the form of C_e/T v.s. T.

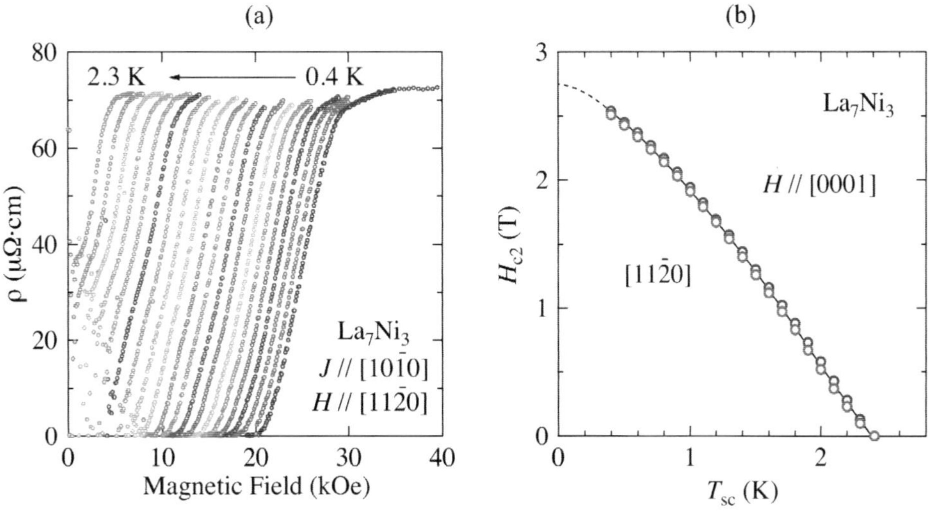

Figure 4. Temperature dependences of (a) eld dependences of electrical resistivity and (b) temperature dependence of upper critical elds H_{c2} for $H \parallel [0001]$ and $[11\bar{2}0]$ in La_7Ni_3.

the electrical speci c heat coefficient is $\gamma = 44$ mJ/(K^2·f.u.-mol). The γ-value approximately follows the universal value obtained by the Kadowaki–Woods ratio. The speci c heat jump is $\Delta C_e/\gamma T_c = 1.2$, which is slightly lower than the weak coupling BCS-value, 1.43. The low temperature data below 0.7 K follow the BCS asymptotic formula, $\Delta C_e/T = \gamma_0 + 3.15(\Delta/1.76k_BT)^{5/2}\gamma\exp[-\Delta/k_BT]$. The obtained gap energy is $\Delta = 4.0$ K and $2\Delta/k_BT_{sc} = 3.34$, which is also consistent with the BCS value. The thermodynamical critical eld H_c is calculated from the relation $\mu_0H_c^2/2 = \int_{T_c}^{T}[S_s(T) - S_n(T)]dT$, where the entropy of superconducting (S_s) and normal state (S_n) is obtained from $S = \int C_e/TdT$. The thermodynamical critical eld H_c at 0 K is $H_c(0) = 0.026$ T.

We also measured the eld dependence of the electrical resistivity in the temperature range from 2.3 to 0.4 K, as shown in Fig. 4(a). The temperature dependence of H_{c2} is quite isotropic between $H \parallel [0001]$ and $[11\bar{2}0]$. The upper critical eld is $H_{c2}(0) \simeq 2.7$ T for both $H \parallel [0001]$ and $[11\bar{2}0]$, as shown in Fig. 4(b). The value of $H_{c2}(0)$ is below the Pauli limit. $H_p = 1.86T_c$, but is relatively large, compared to many La-based superconductors. The coherence length obtained from $H_{c2} = \phi_0/2\pi\xi^2$ is $\xi = 110$ A. It should be noted that the temperature dependence of H_{c2} dose not follows the so-called WHH model based on the dirty limit case. The initial slope of H_{c2} at $H \to 0$ is $-dH_{c2}/dT = 1.4$ T/K. This large initial slope of H_{c2} corresponds to the moderately enhanced γ-value, which is most likely due to the $3d$ electrons from Ni and the characteristic crystal structure without inversion symmetry. The mean free path, ℓ, which can be calculated from $_0$, T_c, and γ-value, is $\ell \sim 110$ A, assuming the spherical Fermi surface [12]. Since the obtained mean free path is smaller or comparable to the coherence length, La_7Ni_3 is conventional BCS-type superconductor in the dirty limit.

4. Summary

We measured the electrical resistivity and speci c heat under magnetic elds using single crystals of Th_7Ni_3 and La_7Ni_3 with the non-centrocymmetric hexagonal structure. The characteristic features are summarized as follows:

1) Bulk-superconductivity is realized below $T_{sc} = 2.4$ K in La_7Ni_3.
2) The electronic speci c heat C_e below T_{sc} approximately follows the conventional BCS single-gap weak-coupling in both compounds.

3) The relatively large $H_{c2}(0)$ and the large initial slope are probably due to the Ni 3d-based and the characteristic crystal structure without inversion symmetry.

Acknowledgments
This work was supported by JSPS KAKENHI Grant Numbers 25247055, 15K05156, 15H05884, 15H05882, and 16H04006 and start-up research costs from the Project to Promote Gender Equality and Support Female Researchers from Tohoku University.

References

[1] Mukuda M, Ishida K, Kitaoka Y and Asayama K 1998 *J. Phys. Soc. Jpn.* **67** 2101
[2] Onuki Y, Settai R, Haga Y, Machida Y, Izawa K, Honda F and Aoki D 2014 *C. R. Phys.* **15** 616
[3] Hedo M, Inada Y, Yamamoto E, Haga Y, Onuki Y, Aoki Y, Matsuda T D, Sato H and Takahashi S 1998 *J. Phys. Soc. Jpn.* **67** 272
[4] Kittaka S, Sakakibara T, Hedo M, Onuki Y and Machida K 2013 *J. Phys. Soc. Jpn.* **82** 123706
[5] Pribulova Z, Klein T, Marcus J, Marcenat C, Levy F, Park M S, Lee H G, Kang B W, Lee S I, Tajima S and Lee S 2007 *Phys. Rev. Lett.* **98** 137001
[6] Xi X X 2008 *Rep. Prog. Phys.* **71** 116501
[7] Bauer E and Sigrist M *Mon-centrosymmetric Superconductors (Springer, Berlin, 2012) Lecture Note in Physics* Vol.847 p.1 and references therein
[8] Settai R, Katayama K, Aoki D, Sheikin I, Knebel G, Flouquet J and Onuki Y 2011 *J. Phys. Soc. Jpn.* **80** 094703
[9] Mühlbauer S, Binz B, Jonietz F, P eiderer C, Rosch A, Neubauer A, Georgii R and Böni P 2009 *Science* **323** 915
[10] Pedrazzini P, Schmer G, Berisso M G, Kappler J P and Sereni J D 2000 *Physica C* **336** 10
[11] Ueda K, Oda T, Iwamoto Y, Kohara T, Lu Q F, Umehara I and Sato K 1996 *Czech. J. Phys.* **46** Suppl. S2 867
[12] Rauchschwalbe U 1987 *Physica* **147B** 1

IOP Conf. Series: Journal of Physics: Conf. Series **807** (2017) 052013 doi:10.1088/1742-6596/807/5/052013

Random phase approximation for superconducting states in multi-orbital Hubbard models with spin-orbit coupling

Kazutaka Nishiguchi
Computational Condensed Matter Physics Laboratory, RIKEN, Saitama 351-0198, Japan
E-mail: kazutaka.nishiguchi@riken.jp

Tomonori Shirakawa
Computational Quantum Condensed Matter Research Team, RIKEN CEMS, Saitama 351-0198, Japan
E-mail: t-shirakawa@riken.jp

Hiroshi Watanabe
Waseda Institute for Advanced Study, Waseda University, Tokyo 169-8050, Japan
E-mail: h.watanabe@aoni.waseda.jp

Seiji Yunoki
Computational Condensed Matter Physics Laboratory, RIKEN, Saitama 351-0198, Japan
Computational Quantum Condensed Matter Research Team, RIKEN CEMS, Saitama 351-0198, Japan
Computational Materials Science Research Team, RIKEN AICS, Kobe, Hyogo 650-0047, Japan
E-mail: yunoki@riken.jp

Abstract. To better understand electronic and magnetic properties of superconducting (SC) states in $5d$ transition metal oxides, we derive an analytical formula of the dynamical correlation function for multi-orbital systems with the spin-orbit coupling (SOC), which is accessible to the experimental responses. Starting with a multi-orbital BCS Hamiltonian including the SC gap function, we introduce the dynamical correlation function in the non-interacting case. Based on the diagrammatic techniques, we derive the dynamical correlation function for the interacting case within the random phase approximation. With the help of a numerical calculation, we also show the temperature dependence of the magnetic susceptibility for two different SC states in the presence of the SOC.

1. Introduction

$5d$ transition metal oxide Sr_2IrO_4 has been focused both experimentally [1, 2, 3, 4, 5, 6, 7, 8] and theoretically [9, 10, 11, 12, 13, 14, 15, 16, 17, 18] because of the unexpected quantum phenomena arising from not only its electron correlation but also highly entangled spin and orbital degrees of freedom due to the large spin-orbit coupling (SOC). Sr_2IrO_4 has a layered perovskite structure where the two-dimensional square lattice is formed by Ir atoms [1], and its electronic properties are attributed to the five $5d$ electrons of Ir occupying the t_{2g} orbitals. In addition, the large SOC splits the sixfold degenerate t_{2g} orbitals into the effective total angular momentum $j_{\text{eff}} = |-\boldsymbol{l}+\boldsymbol{s}|$ states: the fourfold degenerate $j_{\text{eff}} = 3/2$

states and the doubly degenerate $j_{\rm eff} = 1/2$ states. Thus the four electrons fill the $j_{\rm eff} = 3/2$ states and the remaining electron occupies one of the $j_{\rm eff} = 1/2$ states, contributing mostly to determine the electronic properties.

Several experiments have revealed that an antiferromagnetic insulating state appears at low temperatures [3], and very recently, a pseudogap-like structure (Fermi arc) has been observed in the electron-doped Sr_2IrO_4 [6, 7, 8], analogous to high-T_c cuprate superconductors [19, 20, 21]. These similarities naturally expect superconductivity in Sr_2IrO_4 once mobile carriers are introduced. However, this has not been reported experimentally up to date. On the other hand, theoretical studies have proposed that pseudospin $j_{\rm eff} = |-\boldsymbol{l}+\boldsymbol{s}|$ singlet pairing superconductivity is favored due to the large SOC. For example, variational Monte Carlo method, functional renormalization group method, and dynamical mean field theory with continuous time quantum Monte Carlo method are employed to show that a $d_{x^2-y^2}$-wave pairing is favored in the electron-doped region [14, 15, 18] and a $s_\pm$-wave [15, 18] or p-wave [17] pairing in the hole-doped region. To detect the unconventional superconductivities experimentally, it is highly desirable to theoretically derive the experimental responses of the pseudospin singlet pairing.

For this end, it is important to evaluate the dynamical correlation function for superconducting (SC) states in the presence of the SOC because experimental responses are obtained from its linear combinations. To achieve this, we derive the random phase approximation (RPA) for multi-orbital systems with the SOC, where the electron correlation can be incorporated by an infinite summation over all bubble and ladder diagrams connected to the Coulomb interactions. In this paper, we evaluate the magnetic susceptibility in the pseudospin singlet SC states, which can be obtained from the linear combination of the dynamical correlation function. Here we adopt an effective model for the iridium oxide Sr_2IrO_4. However, the formalism derived here is general and is applicable to other systems where the spin and orbital degrees of freedom is entangled due to the SOC.

2. Formulation

Our goal in this paper is to calculate the following dynamical correlation function in SC states:

$$[\hat{\chi}(q)]_{\alpha\beta\gamma\delta} = \frac{1}{N}\int_0^\beta d\tau\, \mathrm{e}^{i\epsilon_m\tau} \sum_{\boldsymbol{k}\boldsymbol{k}'} \langle T_\tau c^\dagger_{\boldsymbol{k}\beta}(\tau) c_{\boldsymbol{k}+\boldsymbol{q}\alpha}(\tau) c^\dagger_{\boldsymbol{k}'\gamma}(0) c_{\boldsymbol{k}'-\boldsymbol{q}\delta}(0)\rangle_{\rm c}, \tag{1}$$

where $q = (\boldsymbol{q}, i\epsilon_m)$ and $\epsilon_m = 2mk_{\rm B}T$ is Matsubara frequency for bosons (at temperature T and $m \in \mathbb{N}$), and $c_{\boldsymbol{k}\alpha}$ ($c^\dagger_{\boldsymbol{k}\alpha}$) with $\alpha = (a,\sigma)$ represents the field operator for an electron annihilation (creation) with wave number $\boldsymbol{k}$, spin σ ($=\uparrow,\downarrow$), and orbital a ($= xy, yz, zx$). The subscript "c" denotes the inclusion of only connected diagrams.

2.1. Tight-Binding Model

Let us start with an effective model for Sr_2IrO_4. We consider its effective t_{2g} three-orbital tight-binding model on the square lattice with a SOC,

$$H_0 = \sum_{\boldsymbol{k}\sigma} \vec{c}^\dagger_{\boldsymbol{k}} \hat{\xi}_{\boldsymbol{k}} \vec{c}_{\boldsymbol{k}}, \tag{2}$$

where

$$\vec{c}^\dagger_{\boldsymbol{k}} = \left(c^\dagger_{\boldsymbol{k}yz\uparrow}\; c^\dagger_{\boldsymbol{k}zx\uparrow}\; c^\dagger_{\boldsymbol{k}xy\uparrow}\; c^\dagger_{\boldsymbol{k}yz\downarrow}\; c^\dagger_{\boldsymbol{k}zx\downarrow}\; c^\dagger_{\boldsymbol{k}xy\downarrow}\right) \tag{3}$$

and

$$\hat{\xi}_{\boldsymbol{k}} = \begin{pmatrix} \xi^{yz}_{\boldsymbol{k}} & \mathrm{i}\lambda/2 & 0 & 0 & 0 & -\lambda/2 \\ -\mathrm{i}\lambda/2 & \xi^{zx}_{\boldsymbol{k}} & 0 & 0 & 0 & \mathrm{i}\lambda/2 \\ 0 & 0 & \xi^{xy}_{\boldsymbol{k}} & \lambda/2 & -\mathrm{i}\lambda/2 & 0 \\ 0 & 0 & \lambda/2 & \xi^{yz}_{\boldsymbol{k}} & -\mathrm{i}\lambda/2 & 0 \\ 0 & 0 & \mathrm{i}\lambda/2 & \mathrm{i}\lambda/2 & \xi^{zx}_{\boldsymbol{k}} & 0 \\ -\lambda/2 & -\mathrm{i}\lambda/2 & 0 & 0 & 0 & \xi^{xy}_{\boldsymbol{k}} \end{pmatrix}. \tag{4}$$

$\xi_{\boldsymbol{k}}^a = \epsilon_{\boldsymbol{k}}^a - \mu$ represents the energy dispersion $\epsilon_{\boldsymbol{k}}^a$ measured from the chemical potential μ and λ is the strength of the SOC. According to the previous studies [11, 14], the band structure of Sr_2IrO_4 obtained from the first-principles calculation [2] can be well reproduced using the following tight-binding energy dispersion for the three t_{2g} orbitals:

$$\begin{aligned}\epsilon_{\boldsymbol{k}}^{yz} &= -2t_5\cos k_x - 2t_4\cos k_y,\\ \epsilon_{\boldsymbol{k}}^{zx} &= -2t_4\cos k_x - 2t_5\cos k_y,\\ \epsilon_{\boldsymbol{k}}^{xy} &= -2t_1(\cos k_x + \cos k_y) - 4t_2\cos k_x\cos k_y\\ &\quad - 2t_3(\cos 2k_x + \cos 2k_y) + \mu_{xy},\end{aligned} \tag{5}$$

with $(t_1, t_2, t_3, t_4, t_5, \mu_{xy}, \lambda) = (0.36, 0.18, 0.09, 0.37, 0.06, -0.36, 0.37)$ eV. Notice that the SOC λ is as large as the hopping parameters.

H_0 can be diagonalized by a unitary matrix $\hat{U}_{\boldsymbol{k}}$ as

$$H_0 = \sum_{\boldsymbol{k}} \vec{c}_{\boldsymbol{k}}^{\dagger}\hat{U}_{\boldsymbol{k}}\left(\hat{U}_{\boldsymbol{k}}^{\dagger}\hat{\xi}_{\boldsymbol{k}}\hat{U}_{\boldsymbol{k}}\right)\hat{U}_{\boldsymbol{k}}^{\dagger}\vec{c}_{\boldsymbol{k}} = \sum_{\boldsymbol{k}} \vec{a}_{\boldsymbol{k}}^{\dagger}\hat{E}_{\boldsymbol{k}}\hat{a}_{\boldsymbol{k}}, \tag{6}$$

where

$$\vec{a}_{\boldsymbol{k}}^{\dagger} = \left(a_{\boldsymbol{k}1\Uparrow}^{\dagger}\ a_{\boldsymbol{k}1\Downarrow}^{\dagger}\ a_{\boldsymbol{k}2\Uparrow}^{\dagger}\ a_{\boldsymbol{k}2\Downarrow}^{\dagger}\ a_{\boldsymbol{k}3\Uparrow}^{\dagger}\ a_{\boldsymbol{k}3\Downarrow}^{\dagger}\right) \tag{7}$$

and

$$\hat{E}_{\boldsymbol{k}} = \mathrm{diag}\left(E_{\boldsymbol{k}}^1,\ E_{\boldsymbol{k}}^1,\ E_{\boldsymbol{k}}^2,\ E_{\boldsymbol{k}}^2,\ E_{\boldsymbol{k}}^3,\ E_{\boldsymbol{k}}^3\right). \tag{8}$$

Here, $E_{\boldsymbol{k}}^l$ represents the l-th energy eigenvalue in descending order, i.e., $E_{\boldsymbol{k}}^1 \geq E_{\boldsymbol{k}}^2 \geq E_{\boldsymbol{k}}^3$, and $a_{\boldsymbol{k}lj}$ is its annihilation operator for the Kramers doublet $j(=\Uparrow, \Downarrow)$. The previous studies have revealed that the quasiparticle $a_{\boldsymbol{k}1j}$ for the highest band $E_{\boldsymbol{k}}^1$ is mostly composed of the effective total angular momentum $j_{\mathrm{eff}} = 1/2$ doublet, where $\boldsymbol{j}_{\mathrm{eff}} = -\boldsymbol{l} + \boldsymbol{s}$ is "pseudospin" operator, and $\boldsymbol{s}$ and $\boldsymbol{l}$ are the spin and orbital angular momentum, respectively. Therefore, $E_{\boldsymbol{k}}^1$ is referred to as the $j_{\mathrm{eff}} = 1/2$ band in the following.

As is obvious in the above, the intrinsic spin σ is no longer a good quantum number, but instead the pseudospin j_{eff} becomes a good quantum number. Thus, when the possible superconductivity is considered in carrier doped Sr_2IrO_4, it is expected that a pseudospin SC pairing is favored to form Cooper pairs with opposite wave number $\pm\boldsymbol{k}$ and opposite (singlet) or parallel (triplet) pseudospins. The pseudospin superconductivity depends on the carrier doping: in the electron-doped regions only the single $j_{\mathrm{eff}} = 1/2$ ($l = 1$) band becomes dominant because the other bands are far below the Fermi level [14, 15, 18], while in the hole-doped regions several j_{eff} bands (especially, the two highest bands with $l = 1$ and 2) contribute to form Cooper pairs [15, 17, 18].

We also consider the Coulomb interaction term including the intra- and inter-orbital on-site Hubbard interactions (U and U', respectively), the Hund's coupling (J), and the pair hopping (J'), described by the following Hamiltonian:

$$\begin{aligned}H_{\mathrm{int}} &= U\sum_i\sum_a n_{i\uparrow}^a n_{i\downarrow}^a + U'\sum_i\sum_{a<b} n_i^a n_i^b\\ &\quad - J\sum_i\sum_{a<b}\left(2\boldsymbol{S}_i^a\cdot\boldsymbol{S}_i^b + \frac{1}{2}n_i^a n_i^b\right) + J'\sum_i\sum_{a<b} c_{ia\uparrow}^{\dagger}c_{ia\downarrow}^{\dagger}c_{ib\downarrow}c_{ib\uparrow} + \mathrm{h.c.},\end{aligned} \tag{9}$$

where $n_{i\sigma}^a = c_{ia\sigma}^{\dagger}c_{ia\sigma}$, $n_i^a = n_{i\uparrow}^a + n_{i\downarrow}^a$, and $\boldsymbol{S}_i^a = \sum_{\sigma\sigma'} c_{ia\sigma}^{\dagger}\boldsymbol{\tau}_{\sigma\sigma'}c_{ia\sigma'}$ ($\boldsymbol{\tau}_{\sigma\sigma'}$ is the elements of the Pauli matrix $\boldsymbol{\tau}$). We impose that $U = U' + 2J$ and $J = J'$ for simplicity [22].

2.2. BCS Hamiltonian

Next, we consider a multi-orbital BCS Hamiltonian H^{BCS} in the presence of the SOC by introducing a SC gap function $\hat{\Delta}_{\boldsymbol{k}}$ into H_0 (apart from a constant term),

$$\begin{aligned} H^{\mathrm{BCS}} &= \frac{1}{2}\sum_{\boldsymbol{k}} \vec{C}^{\dagger}_{\boldsymbol{k}} \hat{H}^{\mathrm{BCS}}_{\boldsymbol{k}} \vec{C}_{\boldsymbol{k}} = \frac{1}{2}\sum_{\boldsymbol{k}} C^{\dagger}_{\boldsymbol{k}\bar{A}} [\hat{H}^{\mathrm{BCS}}_{\boldsymbol{k}}]_{\bar{A}\bar{B}} C_{\boldsymbol{k}\bar{B}} \\ &= \frac{1}{2}\sum_{\boldsymbol{k}} \left(\vec{c}^{\dagger}_{\boldsymbol{k}} \,\middle|\, \vec{c}^{\mathrm{T}}_{-\boldsymbol{k}} \right) \left(\begin{array}{c|c} \hat{\xi}_{\boldsymbol{k}} & \hat{\Delta}_{\boldsymbol{k}} \\ \hline \hat{\Delta}^{\dagger}_{\boldsymbol{k}} & -\hat{\xi}^{\mathrm{T}}_{-\boldsymbol{k}} \end{array} \right) \left(\begin{array}{c} \vec{c}_{\boldsymbol{k}} \\ \hline \vec{c}^{\dagger\mathrm{T}}_{-\boldsymbol{k}} \end{array} \right), \end{aligned} \tag{10}$$

where $\vec{C}^{\dagger}_{\boldsymbol{k}}$ is a matrix representation of the field operators,

$$\vec{C}^{\dagger}_{\boldsymbol{k}} = (\vec{c}^{\dagger}_{\boldsymbol{k}}, \vec{c}^{T}_{-\boldsymbol{k}}) \equiv (C^{\dagger}_{\boldsymbol{k}1}, C^{\dagger}_{\boldsymbol{k}2}, \cdots, C^{\dagger}_{\boldsymbol{k}6}, C^{\dagger}_{\boldsymbol{k}7}, \cdots, C^{\dagger}_{\boldsymbol{k}12}). \tag{11}$$

The capital subscript A (B) is used to represent the A-th element $C_{\boldsymbol{k}A}$ of $\vec{C}_{\boldsymbol{k}}$, and $\bar{A}$ indicates the summation over A (= $1, 2, \cdots, 12$). Here the dimension of $\vec{C}^{(\dagger)}_{\boldsymbol{k}}$ is specified as $12 = 3(\text{orbital}) \times 2(\text{spin}) \times 2(\text{Nambu})$ since the effective model for Sr_2IrO_4 is considered. The H^{BCS} can be diagonalized by the unitary matrix $\hat{\mathcal{U}}_{\boldsymbol{k}}$ as

$$\begin{aligned} H^{\mathrm{BCS}} &= \frac{1}{2}\sum_{\boldsymbol{k}} \vec{C}^{\dagger}_{\boldsymbol{k}} \hat{\mathcal{U}}_{\boldsymbol{k}} \big(\hat{\mathcal{U}}^{\dagger}_{\boldsymbol{k}} \hat{H}^{\mathrm{BCS}}_{\boldsymbol{k}} \hat{\mathcal{U}}_{\boldsymbol{k}} \big) \hat{\mathcal{U}}^{\dagger}_{\boldsymbol{k}} \vec{C}_{\boldsymbol{k}} \\ &= \frac{1}{2}\sum_{\boldsymbol{k}} \vec{B}^{\dagger}_{\boldsymbol{k}} \hat{\mathcal{E}}_{\boldsymbol{k}} \vec{B}_{\boldsymbol{k}} = \frac{1}{2}\sum_{\boldsymbol{k}} B^{\dagger}_{\boldsymbol{k}\bar{A}} \mathcal{E}_{\boldsymbol{k}\bar{A}} B_{\boldsymbol{k}\bar{A}}, \end{aligned} \tag{12}$$

where $\vec{B}_{\boldsymbol{k}}$ represents the field operator of the Bogoliubov quasiparticles (in the vector form) in the Nambu representation, and $\hat{\mathcal{E}}_{\boldsymbol{k}}$ is its quasiparticle energy. The unitary transformation from the $\vec{B}_{\boldsymbol{k}}$ to $\vec{C}_{\boldsymbol{k}}$ is given as

$$\begin{cases} \vec{C}_{\boldsymbol{k}} = \hat{\mathcal{U}}_{\boldsymbol{k}} \vec{B}_{\boldsymbol{k}} \\ \vec{C}^{\dagger}_{\boldsymbol{k}} = \vec{B}^{\dagger}_{\boldsymbol{k}} \hat{\mathcal{U}}^{\dagger}_{\boldsymbol{k}} \end{cases} \Leftrightarrow \begin{cases} C_{\boldsymbol{k}A} = [\mathcal{U}_{\boldsymbol{k}}]_{A\bar{A}} B_{\boldsymbol{k}\bar{A}} \\ C^{\dagger}_{\boldsymbol{k}B} = B^{\dagger}_{\boldsymbol{k}\bar{B}} [\hat{\mathcal{U}}^{\dagger}_{\boldsymbol{k}}]_{\bar{B}B} \end{cases}. \tag{13}$$

In Eq. (10), the SC gap function $\hat{\Delta}_{\boldsymbol{k}}$ is introduced as an input parameter. When we consider the pseudospin SC pairing for Sr_2IrO_4, $\hat{\Delta}_{\boldsymbol{k}}$ can be obtained from the unitary transformation of the pseudospin SC gap function $\hat{D}_{\boldsymbol{k}}$ via $\vec{c}^{\dagger}_{\boldsymbol{k}} \hat{\Delta}_{\boldsymbol{k}} \vec{c}^{\dagger\mathrm{T}}_{-\boldsymbol{k}} = \vec{a}^{\dagger}_{\boldsymbol{k}} \hat{D}_{\boldsymbol{k}} \vec{a}^{\dagger\mathrm{T}}_{-\boldsymbol{k}}$. For instance, the specific form of the $d_{x^2-y^2}$-wave pseudospin $j_{\mathrm{eff}} = 1/2$ singlet pairing is given by $[\hat{D}_{\boldsymbol{k}}]_{12} = -[\hat{D}_{\boldsymbol{k}}]_{21} = D_d(\cos k_x - \cos k_y)/2$ and the other elements are zero [14, 15, 18], while that of the $s_{\pm}$-wave pseudospin singlet pairing is given by $[\hat{D}_{\boldsymbol{k}}]_{12} = -[\hat{D}_{\boldsymbol{k}}]_{21} = +D_s$, $[\hat{D}_{\boldsymbol{k}}]_{34} = -[\hat{D}_{\boldsymbol{k}}]_{43} = -D_s$ and the other elements are zero [15, 17, 18]. Here, D_d and D_s denote the amplitudes of the $d_{x^2-y^2}$- and $s_{\pm}$-wave pseudospin singlet pairing, respectively.

2.3. Green's Functions

To formulate the RPA scheme for SC states, we here introduce a Green's function defined as

$$\begin{aligned} \hat{\mathcal{G}}(k) &\equiv -\int_0^{\beta} d\tau\, \mathrm{e}^{i\omega_n\tau} \langle T_\tau \vec{C}_{\boldsymbol{k}}(\tau) \vec{C}^{\dagger}_{\boldsymbol{k}}(0) \rangle \\ &= \left(\begin{array}{c|c} \hat{G}(k) & \hat{F}(k) \\ \hline \hat{\bar{F}}(k) & \hat{\bar{G}}(k) \end{array} \right), \end{aligned} \tag{14}$$

where $\beta = 1/k_{\mathrm{B}}T$, $k = (\boldsymbol{k}, i\omega_n)$, and $\omega_n = (2n+1)k_{\mathrm{B}}T$ is a Matsubara frequency for fermions with $n \in \mathbb{N}$. The elements of the 6×6 matrix G ($\bar{G}$) and F ($\bar{F}$) are the normal and anomalous Green's functions, respectively. In the non-interacting case, the Green's function $\hat{\mathcal{G}}_0(k)$ can be obtained immediately from the BCS Hamiltonian matrix, i.e.,

$$\begin{aligned}\hat{\mathcal{G}}_0(k) &= \left(i\omega_n - \hat{H}^{\rm BCS}_{\boldsymbol{k}}\right)^{-1} = \hat{\mathcal{U}}_{\boldsymbol{k}}\left(i\omega_n - \hat{\mathcal{E}}_{\boldsymbol{k}}\right)^{-1}\hat{\mathcal{U}}^\dagger_{\boldsymbol{k}} \\ &= \left(\begin{array}{c|c} \hat{G}_0(k) & \hat{F}_0(k) \\ \hline \hat{\bar{F}}_0(k) & \hat{\bar{G}}_0(k)\end{array}\right).\end{aligned} \tag{15}$$

2.4. *Dynamical Correlation Function in the Non-interacting Case*

Let us consider the dynamical correlation function in the non-interacting case,

$$[\hat{\chi}_0(q)]_{\alpha\beta\gamma\delta} = \frac{1}{N}\int_0^\beta d\tau\, e^{i\epsilon_m\tau}\sum_{\boldsymbol{k}\boldsymbol{k}'}\langle T_\tau c^\dagger_{\boldsymbol{k}\beta}(\tau)c_{\boldsymbol{k}+\boldsymbol{q}\alpha}(\tau)c^\dagger_{\boldsymbol{k}'\gamma}(0)c_{\boldsymbol{k}'-\boldsymbol{q}\delta}(0)\rangle_{0\,\mathrm{c}}, \tag{16}$$

where the subscript "0" denotes the statistical average over the one-body Hamiltonian $H_{\rm BCS}$. With Wick's theorem, $\hat{\chi}_0(q)$ can be represented by using both normal and anomalous Green's functions as

$$[\hat{\chi}_0(q)]_{\alpha\beta\gamma\delta} = \frac{1}{N\beta}\sum_k\left[\hat{G}_0(q+k)\right]_{\alpha\gamma}\left[\hat{\bar{G}}_0(-k)\right]_{\beta\delta} - \frac{1}{N\beta}\sum_k\left[\hat{F}_0(q+k)\right]_{\alpha\delta}\left[\hat{\bar{F}}_0(-k)\right]_{\beta\gamma}, \tag{17}$$

where the summation over k is the short hand notation of the summation over $\boldsymbol{k}$ and $i\omega_n$. The diagrammatic representation of Eq. (17) is depicted in Fig. 1. In addition to the first term in Eq. (17) composed of the normal Green's functions (G and $\bar{G}$), it is necessary to include the second one composed of the anomalous Green's functions (F and $\bar{F}$) in SC states.

Furthermore, one can perform the summation over $i\omega_n$ in Eq. (17) by considering the Bogoliubov quasiparticle basis. To achieve this, it is convenient to consider the dynamical correlation function in the Nambu representation

$$[\hat{X}_0(q)]_{ABCD} = \frac{1}{N}\int_0^\beta d\tau\, e^{i\epsilon_m\tau}\sum_{\boldsymbol{k}\boldsymbol{k}'}\langle T_\tau C^\dagger_{\boldsymbol{k}B}(\tau)C_{\boldsymbol{k}+\boldsymbol{q}A}(\tau)C^\dagger_{\boldsymbol{k}'C}(0)C_{\boldsymbol{k}'-\boldsymbol{q}D}(0)\rangle_{0\,\mathrm{c}}. \tag{18}$$

By using the unitary matrix $\hat{\mathcal{U}}_{\boldsymbol{k}}$ and the quasiparticle dispersion $\hat{\mathcal{E}}_{\boldsymbol{k}}$, we obtain the following form of $\hat{X}_0(q)$:

$$\begin{aligned}[\hat{X}_0(q)]_{ABCD} &= \frac{1}{N}\sum_{\boldsymbol{k}}[\mathcal{U}_{\boldsymbol{k}+\boldsymbol{q}}]_{A\bar{A}}[\mathcal{U}^\dagger_{\boldsymbol{k}+\boldsymbol{q}}]_{\bar{A}C}[\mathcal{U}^\dagger_{\boldsymbol{k}}]_{\bar{B}B}[\mathcal{U}_{\boldsymbol{k}}]_{\bar{B}D}\frac{f(\mathcal{E}_{\boldsymbol{k}+\boldsymbol{q}\bar{A}}) - f(\mathcal{E}_{\boldsymbol{k}\bar{B}})}{i\epsilon_m - (\mathcal{E}_{\boldsymbol{k}+\boldsymbol{q}\bar{A}} - \mathcal{E}_{\boldsymbol{k}\bar{B}})} \\ &- \frac{1}{N}\sum_{\boldsymbol{k}}[\mathcal{U}_{\boldsymbol{k}+\boldsymbol{q}}]_{A\bar{A}}[\mathcal{U}_{-(\boldsymbol{k}+\boldsymbol{q})}]_{D\bar{A}'}[\mathcal{U}^\dagger_{\boldsymbol{k}}]_{\bar{B}B}[\mathcal{U}^\dagger_{-\boldsymbol{k}}]_{\bar{B}',C}\frac{f(\mathcal{E}_{\boldsymbol{k}+\boldsymbol{q}\bar{A}}) - f(\mathcal{E}_{\boldsymbol{k}\bar{B}})}{i\epsilon_m - (\mathcal{E}_{\boldsymbol{k}+\boldsymbol{q}\bar{A}} - \mathcal{E}_{\boldsymbol{k}\bar{B}})},\end{aligned} \tag{19}$$

where $f(x) = \left(e^{\beta x} + 1\right)^{-1}$ is the distribution function for fermions, and $\bar{A}'(\bar{B}') = \mathrm{Mod}(6 + \bar{A}(\bar{B}) - 1, 12) + 1$. For the derivation, we have used the following identity for the summation over the Matsubara frequencies:

$$\frac{1}{\beta}\sum_{i\omega_n}\frac{1}{i\omega_n + i\epsilon_m - \mathcal{E}_{\boldsymbol{k}+\boldsymbol{q}A}} \times \frac{1}{-i\omega_n + \mathcal{E}_{\boldsymbol{k}B}} = \frac{f(\mathcal{E}_{\boldsymbol{k}+\boldsymbol{q}A}) - f(\mathcal{E}_{\boldsymbol{k}B})}{i\epsilon_m - (\mathcal{E}_{\boldsymbol{k}+\boldsymbol{q}A} - \mathcal{E}_{\boldsymbol{k}B})}. \tag{20}$$

Notice that the elements of $\hat{X}_0(q)$ for A, B, C, and $D = 1, 2, \cdots 6$ correspond to those of $\hat{\chi}_0(q)$ as $\vec{C}^\dagger_{\boldsymbol{k}} = (\vec{c}^{\,\dagger}_{\boldsymbol{k}}, \vec{c}^{\,T}_{-\boldsymbol{k}})$.

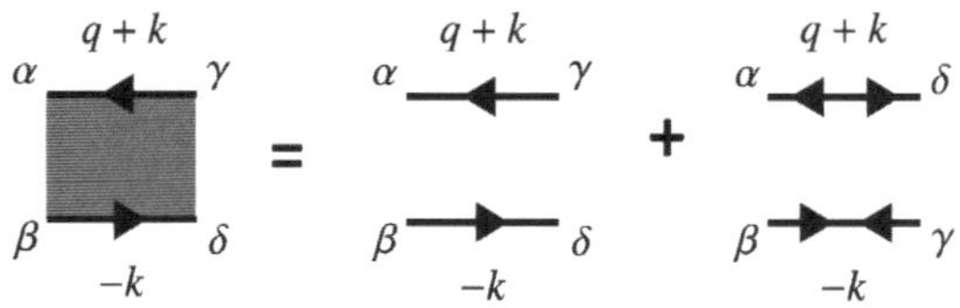

Figure 1. Schematic representation of the dynamical correlation function $\hat{\chi}_0(q)$ in the non-interacting case in SC states. In addition to the first term composed of the normal Green's functions (G and $\bar{G}$), the second one composed of the anomalous Green's functions (F and $\bar{F}$) is also taken into account because of SC states. Notice that the order of the vertex indices is different in the first and second terms.

2.5. Dynamical Correlation Function in the RPA

The dynamical correlation function $\hat{\chi}(q)$ in the RPA can be obtained diagrammatically with the infinite summation over all bubble and ladder diagrams of $\hat{\chi}_0(q)$ connected by the Coulomb interactions [23, 24, 25]. In a one-band Hubbard model with the on-site Hubbard interaction U, the bubble diagrams can be interpreted as the charge and longitudinal spin fluctuations, while the ladder ones as the transverse spin fluctuations. In the presence of the SOC, however, the bubble and ladder diagrams can not be separated from each other due to the entanglement of the spin and orbital degrees of freedom. Therefore, one needs to introduce a general expression for $\hat{\chi}_0(q)$ being taken account of the spin and orbital indices, as depicted in Figure 1.

To obtain the analytical expression for $\hat{\chi}(q)$, let us consider the Bethe-Salpeter equation for $\hat{\chi}(q)$, i.e.,

$$\hat{\chi}(q) = \hat{\chi}_0(q) + \hat{\chi}_0(q)\hat{V}\hat{\chi}(q), \tag{21}$$

which is schematically represented in Figure 2. Here, the product of the tensors is defined as $[\hat{A}\hat{B}]_{\alpha\beta\gamma\delta} = \sum_{\mu\nu} A_{\alpha\beta\mu\nu}B_{\mu\nu\gamma\delta}$. $\hat{V}$ is the Coulomb interaction in the tensor form whose elements contain the intra- and inter-orbital on-site Hubbard interactions U and U', respectively, the Hund's coupling J, and the pair hopping J in $H_{\rm int}$ of Eq. (9). The solution for $\hat{\chi}(q)$ can be obtained immediately as

$$\hat{\chi}(q) = \left(1 - \hat{\chi}_0(q)\hat{V}\right)^{-1} \hat{\chi}_0(q). \tag{22}$$

The explicit expression of the interaction tensor $\hat{V}$ is given as

$$\begin{aligned}
V_{a\uparrow,b\uparrow,c\uparrow,d\uparrow} = V_{a\downarrow,b\downarrow,c\downarrow,d\downarrow} &= \begin{cases} U' - J & (a = c \neq b = d) \\ -(U' - J) & (a = b \neq c = d) \\ 0 & (\text{others}) \end{cases}, \\
V_{a\uparrow,b\uparrow,c\downarrow,d\downarrow} = V_{a\downarrow,b\downarrow,c\uparrow,d\uparrow} &= \begin{cases} -U & (a = b = c = d) \\ -J & (a = c \neq b = d) \\ -U' & (a = b \neq c = d) \\ -J & (a = d \neq b = c) \\ 0 & (\text{others}) \end{cases}, \\
V_{a\uparrow,b\downarrow,c\uparrow,d\downarrow} = V_{a\downarrow,b\uparrow,c\downarrow,d\uparrow} &= \begin{cases} U & (a = b = c = d) \\ U' & (a = c \neq b = d) \\ J & (a = b \neq c = d) \\ J & (a = d \neq b = c) \\ 0 & (\text{others}) \end{cases},
\end{aligned} \tag{23}$$

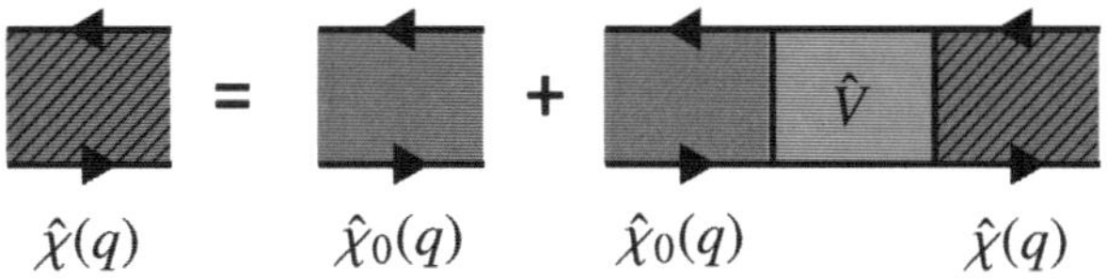

Figure 2. Schematic representation of the dynamical correlation function $\hat{\chi}(q)$ in the RPA. This can be obtained from the Bethe-Salpeter equation, where all the diagrams connected by the interaction tensor are summed until the infinite order.

and the other components are zero [26, 27]. Although we have started with the effective model for Sr_2IrO_4, the obtained expression for $\hat{\chi}(q)$ is general in the RPA scheme and can also be applied to the SC states for other multi-orbital systems in the presence of a SOC.

3. Results

We now show the numerical results for the magnetic susceptibility in the pseudospin singlet SC states. Let us first consider the theoretical expression for the magnetic susceptibility, which can be derived from the linear combination of the dynamical correlation function $\hat{\chi}(q)$.

The magnetic susceptibility $\chi_M^{uv}(q)$ can be written as,

$$\chi_M^{uv}(q) = \frac{1}{N}\int_0^\beta d\tau\, e^{i\epsilon_m\tau}\langle T_\tau M_{\boldsymbol{q}}^u(\tau)M_{-\boldsymbol{q}}^v(0)\rangle_c, \tag{24}$$

where the magnetic moment operator $M_{\boldsymbol{q}}^u \equiv L_{\boldsymbol{q}}^u + 2S_{\boldsymbol{q}}^u$ with wave number $\boldsymbol{q}$ and $u\ (=\pm, z)$ is the sum of the spin and orbital angular momentum operator $S_{\boldsymbol{q}}^u$ and $L_{\boldsymbol{q}}^u$, respectively. $M_{\boldsymbol{q}}^{\pm}$, $L_{\boldsymbol{q}}^{\pm}$, and $S_{\boldsymbol{q}}^{\pm}$ represent the raising and lowering operators, and $M_{\boldsymbol{q}}^z$, $L_{\boldsymbol{q}}^z$, and $S_{\boldsymbol{q}}^z$ are the z component of the operators. Thus, $\chi_M^{-+}(q)$ and $\chi_M^{zz}(q)$ represent the transverse and longitudinal magnetic susceptibilities, respectively. $M_{\boldsymbol{q}}^u$ can be expressed by the field operators as

$$M_{\boldsymbol{q}}^u = \sum_{\alpha\beta}\sum_{\boldsymbol{k}} c_{\boldsymbol{k}\alpha}^\dagger [\hat{M}^u]_{\alpha\beta} c_{\boldsymbol{k}+\boldsymbol{q}\beta}, \tag{25}$$

where $\hat{M}^u$ denotes the matrix representation of the magnetic moment operator. From the definition of $M_{\boldsymbol{q}}^u$, $\hat{M}^u$ can be obtained from the sum of the matrix representation for the spin and orbital angular momentum operator $\hat{S}^u$ and $\hat{L}^u$, respectively, as

$$\hat{M}^u = \hat{L}^u + 2\hat{S}^u, \tag{26}$$

where the explicit forms of $\hat{S}^u$ and $\hat{L}^u$ are given as

$$\begin{aligned} \hat{S}^z &= \frac{1}{2}\left(\begin{array}{c|c} +\hat{1} & \hat{0} \\ \hline \hat{0} & -\hat{1} \end{array}\right), \\ \hat{S}^+ &= \left(\begin{array}{c|c} \hat{0} & \hat{1} \\ \hline \hat{0} & \hat{0} \end{array}\right),\ \hat{S}^- = \left(\begin{array}{c|c} \hat{0} & \hat{0} \\ \hline \hat{1} & \hat{0} \end{array}\right), \end{aligned} \tag{27}$$

and

$$\hat{L}^z = \left(\begin{array}{ccc|ccc} 0 & i & 0 & & & \\ -i & 0 & 0 & & \hat{0} & \\ 0 & 0 & 0 & & & \\ \hline & & & 0 & i & 0 \\ & \hat{0} & & -i & 0 & 0 \\ & & & 0 & 0 & 0 \end{array}\right),\quad \hat{L}^\pm = \left(\begin{array}{ccc|ccc} 0 & 0 & \pm 1 & & & \\ 0 & 0 & i & & \hat{0} & \\ \mp 1 & -i & 0 & & & \\ \hline & & & 0 & 0 & \pm 1 \\ & \hat{0} & & 0 & 0 & i \\ & & & \mp 1 & -i & 0 \end{array}\right). \tag{28}$$

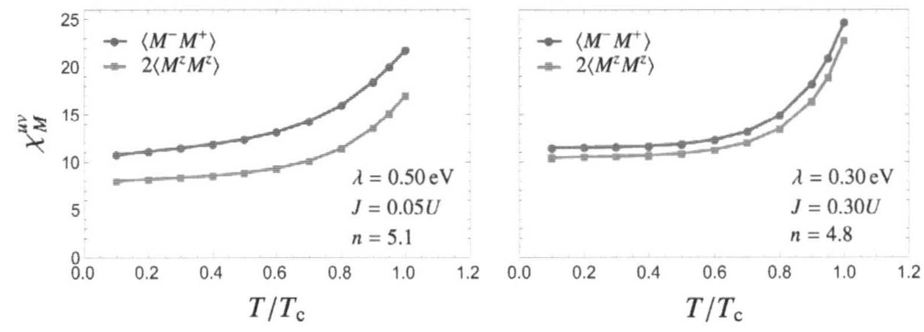

Figure 3. Magnetic susceptibility $\chi_M^{uv} \equiv \chi_M^{uv}(q=0)$ in the pseudospin singlet SC states. The $d_{x^2-y^2}$ and $s_\pm$-wave pseudospin singlet pairings are assumed in the left and right panels, respectively.

Finally, one can obtain the magnetic susceptibility $\chi_M^{uv}(q)$ as the linear combination of the dynamical correlation function $\hat{\chi}(q)$,

$$\chi_M^{uv}(q) = \sum_{\alpha\beta\gamma\delta} [\hat{M}^u]_{\beta\alpha}[\hat{M}^v]_{\gamma\delta}[\hat{\chi}(q)]_{\alpha\beta\gamma\delta}. \tag{29}$$

Figure 3 displays the temperature dependence of the uniform magnetic susceptibility $\chi_M^{uv} \equiv \chi_M^{uv}(q=0)$ in the pseudospin singlet SC states. Here, the $d_{x^2-y^2}$- and $s_\pm$-wave pseudospin singlet pairing are assumed in the left and right panels, respectively, in Figure 3. We assume that the temperature dependence of the amplitude $D_{d(s)}(T)$ of the SC gap function is determined by the gap equation

$$D_{\boldsymbol{k}} = -\frac{1}{N}\sum_{\boldsymbol{k}'} V_{\boldsymbol{k},\boldsymbol{k}'}\frac{D_{\boldsymbol{k}'}}{2E_{\boldsymbol{k}'}}\tanh\frac{E_{\boldsymbol{k}'}}{2T} \tag{30}$$

with the pairing interaction $V_{\boldsymbol{k},\boldsymbol{k}'} = -|V|(\cos k_x - \cos k_y)(\cos k'_x - \cos k'_y)$ for the $d_{x^2-y^2}$-wave pairing and $V_{\boldsymbol{k},\boldsymbol{k}'} = -|V|$ for the $s_\pm$-wave pairing. We set the single-particle parameters $(t_1, t_2, t_3, t_4, t_5, \mu_{xy}) = (0.36, 0.18, 0.09, 0.37, 0.06, -0.36)$ eV and the intra-orbital on-site Hubbard interaction $U = 0.75$ eV. The two-dimensional k-point mesh in the first Brillouin zone is chosen to be $(k_x, k_y) = (64, 64)$. The amplitude of the SC gap at $T = 0$ is also set to be $D_{d(s)}(0) = 4k_BT_c$ and the transition temperature T_c to be $k_BT_c = 0.1t_1$. In addition to this, we set the SOC $\lambda = 0.5$ (0.3) eV, the Hund's coupling $J = 0.05U$ (0.3U), and the electron density $n = 5.1$ (4.8) for the $d_{x^2-y^2}$ ($s_\pm$)-wave pseudospin singlet pairing superconductivity. These parameters are selected simply because the previous studies [14, 15, 18] have shown that the $d_{x^2-y^2}$- ($s_\pm$-) wave pseudospin singlet superconducting pairing is dominant in the electron- (hole-) doped region with large (small) SOC λ and small (large) Hund's coupling J. The effective model of the electron-doped Sr_2IrO_4 with large λ becomes a one-band system and thus the $d_{x^2-y^2}$-wave pairing is favored by the intra-band repulsive (k-dependent) pairing interaction in analogy to cuprate superconductors. On the other hand, the effective model of the hole-doped Sr_2IrO_4 with small λ becomes a two-band system and therefore the $s_\pm$-wave pairing is favored by the inter-band repulsive pairing interaction in analogy to iron-based superconductors.

Figure 3 shows clearly that the transverse and longitudinal magnetic susceptibilities are different from each other. This difference is originated from the SOC and Hund's coupling J, which induce magnetic anisotropy. One can also notice that these magnetic susceptibilities decrease below T_c with decreasing T but they do not become zero at $T = 0$. This is understood because neither spin nor orbital angular momentum is good quantum number in the presence of the SOC and therefore the formation of singlet Cooper pairs does not completely freeze the magnetic fluctuations even at $T = 0$.

4. Summary

To summarize, we have formulated the RPA for multi-orbital systems with the SOC in order to understand the electronic and magnetic properties for SC states in 5d transition metal oxides such as Sr_2IrO_4. Starting with the effective model for iridium oxide insulator Sr_2IrO_4, we have considered the BCS Hamiltonian by introducing the SC gap function, from which the normal and anomalous Green's functions can be obtained. The dynamical correlation function for the non-interacting case can be obtained from the Green's functions, and the electron correlation can be incorporated by an infinite summation over all bubble and ladder diagrams connected via the Coulomb interactions. The formalism given here is general and is applicable to other multi-orbital systems with a SOC. Finally, we have evaluated the magnetic susceptibility in the pseudospin singlet SC states with $d_{x^2-y^2}$- and $s_\pm$-wave pairing symmetries and demonstrated that the transverse and longitudinal magnetic susceptibilities are anisotropic due to the SOC. We have also shown that these magnetic susceptibilities decrease below T_c but they are finite even at $T = 0$. This is because the magnetic fluctuations are not completely frozen in the SC states with the pseudospin singlet pairing.

References

[1] J. J. Randall, L. Katz, and R. Ward, Journal of the American Chemical Society **79**, 266 (1957).
[2] B. J. Kim, H. Jin, S. J. Moon, J.-Y. Kim, B.-G. Park, C. S. Leem, J. Yu, T.W. Noh, C. Kim, S.-J. Oh, J.-H. Park, V. Durairaj, G. Cao, and E. Rotenberg, Phys. Rev. Lett. **101**, 076402 (2008).
[3] B. J. Kim, H. Ohsumi, T. Komesu, S. Sakai, T. Morita, H. Takagi, and T. Arima, Science **323**, 1329 (2009).
[4] J. Kim, D. Casa, M. H. Upton, T. Gog, Y.-J. Kim, J. F. Mitchell, M. van Veenendaal, M. Daghofer, J. van den Brink, G. Khaliullin, and B. J. Kim, Phys. Rev. Lett. **108**, 177003 (2012).
[5] S. Fujiyama, H. Ohsumi, T. Komesu, J. Matsuno, B. J. Kim, M. Takata, T. Arima, and H. Takagi, Phys. Rev. Lett. **108**, 247212 (2012).
[6] Y. K. Kim, O. Krupin, J. D. Denlinger, A. Bostwick, E. Rotenberg, Q. Zhao, J. F. Mitchell, J. W. Allen, and B. J. Kim, Science 345, **187** (2014).
[7] Y. K. Kim, N. H. Sung, J. D. Denlinger, and B. J. Kim, Nature Physics , 10.1038/nphys3503 (2015).
[8] Y. J. Yan, M. Q. Ren, H. C. Xu, B. P. Xie, R. Tao, H. Y. Choi, N. Lee, Y. J. Choi, T. Zhang, and D. L. Feng, Phys. Rev. X **5**, 041018 (2015).
[9] G. Jackeli and G. Khaliullin, Phys. Rev. Lett. **102**, 017205 (2009).
[10] H. Jin, H. Jeong, T. Ozaki, and J. Yu, Phys. Rev. B **80**, 075112 (2009).
[11] H. Watanabe, T. Shirakawa, and S. Yunoki, Phys. Rev. Lett. **105**, 216410 (2010).
[12] F. Wang and T. Senthil, Phys. Rev. Lett. **106**, 136402 (2011).
[13] R. Arita, J. Kuneš, A. V. Kozhevnikov, A. G. Eguiluz, and M. Imada, Phys. Rev. Lett. **108**, 086403 (2012).
[14] H. Watanabe, T. Shirakawa, and S. Yunoki, Phys. Rev. Lett. **110**, 027002 (2013).
[15] Y. Yang, W.-S. Wang, J.-G. Liu, H. Chen, J.-H. Dai, and Q.-H. Wang, Phys. Rev. B **89**, 094518 (2014).
[16] H. Watanabe, T. Shirakawa, and S. Yunoki, Phys. Rev. B **89**, 165115 (2014).
[17] Z. Y. Meng, Y. B. Kim, and H.-Y. Kee, Phys. Rev. Lett. **113**, 177003 (2014).
[18] K. Nishiguchi, T. Shirakawa, H. Watanabe, R. Arita, and S. Yunoki (in preparation).
[19] D. J. Scalapino, Rev. Mod. Phys. 84, 1383 (2012).
[20] Y. J. Uemura, Nat. Mater. 8, 253 (2009).
[21] H. Aoki, J. Supercond. Nov. Magn. 25, 1243 (2012).
[22] J. Kanamori, Prog. Theor. Phys. **30**, 275 (1963).
[23] N. E. Bickers, D. J. Scalapino, and S. R. White, Phys. Rev. Lett. **62**, 961 (1989).
[24] N. Bickers and D. Scalapino, Ann. Phys **193**, 206 (1989).
[25] T. Dahm and L. Tewordt, Phys. Rev. Lett. **74**, 793 (1995).
[26] K. Yada and H. Kontani, J. Phys. Soc. Jpn. **74**, 2161 (2005).
[27] T. Takimoto, T. Hotta, and K. Ueda, Phys. Rev. B 69, 104504 (2004).
[28] K. Nishiguchi, H. Watanabe, and S. Yunoki, JPS Conf. Proc. **3**, 015037 (2014).

IOP Conf. Series: Journal of Physics: Conf. Series **807** (2017) 052014 doi:10.1088/1742-6596/807/5/052014

Fractional vortices in a nano-scaled superconducting composite structure (d-dot) with a twin boundary

Norio Fujita[1,3], Masaru Kato[1] and Takekazu Ishida[2]

[1]*Department of Mathematical Sciences, Osaka Prefecture University*
1-1 Gakuencho, Nakaku, Sakai, Osaka 599-8531, Japan
[2]*Department of Physics and Electronics, Osaka Prefecture University*
1-1 Gakuencho, Nakaku, Sakai, Osaka 599-8531, Japan
[3]*Research Fellow of Japan Society for the Promotion of Science*

E-mail: su104021@edu.osakafu-u.ac.jp

Abstract. In order to investigate effects of twin boundary (TB) on structure of spontaneous half-quantized vortices (SHQVs) in a nano-scaled composite structure that is consisted of *d*- and *s*-wave superconductors (d-dot), we investigate magnetic field distribution with a d-dot model which has a single TB using anisotropic two-component Ginzburg-Landau equations and finite element method. We find that fractional vortices appear at the edges of TB and that the fractional vortices merge with SHQVs when effective interaction between *d*- and *s*-wave SCs is small.

1. Introduction

Corner junctions made with *d*- and *s*-wave superconductors (SCs) are called π-junctions, because the phase between *d*- and *s*-wave SCs is different by due to symmetry of order parameter in *d*-wave SC [1]. With quantization of fluxoids including this phase differences, half-quantized vortices spontaneously appear around π-junctions with zero magnetic field [2].

A d-dot, which is a nano-scaled composite structure that consists of a *d*-wave SC embedded in an *s*-wave matrix [3,4] has four -junctions and then four spontaneous half-quantized vortices (SHQVs). If the size of the *d*-wave SC in d-dot is sufficiently small than penetration depth of the *s*-wave SC, SHQVs form anti-ferromagnetic configuration. So d-dot has two arrangements of SHQVs, which are doubly degenerated stable states controlling by an external field or current [5,6] we can use it as a classical bit with. It is also predicted that we can use superposition of these stable states as quantum bit by reducing the size of d-dot's to increase transition probability between both states [7].

Fujii et al. made d-dot's that consist of $YBa_2Cu_3O_{7-\delta}$ (YBCO) and Pb, but SHQVs were not observed [8]. One of reasons of this result seems to be defects in YBCO, especially twin boundaries (TBs). Smilde et al. reported TBs suppress the SHQVs [9].

In this paper, to clarify effects of TB on structures of SHQVs in a d-dot, we investigate magnetic field distribution with a d-dot model that has a single TB using anisotropic two-component Ginzburg-Landau (GL) equations and finite element method (FEM). We show that fractional vortices appear at the edges of TB and that the fractional vortices merge with SHQVs when effective interaction between *d*- and *s*-wave SCs is small.

2. Methods

In order to incorporate effects of TBs into the GL equations, we introduce an anisotropic effective mass into the Gor'kov equations [10] as follows.

$$\left\{ i\hbar\omega_n - \frac{1}{2m_x}\left(-i\hbar\frac{\partial}{\partial x}+\frac{e}{c}A_x\right)^2 - \frac{1}{2m_y}\left(-i\hbar\frac{\partial}{\partial y}+\frac{e}{c}A_y\right)^2 + \mu \right\}\tilde{G}(\boldsymbol{x},\boldsymbol{x}';\omega_n) + \int d\boldsymbol{x}''\Delta(\boldsymbol{x},\boldsymbol{x}'')F^{\dagger}(\boldsymbol{x}'',\boldsymbol{x}';\omega_n) = \delta(\boldsymbol{x}-\boldsymbol{x}') \quad (1)$$

$$\left\{ -i\hbar\omega_n - \frac{1}{2m_x}\left(i\hbar\frac{\partial}{\partial x}+\frac{e}{c}A_x\right)^2 - \frac{1}{2m_y}\left(i\hbar\frac{\partial}{\partial y}+\frac{e}{c}A_y\right)^2 + \mu \right\}F^{\dagger}(\boldsymbol{x},\boldsymbol{x}';\omega_n) - \int d\boldsymbol{x}''\Delta^{*}(\boldsymbol{x},\boldsymbol{x}'')\tilde{G}(\boldsymbol{x}'',\boldsymbol{x}';\omega_n) = 0 \quad (2)$$

where $\tilde{G}$ and F are the single particle normal and anomalous Green's functions, respectively. m_x and m_y are the effective mass along x- and y-axis, respectively. $\omega_n = (2n+1)\pi T$ is the Matsubara frequency. The order parameter in real space is given

$$\Delta^{*}(\boldsymbol{x},\boldsymbol{x}') = V(\boldsymbol{x}-\boldsymbol{x}')T\sum_{\omega_n} F^{\dagger}(\boldsymbol{x},\boldsymbol{x}';\omega_n) \quad (3)$$

$$V(\boldsymbol{k},\boldsymbol{k}') = V_s + V_d(\hat{k}_x^2 - \hat{k}_y^2)(\hat{k}_x'^2 - \hat{k}_y'^2) \quad (4)$$

where $V(\boldsymbol{k},\boldsymbol{k}')$ is an effective pairing interaction between two electrons in momentum space, and V_s and V_d are interaction constants for s- and d-wave pairings [11], respectively.

Substituting equations (1), (2) and (4) into equation (3), we derive self-consistent equations for s- and d-wave order parameters [12], Δ_s and Δ_d,

$$\Delta_s^{*} = 2\lambda_d\left(\frac{V_s}{V_d}\right)\Delta_s^{*}\ln\left(\frac{2e^{\gamma}\hbar\omega_D}{\pi k_B T}\right) - 2\lambda_d\left(\frac{V_s}{V_d}\right)\alpha\left[\left(\frac{\varepsilon_F}{2m_x}\Pi_x^2+\frac{\varepsilon_F}{2m_y}\Pi_y^2\right)\Delta_s^{*}+\left(\frac{\varepsilon_F}{4m_x}\Pi_x^2-\frac{\varepsilon_F}{4m_y}\Pi_y^2\right)\Delta_d^{*}+|\Delta_s|^2\Delta_s^{*}+|\Delta_d^{*}|^2\Delta_s^{*}+\frac{1}{2}\Delta_d^{*2}\Delta_s^{*}\right] \quad (5)$$

$$\Delta_d^{*} = 2\lambda_d\Delta_d^{*}\ln\left(\frac{2e^{\gamma}\hbar\omega_D}{\pi k_B T}\right) - 2\lambda_d\alpha\left[\left(\frac{\varepsilon_F}{4m_x}\Pi_x^2+\frac{\varepsilon_F}{4m_y}\Pi_y^2\right)\Delta_d^{*}+\left(\frac{\varepsilon_F}{4m_x}\Pi_x^2-\frac{\varepsilon_F}{4m_y}\Pi_y^2\right)\Delta_s^{*}+|\Delta_s|^2\Delta_d^{*}+\frac{3}{8}|\Delta_d^{*}|^2\Delta_d^{*}+\frac{1}{2}\Delta_s^{*2}\Delta_d\right] \quad (6)$$

Here γ is the Euler constant, and ω_D is a cutoff frequency for interactions. Also $\alpha = 7\zeta(3)/8(\pi k_B T)^2$, $\varepsilon_F = \hbar^2 k_{Fx}^2/2m_x = \hbar^2 k_{Fy}^2/2m_y$ and $\lambda_d = N(0)V_d/2$ where $N(0)$ is a density of states at the Fermi surface. And $\mathbf{\Pi} = -i\hbar\nabla_{\mathbf{R}} - (2e/c)\mathbf{A}_{\mathbf{R}}$ where $\mathbf{R} = (\boldsymbol{x}+\boldsymbol{x}')/2$.

Note that gradient terms and anisotropic coupling terms between Δ_s and Δ_d in equations (5) and (6) contain anisotropic mass. In order to analyze how these terms affect d-dot's, we solve equations (5) and (6) using the FEM [3].

In this calculation, we use material parameters of YBCO and Pb for d- and s-wave SCs, respectively.

Figure1 shows (a) a schematic diagram and (b) a calculation model of a d-dot with a single TB. We consider following system; an s-wave matrix is $L\times L$ $(L=\xi_{s0}=50\xi_{d0}\simeq 80nm)$ and an embedded d-wave region is $L/2\times L/2$ $(L/2\simeq 40nm)$. The single TB is shown by a straight line (yellow line) in the d-wave SC region from the middle point of the bottom to the middle point of the left side, which divides the d-wave SC region into a left lower domain (orange area) and a right upper domain (purple area). Effective masses of electrons in the left lower domain are switched in the right upper domain. Here, ξ_{d0} and ξ_{s0} are the coherent length of d- and s-wave SCs at 0 K.

We assume that the s-wave order parameter exists in the s-wave SC region and penetrates into d-wave SC region. On the other hand, the d-wave order parameter exists in the d-wave SC region and

IOP Conf. Series: Journal of Physics: Conf. Series **807** (2017) 052014 doi:10.1088/1742-6596/807/5/052014

penetrates into the s-wave SC region. So, interactions between s- and d-wave order parameters around interface of d- and s-wave SCs are important for appearance of SHQVs.

Note that in this calculation, ratio of interaction constants for s- and d-wave pairings V_s, V_d are different between d- and s-wave SC regions.

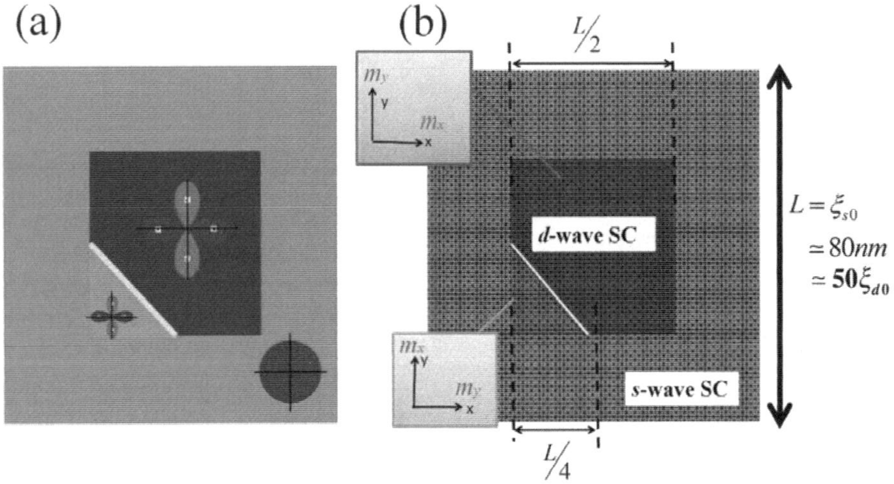

Figure 1. (a) A Schematic diagram and (b) a calculation model of a d-dot with a single TB. The s-wave matrix is $L \times L$ $(L = \xi_{s0} = 50\xi_{d0} \simeq 80nm)$ and the embedded d-wave region is $L/2 \times L/2$ $(L/2 \simeq 40nm)$. The TB is shown by a straight yellow line which divides a central d-wave square into a left lower domain (orange area) and a right upper domain (purple area). Effective masses of Cooper pairs in the left lower domain are switched in the top right domain. Here, ξ_{d0} are ξ_{s0} are the coherent length of d- and s-wave SCs at 0 K.

3. Results and Discussions

Figure 2 shows magnetic field distributions for isotropic system with m_y/m_x= 1.0 (a), and for anisotropic system with m_y/m_x= 2.0 (b), respectively. In figures 2 (a) and (b), red and blue colors mean $H_z < 0$ and $H_z > 0$, respectively. In this case we set ratios of V_s and V_d in d-wave SC $(V_s/V_d)_d$= 1.0 and in s-wave SC $(V_d/V_s)_s$= 1.0, respectively.

In the isotropic case, figure 2 (a), only four SHQVs appear on each corners of d-wave domain and form antiferromagnetic configuration [3]. In the anisotropic cases, figure 2 (b), shapes and configuration of magnetic field distributions are almost same as figure 2 (a), but we can also see additional vortices in solid circles (green circles). These additional vortices are fractional vortices, not SHQVs, because peak values of these additional vortices are less than SHQVs' in figure 2 (a).

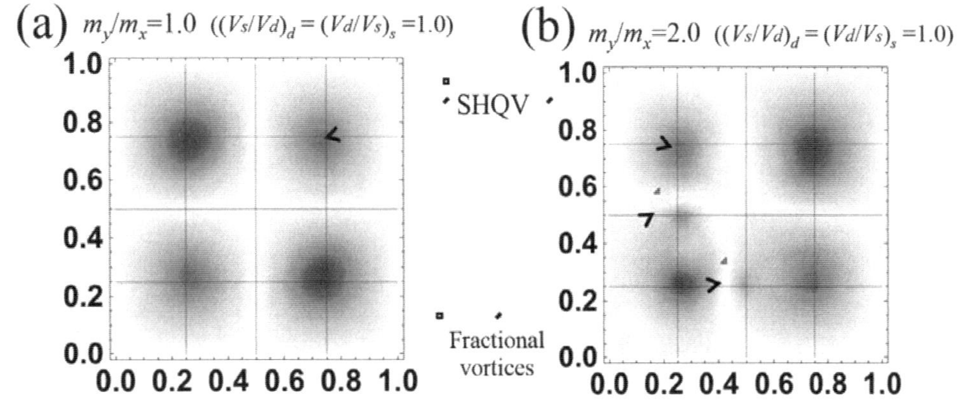

Figure 2. Density plots of magnetic field distributions for (a) isotropic system with $m_y/m_x = 1.0$, and for (b) anisotropic system with $m_y/m_x = 2.0$, respectively. We set ratios of V_s and V_d in d-wave SC $(V_s/V_d)_d = 1.0$ and in s-wave SC $(V_d/V_s)_s = 1.0$, respectively. In (a) and (b), red and blue colors mean $H_z > 0$ and $H_z < 0$, respectively. In (a), only four SHQVs appear on each corners of d-wave domain and form antiferromagnetic order. But in (b), We can see fractional vortices in solid circles (green circles).

One can understand how these fractional vortices appear as below. The distributions of supercurrents around the corners are in an elliptic form, because the penetration depth λ is inversely proportional to the square root of effective mass

$$\lambda = \left(\frac{c^2 m^*}{4\pi n_s e^{*2}} \right)^{1/2} \quad (7)$$

Here e^* and m^* is charge and effective mass of cooper pair i.e. $e^*=2e$, $m^*=m_x$ or m_y. Around the corner junctions, supercurrent $\boldsymbol{j}_s(\boldsymbol{r})$ is large. Then they decrease continuously from the corner to the middle point of the corners. Because penetration depth changes due to the change of m^* across the TB, there is finite imbalance of $\boldsymbol{j}_s(\boldsymbol{r})$ at the edges of the TB. This supercurrent imbalance along the boundary of d- and s-wave SCs, especially in solid circles (green circles), generate fractional vortices.

To analyze dependence of interaction between d- and s-wave SCs, we obtain magnetic field distribution for the case of $(V_s/V_d)_d = (V_d/V_s)_s = 0.1$, respectively. Figure 3 shows density plots of magnetic field distributions for (a) isotropic system with $m_y/m_x = 1.0$, and for (b) anisotropic system with $m_y/m_x = 2.0$, respectively, $((V_s/V_d)_d = (V_d/V_s)_s = 0.1)$. In figure3 (a), only four SHQVs also appear on each corners of d-wave domain and form antiferromagnetic order as same as figure 2 (a). But the distribution of each SHQV concentrate more inside of d-wave region as dashed lines indicating. This is because efficient penetration depth in d-wave region becomes larger than the case of figure 2. In figure 3 (b), fractional vortices cannot be distinguished from SHQVs because SHQVs more spread inside d-wave region and merge with the nearest parallel fractional vortices.

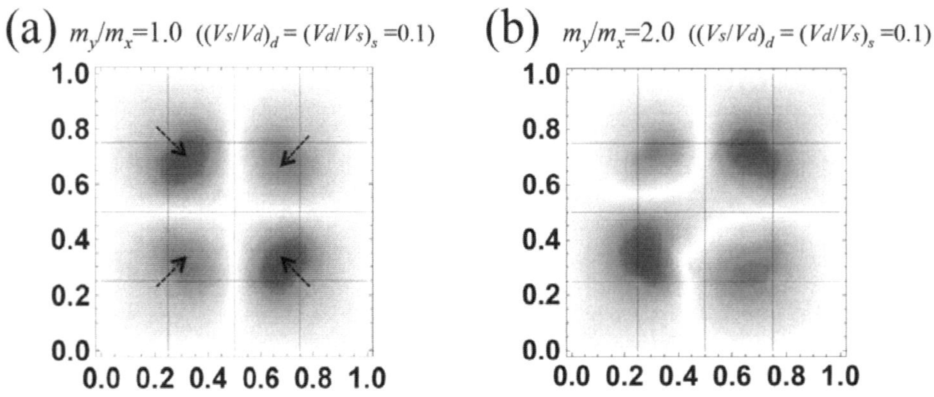

Figure 3. Density plots of magnetic field distributions for (a) isotropic system with $m_y/m_x = 1.0$, and for (b) anisotropic system with $m_y/m_x = 2.0$, respectively, $((V_s/V_d)_d = (V_d/V_s)_s = 0.1)$. In (a) and (b), red and blue colors mean $H_z > 0$ and $H_z < 0$, respectively. In (a), only four SHQVs appear on each corners of *d*-wave domain and form antiferromagnetic order as same as figure 2 (a), but the distribution of each SHQV concentrate more inside *d*-wave region as dashed lines indicating. In (b), we can't distinguish fractional vortices from SHQVs because SHQVs more spread inside *d*-wave region and merge with the nearest parallel fractional vortices.

4. Summary

In conclusion, in order to analyze the effects of TBs on structure of magnetic field distribution in d-dot, phenomenologically, we calculated magnetic field distribution in a d-dot model with a single TB using modified two-component GL equations that include anisotropic mass terms and the FEM. We find that fractional vortices appear at the edges of TB beside SHQVs and that the fractional vortices merge with the nearest parallel SHQVs when effective interaction between *d*- and *s*-wave SCs is small.

Acknowledgments

We would like to thank Y. Higashi, S. Fukui and M. Umeda for useful discussions. This work is supported by the Program for Leading Graduate Schools of Ministry of Education, Culture, Sports, Science and Technology in Japan (MEXT), JSPS KAKENHI Grant Number JP26400367 and Grant-in-Aid for JSPS Fellows 1681004900.

References

[1] Tsuei C C and Kirtley 2000 Rev. Mod. Phys. **72** 969
[2] Hilgenkamp H, Ariando, Smilde H, Blank D H A, Rijnders G, Rogalla H, Kirtley J R and Tsuei C C 2003 Nature **422** 50
[3] Kato M, Ishida T, Koyama T and Machida M, 2012 *Superconductors -Materials, Properties and Applications-* (InTech, Croatia) Chap.13, p. 319
[4] Ishida T, et al. 2006 Physica C **437** 104
[5] Ako M, Machida M, Koyama T, Ishida T and Kato M 2004 Physica C **412-414** 544
[6] Ako M, Machida M, Koyama T, Ishida T and Kato M 2005 Physica C **426-431** 122
[7] Koyama T, Machida M, Kato M and Ishida T 2005 Physica C **426-431** 1561
[8] Fujii M, et al. 2005 Physica C **426** 104
[9] Smilde H J H, Ariando, Blank D H A, Gerritsma G J, Hilgenkamp H and Rogalla H 2002 Phys. Rev. Lett. **88** 057004
[10] Gor'kov L P 1960 Sov. Phys. JETP **9** 1364
[11] Xu J H, Ren Y and Ting C S 1995 Phys. Rev. B **52** 7663

[12] Fujita N, Kato M and Ishida T 2015 Physica C **518** 44

IOP Conf. Series: Journal of Physics: Conf. Series **807** (2017) 052015 doi:10.1088/1742-6596/807/5/052015

^{9}Be-NMR studies on anomalous superconducting phase diagram in UBe_{13}

Haruki Matsuno [1,*], Kyohei Morita [1], Hisashi Kotegawa [1], Hideki Tou [1,†], Yoshinori Haga [2], Etsuji Yamamoto [2] and Yoshichika Ōnuki [3]

[1] Department of Physics, Kobe University, Kobe, Hyogo 675-8501, Japan
[2] Advanced Science Research Center, Japan Atomic Energy Agency, Tokai 319-1195, Japan
[3] Faculty of Science, University of Ryukyus, Nishihara, Okinawa 903-0213, Japan

E-mail: *125s118s@stu.kobe-u.ac.jp
†tou@crystal.kobe-u.ac.jp

Abstract. In order to clarify unusual superconducting properties, we have carried out ^{9}Be NMR measurements for a single crystal of UBe_{13} with $T_c \cong 0.85$ K. The NMR spectra under 5 T parallel to [001] crystal axis show no change between $T_c(H = 5\text{ T}) = 0.41$ K and $T_a(H = 5\text{ T}) = 0.35$ K. Below T_a, however, the reduction of the Knight shift was observed. The reduction of the Knight shift of Be(II) is amount to ~ 0.02 % ,which is much smaller than spin part of the Knight shift, $K_\gamma \sim 0.107$ % estimated from electronic specific coefficient $\gamma_e \cong 1000$ $mJ/(K^2 \cdot$ mol). The origin of reduction of the Knight shift cannot be explained by spin singlet superconductivity.

1. Introduction

Since the discovery of heavy fermion superconductivity in UBe_{13}, extensive experimental and theoretical works have been carried out in order to clarify the nature of the unusual normal and superconducting(SC) states of UBe_{13}. The SC transition occurs at around $T_c \cong 0.85$ K with large specific heat jump, $\Delta\, C/\gamma_e T_c \cong 1$ [1, 2], indicating that the heavy-quasiparticles are responsible for the superconductivity. In the SC state, the power law temperature dependence of the specific heat, the NMR spin-lattice relaxation rate, magnetic field penetration depth, etc., suggest that the SC energy gap vanishes at points and/or lines on the Fermi surface [2, 3, 4, 5]. These features have been interpreted as evidence for an anisotropic pairing state, most likely a p-wave triplet state in UBe_{13}. Furthermore, the anomalies have been observed at B^* in several measurements, specific heat, magnetization, surface impedance etc.[6, 7, 8], which have been discussed the change of the SC order parameters like UPt_3[9], related to second phase transition in $U_{1-x}Th_xBe_{13}(x = 0.02 \sim 0.04)$ [10]. However, the reduction of the Knight shift below T_c have been observed in μ^+SR studies and suggests that the spin singlet state is realized in UBe_{13} [11]. Moreover, recent specific heat measurements suggest multi-band full-gap like symmetry [12]. Therefore, the parity of the superconductivity of UBe_{13} has been controversial and more information is needed in order to clarify the nature of superconductivity.

In order to gain a further understanding of the unconventional SC state, we have carried out ^{9}Be NMR measurements on a single crystal UBe_{13} applied field 5 T parallel to [001] crystal axis. In an applied field of 5 T, UBe_{13} shows the SC transition and anomalous behavior at $T_c =$

0.45 K and T_a = 0.32 K respectively, which have been obtained by surface impedance studies as shown in Fig.1.

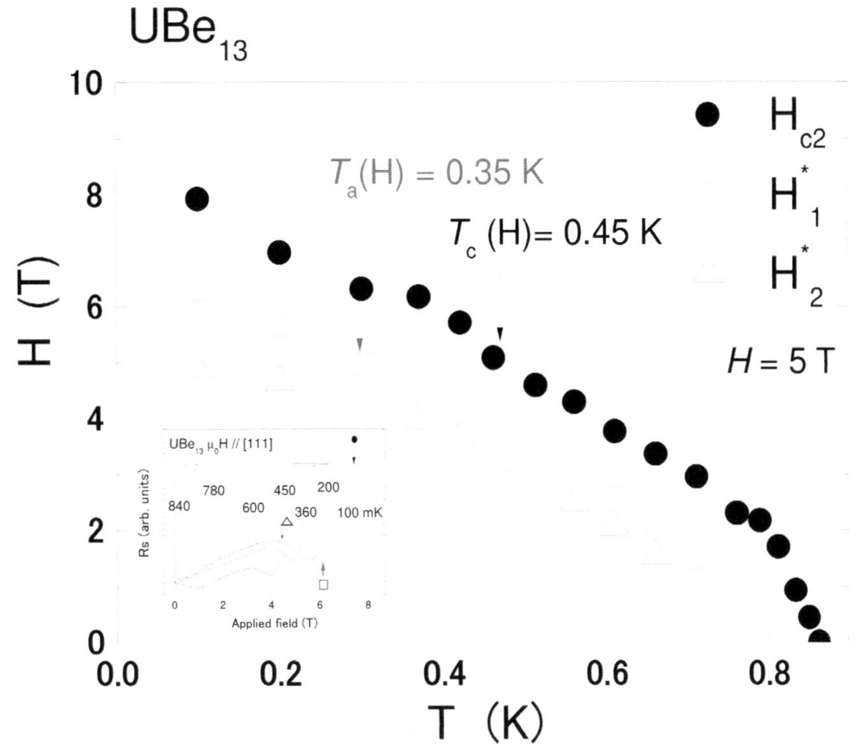

Figure 1. The T-H phase diagram of UBe_{13} obtained from surface impedance measurements. H_1^*(open square) and H_2^*(open triangle) are indicate maximal and minimal point of the surface resistance respectively. Closed circle indicates the H_{c2}. T_c and T_a indicate SC transition and appearance of anomaly temperature under applied field of 5 T respectively. NMR measurements were carried out under 5 T, which is shown by the solid line. The inset shows field dependence of the surface resistivity R_S [8].

2. Experimental procedures

Single crystals of UBe_{13} were grown by the Al-flux method. Details of sample preparation techniques for single crystals of UBe_{13} was reported elsewhere [13]. Samples were characterized to be a single phase by a Laue photograph as well as the X-ray diffraction. The SC phase transition was confirmed to occur at $T_c \cong 0.85$ K by DC electrical resistivity and surface impedance measurements [8, 13]. NMR measurements have been carried out using a conventional pulsed NMR spectrometer and a 12 T solenoid superconducting magnet. The sample was fixed on a Bakelite holder and installed a top loading type ^{3}He-^{4}He dilution refrigerator down to 0.12 K. The sample temperature was monitored using a calibrated RuO_2 resistor at low T's (50 mK - 20 K).

3. Results and Discussion

The temperature dependence of ^{9}Be-NMR spectra of UBe_{13} under applied field of 5 T || [001] axis is shown in Fig.2. The NMR spectra show no change between 1 K and 0.37 K, even though temperature is below $T_c = 0.45$ K. On the other hand, the change of the spectra can be observed below $T_a = 0.35$ K and the change become more clearly with decreasing the temperature. The T_a at which the anomalous spectra change was observed is equivalent to the minimal point in surface impedance as shown in Fig.1 by open square.

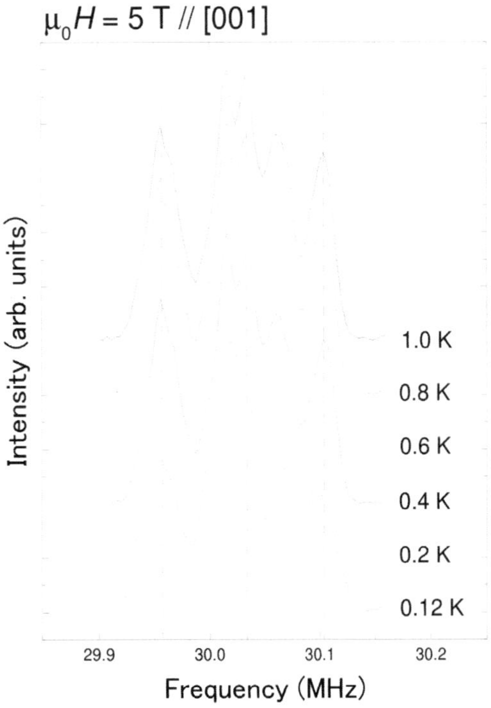

Figure 2. The temperature dependence of ^{9}Be NMR spectra at $\mu_0 H = 5$ T parallel to [001]. The dashed line is guide to the eye.

In order to clarify the origin of the change of the spectra below T_a, we tried the spectral simulation calculation. The resonance frequencies for each Be sites were calculated by exact diagonalization of the 4 × 4 nuclear spin Hamiltonian matrix of ^{9}Be as follow,

$$\begin{aligned}\mathcal{H} = & - \gamma_n \hbar I \cdot H_0 - \gamma_n \hbar [I \cdot \tilde{K} \cdot H_0] - \gamma_n \hbar I \cdot H_d \\ & + \frac{1}{6} h \nu_Q [3I_Z^2 - I(I+1) + \frac{\eta}{2}(I_+^2 + I_-^2)]\end{aligned} \quad (1)$$

The first term represents to the Zeeman interaction. The second term is due to the hyperfine interaction described by the second rank Knight shift tensor $\tilde{K}$. Here, $\tilde{K}$ is composed of the isotropic part K_{iso} and the anisotropic part $K_i (i = 1, 2, 3)$. The third term is the classical dipolar interaction between nuclear spin and the dipole field arising from the localized U spin moment. The last term corresponds to the nuclear quadrupole interaction. Here, h is Planck constant and $\hbar = h/2\pi$. $\nu_Q = 3e^2 qQ/2hI(2I-1)$ denotes the NQR frequency where $eq = V_{ZZ}$ is the maximum principal value of the electric field gradient (EFG) tensor at each sites. $\eta = (V_{XX} - V_{YY})/V_{ZZ}$ exhibits the asymmetry parameter. The notation described here follows the previous report and the detail of the calculation has been reported in ref [13]. These simulation calculations were carried out through a trial and error process.

Fig.3(a) displays the observed spectrum(solid line) and its simulation (dotted line) at $T =$ 1 K for $H \|$ [001] at 5 T. The simulation calculation using previous NMR/NQR parameters reproduces well the experimental results. The best fit parameters are obtained as $\nu_Q = 0.8 \pm 0.02$

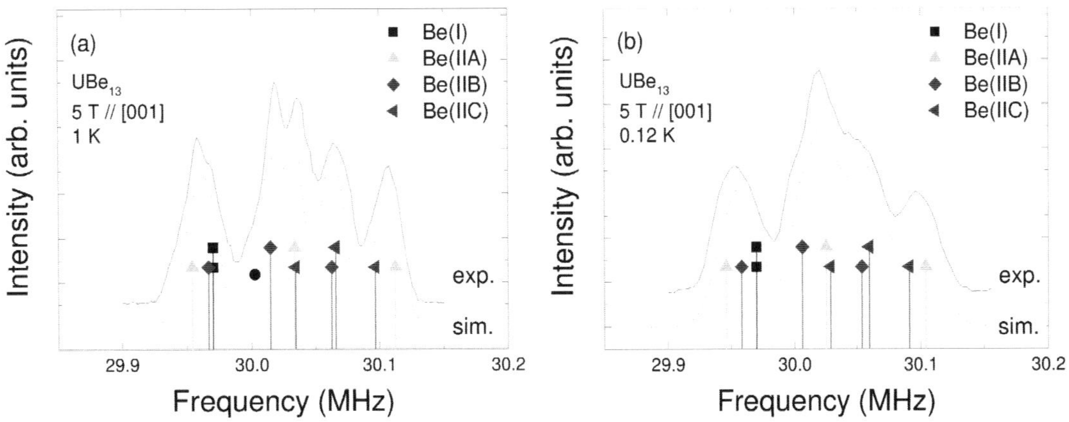

Figure 3. The comparison of the spectra between experimental data (solid line) and simulation (dotted line) at (a) 1K and (b) 0.12K.

kHz, $\eta = 0.21 \pm 0.01$, $K_{iso} = 0.13 \pm 0.01$ %, $(K^1, K^2, K^3) = (0.04, 0.015, -0.055) \pm 0.01$ % and line width $\sigma = 9 \pm 1$ kHz at 1 K.

Fig.3(b) shows the spectrum observed at 0.12 K. Apparently, it is difficult to reproduce the whole line profile using the parameters at 1 K. Therefore, we try to determine the parameters which can explain NMR the spectrum at 0.12 K. The calculated spectrum is shown by dotted line in Fig.3(b). At 0.12 K, the NMR spectrum can be reproduced by changing only the isotropic Knight shift and line width with $K_{iso} = 0.11 \pm 0.01$ %, and $\sigma = 13 \pm 1$ kHz. The reduction of the K_{iso} is $\Delta K_{iso} \sim 0.02$ %, which is 15 % of the K_{iso}. At the present stage, the change of K_i cannot be determined. Because, the amount of the change of K_i is smaller than $\Delta K_i = 0.01$ %. Even if we assume the reduction of K_i the same ratio (~ 15 %) as K_{iso}, the NMR spectra can be also reproduced in anomaly phase.

The temperature dependence of K_{iso} are obtained and shown in Fig.4. The temperature dependence of K_{iso} does not change between 1 K and 0.37 K passing through T_c, whereas it decreases below 0.37 K and reaches constant value at 0.32 K. The observed reduction of K_{iso} is very small amount to $\Delta K_{iso} \sim 0.02$ %, as mentioned before. If we estimate the Knight shift of the quasiparticle spin part from the specific heat coefficient γ_e, $K_\gamma \sim 0.107$ % is obtained assuming the conventional Fermi liquid picture [14]. This estimated value K_γ is almost comparable to the experimental value K_{iso} in the normal state. Thus the Knight shift would be almost constituted by spin part K_s contribution.

If the reduction of the Knight shift is related with superconductivity, several possibilities are expected. One is that the spin singlet state is realize in UBe_{13}, another is existence of degenerated order parameters like UPt_3 [9]. In the former case which the spin singlet state is realized below T_a, Knight shift should decrease to $K \sim 0.01$ % at $T \ll T_c$. This expected value is markedly different from observed value $K_{iso} \sim 0.1$ %. Hence it is difficult that the origin of the reduction of the Knight shift is ascribed to spin singlet superconductivity. Whereas, in the latter case that UBe_{13} has the degenerated order parameters of SC state, the reduction of Knight shift can be explained by the change of the order parameter below T_a. However, recent several measurements suggest that a short range ordering occurs below $B^*(T_a)$ [15, 16]. If such a short range ordering occurs, the NMR spectra are expected to change. Unfortunately, our data are insufficient to conclude that such a short range ordering occurs. To confirm whether or not such a short range ordering occurs below T_a, we have to measure NMR spectra in other

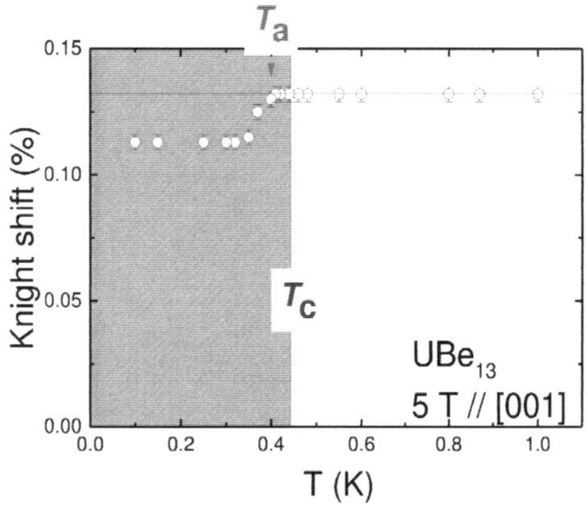

Figure 4. The temperature dependence of the isotropic part of Knight shift.

field directions and compare with them.

4. Summary

We have performed the NMR studies and observed the anomaly in the Knight shift which decrease below T_a. The reduction of the Knight shift is very small ~ 0.02 %. This reduction of the Knight shift cannot be explained by spin singlet state. On the other hand, the degenerated order parameters can be exist in spin triplet SC state, which can cause the change of the Knight shift. Hence the reduction of Knight shift would be related to spin triplet SC state. However, the possibility of short range order have not been excluded. To obtain more information of the anomaly below T_a, the measurements in various field and direction are needed.

Acknowledgments

This work has been supported by JSPS KAKENHI Grant Numbers 15H05885, (J-Physics), 26400359, and 15H05745. This work was partly supported by JSPS KAKENHI Grant Numbers, 15H03689, 15K05882 and 15K21732.

References

[1] Ott H R, Rudigier H, Fisk Z and Smith J L 1983 *Phys. Rev. Lett.* **50** 1595
[2] Ott H R, Rudigier H, Rice T M, Ueda K, Fisk Z and Smith J L 1984 *Phys. Rev. Lett.* **52** 1915
[3] Tien C and Jiang I M 1989 *Phys. Rev.* B **40** 229.
[4] Golding B, Bishop D J, Batlogg B, Haemmerle W H, Fisk Z, Smith J L and Ott H R 1985 *Phys. Rev. Lett.* **55** 2479
[5] Einzel D, Hirschfeld P J, Gross F, Chandrasekhar B S, Andres K, Ott H R, Beuers J, Fisk Z and Smith J L 1986 *Phys. Rev. Lett.* **56** 2513
[6] Ellman B, Rosenbaum T F, Kim J S and Stewart G R 1991 *Phys. Rev.* B **44** 12074
[7] Shimizu Y, Haga Y, Ynagisawa T and Amitsuka H 2016 *Phys. Rev.* B **93** 024502
[8] Matsuno H, Morita K, Kotegawa H, Tou H, Haga Y, Yamamoto E and Onuki Y 2015 *J. Phys.: Conf. Series* **592** 012067
[9] Machida K, Itoh A, So Y, Izawa K, Haga Y, Yamamoto E, Kimura N, Onuki Y, Tsutsumi Y and Machida K 2012 *Phys. Rev. Lett.* **108** 157002
[10] Ott H R, Rudigier H, Felder E, Fisk Z and Smith J L 1986 *Phys. Rev.* B **33** 126
[11] Sonier J E, Heffner R H, Morris G D, MacLaughlin D E, Bernal O O, Cooley J, Smith J L and Thompson J D 2007 *Physica* B **326** 414-417
[12] Shimizu Y, Kittaka S, Sakakibara T, Haga Y, Yamamoto E, Amitsuka H, Tsutsumi Y and Machida K 2015 *Phys. Rev. Lett.* **114** 147002
[13] Tou H, Tsugawa N, Sera M, Harima H, Haga Y and Onuki Y 2007 *J. Phys. Soc. Jpn.* **76** 024705
[14] Tou H, Ishida K and Kitaoka Y 2005 *J. Phys. Soc. Jpn.* **74** 1245

[15] Kromer F, Helfrich R, Lang M, Steglich F, Langhammer C, Bach A, Michels T, Kim J S and Stewart G R 1998 *Phys. Rev. Lett.* **81** 4476
[16] Shimizu Y, Haga Y, Ikeda Y, Ynagisawa T and Amitsuka H 2012 *Phys. Rev. Lett.* **109** 217001

^{51}V-NMR study of charge order induced by cation order in δ-$Ag_{2/3}V_2O_5$

Y. Kawasaki[1], R. Morioka[1], Y. Kishimoto[1], K. Nakamura[1], K. Nishiyama[2], T. Koyama[2], T. Mito[2], T. Baba[3], T. Yamauchi[3], M. Isobe[4] and Y. Ueda[5]

[1] Graduate School of Science and Technology, Tokushima University, Tokushima 770-8506, Japan
[2] Graduate School of Material Science, University of Hyogo, Hyogo 678-1297, Japan
[3] Institute for Solid State Physics, University of Tokyo, Kashiwa 277-8581, Japan
[4] Max Planck Institute for Solid State Research, D-70569 Stuttgart, Germany
[5] Toyota Physical and Chemical Research Institute, Nagakute, Aichi 480-1192, Japan

E-mail: kawasaki.yu@tokushima-u.ac.jp

Abstract. ^{51}V-NMR experiments have been performed to investigate the local magnetic and electronic properties of the mixed-valence oxide δ-$Ag_{2/3}V_2O_5$, which shows the novel V^{4+}/V^{5+} charge ordering trigged by Ag ion ordering. We have observed the abrupt loss of ^{51}V-NMR signal coming from the non-magnetic V^{5+}-like ions above T_c = 225 K. It indicates that the majority of V ions are magnetic above T_c and that the phase transition is accompanied by the charge separation and the charge ordering of 3*d* electrons on V sites. In the low-temperature phase below T_c, the nuclear spin-lattice relaxation rate $1/T_1$ shows the thermal activation-type temperature dependence with the activation energy of about 130 K. It has been clarified that the ground state of this material is the charge ordered one with the spin singlet of 3*d* electrons on V^{4+}-V^{4+} pair from a microscopic point of view.

1. Introduction

Various exotic phenomena due to spin, charge, lattice and orbital degrees of freedom have attracted much attention in strongly correlated electron systems [1]. Among them, the charge ordering is one of the most intriguing and extensively studied phenomena related to the charge degrees of freedom. In the vanadium bronzes $A_xV_2O_5$ (A: alkali, alkaline-earth or other metals), mixed-valence oxides of V^{4+} ($3d^1$) and V^{5+} ($3d^0$), the charge ordering may appear in its stoichiometric composition, where the number of 3*d* electrons could be controlled as a function of x. The first discovery of the charge ordering in this system has been made on α'-NaV_2O_5 [2, 3]. For the last decade, the charge ordering in β-$A_{1/3}V_2O_5$ has been intensively investigated in connection with the superconductivity that appears when the charge ordering is suppressed by applying pressure for A = Li, Na and Ag [4, 5]. For example, β-$Na_{1/3}V_2O_5$ shows the successive phase transitions at ambient pressure; the ordering of Na atoms at 242 K, the metal-insulator transition at 135 K and the antiferromagnetic one at 24 K. Various experiments have clarified that the metal-insulator transition is accompanied by the charge ordering or the charge disproportionation [6, 7, 8, 9].

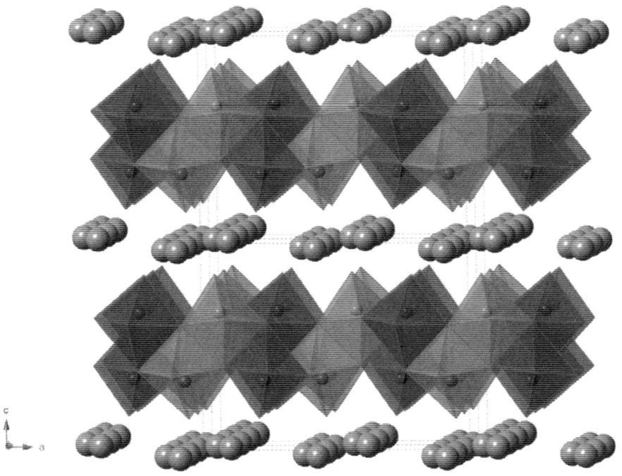

Figure 1. Crystal structure of δ-$Ag_{2/3}V_2O_5$, where mauve octahedra, blue octahedra and orange balls represent $V1O_6$, $V2O_6$ and Ag, respectively. (from Ref.[10])

Now, we focus on δ-$Ag_{2/3}V_2O_5$, which shows the novel charge ordering induced by cation ordering [11]. The crystal structure consists of V_2O_5 double-torellis layers and Ag ions located between the double-trellis layers (Fig. 1) [11, 12, 13]. There are two V sites denoted as V1 and V2, and each V forms a distorted octahedron at room temperature, $V1O_6$ (mauve octahedra in the figure) and $V2O_6$ (blue ones). The V_2O_5 double-trellis layers are formed by edge/corner-shared $V1O_6$ and $V2O_6$ octahedra. Ag ions (orange balls) partly occupy one of two nearest-neighboring sites between the double-trellis layers.

Regarding its electromagnetic properties, this oxide shows the first-order transition at around 220 K, accompanied by the jumps in electric resistivity and magnetic susceptibility [11, 13]. The transport properties become more insulating below T_c, which may be associated with the V^{4+}/V^{5+} charge ordering in the formal valence of $V^{4+}/V^{5+} = 1/2$. Recently, Baba et al. have revealed that the phase transition is caused by the Ag ion ordering, which triggers the charge ordering [11], where this phenomenon seems to be closely related to the fact that the ratio of $V^{4+}/V^{5+} = 1/2$ is identical to the ratio of Ag vacancies/Ag = 1/2. Interestingly, the phase transition can be supercooled to the lowest temperature by rapid cooling, which may prevent the Ag ion ordering. Below the transition temperature, the magnetic susceptibility shows a broad maximum at around 110 K followed by spin-gapped behavior, in contrast to Curie-Weiss behavior above the transition temperature. These results indicate that the magnetic V^{4+} ions have a specific geometrical arrangement required for the spin-gap state and that the geometrical arrangement is closely related to the Ag ion ordering. In this study, we have performed ^{51}V-NMR measurements to investigate the electromagnetic properties of δ-$Ag_{2/3}V_2O_5$, focusing on the spin-gap behavior in the low-temperature phase below T_c.

2. Experimental

A powder sample of δ-$Ag_{2/3}V_2O_5$ was prepared by a solid state reaction method. The detailed preparation method has been reported elsewhere [11]. X-ray diffraction and magnetic susceptibility confirm that the structural and magnetic properties are consistent with the published data [11, 13]. The transition temperature T_c on warming process for our sample was determined to be 225 K from the inflection point on the ^{51}V-NMR signal intensity vs. temperature curve, where the signal intensity abruptly decreases as mentioned later.

For NMR, we cooled the sample down to 5.0 K with sufficiently slow rate of 2 K/min and subsequently performed measurements at elevated temperature. The spin-echo signal of ^{51}V nuclei (the nuclear spin of $I = 7/2$) was observed by the spin-echo method, using a phase-coherent pulsed NMR spectrometer. The ^{51}V-NMR spectrum was obtained by tracing the

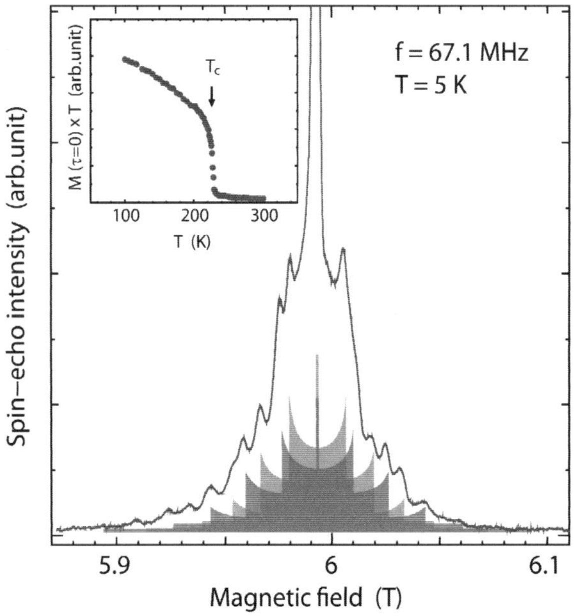

Figure 2. Field-swept ^{51}V-NMR spectrum of δ-$Ag_{2/3}V_2O_5$ at 5.0 K. Inset: Temperature dependence of spin-echo intensity multiplied by temperature, $M(\tau = 0) \times T$.

spin-echo intensity as a function of external field at the fixed frequency of 67.1 MHz. For the measurements of the nuclear spin-lattice relaxation rate $1/T_1$, we used the saturation recovery method. The spin-spin relaxation rate $1/T_2$ was obtained from the decaying rate of the spin-echo intensity $M(\tau)$ against the pulse separation τ in the spin-echo method. Here, the spin-echo intensity $M(\tau)$ is expressed by the equation, $M(\tau) = M(\tau = 0)\exp(-2\tau/T_2)$.

3. Results and discussion

In Fig.2, we show the field-swept ^{51}V-NMR spectrum measured at the fixed frequency of 67.1 MHz at 5.0 K in the low-temperature phase. The spectrum consists of two typical powder patterns (S1 and S2) for ^{51}V nuclei with different NQR frequencies, which are indicated by the green and the purple areas, respectively, in the figure. The NQR frequencies for S1 and S2 are estimated to be $\nu_{Q1} = 0.29$ MHz and $\nu_{Q2} = 0.43$ MHz, respectively, from the 1st satellite peak positions, where the respective asymmetric parameter is nearly zero for both.

One may consider that S1 and S2 are associated with non-magnetic V^{5+}-like ions located at the V1 and V2 sites ($V1^{5+}$ and $V2^{5+}$), although it is unclear whether S1 (S2) comes from $V1^{5+}$ or $V2^{5+}$ at present. The non-magnetic nature of V ions associated with S1 and S2 is indicated by the facts: (i) the positions of central peak for S1 and S2 are almost same and the corresponding Knight shifts are nearly zero, (ii) T_2's for S1 and S2 are comparable, as described in the next paragraph, which indicates that the V ions associated with S1 and S2 are magnetically similar, (iii) T_2's for S1 and S2 are long and these values are about a few hundred microseconds, which are comparable to the value of $T_2 \sim 100$ μs for the V^{5+}-like site in the charge ordered phase of β-$Ag_{1/3}V_2O_5$ [14] and (iv) for both S1 and S2, the signal intensity abruptly decreases at $T_c = 225$ K (see the inset of Fig. 2, showing the temperature dependence of the spin-echo intensity multiplied by temperature, $M(\tau = 0) \times T$), above which the V^{5+}-like ions are expected to be almost absent due to the collapse of the charge separation and charge order. A short nuclear spin-spin relaxation time T_2 may prevent us from clearly observing the ^{51}V-NMR signal of magnetic V^{4+}-like ions.

To examine T_2's for S1 and S2, we show the ^{51}V-NMR spectrum with typical values of τ in

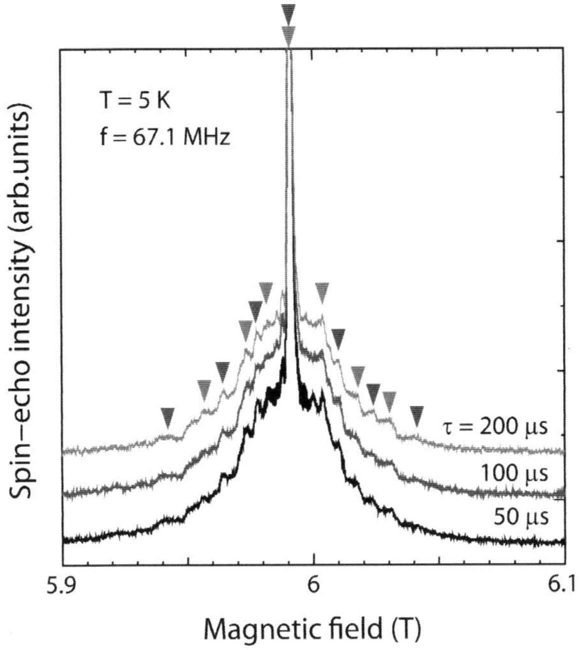

Figure 3. Field-swept ^{51}V-NMR spectra of δ-$Ag_{2/3}V_2O_5$ with pulse separation of $\tau = 50\mu s$ (black), 100 μs (blue) and 200 μs (red) at 5 K.

Fig. 3, where the black, blue and red lines represent the data with $\tau = 50$ μs, 100 μs, and 200 μs, respectively. The green and purple down triangles indicate the peak positions in powder pattern for S1 and S2, respectively, in the figure. The relative peak intensity for S1 and S2 is independent of τ, which indicates that T_2's for S1 and S2 are comparable. The comparable values of T_2's for S1 and S2 are also confirmed by the fact that the decay of the spin-echo intensity $M(\tau)$ against τ, measured at the center peak in the spectrum, has a single exponential form.

Next, we go on to the experimental results on the nuclear spin-lattice relaxation rate $1/T_1$ in δ-$Ag_{2/3}V_2O_5$. In Fig. 4, we show $(M_0 - M(t))/M_0$ against t at 105 K as a typical example. Here, $M(t)$ is the spin-echo intensity, which is proportional to nuclear magnetization t seconds after the saturation pulse. M_0 is $M(t)$ in the thermal equilibrium state ($t \to \infty$). The recovery of $M(t)$ to the thermal equilibrium state is expressed by

$$\frac{M_0 - M(t)}{M_0} = C_1 \exp\left(-\frac{t}{T_1}\right) + C_2 \exp\left(-\frac{6t}{T_1}\right) + C_3 \exp\left(-\frac{15t}{T_1}\right) + C_4 \exp\left(-\frac{28t}{T_1}\right) \quad (1)$$

for the transition between $I_z = -1/2$ and $1/2$ of the nuclear spin $I = 7/2$. Here, $C_1 = 0.019$, $C_2 = 0.0682$, $C_3 = 0.206$, and $C_4 = 0.714$ [15]. The solid line in Fig. 4 represents the best fit by Eq.(1) to the data. We note that Eq.(1) reproduces the data for all the measured temperature. It indicates that the observed nuclear spin relaxation is expressed by a single component of $1/T_1$, that is to say, the local environments of V1 and V2 sites are magnetically similar, which is consistent with comparable T_2's for S1 and S2. The fit enables us to estimate the nuclear spin-lattice relaxation rate $1/T_1$ at each temperature.

Fig. 5 shows the temperature dependence of the nuclear spin-lattice relaxation rate $1/T_1$. The value of $1/T_1$ decreases exponentially with decreasing temperature below T_c, followed by the temperature independent behavior below 20 K. The exponential temperature dependence of $1/T_1$ is confirmed by the Arrhenius plot of the data (inset of Fig. 5). This result indicates the existence of energy gap in magnetic excitation, which is consistent with the spin-singlet state of $3d$ electrons on V^{4+}-V^{4+} pairs. One may fit the data by the thermal activation-type relation in

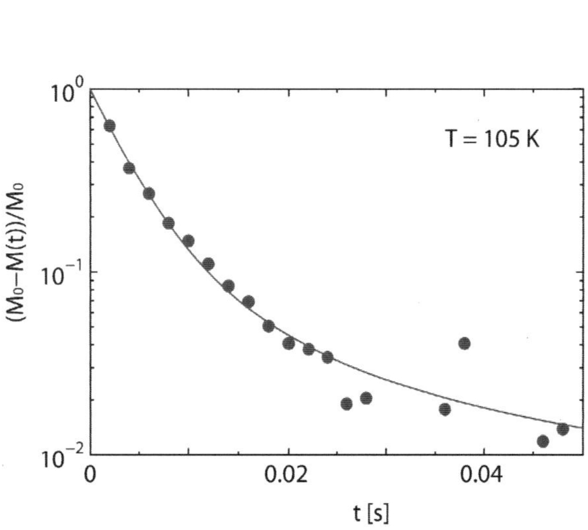

Figure 4. $(M_0 - M(t))/M_0$ vs. t at 105 K. The solid line indicates the best fit to the data by Eq.(1) (see text).

the temperature range between $T_c = 225$ K and 30 K, represented by solid lines in Fig. 5 and its inset. This relation is given by

$$\frac{1}{T_1} \propto \exp\left(-\frac{\Delta}{k_B T}\right), \tag{2}$$

where Δ is the size of the energy gap and k_B is the Boltzmann's constant. The size of the energy gap Δ/k_B is estimated to be 130 K from the present data, which is comparable to 150 K obtained from the magnetic susceptibility [11]. The deviation from the exponential temperature dependence below 20 K may be due to local fluctuation of V^{4+} magnetic moments without forming spin singlet. Above T_c, the value of $1/T_1$ is slightly smaller than that just below T_c, which may be related to the small jump in the magnetic susceptibility at T_c.

4. Conclusion

We have investigated the mixed-valence vanadate δ-$Ag_{2/3}V_2O_5$, which shows the novel V^{4+}/V^{5+} charge ordering triggered by the Ag ion ordering, from ^{51}V-NMR measurements. It has been microscopically confirmed that the phase transition is accompanied by the charge separation on V sites at T_c, above which the ^{51}V-NMR signal coming from the non-magnetic V^{5+}-like ions is significantly reduced. In the low-temperature phase below T_c, the nuclear spin-lattice relaxation rate $1/T_1$ indicates the existence of energy gap in magnetic excitation, which is consistent with the formation of spin singlet on V^{4+}-V^{4+} pair. The size of the energy gap is estimated to be about 130 K.

References

[1] Imada M, Fujimori A and Tokura Y 1998 *Rev. Mod. Phys.* **70** 1039
[2] Isobe M and Ueda Y 1996 *J. Phys. Soc. Jpn.* **65** 1178
[3] Seo H and Fukuyama H 1998 *J. Phys. Soc. Jpn.* **67** 2602
[4] Yamauchi T, Ueda Y and Môri N 2002 *Phys. Rev. Lett.* **89** 057002
[5] Yamauchi T and Ueda Y 2008 *Phys. Rev.* B **77** 104529
[6] Yamaura J, Isobe M, Yamada H, Yamauchi T and Ueda Y 2002 *J. Phys. Chem. Solids* **63** 957
[7] Heinrich M, Krug von Nidda H A, Eremina R M, Loidl A, Helbig Ch, Obermeier G and Horn S 2004 *Phys. Rev. Lett.* **93** 116402
[8] Nagai S, Hishi M, Kakurai K, Oohara Y, Yoshizawa H, Kimura H, Noda Y, Grenier B, Yamauchi T, Yamaura J, Isobe M, Ueda Y and Hirota K 2005 *J. Phys. Soc. Jpn.* **74** 1297
[9] Itoh M, Yamauchi I, Kozuka T, Suzuki T, Yamauchi T, Yamaura J and Ueda Y 2006 *Phys. Rev.* B **74** 054434

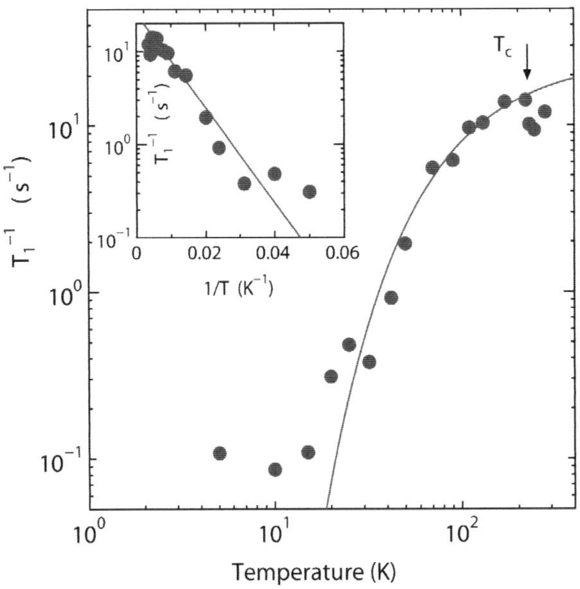

Figure 5. The temperature dependence of $1/T_1$. Inset: Arrhenius plot of the data shown in the main panel. The solid lines indicate the best fit to the data by the thermal activation-type relation between 20 K and T_c.

[10] Kawasaki Y, Morioka R, Nakamura K, Nishiyama K, Koyama T, Mito T, Baba T, Yamauchi T, Isobe M and UedaY 2015 *J. Phys.: Conf. Ser.* **592** 012042
[11] Baba T, Yamauchi T, Yamazaki S, Ueda H, Isobe M, Matsushita Y and Ueda Y 2015 *J. Phys. Soc. Jpn.* **84** 024718
[12] Andersson S 1965 *Acta Chem. Scand.* **19** 1371
[13] Onoda M and Arai R 2001 *J. Phys.: Condens. Matter* **13** 10399
[14] Hisada A, Fujiwara N, Yamauchi T and Ueda Y 2009 *J. Phys. Soc. Jpn.* **78** 094705
[15] Narath A 1967 *Phys. Rev.* **162** 320

IOP Conf. Series: Journal of Physics: Conf. Series **807** (2017) 062002 doi:10.1088/1742-6596/807/6/062002

NMR studies of coexistence of superconductivity and CDW in $LaPt_2Si_2$

Taisuke Aoyama[1A], Tetsuro Kubo[1], Haruki Matsuno[1B], Hisashi Kotegawa[1], Hideki Tou[1], Akihiro Mitsuda[2], Yutaro Nagano[2], Nobutaka Araoka[2], Hirohumi Wada[2] and Yuh Yamada[3]

[1] Department of Physics, Kobe University, Kobe, Hyogo 657-8501, Japan
[2] Department of Physics, Kyushu University, Fukuoka 819-0395, Japan
[3] Department of Physics, Niigata University, Niigata 950-2181, Japan

E-mail: [A]`t_aoyama@stu.kobe-u.ac.jp`

E-mail: [B]`tou@crystal.kobe-u.ac.jp`

Abstract. We carried out ^{139}La-, ^{29}Si- and ^{195}Pt-NMR measurements of $LaPt_2Si_2$ which shows coexistence of a superconducting state and a charge density wave (CDW) state. Slight gradual decreasing ~0.006% are observed in Knight shifts of ^{139}La and ^{29}Si-line C from 200K to 5K. The decreasing are much less than the previous ^{195}Pt-NMR spectra PeakD's shifts reported in T. Kubo *et al*, so the decreasing can be expected following ^{195}Pt-Knight shift decreasing. $LaPt_2Si_2$ has two types of Pt-sites: Pt(1)2a(Wycoff position) and Pt(2)2c. By carrying out single crystalline ^{195}Pt-NMR, the site assignments of ^{195}Pt-NMR spectra to each respective Pt sites are succeeded for both $H \| a$ and $\| c$. Associating for $1/T_1T$ results for polycrystalline powder reported by T. Kubo *et al*, Pt(1)-site may contributes to the CDW order below T^*. It is necessary to measure the temperature dependences of line1's $1/T_1T$ by using single crystalline $LaPt_2Si_2$.

1. Introduction

Recently some RPt_2Si_2-type compounds (R=rare-earth elements) have been attracted attention because RPt_2Si_2 exhibits various transitions by changing R-elements[1]. Especially among them, $LaPt_2Si_2$ exhibits successive phase transitions. Electric resistivity, magnetic susceptibility, X-ray diffraction (XRD) and selected area electron diffraction (SAED) measurements using polycrystalline sample show that $LaPt_2Si_2$ occurs a structural phase transition from tetragonal structure to orthorhombic one at $T^* \sim 112$K, and successively transitions to superconducting state at $T_c \sim 1.8$K[2]. It is reported that the structural phase transition at T^* is accompanied by the charge-density-wave (CDW) state with a nesting vector $\boldsymbol{q}_{CDW} = (1/3, 0, 0)$ or $(2/3, 0, 0)$. Therefore it has been discussed that superconductivity (SC) coexists with the CDW state below T_c in $LaPt_2Si_2$. Though the coexistence of SC and CDW was reported in various compounds such as transition metal dichalcogenides (2H-$NbSe_2$[3], 2H-$TaSe_2$[4], etc.) and APt_2X_2(A=La, Sr; X=As, Ge) compounds ($SrPt_2As_2$[5], $LaPt_2Ge_2$[6], etc.), the origin of the coexistence has been still unrevealed.

$LaPt_2Si_2$ has the quasi-two dimensional $CaBe_2Ge_2$-type structure with the space group P4/nmm and two types of Pt_2Si_2 layers as shown in Fig. 1, where a Pt(1)-layer consists of $Pt(1)Si(1)_4$ tetrahedra and a Pt(2)-layer, $Si(2)Pt(2)_4$ one (Table 1). I. Hase *et al* and S. Kim *et al* independently carried out first principle band structure calculations at tetragonal state[7][8]. The band calculations suggest that the Pt(1)-5d bands have main contribution to the electronic density of states at Fermi level ($N(E_F)$). So they predict that the CDW occurs in Pt(1)-layers. However there are no experimental evidences to confirm the prediction. It has been required to perform separable measurements so as to resolve two kinds of Pt_2Si_2 layers. Then we carried out nuclear magnetic resonance (NMR) measurements which are able to measure local $N(E_F)$ on atomic site. In our previous study, we measured ^{139}La and ^{195}Pt-NMR for polycrystalline powdered sample[9]. The ^{195}Pt-NMR line profile is distinct from the typical powder pattern spectrum for $I = 1/2$ and consists of four peaks originating from the Pt(1) and Pt(2) site. However, the line width is too broad to assign ^{195}Pt NMR lines to their respective Pt sites, so that the assignment of the ^{195}Pt NMR lines is still insufficient.

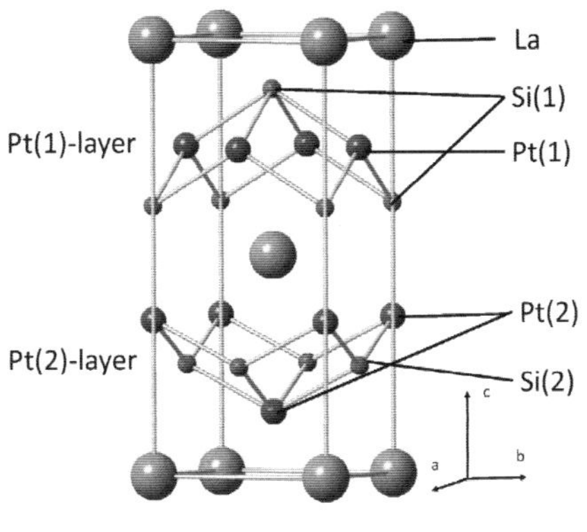

Figure 1. Structure of $LaPt_2Si_2$.

Table 1. Local inversion symmetries at each atomic sites.

atomic site	local symmetry	Wycoff position
Pt(1)	$\bar{4}$m2	2b
Si(1)	4mm	2c
Pt(2)	4mm	2c
Si(2)	$\bar{4}$m2	2a
La	4mm	2c

2. Experimental

The polycrystalline sample was prepared by arc-melting stoichiometric amounts of the constituent elements on a water-cooled copper hearth in an argon atmosphere at Kyushu University. The single crystalline sample was grown by a tetra-arc Czochralski Furnace at Niigata University. The dimensions of the single crystal sample are $a \times b \times c \approx 3 \times 3 \times 7$ mm^3. We carried out ^{139}La(I=7/2, γ_n=6.0146MHz/T)-, ^{29}Si(I=1/2, γ_n=8.4577MHz/T)- and ^{195}Pt(I=1/2, γ_n=9.094MHz/T)-NMR measurements by using a single crystalline sample and polycrystalline powder sample. NMR measurements were carried out by using a conventional pulsed spectrometer and a 9 T superconducting magnet in the temperature range of 5K-200 K. Knight shift was extracted from the frequency of each resonance lines ω_{res} by using the relation of $K = (\omega_{res} - \omega_0)/\omega_0$, where ω_0 is a reference frequency defined by $\omega_0 = \gamma_n H$ in the external magnetic field H. In simple metals showing weak electron-electron interactions, Knight shift is proportional to $N(E_F)$. When a CDW order occurs, Knight shift is expected to decrease below the CDW transition temperature because of the formation of the energy gap at Fermi level due to the CDW.

3. Results and Discussions

Figure 2(a) shows the central line (m=1/2↔-1/2) of ^{139}La-NMR spectra for the single crystalline sample for $H \| c$ in order to investigate the peak shift associated with the reduction of the $N(E_F)$ due to the CDW transition. ^{139}La-satellite lines cannot be detected because of the weak signal to noise ratio for the single crystal sample. No clear change was observed for the line profile passing

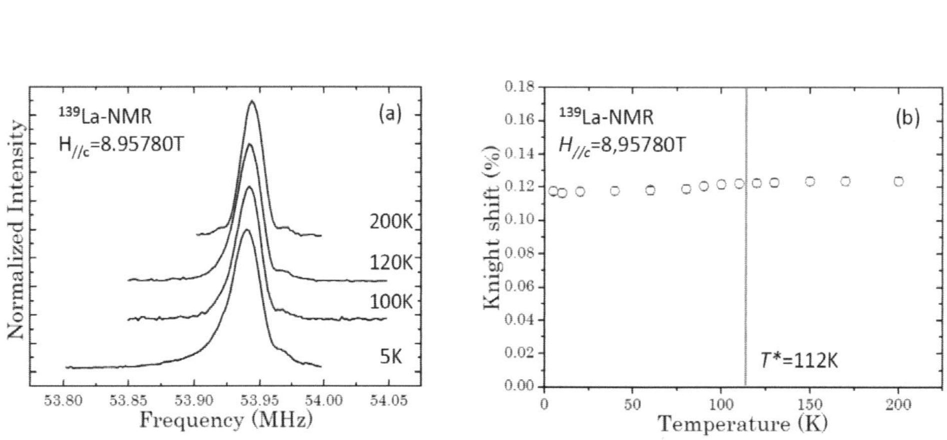

Figure 2. (a)^{139}La-NMR spectra by using single crystal oriented H//c. (b)Temperature dependences of ^{139}La-Knight shift.

through T^*. Figure 2(b) displays the temperature dependence of the ^{139}La-NMR Knight shift (^{139}K), which is determined from the peak positions at each temperature. A slight reduction, $\Delta^{139}K \sim 0.006\%$, is observed below T^*. This reduction of ^{139}K is much smaller than one of ^{195}Pt-NMR shift reported by T. Kubo *et al.* Since the Knight shift relates to the local density of states at Fermi level, this result implies the La-layer is hardly affected by the CDW transition. In this context, the CDW transition in $LaPt_2Si_2$ occurs in the Pt_2Si_2-layers. Therefore, ^{195}Pt- and ^{29}Si-NMR is expected to bring about important information on the CDW transition in $LaPt_2Si_2$.

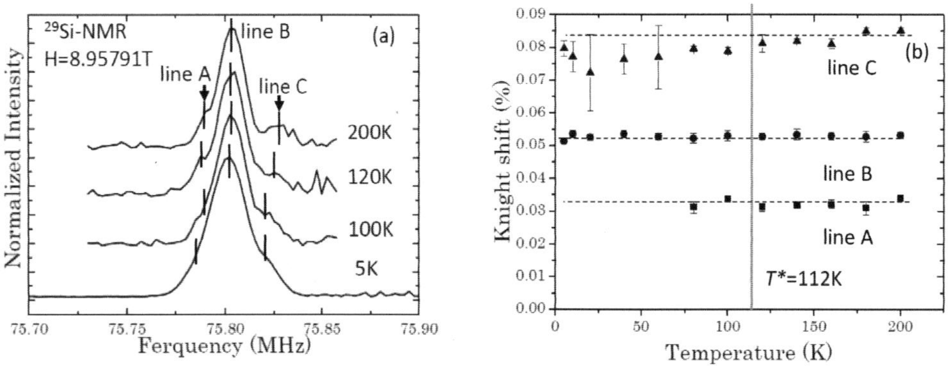

Figure 3. (a)^{29}Si-NMR spectra by using polycrystalline powder. (b)Temperature dependences of ^{29}Si-Knight shifts for line A, B and C.

Figure 3(a) shows the temperature dependence of ^{29}Si-NMR spectra for the polycrystalline powder sample. Although ^{29}Si-NMR spectrum for the single crystal sample can be observed at 5 K, the signal to noise ratio is too weak to detect ^{29}Si-NMR signals at hight temperatures above 100 K. Therefore we measured the temperature dependence of ^{29}Si-NMR spectra for the c-axis oriented powder sample. At high temperature region, we found that the ^{29}Si-NMR line profile can be fitted by at least three Gaussian lines of line A, line B and line C as show in the Fig. 3(a). With decreasing temperature, the line profile does not change but the line width of each line slightly broadens. Temperature dependences of the ^{29}Si-NMR Knight shifts (^{29}K) from the frequencies of line A, B and C are shown in Fig. 3(b), here the ^{29}K of the line A can not be detected below 60K because of the overlap of line A with the other lines. While the temperature dependences of ^{29}K for line A and B stay constant within the experimental accuracy with decreasing temperature, ^{29}K for the line C decreases gradually by $\Delta^{29}K \sim 0.006\%$. However this reduction is same order as that of ^{139}K. These results suggest that the local density of states at Si-sites are small and it is difficult to extract the characteristic feature for the CDW transition.

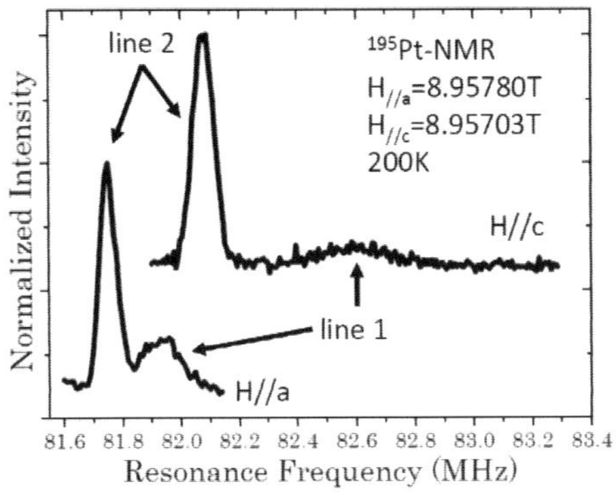

Figure 4. ^{195}Pt-NMR spectra at 200K for single crystal at $H \| a$ and $\| c$.

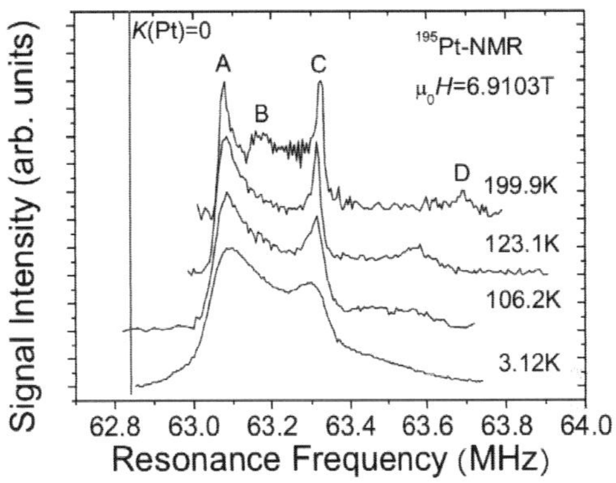

Figure 5. Temperature dependences ^{195}Pt-NMR spectra for polycrystalline powder[9].

Figure 4 displays ^{195}Pt-NMR spectra for the single crystal sample for $H \| a$ and $\| c$ measured at 200 K. The spectrum for each field directions is consists of two lines with a sharp line (line 2) and a broad line (line 1). In the previous paper, we could not deconvolute the characteristic line for two kinds of Pt sites as shown Fig. 5. By comparing line profile in Fig. 4 with Fig. 5, we interpret that Peak A and C in Fig. 5 correspond to line 2 in Fig. 4 and Peak B and D, to line 1. From $1/T_1T$ results by using polycrystalline powder[9] and band calculations[7][8], we can deconvolute line 1 arises from the Pt(1)-site and line 2, from Pt(2)-site in the previous paper[9]. The temperature dependence of $1/T_1T$ at Peak D could be measured at only high temperature region above 112K so that it is necessary to measure the temperature dependences of Knight shift and $1/T_1T$ for line 1 for single crystal sample investigate the temperature dependence of $N(E_F)$ in detail.

4. Conclusions

We performed ^{139}La-, ^{29}Si- and ^{195}Pt-NMR measurements by using single crystal and polycrystalline powder $LaPt_2Si_2$. From the ^{139}La- and ^{29}Si-NMR results, La- and Si-sites are nearly simple metallic behavior at 5K-200K. From ^{195}Pt-NMR spectra by using single crystal, we can deconvolute two Pt-sites as is expected from the crystal structure: line 1 corresponding to Pt(1)-site and line 2, Pt(2)-site. For future studies, we will measure the temperature dependences of Knight shift and $1/T_1T$ at around T^* by using single crystal in order to investigate the temperature dependence of $N(E_F)$ for line 1 in detail. .

5. Acknowledgments

This work has been supported byJPSJ KAKENHI Grant numbers 15H05885(J-Physics), 15H05882(J-Physics), 15K21732(J-Physics), 26400359, 15H05745 and 15H03689.

IOP Conf. Series: Journal of Physics: Conf. Series **807** (2017) 062002 doi:10.1088/1742-6596/807/6/062002

[1] K. HIEBL and P. ROGL, *J. Magn. Magn. Mat.* **50** (1985) 39-48.
[2] Y. Nagano *et al.*, *J. Phys. Soc. Jpn.* **82** (2013) 064715.
[3] M. Nagano and S. Tanaka, *J. Phys. Soc. Jpn.* **51** (1982) 219-227.
[4] T. Kumakura *et al.*, *Chech. J. Phys.* **46** (1996). Suppl. S5.
[5] K. Kudo, Y. Nishikubo and M. Nohara, *J. Phys. Soc. Jpn.* **79** (2010) 123710.
[6] S. Maeda *et al.*, *Phys. Rev.* B **91**, 174516 (2015).
[7] I. Hase *et al* and T. Yanagisawa, *et al Physica* C **484** (2013) 59-61.
[8] S. Kim, K. Kim and B. I. Min, *Sci Rep.* 2015; 5: 15052.
[9] T. Kubo *et al.*, *JPS Conf. Proc.* **3**, 017031 (2014)

Change in crystal structure and physical properties of the Multiferroics $YMnO_3$ single crystals by Strong gravitational field

M Tokuda[1], M Weijian[1], S Hayami[2], A Yoshiasa[2], T Mashimo[1]

[1]Institute of Pulsed Power Science, Kumamoto University, Kumamoto 860-8555, Japan
[2]Faculty of Science, Kumamoto University, Kurokami 2-39-1, Kumamoto 860-8555, Japan

Abstract. Many researchers have studied the multiferroicity of the hexagonal $RMnO_3$ (R: rare-earth element) for both applications and fundamental studies. To investigate the relationship between the structure and physical properties of materials, some people apply the chemical pressure effect. The procedure of chemical pressure effect involves substituting rare-earth elements for ones which have a different ionic radius. Mashimo et al. have developed a high-temperature ultracentrifuge apparatus that can generate extended duration strong gravitational field in excess of 10^6 G under a wide range of temperatures (up to 500°C). Strong gravitational fields directly act on each atom as a different body force. This can cause the change in crystal structure. Thus, we subjected $YMnO_3$ single crystal to strong gravity experiments (0.78×10^6 G, 400°C, 2 h) and investigated the resulting changes in the crystal structure and physical properties of the gravity sample. The single crystal four-circle X-ray diffraction measurements revealed the change in the nearest neighboring Mn-Mn and M-O bond distances. The temperature dependence of magnetic susceptibility by SQUID showed the change in the magnetic anisotropy of gravity sample.

1. Introduction

$RMnO_3$ (R is rare earth) crystallizes in a hexagonal structure (Space group: $P6_3cm$) when the ionic radius of R is small (R = Ho-Lu, Y, and Sc). These compounds have been investigated due to showing the typical multiferroic effects [1,2]. In this hexagonal structure, each Mn^{3+} ions with $S = 2$ is surrounded by three in-plane and two apical oxide ions, and thus subject to a trigonal crystal field. These MnO_5 trigonal bipyramids are two-dimensionally connected with each other on their corners on *ab*-plane and each MnO_5 layers are separated by RO_7 layers [3]. Mn^{3+} ions forms two-dimensional triangular lattice in *ab*-plane with stacking along *c*-axis. This two-dimensional triangular lattice with an antiferromagnetic interaction is a good canonical example of a geometrical frustrated system. Actually, frustration parameter f which is defined as the ratio of $|\theta_{CW}|$ to T_N is very large (~ 10) compared with other triangular lattice compounds. The magnetic properties of $RMnO_3$ single crystals also have been studied [4, 5, 6]. Hexagonal $RMnO_3$ single crystals have a magnetic anisotropy, which causes the difference of susceptibility of parallel and perpendicular to *c*-axis. The susceptibility parallel to *c*-axis $\chi_c(T)$ shows the cusp of susceptibility curve at Néel temperature (T_N ~ 70 K); on the other hand, the susceptibility perpendicular to *c*-axis $\chi_{ab}(T)$ shows no cusp at Néel temperature. Curie-Weiss temperature θ_{CW} calculated from inverse susceptibility $1/\chi_{ab}(T)$ is about −700 K. The study of structural change for $YMnO_3$ has been investigated for fundamental understanding of physical properties. According to high-pressure study in temperature range 10-295 K, Kozlenko, et al., revealed that the triangular lattice of Mn ions at high-pressure enhances the effects of geometrical magnetic frustration [7]. By study of high-temperature and -pressure for $YMnO_3$, Waintal, et al., found orthorhombic $YMnO_3$, which is perovskite-type structure [8]. Park, et al. investigated the relationship between the structure and physical properties used by the chemical pressure effect on $Y_{1-x}Lu_xMnO_3$ [9]. Both Y^{3+} and Lu^{3+} ions are nonmagnetic elements therefore there is no magnetic contribution and no disturbing the Mn magnetic moments.

Mashimo et al. have developed a high-temperature ultracentrifuge apparatus that can generate extended duration strong gravitational field in excess of 10^6 G under a wide range of temperatures (up to 500°C) [10]. Gravity is a field state variable, whereas pressure and temperature are thermodynamic state variables. Thus, gravitational fields directly act on the respective atoms by body force in alloys and compounds, while pressure or temperature influence atoms isotopically. Under a strong gravitational field (~10^6 G), heavy atoms are displaced in the gravitational direction, while light atoms are shifted towards the opposite direction by the different body forces relative to their respective atomic weight in the crystal compound [11, 12]. Under a strong gravitational field, unique structural changes can be generated in compounds by the one-dimensional displacement of atoms caused by different body forces. Thus, the applying strong gravitational field is a clean way of exploring the relationship between the structure and physical properties.

In this work, we performed strong-gravity experiments on $YMnO_3$ single crystals to examine the changes in crystal structure and physical properties.

2. Experimental Procedure

$YMnO_3$ single crystal was grown by floating-zone method using the desktop-type single crystal growth equipment (Canon Machinery, Japan) at a feeding speed of 5 mm/h in air. A stoichiometric mixture of Y_2O_3(99.99%) and Mn_3O_4(99.9%) powder was ground and clacined three times at 1000°C in air for 12 h with an intermittent grinding procedure. Then resulting powder was cold-pressed under 400 kg cm^{-2} and sintered in air at 1400°C for 48 h. The direction of obtained crystal was confirmed by observation of a Laue X-ray backscattering photograph (TRY SE TRY-YGR) obtained using a W target and radiation at 35kV-15mA. We cut to the plate sample with 2×1×1 mm^3 whose plane (2×1 mm^2) is parallel to (0001) plane. The sample was set on a Si_3N_4 plate in a stainless steel SUS304 capsule, as shown in Fig 1. A strong-gravitational field was applied along the *c*-axis of the crystal where the maximum distance from the rotor axis was about 35.5 mm. The rotor was radiatively heated to 400°C by a follow carbon cylinder, which in turn was heated by a high-frequency heating system.

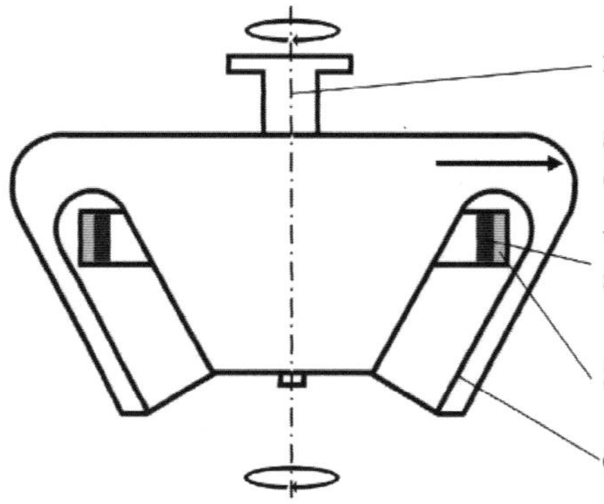

Figure 1. the schematic illustration of the rotor assembly. The sample was fixed on a Si_3N_4 plate. A gravitational field was applied along the

c-axis of the crystal where the maximum distance from the rotor axis was about 35.5 mm.

The duration of each experiment was 2 h. At the end of the experiment, the sample was quenched down to room temperature, and then strong gravitational field was removed. The rotor was radiatively heated to 400°C by a follow carbon cylinder, which in turn was heated by a high-frequency heating system. The duration of each experiment was 2 h. At the end of the experiment, the sample was quenched down to room temperature, and then strong gravitational field was removed.

The magnetic properties were measured using a superconducting quantum interference devise magnetometer (Magnetic Property Measurement System MPMS-XL, Quantum Design) under magnetic field of 0.1 T and temperature range between 5 and 300 K.

Single crystal X-ray diffraction measurements for initial and gravity samples were performed with Rigaku AFC-7R automated four-circle diffractometer. The radiation source Mo $K\alpha$ (λ = 0.7102 Å) is near the absorption K edge of Y atom (~ 0.7277 Å) therefore $YMnO_3$ compounds have a large linear absorption coefficient μ = 28.118 mm^{-1}. To carry out the accurate structure refinement, we should compute the absorption coefficient $\mu \cdot r$, where r is the radius of sample shaped into a sphere. Structure refinements were carried out by minimizing the function using Shelx97 program (Sheldrick, 1997).

3. Results and Discussion

We examined the crystal structures of initial and gravity samples at room temperature. The least-squares calculation for initial and gravity samples was converged at R_1 (I > 2.0σ(I)) 2.28, and 2.53%, respectively. The obtained crystal structure for initial sample is in agreement with previous reported [13]. The selected bond distances are summarized in Table 1. In all hexagonal $RMnO_3$ compounds, the two apical bond distances (Mn-O1 and Mn-O2) are shorter than the three equatorial bond distances (Mn-O3 and Mn-O4). The difference in apical bond distances for gravity sample increased. According to previous high-pressure study, the difference in apical bond distances increases with increasing pressure [8]. By analogy with this, the increase in difference in apical bond distances was caused by compressed uniaxially. The Mn-O bond distance in *ab*-plane splits the inequivalent bond distances Mn-O3 and Mn-O4. There are two ways of the direction of Mn trimerization, namely around O3 site and O4 site. The Mn-O bond distances in *ab*-plane of initial and gravity sample has a significant difference. It means that Mn trimerization changed after strong -gravity experiment.

Table 1. Results of least-squares calculation lattice parameters, and selected bond distances for initial and gravity samples.

	initial	gravity
R1 and S values		
*R*1 (I > 2.0σ(I))	2.28	2.53
S	1.018	1.015
lattice parameter		
a (Å)	6.1352(4)	6.1421(4)
c (Å)	11.4137(8)	11.4160(10)
MnO_5 bipyramid		
Mn-O1 (Å)	1.869(6)	1.886(14)
Mn-O2 (Å)	1.865(5)	1.854(11)
Mn-O3 (Å)	2.0631(15)	2.053(4)
Mn-O4 (Å) [× 2]	2.0542(9)	2.0636(16)

IOP Conf. Series: Journal of Physics: Conf. Series **807** (2017) 072001 doi:10.1088/1742-6596/807/7/072001

mean Mn-O (Å)	1.9811(16)	1.9840(37)

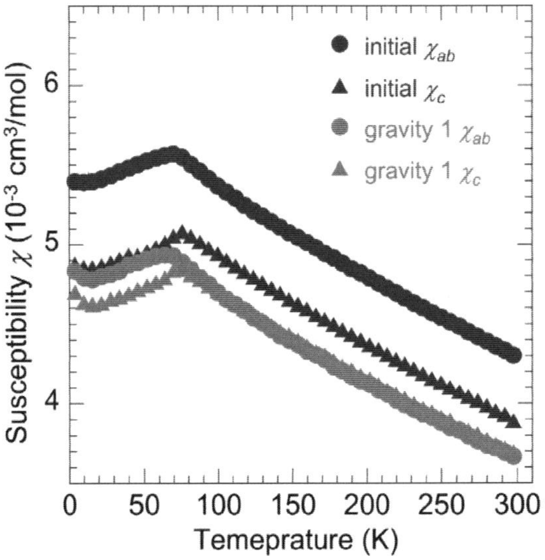

Figure 2. Temperature dependence of magnetic susceptibility of initial and gravity samples. The closed circle in blue and red represent the susceptibility with magnetic field perpendicular to *c*-axis (χ_{ab}) for initial and gravity samples, respectively. The closed triangle in blue and red represent the susceptibility with magnetic field parallel to *c*-axis (χ_{ab}) for initial and gravity samples, respectively.

Figure 3 shows temperature dependence of magnetic susceptibility of initial and gravity samples. In $YMnO_3$, the susceptibility of $\chi_c(T)$ and $\chi_{ab}(T)$ are parallel to and perpendicular to *c*-axis, respectively. The susceptibility of initial sample is consistent with previous reports, which state that $\chi_{ab}(T)$ is larger than $\chi_c(T)$ in whole temperature range [5, 6]. However, the susceptibility after gravity experiment is different from the one of initial and previous reports. The $\chi_c(T)$ and $\chi_{ab}(T)$ of gravity sample is close to each other. In other ward, the anisotropy of $\chi(T)$ for gravity sample is suppressed. We should note that the anomaly at T_N still remain in gravity samples. The Curie-Weiss temperature θ_{CW} is derived from the inverse susceptibility $1/\chi$. The obtained θ_{CW} of initial and gravity samples were −708 and −602 K, respectively. The frustration parameter *f* of initial and gravity samples were 9.3 and 8.2, respectively. To estimate the value of nearest neighboring interaction *J*, we use the theoretical Curie-Weiss temperature using a high-temperature expansion of spin susceptibility. The obtained *J* of initial and gravity samples were 2.5 and 2.2 meV, respectively, which is consistent with the theoretical calculation of recently reported [9].

Now, we consider that the correlation with Mn-O bond length in *ab*-plane and Curie-Weiss temperature θ_{CW}. Above mentioned before, The Mn-O bond distance in *ab*-plane splits the inequivalent bond distances Mn-O3 and Mn-O4. This leads to two different magnetic exchange interactions between the neighboring Mn^{3+} ions in *ab*-plane [14] and to relieve a geometric frustration in the system. We note that observed Néel temperature T_N are ~ 72 K for initial and gravity samples. Therefore the ratio of Mn-O3 and Mn-O4 bond distances in room temperature range is the major factor controlling the Curie-Weiss temperature $|\theta_{CW}|$ and frustration parameter *f*. Mn trimers of initial connect to each other organically over *ab*-plane, which means the strong two-dimensional character.

On the other hand, each Mn trimers of gravity sample are isolated. We assume that it leads to change in the susceptibility $\chi_c(T)$ and $\chi_{ab}(T)$ and Curie-Weiss temperature $|\theta_{\mathrm{CW}}|$ for gravity sample.

4. Acknowledgement

The authors would like to thank N. Saigo, graduated school of Science and Technology Kumamoto University, for providing us details of techniques about measurement of the temperature dependence of magnetic susceptibility.

5. References

[1] Cheong S.-W and Mostovoy M 2007 *Nature Mater.* **6** 13
[2] Lee S, Pirogov A, Kang M, Jang K. H, Yonemura M, Kamiyama T, Cheong S. W, Gozzo F, Shin N, Kimura H 2008 *Nature* **451** 805
[3] Yakel H. L, Koehler W. C, Bertaut E. F, and Forrat E. F 1963 *Acta Crystallogr.* **16** 957
[4] Van Aken B. B, Bos J. W. G, de Groot R. A, and Palstra T. T. M 2001 *Phys. Rev.* B **63** 125127
[5] Katsufuji T, Mori S, Masaki, Moritomo Y, Yamamoto N, and Takagi H 2001 *Phys. Rev.* B **64** 104419
[6] Aikawa Y, Katsufuji T, Arima T, and Kato K 2005 *Phys. Rev.* B **71** 184418
[7] Kozlenko D. P, Kichanov S. E. Lee S, Park J.-G, Glazkov V. P, and Savenko B. N 2005 *JETP Letters,* **82** 193
[8] Waintal A and Chenavas J 1967 *C. R. Seances Acad. Sci. Ser.* B **264**, 168
[9] Park J, Lee S, Kang M, Jang K.-H, Lee C, Streltsov S. V, Mazurenko V. V, Valentyuk M. V, Medvedeva J. E, Kamiyama T, and Park J.-G 2010 *Phys. Rev.* B **82** 054428
[10] Mashimo T, Huang X. S, Osakabe T, Ono M, Nishibara M, Ihara H, Sueyoshi M, Shibasaki K, Shibasaki S, Mori N 2003 *Rev. Sci. Inst.r* **74** 160
[11] Mashimo T 1988 *Phys. Rev.* A **38** 4149
[12] Mashimo T, Ikeda T, Minato I 2001 *J. Appl. Phys.* **90** 741
[13] Van Aken B. B, Palstra T. T. M, Filippetti A, and Spaldin N. A 2004 *Nat. Mater* **3** 164
[14] Sato T. J, Lee S. H, Katsufuji T, Masaki M, Park S, Copley J. R. D, Takagi H 2003 *Phys. Rev.* B **68** 014432

Spin polarization measurements in ferromagnetic $SrRuO_3$ using point-contact Andreev reflection technique

Masanobu Shiga[1], Naoto Nishimura[1], Yuji Inagaki[1], Tatsuya Kawae[1], Hiroshi Kambara[2] and Kenichi Tenya[2]

[1]*Department of Applied Quantum Physics, Faculty of Engineering, Kyushu University, Motooka, Fukuoka 812-8581, Japan*

[2]*Faculty of Education, Shinshu University, Nagano, 380-8544, Japan*

E-mail: 2TE15295Y@s.kyushu-u.ac.jp

Abstract. Using the point-contact Andreev reflection (PCAR) technique, we have performed spin polarization measurements in a polycrystalline $SrRuO_3$ to study the difference of the spin polarization between the ballistic and diffusive transport samples. PCAR spectra were measured with Nb tips for $T = 4.2$ K with changing the contact area mechanically. The results were well fitted by the modified Blonder-Tinkham-Klapwijk (mBTK) model. We estimate the spin polarization $P \approx 0.59$, which is slightly larger than that of a single crystal film with $P \approx 0.51$ due to the change of transport property.

1. Introduction

After the discovery of the giant magnetoresistance (GMR), intensive efforts have been devoted to fabricate spintronics devices. Since magnetoresistance is sensitive to the spin polarization of ferromagnetic electrode, studying the spin-polarized transport properties of ferromagnetic materials is crucial for the development of spintronics devices. Spin polarization P_n is defined by difference between minority-spin and majority-spin bands at the Fermi level E_F. Hence, it is possible to define P_n by the following equation [1],

$$P_n = \frac{<N_\uparrow(E_F)v_{F\uparrow}^n> - <N_\downarrow(E_F)v_{F\downarrow}^n>}{<N_\uparrow(E_F)v_{F\uparrow}^n> + <N_\downarrow(E_F)v_{F\downarrow}^n>}, \tag{1}$$

where $N_\uparrow(E_F)$ and $N_\downarrow(E_F)$ are the density of states (DOS) of majority and minority spins at the Fermi level, and $v_{F\uparrow}$ and $v_{F\downarrow}$ are the Fermi velocity, respectively. Spin-resolved photoemission experiments measure the spin polarization P_0, which is determined only by the difference of DOS between the majority and minority spins at the Fermi level. On the other hand, transport experiments measure a different spin polarization including the Fermi velocities. When the contact size d at the surface is smaller than mean free path L ($d << L$), electron transport is ballistic. In this case, the DOS is weighted linearly with v_F ,and P_1 ($n = 1$) is measured. In diffusive electron transport ($L < d$), the weighting is quadratic in v_F ($n = 2$) and P_2 is measured.

Therefore, the spin polarization P_n depends on the transport properties of samples; if $v_{F\uparrow}^n$ is larger than $v_{F\downarrow}^n$, P_2 is larger than P_1.

Owing to the structural simplicity and remarkable chemical stability, $SrRuO_3$ is a good candidate to investigate the transport spin polarization P_n including the Fermi velocities. $SrRuO_3$ has a ferromagnetic transition temperature at $Tc = 160$ K and a saturation magnetization of 1.6 μ_B [2]. To measure the spin polarization, the point contact Andreev reflection (PCAR) technique has been used because the $I-V$ characteristics due to Andreev reflection at a ferromagnet-superconductor interface changes significantly depending on the spin polarization of the ferromagnet.

In this paper, we report the spin polarization of a polycrystalline $SrRuO_3$ in the diffusive transport region with changing the contact area mechanically using PCAR technique where Nb is utilized for a probe tip. By fitting the experimental results to the modified Blonder-Tinkham-Klapwijk (mBTK) model [3], the spin polarization is estimated to be $P \approx 0.59$, which is slightly larger than that of a single crystal film with $P \approx 0.51$ [4].

2. Experiments

The polycrystalline sample of $SrRuO_3$ is synthesized by the conventional solid-state reaction method. The powders of $SrCO_3$ and RuO_2 with stoichiometric compositions were carefully mixed and calcined in air at 1100°C for 24 hours. After careful mixing of the calcined samples, they were shaped into pellets and then sintered at 1100 °C for 24 hours [5].

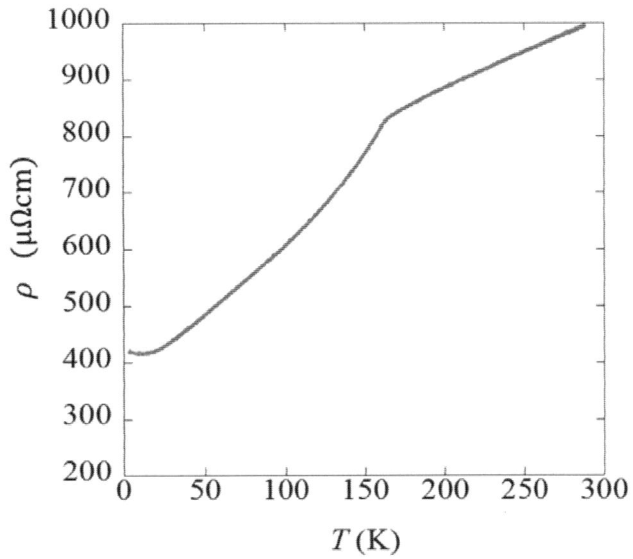

Fig. 1. Temperature dependence of electrical resistivity ρ for polycrystalline $SrRuO_3$.

The conductance versus voltage (*G*-*V*) characteristics of the Nb-$SrRuO_3$ point contact were measured at 4.2 K in vacuum with conventional lock-in technique with modulation frequency of 2 kHz. The Nb tip was fabricated by mechanically polishing a wire with the diameter of 0.2 mm. The sample was mounted in ^{4}He cryostat after polishing its surface. The point contact was prepared by using a stacked piezo actuator, enabling the precise control of the contact size. The spin polarization P_n, interface scattering parameter Z and superconducting energy gap Δ are obtained from fitting the differential conductance d*I*/d*V* to the mBTK function.

3. Results and discussion

Fig. 1 shows the temperature dependence of electrical resistivity ρ for the polycrystalline $SrRuO_3$. An anomaly due to the ferromagnetic transition is visible at $T \sim 160$ K. The residual resistivity of ~ 420 μΩcm is about 8 times larger than that in a single crystal, suggesting the existence of many grain boundaries.

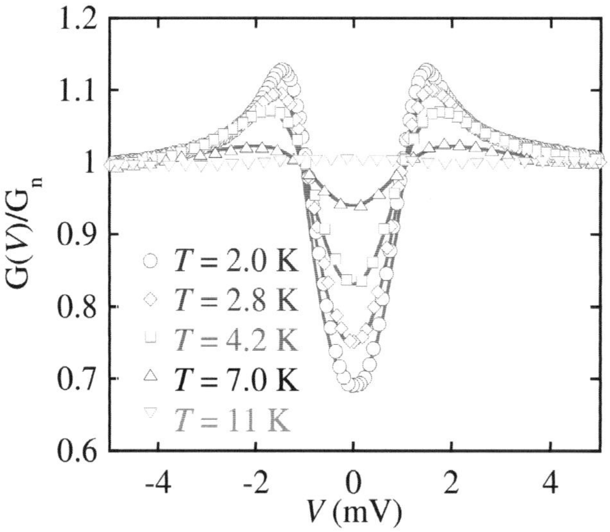

Fig. 2. The temperature dependence of the differential conductance *dI*/*dV* (red line) at the resistance of 12 Ω. The solid lines show fittings by mBTK model.

Fig. 2 illustrates the temperature dependence of the differential conductance spectra *G*(*V*), which is normalized by the conductance at the normal region (G_n). Here, the contact resistance at the interface is controlled to be the same for all the measurements. When the temperature is higher than the Nb superconducting transition temperature $T_C = 9.5$ K, the spectra shows no anomaly. In contrast, as the temperature is decreased, a dip at around the zero bias voltage is grown rapidly because of the increase of the superconducting energy gap at E_F. It is seen that the anomaly is well reproduced by the fitting curve calculated by the mBTK function. This indicates that the anomaly is understood by the Andreev reflection at the superconductor/ ferromagnet interface.

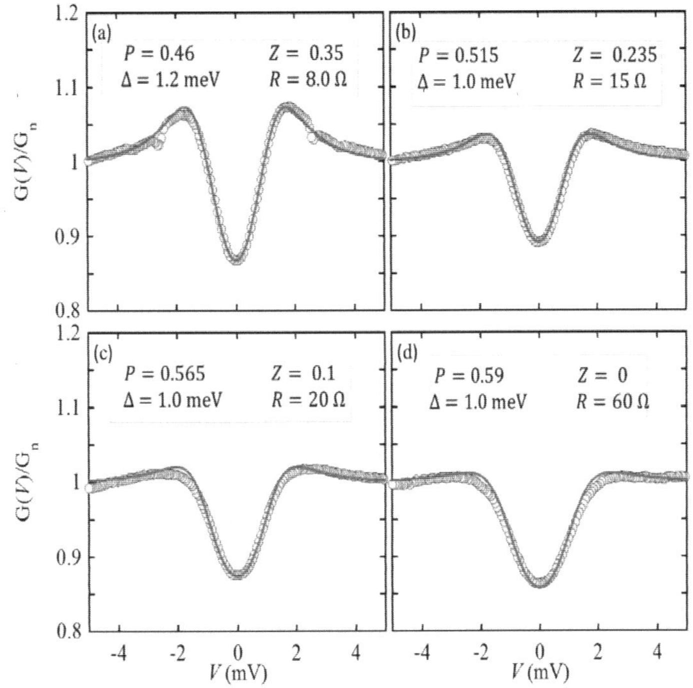

Fig. 3. The representative spectra of differential conductance at T = 4.2 K. The spectra are measured with increasing the resistance at the interface without breaking the contact after preparing it. The solid line shows fitting by mBTK model.

We measure the evolution of the differential conductance spectra with increasing the resistivity at the interface, which corresponds to decrease of the contact size. It is important that the measurements are performed without breaking the contact after preparing it, which suppresses the scattering of data. We depict the representative spectra at T = 4.2 K for various resistances in Fig. 3(a)-3(d), where the fitting curve by the mBTK function and the parameters for the fitting are also shown. The superconducting gap of bulk Nb is 1.5 meV, which is larger than that estimated from the present experiments. The gap is suppressed by magnetic scattering at the superconductor/ferromagnet interface. Moreover, the scattering parameter Z increases with decreasing the resistance R at the contact as shown in the inset of Fig. 4. This implies that the effect of ferromagnetic scattering, which suppresses the superconducting current, is increased with increasing the contact size.

We plot P_n versus Z in Fig. 4, indicating that P_n is suppressed with increasing Z [6]. From the results, the value of spin polarization at $Z = 0$ is evaluated to be $P_n \approx 0.59$. The spin polarization of a single-crystal thin film, which is measured in the ballistic transport region, is reported to be $P_l = 0.51$ at $T = 4.2$ K [4]. The spin polarization in the present experiments is slightly larger than the single-crystal thin film. The origin is likely due to the difference of transport properties between the samples as follows.

The electrical resistivity of the present sample is $\rho \sim 420$ μΩcm at $T = 4.2$ K as shown in Fig. 1. We can estimate the mean free path from the electrical resistivity ρ to be $L \sim 0.8$ and 0.45 nm for minority and majority carriers, respectively [7]. Furthermore, the contact resistance R_n is approximated by the following equation,

$$R_n \approx \frac{4}{3\pi}\frac{\rho L}{d^2} + \frac{\rho}{2d}, \tag{2}$$

where ρ is electrical resistivity [8]. When the contact is assumed to have a circular shape, the contact diameter of this sample is larger than several tens of nanometers, e.g., $d \sim 36$ nm at $R = 60\ \Omega$, representing that the measurements are done in the diffusive transport ($L < d$) region. Thus, the spin polarization in the present experiments should be larger than that in the single-crystal as mentioned in Sec. 1 and we can conclude that the effect of the Fermi velocity in P_n is different between the ballistic and diffusive transports for the Nb-$SrRuO_3$ interface. The similar results are found in the spin polarization of $La_{0.7}Sr_{0.3}MnO_3$ [9].

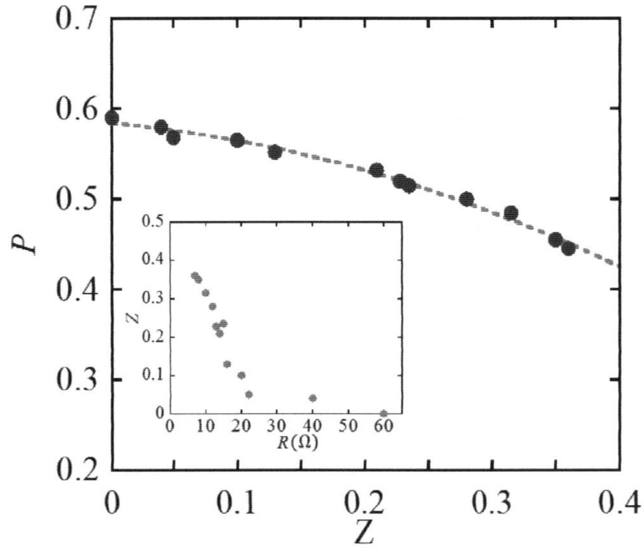

Fig. 4. Spin polarization P as a function of interface scattering parameter Z. Dotted lines were extrapolated to each data by a quadratic function. Inset: R dependence of the interface scattering parameter Z.

4. Conclusion

We have studied the spin polarization of a polycrystalline $SrRuO_3$ using point-contact Andreev reflection technique. The spin polarization P_n at $T = 4.2$ K is obtained as a function of the interface scattering parameter Z, giving $P_n \approx 0.59$ at $Z = 0$. This value is larger than that in the single-crystal, which can be understood by the difference of the transport property between the ballistic and diffusive transports for the Nb-$SrRuO_3$ interface.

5. Acknowledgements

This work was supported by a Grant-in-Aid for Scientific Research (Grant Nos. 25220605, 25287076, and 26600102) from the Japan Society for the Promotion of Science.

References

[1] Mazin I I 1999 *Phys. Rev. Lett.* **83** 1427
[2] Gao G, McCall S, Shepard M, Crow E J and Guertin P R 1997 *Phys. Rev.* B **56** 321
[3] Mazin I I, Golubov A A and Nadogorny B 2001 *J. Appl. Phys.* **89** 7576
[4] Raychaudhuri P, MackenZie P A, Reiner W J and Beasley R M 2003 *Phys. Rev.* B **67** 020411(R)
[5] Kawasaki I, Yokoyama M, Nakano S, Fujimura K, Netsu N, Kawanaka H and Tenya K 2014 *J. Phys. Soc. Jpn.* **83** 0647712
[6] Ji Y, Strijkers J G, Yang Y F, Chien L C, Byers M J, Anguelouch A, Xiao Gang and Gupta A 2001 *Phys. Rev. Lett.* **86** 5585
[7] Nadgory B, Osofsky S M, Singh J D Woods T G, Soulen J R, Jr, Lee K M and Eom B C 2003 *Appl. Phys. Lett.* **82** 427
[8] Nikolic B and Allen B P 1999 *Phys. Rev.* B **60** 3963
[9] Nadogorny B, Mazin I I, Osofsky M, Soulen J R, Jr, Broussard P, stroud M R, Singh J D and Harris G V 2001 *Phys. Rev.* B **63** 184433

Kondo-Fano resonance in atomic-scale contacts for ferromagnetic metals

MS Islam[1], H Takata[1], Y Ueno[1], K Ienaga[2], Y Inagaki[1], H Tsujii[3] and T Kawae[1]

[1]Department of Applied Quantum Physics, Kyushu University, Fukuoka, Japan

[2]Department of Physics, Tokyo Institute of Technology, Tokyo, Japan

[3]Division of Science Education, Kanazawa University, Kanazawa, Japan

E-mail: sislamru@gmail.com

Abstract. The electrical conductance of cobalt (Co) nano-scale contacts prepared by a Mechanically Controllable Break Junction (MCBJ) technique has been studied to understand the origin of Fano resonance in ferromagnetic atomic-sized contacts. In an atomic-scale contacts, the zero-bias anomaly is well-fitted by the Fano formula. The characteristic temperature, which is introduced as the fitting parameter of that formula, exhibits the log-normal distribution. These indicate that the anomaly is likely caused by Kondo effect. Moreover, the zero-bias anomaly is observed in large scale contacts where the bulk ferromagnetic properties should be retained. This suggests the coexistence of Kondo effect and ferromagnetism.

Keywords: Kondo effect, Fano resonance, MCBJ, Electrical properties

1. Introduction

The Kondo effect is understood as the quenching of a magnetic impurity by the bath of conduction electrons and forming a singlet ground state [1] and is observed in a variety of systems ranging from bulk magnetic materials to nanosystems such as single magnetic adatoms on nonmagnetic surfaces, quantum dots, single-electron and -molecule transistors [2-7]. In nanosystems, the Kondo resonance has been observed directly through the Fano anomalies using spectroscopy measurements [8].

During the last few decades, a lot of sustainable effort has been devoted to form and analyze atomic-sized contacts in 3*d* ferromagnetic materials theoretically and experimentally to study the magnetoresistance at the domain wall in nano-sized constrictions. Calvo *et al.* first recorded a zero-bias anomaly in atomic-scale contacts of ferromagnets such as Fe, Co and Ni prepared by scanning tunneling microscopy (STM) and electromigration technique. It is surprising that the anomaly is well-reproduced by the Fano resonance and the amplitude shows logarithmic temperature dependence because these suggest that the anomaly is caused by the Kondo resonance [9]. In addition, Ienaga *et al.* observed a zero-bias anomaly due to the Fano resonance in Ni nanoconstrictions made by a MCBJ technique and showed the anomaly retains in large scale constrictions exhibiting the ferromagnetic properties [10].

In the present study, we measure the electrical conductance of Co constrictions with changing its size by a MCBJ technique to examine the size dependence of the Fano resonance in Co atomic-scale contact. A zero-bias anomaly is also found in Co atomic-scale contacts as well as in large constrictions like Ni nanoconstrictions [10].

2. Experimental details

The sample wire, a ferromagnetic Co wire, which has a typical diameter of 0.25 mm with a purity of 99.99%, was intimately fixed on the top of the flexible phosphor-bronze substrate after making notch at the center by a commercial blade under a microscope and mounted in a triple point bending configuration in a vacuum chamber. After cooling at liquid helium temperature, the sample wire was stretched by bending the substrate mechanically.

Nanometer-scale contacts of Co were made by employing a homemade MCBJ technique [11]. A piezo element was then used to control the constriction size (a few 100 nm ~ atomic-sized) precisely. Thereafter, we have measured the spectra of differential conductance d*I*/d*V* as a function of the bias voltage using a lock-in technique with a frequency of 1 kHz. The whole experiment was performed at cryogenic environment, enabling to keep a stable contact for a long time by diminishing the thermal fluctuation. Besides, any contaminants at the contact are prevented in a cryogenic vacuum environment.

3. Results and Analysis

By the repetition of making and breaking the contacts around two thousand times in a controlled way, the conductance histogram is built to study the atomic contacts as illustrated in Fig. 1(a). The position of the first peak at around $1.6G_0$ corresponds to the conductance of a single-atom contact, where $G_0 = 2e^2/h$ is the quantized unit of electrical conductance. The result is good agreement with previous theory and experiment done under similar conditions [9,12]. Further, the conductance histogram shows no fractional quantization which observed in atomic contacts of pure magnetic (Fe, Co and Ni) and nonmagnetic (Pt) metals by C Untiedt *et al.* [13].

To confirm the emergence of zero-bias anomaly in Co atomic-scale contacts reported by Calvo *et al.*, we have measured the zero-bias conductance and d*I*/d*V* spectra for hundreds of atomic-scale contacts from $5G_0$ to the final stage of breaking such as those shown in Fig. 1(b) and (c), respectively. It is significant that most of the spectra show Fano-like features such as dip, tip or asymmetric shapes at around zero bias. Similar features are observed in spectroscopy measurements for single magnetic adatoms on nonmagnetic surfaces [7], in molecular devices [14] and in Ni nanoconstrictions [10]. As the constriction is stretched, the zero-bias anomaly changes gradually in the same conductance plateau, while the shape of the d*I*/d*V* spectra changes remarkably at the jump of the conductance. Generally, the stretching induces a shift of the electron energy level in the contact due to a change of the overlap of the wave function, implying that the motion of the electrodes gives rise to a similar behavior as that observed in the gate-voltage effect in quantum dot and molecular device systems [3,14].

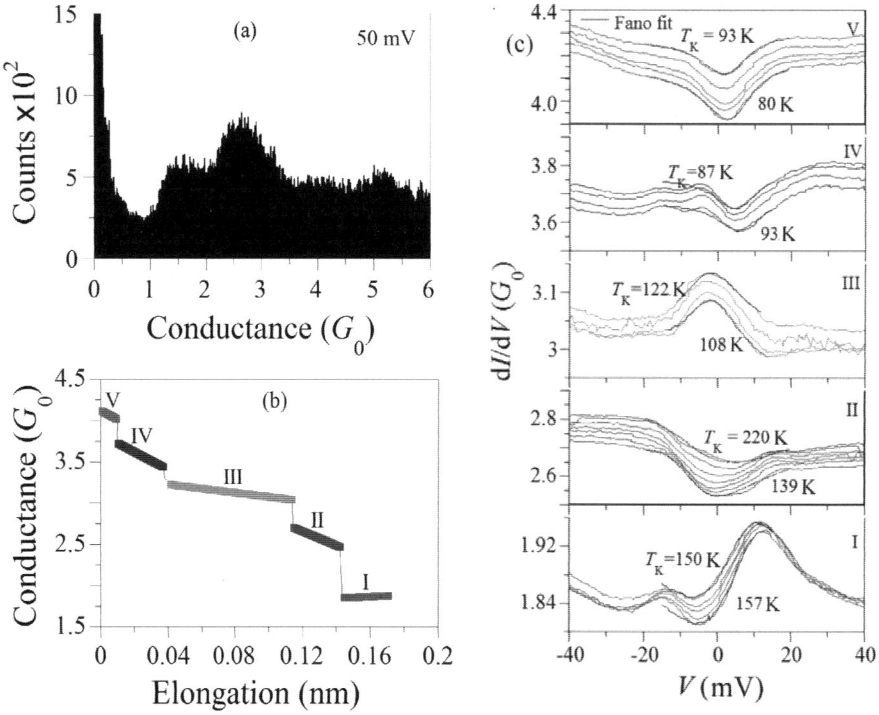

Fig. 1. (colour online) (a) The histogram depicted for around two thousands of conductance traces at 4.2 K; (b) Illustration of the zero-bias conductance for a monotonic elongation of Co nanocontact until the final stage of breaking; (c) Corresponding d*I*/d*V* spectra measured as a function of bias voltage according to the five regions observed in (b). The black lines obtained by fitting with the Fano formula (1).

It is well-known that a Fano resonance originates from the quantum interference between a discrete state and the continuum which can be represented as a mixing of two different configurations [8]. To analyze such resonance, it is important to fit the low bias conductance ($|eV| \leq \Gamma_K$) by the Fano formula:

$$G(V) = \frac{1}{G_0}\frac{dI}{dV} = g_{\mathrm{off}} + \frac{A}{1+q^2}\frac{(q+\varepsilon)^2}{1+\varepsilon^2} \qquad (1),$$

where $\varepsilon = (eV - \varepsilon_K)/k_B T_K$ is the bias shifted with respect to the center of the resonance ε_K and normalized by the natural width of the resonance $k_B T_K$; k_B is the Boltzmann's constant, T_K is the fitting parameter; q is the dimensionless Fano parameter that determines the shape of the Fano resonance and A is the amplitude of such resonance. By fitting these spectra at around $V = 0$ mV, we obtained the variation of T_K from 80 to 220 K that is almost similar agreement in the previous work performed by STM [7,9].

Additionally, the statistical analysis of the data is effective to obtain the origin of T_K for the resonances in atomic-scale contacts. We have constructed the histogram of T_K from fittings of more than two hundreds of atomic contacts, which is shown in Fig. 2. Obviously, the histogram follows a log-normal distribution, namely the natural logarithm of T_K is normally distributed. This indicates that the distribution is understood by assuming the Kondo model, where the Kondo temperature T_K varies exponentially with the coupling parameter J_{cd} between conduction electrons and d electrons and the density of states at Fermi energy $D(E_F)$ as shown in the following formula:

$$T_K = \frac{W}{k_B}\exp\left[-\frac{1}{J_{cd}\,D(E_F)}\right] \quad (2),$$

where W is the bandwidth. These parameters J_{cd} and $D(E_F)$ are determined by the local properties of the constrictions, which show normal distributions. The fitting of T_K histogram yields a most frequent value for the resonance width corresponding to $T_K = 90$ K, which is reasonable agreement with the previous results reported in [7,9]. As described above, it is important to consider that the origin of the Fano resonance in Co atomic-scale contacts is caused by the Kondo effect as pointed out by Calvo *et al.*

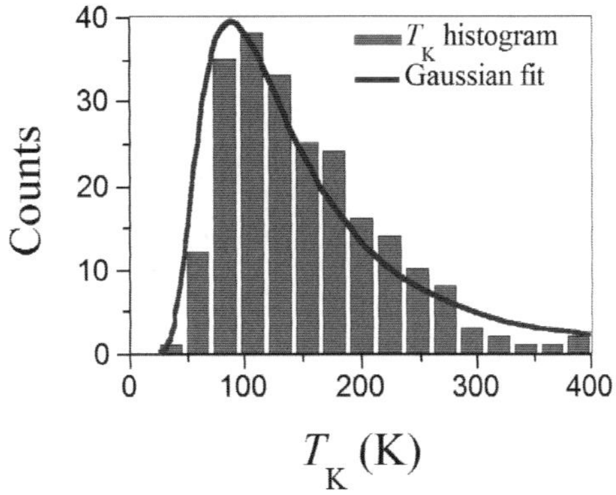

Fig. 2. (colour online) Histogram of T_K for more than two hundred fittings of Co atomic contacts. The solid (blue) line indicates the fit of the data to the log-normal distributions of T_K estimated below $6G_0$.

As mentioned above, the Fano resonance is survived in Ni nanoconstrictions where the bulk ferromagnetic properties are expected. To identify whether the similar feature appears in Co, we measure d*I*/d*V* spectra in large scale constrictions. Fig. 3 illustrates the size dependence of d*I*/d*V* spectra up to $85G_0$. Although the depth of the anomaly reduces by increasing the size of the constrictions, a dip shape anomaly is fitted by the Fano formula as shown by the solid black lines in Fig. 3. Due to the suppression of the depth of anomaly, we have enlarged the d*I*/d*V* spectra at $V \leq 20$ mV. Indeed, the Kondo temperature T_K gradually decreases and becomes constant because of reducing effect of both the parameters J_{cd} and $D(E_F)$ by increasing the contact size. The spectra imply that the ferromagnetic order and Kondo effect can coexist in Co nanoconstrictions as in the case of Ni ones.

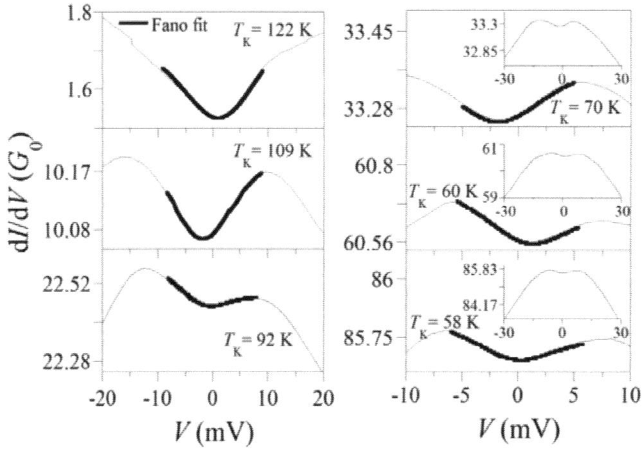

Fig. 3. (colour online) Spectra of d*I*/d*V* for large sized constrictions until $85G_0$. The insets in the right panel show the whole spectra. Solid lines indicate calculated Fano profiles with T_K.

Finally, we discuss the origin of the Kondo effect. Fig. 1 represents that a slight variation of the spatial geometry of the Co constrictions induces a change of the Kondo resonance spectra markedly. This suggests that the geometry is strongly associated with the appearance of the Kondo effect as in the case of Ni nanoconstrictions, in which a bridge-like structure where a Ni atom is placed on the Ni chain or surface like a bridge plays a crucial role for the existence of Kondo effect [10]. Generally, the magnetic moment of the bridge atoms should be varied from the bulk value due to the formation of an anisotropic structure in the crystal environment. Accordingly, a bridge-like atom forms a localized level, leading to the antiferromagnetic Kondo coupling J_{cd} due to the hybridization with the conduction electrons. The importance of the spatial geometry for the conductance is also pointed out by theoretical studies [15,16]. For instance, in Co adatom on the Cu(100) surface, the zero-bias anomaly due to Fano resonance appears in terms of the interference between the s and d_z^2 orbitals [16].

4. Conclusion

The spectra of d*I*/d*V* for Co monatomic contacts have been studied using MCBJ by the successive change of the geometry due to the stretching. Most of the spectra exhibit Fano-like features at around zero bias. The histogram of Kondo temperature follows a log-normal distribution that satisfies the normal distribution of their electronic properties in monatomic contacts. The size dependence of d*I*/d*V* spectra obtained in Co constrictions, suggest the existence of Kondo effect in ferromagnetic phase.

Acknowledgement

This work was supported by a Grant-in-Aid for Scientific Research (Grant Nos. 25220605, 25287076 and 26600102) from the Japan Society for the Promotion of Science.

References

[1] Kondo J 1964 *Prog. Theor. Phys.* **32** 37
[2] Madhavan V, Chen W, Jamneala T, Crommie M F and Wingreen N S 1998 *Science* **280** 567
[3] Cronenwett S M, Oosterkamp T H and Kouwenhoven L P 1998 *Science* **281** 540
[4] Goldhaber-Gordon D, Shtrikman H, Mahalu D, Abusch-Magder D, Meirav U and Kastner M A 1998 *Nature* **391** 156
[5] Park J, Pasupathy A N, Goldsmith J I, Chang C, Yaish Y, Petta J R, Rinkoski M, Sethna J P, Abruña H D, McEuen P L and Ralph D C 2002 *Nature* **417** 722
[6] Yu L H and Natelson D 2004 *NanoLett.* **4** 79
[7] Néel N, Kröger J, Limot L, Palotas K, Hofer W A and Berndt R 2007 *Phys. Rev. Lett.* **98** 016801
[8] Fano U 1961 *Phys. Rev.* **124** 1866
[9] Calvo M R, Fernández-Rossier J, Palacios J J, Jacob D, Natelson D and Untiedt C 2009 *Nature* **458** 1150
[10] Ienaga K, Nakashima N, Inagaki Y, Tsujii H, Honda S, Kimura T and Kawae T 2012 *Phys. Rev.* B **86** 064404
[11] Agrait N, Yeyati A L and van Ruitenbeek J M 2003 *Phys. Rep.* **377** 81
[12] Häfner M *et al.* 2008 *Phys. Rev.* B **77** 104409
[13] Untiedt C *et al.* 2004 *Phys. Rev.* B **69** 081401(R)
[14] Parks J J *et al.* 2010 *Science* **328** 1370
[15] Lucignano P *et al.* 2009 *Nat. Mater.* **8** 563
[16] Baruselli P P *et al.* 2015 *Phys. Rev.* B **92** 045119

IOP Conf. Series: Journal of Physics: Conf. Series **807** (2017) 082003 doi:10.1088/1742-6596/807/8/082003

Spin Hall conductance in Y-shaped junction devices

Sudin Ganguly[a] **and Saurabh Basu**[b]

Indian Institute of Technology Guwahati, Assam-781039, India

E-mail: [a]`sudin@iitg.ernet.in`, [b]`saurabh@iitg.ernet.in`

Abstract. We study the spin Hall effect in Y-shaped junction devices in presence of Rashba spin orbit coupling (RSOC). The voltage and the net spin current registered at one of the arms of the Y-junction are seen to increases although the spin Hall conductance (SHC) diminishes as the strength of the RSOC is increased. This implies the voltage increases further than that of the current, thereby causing a loss of the RSOC. Various other characteristic features obtained from our study include, a perfectly antisymmetric behaviour of the spin current and the SHC with respect to the zero bias, while the voltage shows a symmetric character. Finally a large RSOC completely destroys the SHC, owing to a complete disappearance of the local density of states, thereby reinforcing our earlier claim that RSOC emulates the effect of disorder on the quantum conductance of junction device.

1. Introduction

Nowadays dissipationless spin current [1] has been a central focus of active investigation in mesoscopic physics. Various mechanisms, such as spin field effect transistor (spin-FET) [2], spin-transfer torque [3, 4], spin pumping [5] and (inverse) spin Hall effect [6, 7] have been proposed to generate, control and detect the spin current. The spin current generated by these means is in general dissipative in nature. On the other hand, a dissipationless spin current can be generated by an external electric field via intrinsic spin Hall effect [1, 8]. The intrinsic spin Hall effect can be achieved in presence of spin orbit coupling such as Rashba spin orbit coupling (RSOC).

The spin-orbit coupling produces a transverse force on a moving electron and this force tends to form a transverse spin current. In the simplest form, a spin current is the flow of spin-up electrons in one direction, say $+x$, accompanied by the flow of an equal number of spin-down electrons in the opposite direction, $-x$. The total charge current in the x direction is therefore zero, that is $I^q = e\,(I_\uparrow + I_\downarrow) = 0$. The total spin current is finite, that is $I^s = e\,(I_\uparrow - I_\downarrow) \neq 0$.

In a two-dimensional electron gas (2DEG) lacking structure inversion symmetry of the confining potential, the Rashba spin orbit coupling (RSOC)can be important. The Hamiltonian for the system in presence of RSOC is written as,

$$H = \frac{\mathbf{p}^2}{2m^*} + \alpha\,(\sigma^x p_y - \sigma^y p_x) \tag{1}$$

where the second term is the Rashba spin-orbit coupling (RSOC). $\sigma^{x/y}$ is the x/y component of the Pauli matrices, and α is the coupling parameter which denotes the strength of RSOC.

A three terminal structure is believed to be a suitable candidate for studying the spin Hall effect [9]. In this paper, we have studied the behaviour of a Y-shaped three terminal structure with different angular separation between the arms of the Y-shaped geometry as shown in Fig.1. These Y-shaped junction devices can be fabricated experimentally. The technique by which we make these devices for a theoretical investigation is presented in the following section. Specifically, we have studied the conductance properties in order to observe the effects of the angular separation between the two arms of the Y-shaped junction.

We organize our paper as follows. The theoretical formalism leading to the expressions for the spin Hall and longitudinal conductances using Landauer Büttiker formula are presented in section II. Section III includes an elaborate discussion of the results obtained for the spin Hall and longitudinal conductances in presence of spin orbit interaction.

2. Theoretical formulation

We begin our discussion by a prescription of fabricating a Y-shaped junction device which can be interesting from an experimental perspective.

We choose a three-probe measuring set-up as shown in Fig.1 to observe the spin Hall effect. Here the three ideal semi-infinite leads are attached to the central conducting region, which in our case is the Y-shaped device formed via a square lattice geometry and includes spin orbit interaction. In this work, we have taken three different Y-shapes by changing the angle between the two arms of the Y, we call it θ_Y as shown in Fig.1. The leads shown as 1,2 and 3 are semi-infinite in nature. The voltages applied at the leads are V_1, V_2 and V_3 respectively. The width of the scattering region is d, while the arms has width $d/2$.

Fig.2 provides a technique how we can fix the angle, θ_Y. According to Fig.1, θ_Y is twice the angle as shown in Fig.2. For the Y-shape shown in Fig.1(a), we add the lattice sites in the arms of Y as depicted in Fig.2(a). First we add 3 sites along x-axis with spacing 'a' and then add another site along y-axis just above the third site along x. We repeat the same procedure to build up the rest of the arm of the Y-geometry. Since we need 3 sites along x-axis and 2 sites along y-axis we call it a (3,2) scheme. The calculation of the angle in now straight forward from the geometry. From Fig.2, b is the base of the triangle and h is the height. In the (3,2) scheme, $b = 2a$ and $h = a$. Hence, the angular separation between the arms of the Y will be twice the calculated angle and the angle is, $\theta_Y = 2\tan^{-1}\frac{a}{2a} = 53.13°$.

Similarly, corresponding to $\theta_Y = 90°$, we need the (2,2) scheme, for which 2 sites along x-axis and 1 site along y-axis above the second site along x are required, as shown in Fig.2(b). In the given case, $\theta_Y = 2\tan^{-1}\frac{a}{a} = 90°$. Again from Fig.2(c), we should now have (2,3) scheme. The angle is, $\theta_Y = 2\tan^{-1}\frac{2a}{a} = 128.87°$.

Now an unpolarized charge current is allowed to pass through the longitudinal leads along lead-1 and lead-3 (see Fig.1) inducing spin Hall current along lead-2. Assuming a tight binding approximation with nearest neighbour hopping, Eq.(1) reads as,

$$H = \epsilon \sum_{i,\sigma} c^{\dagger}_{i\sigma} c_{i\sigma} + t \sum_{\langle ij \rangle,\sigma} c^{\dagger}_{i\sigma} c_{j\sigma} + V_R \sum_{i} \left[\left(c^{\dagger}_{i\uparrow} c_{i+\delta_x\downarrow} - c^{\dagger}_{i\downarrow} c_{i+\delta_x\uparrow} \right) - i \left(c^{\dagger}_{i\uparrow} c_{i+\delta_y\downarrow} + c^{\dagger}_{i\downarrow} c_{i+\delta_y\uparrow} \right) \right] \quad (2)$$

Here ϵ is the on-site potential and t is the hopping strength, $V_R = \frac{\bar{h}\alpha}{a}$ is the modified Rashba coupling strength with a to be the lattice constant. $\delta_{x/y}$ is the unit vector along x/y direction.

For the three terminal case, the longitudinal and spin Hall conductances are defined as [10],

$$G_L = \frac{I_3^q}{V_3 - V_1} \quad \text{and} \quad G_{SH} = \frac{\bar{h}}{2e} \frac{I_2^s}{V_2 - V_1} \quad (3)$$

where I_3^q and I_2^s are the charge and spin currents flowing through the lead-3 and lead-2 respectively. V_m is the potential at the m-th lead.

The calculation of the electric and spin currents is based on the Landauer-Büttiker multi-probe formalism [11]. The charge and spin currents flowing through lead m(=1,2,3) with potential, V_m can be written in terms of the spin resolved transmission probability as [9],

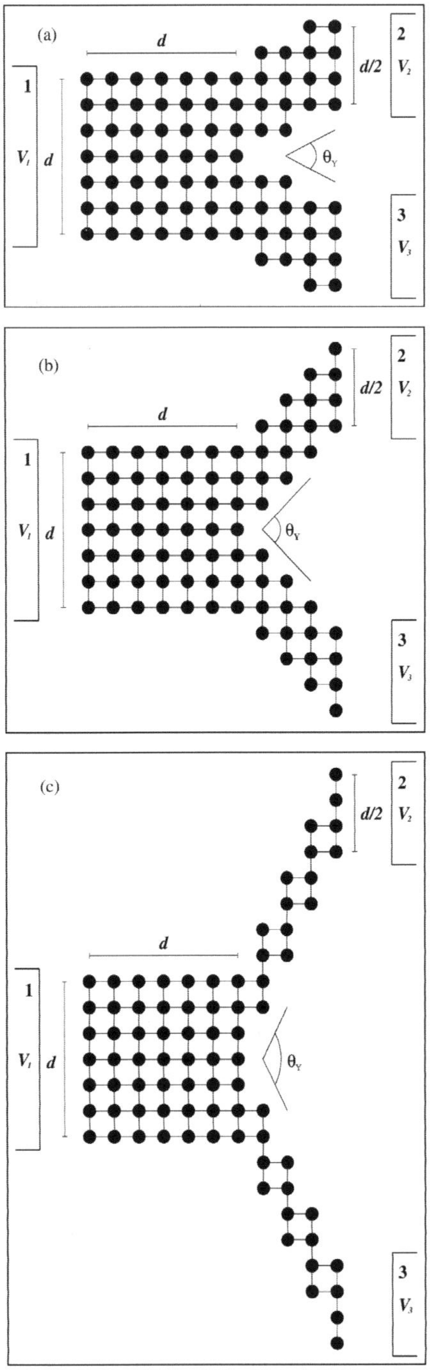

Figure 1: Y-shaped three terminal junction with three different angles (θ_Y). V_1, V_2 and V_3 are the applied voltages at the three terminals. The leads are not shown in the figure.

$$I_m^q = \frac{e^2}{h} \sum_{n \neq m, \sigma, \sigma'} \left(T_{nm}^{\sigma\sigma'} V_m - T_{mn}^{\sigma'\sigma} V_n \right) \quad (4)$$

and,

$$\begin{aligned} I_m^s &= \frac{e^2}{h} \sum_{n \neq m, \sigma'} \left[\left(T_{nm}^{\sigma'\sigma} - T_{nm}^{\sigma'-\sigma} \right) V_m \right. \\ &\quad \left. + \left(T_{mn}^{-\sigma\sigma'} - T_{mn}^{\sigma\sigma'} \right) V_n \right] \\ &= \frac{e^2}{h} \sum_{n \neq m} \left[T_{nm}^{out} V_m - T_{mn}^{in} V_n \right] \end{aligned} \quad (5)$$

where, we have defined two useful quantities as follows,

$$\begin{aligned} T_{pq}^{in} &= T_{pq}^{\uparrow\uparrow} + T_{pq}^{\uparrow\downarrow} - T_{pq}^{\downarrow\uparrow} - T_{pq}^{\downarrow\downarrow} \\ T_{pq}^{out} &= T_{pq}^{\uparrow\uparrow} + T_{pq}^{\downarrow\uparrow} - T_{pq}^{\uparrow\downarrow} - T_{pq}^{\downarrow\downarrow} \end{aligned} \quad (6)$$

Physically, the term $\frac{e^2}{h} \sum_{n \neq m} T_{nm}^{out} V_m$ is the total spin current flowing from the m-th lead with potential V_m to all other n leads, while the term $\frac{e^2}{h} \sum_{n \neq m} T_{mn}^{in} V_n$ is the total spin current flowing into the m-th lead from all other n leads having potential V_n.
The zero temperature conductance, $G_{pq}^{\sigma\sigma'}$ that describes the spin resolved transport measurements, is related to the spin resolved transmission coefficient by [12, 13],

$$G_{pq}^{\sigma\sigma'} = \frac{e^2}{h} T_{pq}^{\sigma\sigma'}(E) \quad (7)$$

The transmission coefficient can be calculated as [14, 15],

$$T_{pq}^{\sigma\sigma'} = \mathrm{Tr}\left[\Gamma_{\mathrm{p}}^{\sigma} \mathrm{G_R} \Gamma_{\mathrm{q}}^{\sigma'} \mathrm{G_A} \right] \quad (8)$$

Γ_p^σ's are the coupling matrices representing the coupling between the central region and the leads, and they are defined by the relation [16],

$$\Gamma_p^\sigma = i \left[\Sigma_p^\sigma - (\Sigma_p^\sigma)^\dagger \right] \quad (9)$$

Here Σ_p^σ is the retarded self-energy associated with the lead p. The self-energy contribution is computed by modeling each terminal as a semi-infinite perfect wire [17].

The retarded Green's function, G_R is computed using

$$G_R = \left(E - H - \sum_{p=1}^{4} \Sigma_p \right)^{-1} \tag{10}$$

where E is the electron Fermi energy and H is the model Hamiltonian for the central conducting region. G_A is the advanced Green's function and is given by, $G_A = G_R^\dagger$.

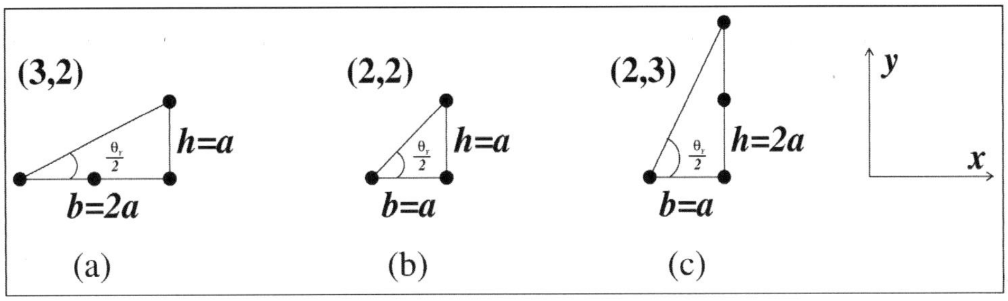

Figure 2: Measurement of the angle between the two arms of the Y-shaped device is shown. b is the base of the triangle and h is the height. a is the lattice constant. According to Fig.1, this angle is half of θ_Y as depicted in the given figure.

Now, following the spin Hall phenomenology, in our set-up since lead-2 is a voltage probe, $I_2^q = 0$. Also, as the currents in various leads depend only on voltage differences among them, we can set one of the voltages to zero without any loss of generality. Here we set $V_1 = 0$ and $V_3 = 1$. With the help of these conditions, from Eq.(4), one can determine the voltage, V_2 and the spin current flowing through terminal 2,

$$V_2 = \frac{T_{23}}{T_{12} + T_{32}} \quad \text{and,} \quad I_2^s = \frac{e^2}{h} \left[\left(T_{12}^{\text{out}} + T_{32}^{\text{out}} \right) V_2 - T_{23}^{\text{in}} \right] \tag{11}$$

Finally, from Eq.(3) the expressions for the longitudinal and spin Hall conductances become,

$$G_L = \frac{e^2}{h} \left[(T_{13} + T_{23}) - \frac{T_{23} T_{32}}{T_{12} + T_{32}} \right] \tag{12}$$

$$G_{SH} = \frac{e}{4\pi} \left[\left(T_{12}^{\text{out}} + T_{32}^{\text{out}} \right) - T_{23}^{\text{in}} \frac{T_{12} + T_{32}}{T_{23}} \right] \tag{13}$$

3. Results and Discussion

We have investigated the effects of the angle variation of the Y-shaped junction in presence of Rashba spin orbit coupling on the experimentally measurable quantities such as the longitudinal conductance (G_L) and the spin Hall conductance (G_{SH}).

We briefly describe the values of different parameters used in our calculation. Throughout our work, we have considered for the Y-shaped system, $d = 20$ (see Fig.1), onsite term, $\epsilon = \epsilon_L = 0$, hopping term, $t = t_L = t_C = 1$. All the energies are measured in unit of t. Further we choose a unit where $c = h = e = 1$. The longitudinal conductance, G_L is measured in unit of $\frac{e^2}{h}$. The spin Hall conductance, G_{SH} is measured in unit of $\frac{e}{4\pi}$. Also the lattice constant is taken to be unity. For most of our numerical calculations we have used KWANT [18]. At the onset, V_R is chosen to have a representative value of 0.5. In this work, we have taken three different angles for the Y-shaped device. However, we have been able to generate a number of other values for the angular separation, θ_Y by following the prescription given above. however to keep our discussion on the angular separation crisp, we have considered θ_Y values to be less than, equal to and greater than $90°$, and as said earlier, in particular we have considered, $\theta_Y = 53.13°, 90°$ and $128.87°$.

Fig.3 shows the variation of the longitudinal conductance as a function of energy, E for three different angles of the Y-shaped device. In Fig.3(a), for $\theta_Y = 90°$, G_L on an average, except for some fluctuations shows step-like nature in absence of the RSOC demonstrating the discreteness of the energy states available for conduction. Another important observation is that, the magnitude of G_L at $E = 0$ decreases with increase in the value of θ_Y. This can be explained in the following way. G_L is measured at terminal-3 and the charge current, I_3^q is flowing from terminals 3 and 1. Now it is easily understandable that, as we increase the angle between terminal-1 and terminal-3, electrons are likely suffer more scattering in reaching terminal-3. Accordingly, G_L decreases as θ_Y is increased.

Fig.3(b) shows the variation of G_L as a function of E in presence of RSOC with strength, $V_R = 0.5$. In presence of RSOC, though the step-like nature is lost but the decreasing feature of G_L continues with increasing θ_Y. An interesting observation from Fig.3(b) is that, for $\theta_Y = 90°$, there exist a zero mode state, that is at $E = 0$ the longitudinal conductance vanishes. This can be explained from the definition of G_L as given in Eq.(3) and Eq.(4). Since we have assumed the voltage at terminal-3, $V_3 = 1$ and the voltage at terminal-1, $V_1 = 0$, by Eq.(3) G_L is nothing but the charge current flowing through terminal-3. Now from Eq.(4) we can say that there is a cancellation for the charge current, I_3^q between the two given quantities, resulting in vanishing of G_L.

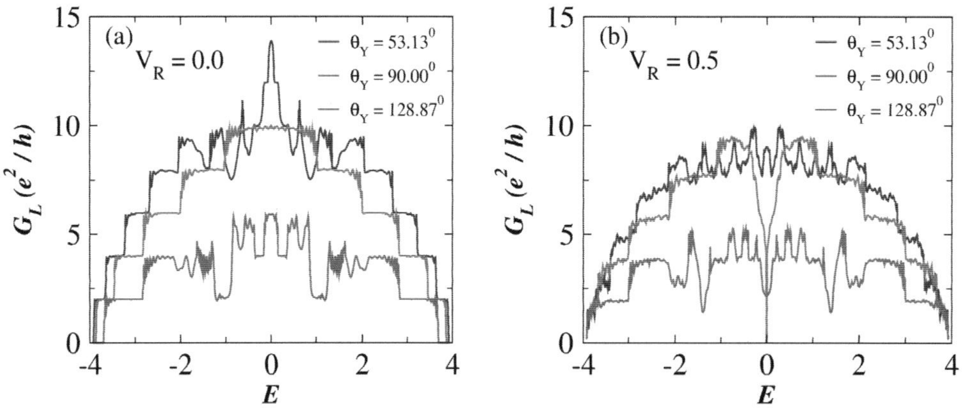

Figure 3: (Color online) (a) G_L is plotted as a function of energy, E for three different angle of the Y-shaped device when RSOC is absent. G_L shows step-like nature. (b) G_L is plotted as a function of E in presence of RSOC with strength $V_R = 0.5$. The step-like nature of G_L is partly lost.

Now let us compute the characteristics of the spin Hall conductance. An useful quantity in this regard, is the spin Hall voltage V_2 measured at terminal-2. Fig.4 shows the variation of the voltage, V_2 at terminal-2 as a function of energy, E for three different θ_Y with RSOC strength, $V_R = 0.5$. In Fig.4(a), (b) and (c), the voltage V_2 in each case shows a symmetric behaviour as a function of the energy, E with respect to $E = 0$. Fig.4(a) is the plot for V_2 for $\theta_Y = 53.13°$, Fig.4(b) for $\theta_Y = 90°$ and Fig.4(c) for $\theta_Y = 128.87°$. V_2 is larger at the edges of the energy band and as we come closer to $E = 0$, G_L decreases. Especially we note that for larger values of θ_Y, namely for $\theta_Y = 90°$ and $128.87°$,G_L almost vanishes around $E = 0$.

Further we have studied the variation of spin current flowing through terminal-2, I_2^s as a

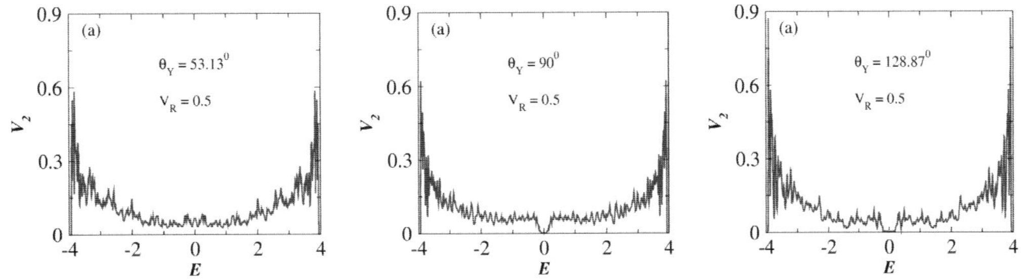

Figure 4: (Color online) The voltage at terminal-2, V_2 is plotted as a function of energy, E for three different θ_Y in presence of RSOC. (a) $\theta_Y = 53.13°$, (b) $\theta_Y = 90°$ and (c) $\theta_Y = 128.87°$. RSOC strength is $V_R = 0.5$.

function of energy, E for three different θ_Y in presence of RSOC as shown in Fig.5. Fig.5(a) shows the variation of I_3^s versus E for $\theta_Y = 53.13°$. Fig.5(b) and Fig.5(c) correspond to $\theta_Y = 90°$ and 128.87° respectively. I_2^s, as expected has an antisymmetric behaviour as a function of E about $E = 0$. As a result, I_2^s is exactly zero at $E = 0$, a fact well known.

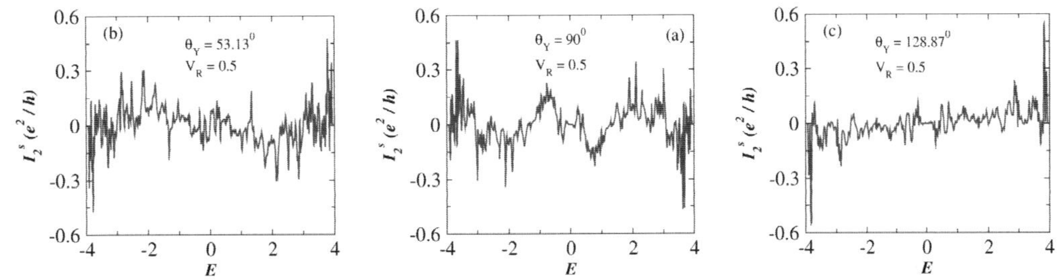

Figure 5: (Color online) The spin current at terminal-2, I_2^s is plotted as a function of energy, E for three different θ_Y in presence of RSOC. (a) $\theta_Y = 53.13°$, (b) $\theta_Y = 90°$ and (c) $\theta_Y = 128.87°$. In the given case, $V_R = 0.5$

Finally, the effect of the angular separation between the arms of the Y-shaped device on the spin Hall conductance, G_{SH} is shown in Fig.6. Here G_{SH} is plotted as a function of energy, E in presence of RSOC. From Fig.4 and Fig.5, we notice that, V_2 has a symmetric nature as a function of E while I_2^s is antisymmetric. As a result, G_{SH} is antisymmetric as a function of E, an important feature of the spin Hall conductance [14, 19, 20]. It can be noticed that for different values of θ_Y the variation of G_{SH} is different, all the while retaining the antisymmetric character.

The main point that would be of interest to the community is that a Y-shaped device can be made, which in presence of RSOC, demonstrates that spin Hall effect is realizable in experiments. It may be noted that in our calculations, the spin Hall conductance (in units of $e/4\pi$) is moderately large (see Fig.6). Thus experiments involving magneto-optical Kerr rotation spectroscopy and other methods [21, 22] should be able to detect the spin Hall effect discussed here. Further, near perfect electrical switching operation is shown to achievable using a CNT Y-shaped junction under an a.c. bias [23]. Also, we have shown that SHE is sensitive to the angular separation between the arms of the Y-shaped device.

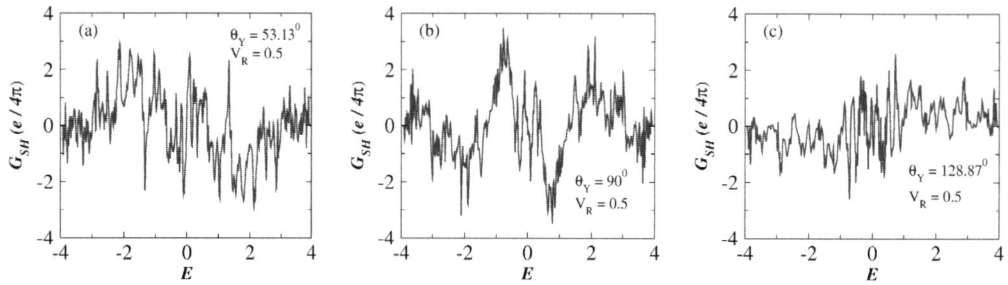

Figure 6: (Color online) (a) G_{SH} is plotted as a function of energy, E for three different values of θ_Y in presence of RSOC. (a) $\theta_Y = 53.13°$, (b) $\theta_Y = 90°$ and (c) $\theta_Y = 128.87°$.

4. Summary and Conclusions

In summary, in the present work we have studied the effect of the angular separation of a three terminal Y-shaped junction device in presence of Rashba spin orbit coupling on the longitudinal and spin Hall conductances by Landauer-Büttiker formalism.

Both in presence of and absence of RSOC, at $E = 0$, G_L decreases as θ_Y increases in a Y-shaped junction device. On the other hand, no systematic variation of the spin Hall properties emerge as a function of the angular separation of the device. In particular, the voltage at terminal-2 has symmetric nature as a function of energy, while the spin current flowing through terminal-2, I_2^s and the spin Hall conductance, G_{SH} having a large fluctuation, have antisymmetric behaviour as a function of energy, E with respect to $E = 0$. For different values of θ_Y the variation of G_{SH} is finite and different, all the while retaining the antisymmetric character. Experimental verification of our results should be possible with improved fabrication techniques of the realizing the mesoscale and nanoscale junction devices.

References

[1] S. Murakami, N. Nagaosa and S. C. Zhang, Science **301**, 1348 (2003).
[2] S. Datta and B. Das, Appl. Phys. Lett. **56**, 665 (1990).
[3] L. Berger, Phys. Rev. B **54**, 9353 (1996).
[4] J. C. Slonczewski, J. Magn. Magn. Mater. **159**, L1 (1996).
[5] Y. Tserkovnyak, A. Brataas, and G. E. W. Bauer, Phys. Rev. Lett. **88**, 117601 (2002).
[6] M. I. D'yakonov and V. I. Perel, JETP Lett. **13**, 467 (1971); Phys. Lett. A **35**, 459 (1971).
[7] J. E. Hirsch, Phys. Rev. Lett. **83**, 1834 (1999).
[8] T. Kimura, Y. Otani, T. Sato, S. Takahashi, and S. Maekawa, Phys. Rev. Lett. **98**, 156601 (2007).
[9] T. P. Pareek, Phys. Rev. Lett. **92**, 076601 (2004).
[10] M. Dey, S. K. Maiti and S. N. Karmakar, J. Appl. Phys. **112**, 024322 (2012).
[11] M. Büttiker, Phys. Rev. Lett. **57**, 1761 (1986).
[12] R. Landauer, IBM J. Res. Dev. **1**, 223 (1957).
[13] R. Landauer, Philos. Mag. 21, 683 (1970).
[14] C. Caroli, R. Combescot, P. Nozieres and D. Saint-James, J. Phys C: Solid State Phys. **4**, 916, (1971).
[15] D. S. Fisher and P. A. Lee, Phys. Rev. B **23**, 6851 (1981).
[16] S. Dutta, *Electronic transport in Mesoscopic systems*, University press (Cambridge), (1995).
[17] B. K. Nikolic, Phys. Rev. B **64**, 014203 (2001).
[18] C. W. Groth, M. Wimmer, A. R. Akhmerov, X. Waintal, *Kwant: a software package for quantum transport*, New J. Phys. **16**, 063065 (2014).
[19] J. Li, L. Hu, S. Q. Shen, Phys. Rev. B **71**, 241305(R) (2005).
[20] C. P. Moca and D. C. Marinescu, Phys. Rev. B **72**, 165335 (2005).
[21] Y. K. Kato, R. C. Myers, A. C. Gossard and D. D. Awschalom, Science **306**, 1910 (2004).
[22] J. Wunderlich, B. Kaestner, J. Sinova and T. Jungwirth, Phys. Rev. Lett. **94**, 047204 (2005).
[23] P. R. Bandaru, C. Daraio, S. Jin and A. M. Rao, Nature Materials **4**, 663 (2005).

Oscillation and suppression of Kondo temperature by RKKY coupling in two-site Kondo systems

Ammar Nejati[1] **and Johann Kroha**[1,2]

[1] Physikalisches Institut and Bethe Center for Theoretical Physics, Universität Bonn, Nussallee 12, D-53115 Bonn, Germany

[2] Center for Correlated Matter, Zhejiang University, Hangzhou, Zhejiang 310058, China

E-mail: kroha@th.physik.uni-bonn.de

Abstract. We apply our recently developed, selfconsistent renormalization group (RG) method [1] to STM spectra of a two-impurity Kondo system consisting of two cobalt atoms connected by a one-dimensional Cu chain on a Cu surface [2]. This RG method was developed to describe local spin screening in multi-impurity Kondo systems in presence of the Ruderman-Kittel-Kasuya-Yosida (RKKY) interaction. Using the RKKY interaction of a one-dimensional chain, we explain the experimentally observed suppression and oscillation of the Kondo temperature, $T_K(y)$, as a function of the length of the chain and the corresponding RKKY interaction parameter y, regardless of the RKKY coupling being ferromagnetic or antiferromagnetic.

1. Introduction

The Kondo effect is a genuine many-body phenomenon which, in its usual magnetic manifestation, appears when a magnetic ion is immersed in a non-magnetic host metal. At high temperatures, the magnetic moment of the ion exhibits a Curie magnetic susceptibility corresponding to uncorrelated moments in a paramagnetic phase. When the temperature is decreased below a characteristic scale, the Kondo temperature T_K, the local moment is collectively screened by the conduction electron spins, leading to a narrow resonance in the electronic spectrum and a temperature-independent Pauli susceptibility [3, 4]. The Kondo effect is observed in a variety of systems, ranging from metallic alloys [5, 6] to artifical semiconductor nanostructures like quantum dots [7, 8, 9, 10] and molecular devices [11, 12, 13].

In a spin-1/2 two-impurity Kondo (2IK) system, two different ground states are possible, the Kondo state where each of the two local spins forms a Kondo singlet with the conduction electrons, and a dimer state where the two local spins are mutually bound into a singlet or triplet. The two states are distinguishable by different behavior of the spin correlation functions between a local spin and the conduction electrons on one hand and between the local spins among each other on the other hand. It has been recognized [1] that it makes a crucial difference whether there is a direct exchange coupling between the impurities or whether the interimpurity coupling is mediated by the conduction electrons, the Ruderman-Kittel-Kasuya-Yosida (RKKY) interaction [14, 15, 16, 17]. With a direct exchange coupling, a quantum phase transition between the Kondo and the dimer phase [18, 19] occurs only for a particular particle-hole symmetry [20], associated with a coupling-decoupling transition of the conduction band. In the case of RKKY-mediated interimpurity coupling, a complete decoupling of the conduction electrons at low energy is not possible. However, it was recently predicted that nevertheless a breakdown of the complete Kondo screening occurs at a critical strength of a dimensionless RKKY coupling parameter y [1], associated with a transition to a phase of partially screened local moments and a universal reduction of the Kondo scale $T_K(y)$. Numerical renormalization group (NRG) studies have recently been performed also for indirect interimpurity coupling [21].

2IK systems have been realized experimentally in semiconductor quantum dot devices [22, 23, 24, 25] and in systems of magnetic adatoms on metal surfaces, studied by scanning tunneling spectroscopy [2, 26, 27, 28]. In the present work we focus on understanding the results of Ref. [2]. In this experiment, Néel *et al.* have assembled atomic chains of up to $n = 6$ Cu atoms on a Cu surface, flanked by a magnetic Co atom on each end. This forms a 2IK system where the indirect RKKY interaction is enhanced by the quasi-one-dimensional (1D) Cu chain. The Kondo temperature was deduced from the width of the Kondo-Fano resonance [29] in the dI/dV signal taken above one of the two Co atoms. It shows an oscillatory suppression as function of the separation of the Co atoms.

This paper is organized as follows: In section 2, we derive the form of the RKKY interaction for the 1D chain and use the result in section 3 to obtain the Kondo temperature via the RKKY-modified RG. In the final section 4, we provide a detailed comparison of the results with the experimental observations.

2. RKKY interaction mediated by the 1D chain

The two-impurity Kondo system is described by the Hamiltonian,

$$H_{2\mathrm{IK}} = \sum_{\mathbf{k}\sigma} \varepsilon_{\mathbf{k}\sigma}\, c^{\dagger}_{\mathbf{k}\sigma} c_{\mathbf{k}\sigma} + J_0 \sum_{i=1,2} \mathbf{S}(x_i) \cdot \mathbf{s}(x_i) \,, \tag{1}$$

where $\mathbf{S}(x_i)$ and $\frac{1}{2}\mathbf{s}(x_i)$ represent the spin operators of the impurities and conduction electrons at the positions of the impurities, $x_{1,2}$, respectively. $(\mathbf{k},\sigma)$ represents a complete basis of the conduction electron system, composed of the substrate and the 1D Cu chain. Density functional theory (DFT) calculations indicate [2] that the RKKY coupling is dominated by the 1D chain (because in the 3D substrate the RKKY interaction decays faster than in 1D) and that the local DOS in the Cu chain is well approximated by an infinitely long chain. Deviations from this behavior are found only for the shortest chains of length $n = 1$, see section 4. We, therefore, approximate the conduction-mediated coupling between the Co spins at positions x_1 and x_2 by the RKKY coupling of two spins separated by $x = x_2 - x_1$ on an infinite chain (infinite chain approximation). In a simple model for the Cu chain we assume a Fermi momentum of $k_F = \pi/(2a)$ corresponding to half filling of the Cu 4s band, where a is the Cu lattice spacing.

In leading order perturbation theory in the local spin exchange coupling J_0, the RKKY coupling reads, $I_{\mathrm{RKKY}}(x) = -\chi_c(x)$, where $\chi_c(x)$ is the free conduction electron susceptibility (or polarization) of the chain [30], as shown in Fig. 1 (a). We calculate $\chi_c(x)$ here directly in position space, as this will be needed for the RG vertex in section 3 and is more concise than the momentum space calculation [30].

$$\chi^R_c(x_2, x_1; \Omega) = \int_{-\infty}^{+\infty} \mathrm{d}\varepsilon \; n_F(-\varepsilon) \Big(G^R_c(x; \varepsilon + \Omega)\, A_c(-x; \varepsilon) + A_c(x; \varepsilon)\, G^A_c(-x; \varepsilon - \Omega) \Big) \,, \tag{2}$$

where $G^{R/A}_c$ is the retarded/advanced Green's function for the free electrons and A_c represents their spectral function. $n_F(\varepsilon)$ denotes the Fermi-Dirac distribution, and $\varepsilon = E - \varepsilon_F$ is the energy with respect to the Fermi level, ε_F. The static susceptibility entering the RG vertex [1] is obtained from the zero-frequency limit $\Omega \to 0$, and $n_F(\varepsilon) = \Theta(\varepsilon)$ for temperature $T = 0$.

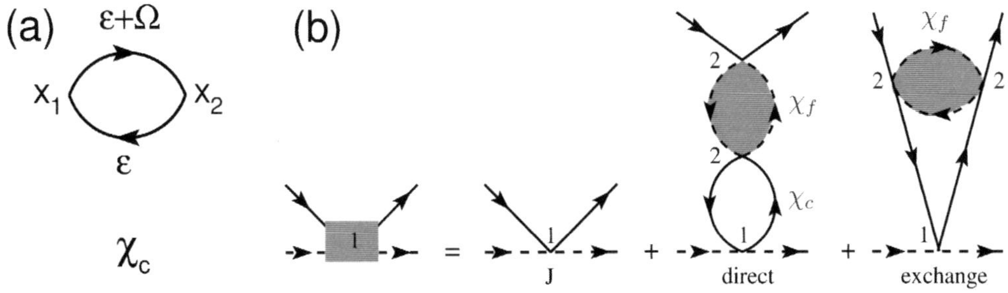

Figure 1: (a) Conduction electron susceptibility or "polarization bubble". (b) Total conduction electron-impurity spin vertex, including non-local RKKY vertex corrections ('direct' and 'exchange' diagrams). The solid (dashed) lines represent conduction electron (impurity spin) propagators, respectively. χ_c (χ_f) is the conduction electron (impurity) susceptibility.

Assuming, for simplicity, a quadratic dispersion, $\varepsilon_k = k^2/2m$, the 1D free Green's function (per spin orientation) reads [31],

$$\begin{aligned} G_c^{0,R/A}(x,\omega) &= \mp i\pi\mathcal{N}(\omega)\exp[\pm ik(\omega)|x|] \, , \\ A_c(x,\omega) &= -\frac{1}{\pi}\mathrm{Im}\, G_c^{0,R}(x,\omega) = \mathcal{N}(\omega)\cos[k(\omega)|x|] \, , \end{aligned} \tag{3}$$

where $\mathcal{N}(\omega) = m/(2\pi k(\omega))$ is the continuum conduction electron DOS and $k(\omega) = \sqrt{2m(\varepsilon_F + \omega)}$ the momentum corresponding to the energy ω. Hence, the static conduction electron susceptibility is real and reads,

$$\chi_c^R(x_2, x_1; \Omega = 0) \equiv \chi_c^R(\varrho; \Omega = 0) = -\frac{1}{2}\mathcal{N}(0)\left(\mathrm{Si}(\varrho) - \frac{\pi}{2}\right) , \tag{4}$$

with $\varrho = 2k_F|x|$, and $\mathrm{Si}(z) = \int_0^z dx \, \frac{\sin(x)}{x}$ the integral sine. Note that the long-distance limit of $\chi_c^R(x; \Omega = 0)$ is

$$\chi_c^R(x; \Omega = 0) \overset{|x|\to\infty}{\longrightarrow} \sim \frac{\cos(\varrho)}{\varrho} , \tag{5}$$

which manifests its slow decay and oscillatory behavior as a function of interimpurity spacing.

3. Renormalization group analysis for symmetrically coupled impurities

In the previous section, we calculated the RKKY interaction between the two Co impurities mediated by the 1D Cu chain connecting the magnetic ions. The RKKY interaction implies that the total spin exchange vertex between conduction electrons and a local impurity spin acquires nonlocal contributions: A conduction electron can scatter from impurity 2, and that spin flip can be transfered to impurity 1 by the itinerant conduction electrons travelling along the Cu chain from the 2nd to the 1st Co impurity. This process is depicted diagrammatically in Fig. 1 (b). As seen from the figure, this transfer process necessarily involves the total, local spin susceptibility χ_f of the 2nd impurity. Its exact behavior is known from the Bethe ansatz solution [32]. At low temperatures, $T < T_K$, it is proportional to $1/T_K$ and crosses over to Curie behavior $\chi_f \sim 1/T$ for $T \gg T_K$.

With the expression for the RKKY-modified Kondo vertex, the RG equations for the Kondo coupling can be obtained by the same procedure as developed in Ref. [1]. The RG equation for the dimensionless Kondo coupling, $g = J\mathcal{N}(0)$, as a function of the running conduction bandwidth, D, is obtained as [1],

$$\frac{dg}{d\ln D} = -2g^2 \left(1 - y\, g_0^2 \frac{D_0}{T_K} \frac{1}{\sqrt{1+(\frac{D}{T_K})^2}}\right) , \tag{6}$$

where D_0 is the full bandwidth of the conduction electrons, $g_0 = J_0\mathcal{N}(0)$ the dimensionless, bare Kondo coupling, $T_K = T_K(y)$ is the RKKY-modified Kondo scale (to be determined), and y is a dimensionless RKKY strength parameter. It is obtained for a 2IK system as (c.f. Ref. [1]),

$$y = -\frac{8W}{\pi^2}\mathrm{Im}\left[\frac{\mathrm{e}^{-ik_F x}}{N(0)^2} G_c^R(x,\Omega=0)\, \chi_c(\varrho,\Omega=0)\right] = \frac{8W}{\pi\mathcal{N}(0)}\cos(\varrho)\, \chi_c(\varrho,\Omega=0) , \tag{7}$$

with $W = 2$ the Wilson ratio of a single Kondo impurity. Note that the term proportional to $\frac{1}{T_K}$ on the right-hand side of Eq. (6) originates from the behavior of the local impurity susceptibility, $\chi_f \propto \frac{1}{T_K}$ [32]. Furthermore, note that the RKKY parameter y of the 2IK system is generically positive, even though $\chi_c^R(x,0)$ has an alternating sign, as shown in Fig. 2, left panel. Technically, this is because in y the oscillatory factors $\exp(-ik_F x)\, G_c^R(x,\Omega=0)$ and $\chi_c(\varrho,\Omega=0)$ combine to form an essentially positive term. It is physically expected because the interimpurity RKKY coupling should reduce the onsite Kondo spin fluctuations irrespective of the RKKY coupling being ferro- or antiferromagnetic.

The RG equation (6) is straight-forwardly integrated by separation of variables,

$$\int_{g(D_0)}^{g(D)} \frac{\mathrm{d}g}{-2g^2} = \int_{D_0}^{D} \frac{\mathrm{d}D}{D}\left(1 - y g_0^2 \frac{D_0}{T_K}\frac{1}{\sqrt{1+(D/T_K)^2}}\right) . \tag{8}$$

To deduce the Kondo temperature T_K, one should use the fact that, as the running cutoff reaches the Kondo scale $D \to T_K$, the effective exchange coupling diverges, $g \to \infty$, indicating that the RG flows to the strong-coupling fixed point with complete Kondo screening. Thus, the Kondo scale, defined in this way, is generally finite even though $g \to \infty$. By equivalence of both Kondo sites, it is the same as the Kondo temperature T_K appearing in the local spin susceptibility χ_f on the second Kondo site (see above). This yields the defining equation for T_K [1],

$$\ln\left(\frac{T_K(y)}{T_K(0)}\right) = -y\,\alpha\, g_0^2 \frac{D_0}{T_K(y)} , \tag{9}$$

where $D_0/T_K(0) \gg 1$ was used and the coefficient $\alpha = \ln(\sqrt{2}+1)$ arises from fixing the integration constant by the condition $D(g \to \infty) = T_K(y)$. Note that this is an implicit or selfconsistent equation, because the β-function itself [right-hand side of Eq. (6)] depends parametrically on T_K via χ_f. The solution of Eq. (9) also indicates that in the 2IK system the Kondo scale is modified by the RKKY interaction [1], i.e., $T_K = T_K(y)$. In the absence of the RKKY interaction, the bare Kondo coupling, $g_0 \equiv g(D_0)$, satisfies

$$-\frac{1}{2g_0} = \ln\left(\frac{T_K(0)}{D_0}\right) . \tag{10}$$

Hence, by defining a rescaled Kondo temperature, $\tau(y) := \frac{T_K(y)}{T_K(0)}$, one obtains the equation for the RKKY-modified Kondo scale as

$$\tau(y) \equiv \frac{T_K(y)}{T_K(0)} = \mathrm{e}^{-y\,\alpha\, g_0^2 \frac{D_0}{T_K(0)}\,\tau(y)} . \tag{11}$$

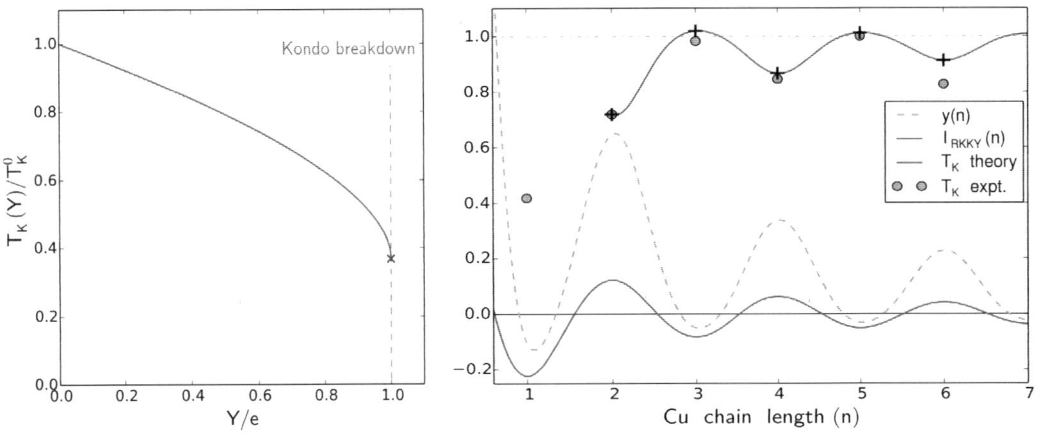

Figure 2: *Left panel:* The universal curve for the RKKY-modified Kondo temperature, $T_K(y)/T_K(0) = \frac{-y/(ey_c)}{W(-y/(ey_c))}$, as a function of the normalized RKKY strength parameter, y/y_c, where the critical RKKY parameter for Kondo breakdown is $y_c = 1/(\mathrm{e}\gamma_0)$ and $\mathrm{e} \approx 2.718$ is Euler's constant (see Eq. (12)). Complete Kondo singlet formation ceases to exist for $y > y_c$. *Right panel:* Oscillation and suppression of the Kondo temperature, $T_K(y(n))/T_K(0)$, for Cu chains of length n: Comparison of theory and experiment Ref. [2]. In the experiment, the single-ion Kondo scale is $T_K(0) = 110$ K. The overall amplitude of $T_K(y(n))/T_K(0)$ is obtained by scaling the theoretical value to the experimental one for $n = 2$. It is remarkable that the 1D infinite chain model accurately reproduces the experimantal data even for short chains of $n \geq 2$. The RKKY coupling, $I_{\mathrm{RKKY}}(n)$ (red solid line) and the RKKY strength parameter, $y(n)$, averaged over one lattice spacing a (dashed green line), are also shown (arbitrary units).

It can be solved in terms of the Lambert W function, [1]

$$\tau(y) = \frac{-\gamma_0\, y}{W(-\gamma_0\, y)}\,, \tag{12}$$

with the parameter,

$$\gamma_0 = g_0^2\, \alpha\, \frac{D_0}{T_K(0)} = \frac{\alpha}{4(\ln(T_K(0)/D_0))^2}\frac{D_0}{T_K(0)}\,. \tag{13}$$

The solution $\tau(y)$ is shown in Fig. 2, left panel [1]. It is seen that the solution of Eq. (12) ceases to exist when y exceeds the critical strength, $y > y_c = 1/(\mathrm{e}\gamma_0)$, indicating that the RG flow does not diverge, i.e. a strong coupling Kondo singlet is not formed for $y > y_c$. As seen from Eq. (13), the critical strength y_c is determined alone by the single-impurity Kondo scale $T_K(0)$. At the breakdown point, the RKKY-induced suppression ratio of the Kondo scale takes the universal, finite value $T_K(y_c)/T_K(0) = 1/\mathrm{e} \approx 0.368$, where e is Euler's constant [1]. We emphasize that, since the solution $1/\mathrm{e} \leq T_K(y)/T_K(0) \leq 1$ for all values of $1 \geq y \geq 0$, the term in brackets on the right-hand side of the RG equation (6) ($\beta-$function) remains always positive and less than one, i.e., the RG flow remains always in the perturbatively controlled regime.

[1] Indeed, the first branch of the Lambert function, W_0, is the solution. More information can be found in Ref. [33].

4. Comparison with experimental results and discussion

Having adapted the RG formalism of Ref. [1] to the Kondo suppression in a 2IK system RKKY-coupled by a 1D Cu chain, we are now in a position to compare with the corresponding experiments of Ref. [2]. All spatial dependence is cast in the distance dependence of the RKKY parameter $y(n)$, Eq. (7). Since the magnetic moment of a Co impurity atom is spread over a Co d−orbital occupying about one unit cell, we average $y(n)$ over one lattice spacing a of the underlying Cu lattice (green dashed curve in Fig. 2, right panel) and insert the result into Eq. (12) to calculate the RKKY-induced Kondo suppression ratio $T_K(y(n))/T_K(0)$ as a function of the Cu chain length. Since the RKKY wave number is fixed to $2k_F = \pi/a$ (half band filling, see section 2), this result has a single scale factor, γ_0, as the only adjustable parameter to fit the experimental results. We adjust it to fit the experimental data for $n = 2$. Note that, in fact, the *relative* suppression, $T_K(y(n))/T_K(0)$, with respect to the single-ion value of experimentally $T_K(0) = 110$ K [2] has no adjustable parameter at all. As seen in Fig. 2 (right panel), the agreement with the experimental data for $n \geq 2$ is quantitatively very good, with only a slight deviation for a Cu chain length of $n = 6$. It is remarkable that the 1D infinite chain model reproduces, without fit parameter, the periodicity as well as the relative amplitude of the $T_K(y(n))$ suppression accurately even for short chain lengths down to $n = 2$, considering the crudeness of this model. It is important to note that our theory explains that the RKKY interaction of either sign (ferro- or antiferromagnetic) suppresses the Kondo scale, as is physically expected and experimentally observed [2]. Only for the shortest Cu chain, $n = 1$, the theory does not fit the experiment. We attribute this to the fact that the simple infinite-chain model fails for this very short cluster, because the local DOS and, hence, the RKKY interaction are strongly influenced by boundary effects, the geometry of the wave functions and chemical bonds. More precise DFT calculations should be employed to calculate the RKKY coupling and the y-parameter in this case. Interestingly, however, the strongest experimentally observed suppression ratio is $T_K(y(n = 1))/T_K(0) = 46/110 \approx 0.42$ (Fig. 2, right panel), which is remarkably close to the theoretical prediction 1/e, considering that the impurity spacing cannot be changed continuously in the experiment.

In summary, we have given a short review of the recently developed renormalization group method to tackle the influence of the RKKY interaction on the Kondo screening in multi-impurity and lattice Kondo systems [1]. The application of this theory to the oscillation and suppression of the Kondo scale in Co-Cu_n-Co chains [2] yield generally good, quantitative agreement. This further supports the validity of this RG method. The deviation of the theory from the experiment for the shortest chain, $n = 1$, is attributed to boundary effects not included in the infinite chain model and may be improved by DFT calculations of the input parameters of the RG.

Acknowledgements

The authors are thankful for discussions with J. Kröger and N. Néel regarding the experimental results. This work was supported in part by DFG through SFB-TR 185.

References

[1] Nejati A and Kroha J 2016 *arXiv:1612.03338*
[2] Néel N, Berndt R, Kröger J, Wehling T O, Lichtenstein A I and Katsnelson M I 2011 *Phys. Rev. Lett.* **107** 106804
[3] Hewson A C 1997 *The Kondo Problem to Heavy Fermions* (Cambridge: Cambridge University Press)
[4] Coleman P 2007 *Handbook of Magnetism and Advanced Magnetic Materials* (Hoboken, NJ: Wiley) chap Heavy fermions: Electrons at the edge of magnetism
[5] De Haas W J, de Boer J and van den Berg G J 1934 *Physica* **1** 1115
[6] Ehm D, Hüfner S, Reinert F, Kroha J, Wölfle P, Stockert O, Geibel C and von Löhneysen H 2007 *PRB* **76** 045117

[7] Goldhaber-Gordon D, Shtrikman H, Mahalu D, Abusch-Magder D, Meirav U and Kastner M A 1998 *Nature* **391** 156
[8] Schmid J, Weis J, Eberl K and Klitzing K 1998 *Physica B* **256-258** 182
[9] van der Wiel W G W G, de Franceschi S, Fujisawa T, Elzerman J M, Tarucha S and Kouwenhoven L P 2000 *Science* **289** 2105
[10] Grobis M, Rau I G, Potok R M and Goldhaber-Gordon D 2007 *Handbook of Magnetism and Advanced Magnetic Materials* (Hoboken, NJ: Wiley) chap The Kondo effect in mesoscopic quantum dots
[11] Nygard J, Cobden D H and Lindelof P E 2000 *Nature* **408** 342
[12] Park J, Pasupathy A N, Goldsmith J I, Chang C, Yaish Y, Petta J R, Rinkoski M, Sethna J P, D H, McEuen P L and Ralph D C 2002 *Nature* **417** 722
[13] Scott G D and Natelson D 2010 *ACS Nano* **4** 3560
[14] Ruderman M A and Kittel C 1954 *Phys. Rev.* **96** 99
[15] Kasuya T 1956 *Prog. Theor. Phys.* **16** 45
[16] Yosida K 1957 *Phys. Rev.* **106** 893
[17] van Vleck J H 1962 *Rev. Mod. Phys.* **34** 681
[18] Jones B A, Varma C M and Wilkins J W 1988 *Phys. Rev. Lett.* **61** 125
[19] Jones B A and Varma C M 1989 *Phys. Rev. B* **40** 324
[20] Affleck I, Ludwig A W W and A J B 1995 *Phys. Rev. B* **52** 9528
[21] Mitchell A K, Derry P G and Logan D E 2015 *Phys. Rev. B* **91** 235127
[22] Craig N J, Taylor J M, Lester E A, Marcus C M, Hanson M P and Gossard A C 2004 *Science* **304** 565
[23] Sasaki S, Kang S, Kitagawa K, Yamaguchi M, Miyashita S, Maruyama T, Tamura H, Akazaki T, Hirayama Y and Takayanagi H 2006 *Phys. Rev. B* **73** 161303(R)
[24] Jeong H, Chang A M and Melloch M R 2001 *Science* **293** 2221
[25] Chang A M and Chen J C 2009 *Rep. Prog. Phys.* **72** 096501
[26] Chen W, Jamneala T, Madhavan V and Crommie M F 1999 *Phys. Rev. B* **60** R8529
[27] Bork J, Zhang Y H, Diekhöner L, Borda L, Simon P, Kroha J, Wahl P and Kern K 2011 *Nature Phys.* **7** 901
[28] Prüser H, Dargel P E, Bouhassoune M, Ulbrich R G, Pruschke T, Lounis S and Wenderot M 2014 *Nature Comm.* **5** 5417
[29] Ujsághy O, Kroha J, Szunyogh L and Zawadowski A 2000 *Phys. Rev. Lett.* **85** 2557
[30] Aristov D N 1997 *Phys. Rev. B* **55** 8064
[31] Economou E N 2006 *Green's functions in quantum physics* (Berlin: Springer)
[32] Andrei N, Furuya K and Lowenstein J H 1983 *Rev. Mod. Phys.* **55** 331
[33] Veberić D 2012 *Computer Phys. Comm.* **183** 2622; arXiv:1209.0735

IOP Conf. Series: Journal of Physics: Conf. Series **807** (2017) 092001 doi:10.1088/1742-6596/807/9/092001

The renormalized superperturbation theory (rSPT) approach to the Anderson model in and out of equilibrium

Enrique Muñoz

Facultad de Física, Pontificia Universidad Católica de Chile, Vicuña Mackenna 4860, Santiago, Chile

E-mail: munozt@fis.puc.cl

Farzaneh Zamani

Max Planck Institute for the Physics of Complex Systems, 01187 Dresden, Germany
Max Planck Institute for Chemical Physics of Solids, 01187 Dresden, Germany

E-mail: farzaneh@pks.mpg.de

Lukas Merker

Peter Grünberg Institut and Institute for Advanced Simulation, Research Centre Jülich, D-52425 Jülich, Germany

E-mail: l.merker@fz-juelich.de

Theo Costi

Peter Grünberg Institut and Institute for Advanced Simulation, Research Centre Jülich, D-52425 Jülich, Germany

E-mail: T.Costi@fz-juelich.de

Stefan Kirchner

Center for Correlated Matter, Zhejiang, University, Hangzhou, China

E-mail: stefan.kirchner@correlated-matter.com

Abstract. The properties of current-carrying steady states of strongly correlated systems away from the linear-response regime are of topical interest. In this article, we review the renormalized perturbation theory (rSPT), or renormalized SPT of reference [1] for the Anderson model. We present an extension to higher orders and compare the higher-order results with NRG calculations. Finally, we elucidate the role of Ward identities in calculating out-of-equilibrium properties and address claims made in the literature.

1. Introduction

Quantum dots as well as other artificial nanostructures, such as single-molecules attached to conducting leads, constitute very versatile devices, since their characteristic energy scales such as Fermi level, charging energy, level spacing can be tuned over a wide range [2, 3, 4]. The high degree of characterization of these systems has made it possible to investigate the properties of strongly-correlated electrons away from equilibrium in a well-defined setting [5, 6]. In particular, the low-temperature, low-bias differential conductance characteristics in these nanostructures often display a nearly universal conductance enhancement due to Kondo screening of the local, *i.e.*, localized on the quantum dot, dynamic degree of freedom by the lead electrons [7]. It is well established, that in equilibrium, i.e. within the linear response regime, the Kondo effect leads to an enhancement of the differential conductance $G = dI/dV|_{V=0}$ on the order of twice the quantum of conductance $2e^2/h$ in the limit $T \to 0$. These experimental results are usually interpreted within the context of the single-level Anderson model [8, 9]

$$H_{\rm AM} = \sum_{\lambda=L/R} H_0^\lambda(c_\lambda^\dagger, c_\lambda) + \sum_{\sigma=\pm} \epsilon_d d_\sigma^\dagger d_\sigma + U d_+^\dagger d_+ d_-^\dagger d_- + \sum_\lambda \sum_\sigma (V d_\sigma^\dagger c_{0\sigma\lambda} + \text{h.c.}) \tag{1}$$

as the effective low-energy model. In Eq. (1), H_0^λ is a free electron Hamiltonian given in terms of $c_\lambda^\dagger$ and c_λ which describes the excitations of lead λ, $c_{0\sigma\lambda}^\dagger$ is the corresponding creation operator of lead λ at the quantum dot site $\mathbf{r} = 0$. V denotes the coupling strength between the lead states and the quantum dot states, described by d_σ and U is the charging energy of the dot.

Correspondingly, a large class of conductance experiments has been accurately fitted by an expression of the form [5, 6, 10, 11, 12]

$$\frac{G(0,0) - G(T,V)}{c_T G(0,0)} = \left(\frac{T}{T_K}\right)^2 + \alpha \left(\frac{eV}{k_B T_K}\right)^2 - \gamma c_T \left(\frac{eVT}{k_B T_K^2}\right)^2, \tag{2}$$

where T_K is some characteristic energy scale which is obtained from the fit to the experimental results. When casting the low-temperature non-linear conductance of the Anderson model, Eq. (1) into the form (2), T_K becomes the Kondo temperature. The coefficients α and γ vary across different experimental systems [5, 6]. Already addressing how α and γ depend on the parameters of $H_{\rm AM}$ is a difficult task as it requires a proper description of the low-energy excitations of the Anderson model in a non-thermal steady state. While there is a rather complete picture of the equilibrium behavior of Eq. (1), such an understanding for the out-of-equilibrium properties is still lacking. This is largely due to the lack of reliable methods that can tackle the strongly correlated nature of the non-thermal steady state associated with $H_{\rm AM}$.

The equilibrium properties of the Anderson model have been studied extensively with the help of *e.g.* numerical renormalization group methods (NRG) [13], pseudoparticle [14] and Monte Carlo methods. Analytical methods have also been developed to study this problem, such as the bare perturbation theory of Yamada and Yosida [15, 16, 17]. An exact analytical solution has been obtained via Bethe Ansatz [18, 19, 20]. Unfortunately, fully reliable generalizations of these methods to the out-of-equilibrium problem are, at least at present, not available. However, a number of perturbative renormalization group (RG) methods have been proposed for non-equilibrium systems [21, 22], including functional perturbative RG [23]. Among the perturbative methods available, the renormalized perturbation theory (RPT) developed by Hewson [7, 24, 25, 26] provides a framework that can in principle be extended out of equilibrium, for instance by relying on exact Ward identities [27, 28]. The RPT provides an accurate description of the equilibrium Fermi-liquid regime in terms of quasi-particle excitations, characterized by effective interaction parameters [7, 24, 25].

Motivated by the above-mentioned experiments, following Refs. [25, 27], and in the spirit of Ref. [29], we recently developed the rSPT [1] - a renormalized superperturbation theory in terms of dual fermions [30, 31] on the Keldysh contour. Using this method, we studied steady-state non-equilibrium transport in the single-level Anderson model beyond particle-hole symmetry, by constructing a perturbative scheme based on a particle-hole symmetric reference system [1]. We constructed the reference system non-equilibrium self-energy at finite bias by extending Oguri's Ward identity approach [27, 32], which has the advantage of providing a current conserving expansion by construction [33, 1, 32]. The dual fermion, an auxiliary fermionic degree of freedom, is used to construct a systematic expansion around the interacting reference system [30].

Within the rSPT, we have studied the role of level asymmetry (gate voltage) and local Coulomb repulsion (charging energy) on the non-linear conductance of the single-level Anderson model in the steady-state regime [1, 34]. A comparison of our analytic results for the linear response transport coefficients with NRG calculations demonstrated an excellent quantitative agreement, even at relatively large level asymmetry [34]. Moreover, our results have recently provided a theoretical framework to describe and interpret magneto-transport experiments in single-molecule transistors [35].

In this article, we review the rSPT and extend the calculation to higher order (in the renormalized interaction) than reported before. In section 5, we compare the higher order calculation with NRG calculations for the linear-response transport coefficients. In the appendix, we exemplify the proof of Oguri's Ward identity [36] and address claims made in the literature regarding the validity of this identity.

2. Path-Integral representation on the Keldysh contour

We consider the single-level Anderson model, Eq. (1), which can be represented by a non-equilibrium functional on the Keldysh contour [1],

$$Z = \int \mathcal{D}[\hat{\psi}^\dagger, \hat{\psi}]\, \mathcal{D}[\hat{\Phi}^\dagger, \hat{\Phi}] e^{iS[\hat{\psi}^\dagger, \hat{\psi}, \hat{\Phi}^\dagger, \hat{\Phi}]}, \tag{3}$$

where the action in the Schwinger-Keldysh matrix representation is

$$\begin{aligned} S[\hat{\psi}^\dagger, \hat{\psi}, \hat{\Phi}^\dagger, \hat{\Phi}] &= \int_{-\infty}^{+\infty} dt \left\{ \sum_{k,\lambda,\sigma} \hat{\psi}^\dagger_{k\lambda\sigma}(t)\,(i\partial_t - \epsilon_{k\lambda})\,\hat{\sigma}_3 \hat{\psi}_{k\lambda\sigma}(t) + \sum_\sigma \hat{\Phi}^\dagger_\sigma(t)\,(i\partial_t - E_{d\sigma})\,\hat{\sigma}_3 \hat{\Phi}_\sigma(t) \right. \\ &\quad \left. + \sum_{k,\lambda,\sigma} \left[V_{k\lambda} \hat{\Phi}^\dagger_\sigma(t) \hat{\sigma}_3 \hat{\psi}_{k\lambda\sigma}(t) + h.c. \right] \right\} + i\, S^{int}[\hat{\Phi}^\dagger, \hat{\Phi}; U]. \end{aligned} \tag{4}$$

Here, the fields are two-component spinors in the index that distinguishes between the forward and backward branch of the Keldysh contour and σ_3 is the 3rd Pauli matrix in this space. The index $\lambda = L, R$ represents the left and right electrodes, U is the Coulomb interaction, and $E_{d\sigma} = \epsilon_{d\sigma} + U/2$ is the shift of the local resonance level with respect to the particle-hole symmetric condition $\epsilon_d = -U/2$. In the presence of a local magnetic field B, the local level has to be shifted by the Zeeman term, *i.e.*, $\epsilon_{d\sigma} = \epsilon_d - \sigma g \mu_B B/2$. The action defined by Eq.(4) is Gaussian in the Grassman fields $\hat{\psi}_{k\lambda\sigma}$ representing the lead electrons by virtue of their non-interacting nature. As a result, the $\hat{\psi}_{k\lambda\sigma}$ fields can be integrated out. This procedure yields an effective action for the localized electron states described by the Grassman fields $\hat{\Phi}_{\sigma\omega}$, which represent the d_σ of Eq. (1). In frequency-space this effective action is given by [1]

$$\begin{aligned} iS[\hat{\Phi}^\dagger, \hat{\Phi}; U, E_{d\sigma}, \Delta] &= i \int_{-\infty}^{+\infty} \frac{d\omega}{2\pi} \sum_\sigma \hat{\Phi}^\dagger_{\sigma\omega}\,(\omega + i\Delta)\,\hat{\sigma}_3 \hat{\Phi}_{\sigma\omega} + iS^{int}[\hat{\Phi}^\dagger, \hat{\Phi}; U] \\ &\quad - i \int_{-\infty}^{+\infty} \frac{d\omega}{2\pi} \sum_\sigma \hat{\Phi}^\dagger_{\sigma\omega} E_{d\sigma} \hat{\sigma}_3 \hat{\Phi}_{\sigma\omega}. \end{aligned} \tag{5}$$

Here, we have defined $\Delta = \Delta_L + \Delta_R$, and

$$i\Delta_\lambda = -\sum_{k,\lambda} \frac{|V_{k\lambda}|^2}{\omega - \epsilon_{k\lambda} + i\eta^+}. \tag{6}$$

In analogy to the equilibrium version of the renormalized theory (RPT) of Hewson [25, 37, 26], we can express the action Eq.(5) in terms of a set of "renormalized" parameters $\tilde{U}$, $\tilde{\epsilon}_{d\sigma}$ and $\tilde{\Delta}$. The fields are rescaled accordingly via the wave-function renormalization factor z given by $z^{-1} = 1 - \partial\Sigma^R_{\sigma\omega}/\partial\omega\Big|_{\omega=0}$.

Here, we seek to construct a renormalized perturbation theory to treat deviations from particle-hole symmetry, represented by the paramater $\tilde{\epsilon}_{d\sigma} = zE_{d\sigma}$ which is assumed to be small compared to the energy scale determined by the renormalized quasi-particle spectral width $\tilde{\Delta} = z\Delta$. Thus,

$$S[\hat{\Phi}^\dagger, \hat{\Phi}; U, E_{d\sigma}, \Delta] \;=\; S[\tilde{\Phi}^\dagger, \tilde{\Phi}; \tilde{U}, \tilde{\epsilon}_{d\sigma}, \tilde{\Delta}] + \delta S[\tilde{\Phi}^\dagger, \tilde{\Phi}; \lambda_1, \lambda_2, \lambda_3]. \tag{7}$$

In order to preserve the original action and thus avoid over counting, one must include counterterms proportional to the parameters λ_1, λ_2 and λ_3. These counterterms are defined as $\lambda_1 = -z\Sigma^R_\sigma(0,0)$, $\lambda_2 = z - 1$ and $\lambda_3 = z^2\left(U - \Gamma_{\uparrow\downarrow}(0,0)\right)$, respectively. Their values are determined in order to satisfy the RG conditions imposed on the retarded Green's function

$$\begin{aligned}
\tilde{\Sigma}^R_{\sigma\omega}\Big|_{\omega=0,T=0,V=0} &= 0, \\
\frac{\partial}{\partial\omega}\tilde{\Sigma}^R_{\sigma\omega}\Big|_{\omega=0,T=0,V=0} &= 0, \\
\tilde{\Gamma}_{\uparrow\downarrow}(0,0) = z^2\Gamma_{\uparrow\downarrow}(0,0) &= \tilde{U},
\end{aligned} \tag{8}$$

at every order in $\tilde{U}$. For the particle-hole symmetric case, the retarded self-energy in equilibrium has been obtained up to second order in $\tilde{u} = \tilde{U}/(\pi\tilde{\Delta})$ [25, 37, 26]. This equilibrium expression can be extended to non-thermal steady states, *e.g.* the one that ensues when the system is experiencing a voltage drop across the interaction region that is constant in time. This extension is based on Ward identities [36, 27, 1, 32]. As a result, one obtains an expression for the selfenergy of the local propagator near the strong-coupling fixed point and at particle-hole symmetry and zero magnetic field

$$\tilde{\Sigma}^R_{\sigma\omega} \;=\; -i\frac{\tilde{\Delta}\tilde{u}^2}{2}\left[\left(\frac{\omega}{\tilde{\Delta}}\right)^2 + \left(\frac{\pi T}{\tilde{\Delta}}\right)^2 + \zeta\left(\frac{V}{\tilde{\Delta}}\right)^2 - \frac{\zeta}{3}\left(\frac{\pi TV}{\tilde{\Delta}}\right)^2\right] + O(T^4, V^4, \tilde{u}^3). \tag{9}$$

At zero magnetic field, it has been shown by Hewson [25, 26] that the counterterm λ_1 cancels the Hartree contribution up to order $\tilde{u}^2$. To the same order, there are no contributions from the counterterms λ_2 and λ_3 [25, 26]. This renormalized action, exact up to terms of order $\tilde{u}^2$ constitutes the reference system around which our perturbative scheme of dual fermions is defined. In the next section, the bare dual fermion Green's function is introduced.

3. The dual Fermion Green's function

The bare dual fermion bare Green's function is given by the expression [1, 32]

$$\mathbf{G}^{f(0)}_{\sigma\omega} \;=\; -\mathbf{g}_{\sigma\omega}\left[\mathbf{g}_{\sigma\omega} - \mathbf{D}^{-1}_{\sigma\omega}\right]^{-1}\mathbf{g}_{\sigma\omega} = \sum_{n=1}^{\infty}\mathbf{g}_{\sigma\omega}\left[\mathbf{D}_{\sigma\omega}\mathbf{g}_{\sigma\omega}\right]^n. \tag{10}$$

Here, we have defined $\mathbf{D}_{\sigma\omega} = \tilde{\epsilon}_{d\sigma}\hat{\sigma}_3$. For convenience we switch from the dynamical index representation $+-$ to the trigonal representation [38] in what follows. Both representations are linked by a similarity transformation

$$\hat{\mathbf{G}}^f_{\sigma\omega} = \hat{L}\hat{\sigma}_3\mathbf{G}^f_{\sigma\omega}\hat{L}^\dagger = \begin{pmatrix} G^{f,R}_{\sigma\omega} & G^{f,K}_{\sigma\omega} \\ 0 & G^{f,A}_{\sigma\omega} \end{pmatrix}, \tag{11}$$

where we defined $\hat{L} = (1 - i\hat{\sigma}_2)/\sqrt{2}$ and $\hat{\sigma}_2$ is the 2nd Pauli matrix. Under this transformation the matrix $\mathbf{D}_{\sigma\omega} \to \hat{L}\hat{\sigma}_3\mathbf{D}_{\sigma\omega}\hat{L}^\dagger = \tilde{\epsilon}_{d\sigma}\mathbf{1}$ becomes proportional to the identity. Moreover, the retarded component of the dual fermion Green's function becomes

$$G^{f(0),R}_{\sigma\omega} = \sum_{n=1}^{\infty} (\tilde{\epsilon}_{d\sigma})^n \left(g^R_{\sigma\omega}\right)^{n+1} = \frac{\tilde{\epsilon}_{d\sigma}\left(g^R_{\sigma\omega}\right)^2}{1 - \tilde{\epsilon}_{d\sigma}g^R_{\sigma\omega}}. \tag{12}$$

The advanced Green's function then follows from the usual relation $G^{f(0),A}_{\sigma\omega} = \left(G^{f(0),R}_{\sigma\omega}\right)^*$. In the steady-state and up to $O(\tilde{u}^4)$, the Keldysh component of the dual fermion Green's function is given by [1, 32]

$$G^{f(0),K}_{\sigma\omega} = \left(1 - 2\tilde{f}(\omega, T, V)\right)\left(G^{f(0),A}_{\sigma\omega} - G^{f(0),R}_{\sigma\omega}\right). \tag{13}$$

Here, up to $O(\tilde{u}^4)$, the non-equilibrium distribution function is given by the expression [1, 32]

$$\tilde{f}(\omega, T, V) = \frac{\tilde{\Delta}_L f_0(\omega + \alpha_L V, T) + \tilde{\Delta}_R f_0(\omega - \alpha_R V, T)}{\tilde{\Delta}_L + \tilde{\Delta}_R}, \tag{14}$$

where $f_0(\omega, T)$ denotes the Fermi distribution and $\alpha_L + \alpha_R = 1$.

Note, that by substituting the definition of the retarded renormalized Green's function of the reference system

$$g^R_{\sigma\omega} = \left(\omega + i\tilde{\Delta} - \tilde{\Sigma}^R_{\sigma\omega}\right)^{-1} \tag{15}$$

into Eq.(12), one concludes that as $\tilde{\epsilon}_{d\sigma} \to 0$ then $G^{f(0),R}_{\sigma\omega} \to 0$, and hence the dual fermion contribution to the perturbed self-energy vanishes in the p-h symmetric limit, as expected.

On the other hand, in the limit $\tilde{\epsilon}_{d\sigma} \to \infty$, we have the non-trivial result [30, 1]

$$\lim_{\tilde{\epsilon}_{d\sigma}\to\infty} G^{f(0),R}_{\sigma\omega} = -g^R_{\sigma\omega}. \tag{16}$$

The relation between the dual fermion Green's function and the local Green's function associated with the $d^\dagger_\sigma, d_\sigma$ operators of Eq. (1) in Keldysh space is

$$\mathbf{G}_{\sigma\omega} = -\tilde{\epsilon}^{-1}_{d\sigma}\hat{\sigma}_3 + \left(\mathbf{g}_{\sigma\omega}\hat{\sigma}_3\tilde{\epsilon}_{d\sigma}\right)^{-1}\mathbf{G}^f_{\sigma\omega}\left(\hat{\sigma}_3\mathbf{g}_{\sigma\omega}\tilde{\epsilon}_{d\sigma}\right)^{-1}. \tag{17}$$

Here, the dressed dual fermion Green's function $\mathbf{G}^f_{\sigma\omega}$ is a solution of the Dyson equation

$$\mathbf{G}^f_{\sigma\omega} = \mathbf{G}^{f(0)}_{\sigma\omega} + \mathbf{G}^{f(0)}_{\sigma\omega}\mathbf{\Sigma}^f_{\sigma\omega}\mathbf{G}^f_{\sigma\omega}, \tag{18}$$

where $\mathbf{\Sigma}^f_{\sigma\omega}$ is the dual fermion self-energy which will be discussed in the next section.

It is convenient to express Eqs. (17) in the trigonal representation, by applying the transformation defined in Eq.(11), which leads to

$$\hat{\mathbf{G}}_{\sigma\omega} = \hat{L}\hat{\sigma}_3\mathbf{G}_{\sigma\omega}\hat{L}^{\dagger} = -\tilde{\epsilon}_{d\sigma}^{-1}\mathbf{I} + \tilde{\epsilon}_{d\sigma}^{-2}\hat{\mathbf{g}}_{\sigma\omega}^{-1}\hat{\mathbf{G}}_{\sigma\omega}^{f}\hat{\mathbf{g}}_{\sigma\omega}^{-1}. \tag{19}$$

Similarly, the Dyson equation becomes in the trigonal representation

$$\hat{\mathbf{G}}_{\sigma\omega}^{f} = \hat{\mathbf{G}}_{\sigma\omega}^{f(0)} + \hat{\mathbf{G}}_{\sigma\omega}^{f(0)}\hat{\mathbf{\Sigma}}_{\sigma\omega}^{f}\hat{\mathbf{G}}_{\sigma\omega}^{f}, \tag{20}$$

where the dual-fermion self-energy adopts the matrix form

$$\hat{\mathbf{\Sigma}}_{\sigma\omega}^{f} = \begin{pmatrix} \Sigma_{\sigma\omega}^{f,R} & \Sigma_{\sigma\omega}^{f,K} \\ 0 & \Sigma_{\sigma\omega}^{f,A} \end{pmatrix}. \tag{21}$$

In particular, solving for the retarded component of the non-equilibrium local Green's function from Eq.(17), we have

$$G_{\sigma\omega}^{R} = -\tilde{\epsilon}_{d\sigma}^{-1} + \tilde{\epsilon}_{d\sigma}^{-2}\left(g_{\sigma\omega}^{R}\right)^{-1}G_{\sigma\omega}^{f,R}\left(g_{\sigma\omega}^{R}\right)^{-1}. \tag{22}$$

This exact expression can be combined with the solution of the Dyson equation for the retarded component of the dual-fermion Green's function obtained from Eq.(18),

$$G_{\sigma\omega}^{f,R} = \frac{G_{\sigma\omega}^{f(0),R}}{1 - G_{\sigma\omega}^{f(0)}\Sigma_{\sigma\omega}^{f,R}}, \tag{23}$$

to obtain an explicit expression for the retarded component of the local non-equilibrium Green's function

$$G_{\sigma\omega}^{R} = \left(\omega - \tilde{\epsilon}_{d\sigma} + i\tilde{\Delta} - \tilde{\Sigma}_{E_d,\sigma\omega}^{R}\right)^{-1}. \tag{24}$$

Here, we have defined the retarded component of the non-equilibrium self-energy by

$$\tilde{\Sigma}_{E_d,\sigma\omega}^{R} = \tilde{\Sigma}_{\sigma\omega}^{R} + \frac{\Sigma_{\sigma\omega}^{f,R}}{1 + g_{\sigma\omega}^{R}\Sigma_{\sigma\omega}^{f,R}}. \tag{25}$$

4. The dual fermion self-energy

In order to obtain an approximation for the dual fermion self-energy, we consider the sum of ladder diagrams, with the effective quasi-particle interaction defined by the renormalized vertex $\tilde{U}$ of Eq. (8). In the trigonal representation, the non-equilibrium dual-fermion self-energy matrix, expressed in terms of the dual fermion vertex $\hat{\mathbf{\Gamma}}^{f}$, is

$$\hat{\mathbf{\Sigma}}_{\sigma\omega}^{f} = \frac{1}{2}\int_{-\infty}^{+\infty}\frac{d\omega'}{2i\pi}\left\{\Gamma_{\sigma,-\sigma}^{f,K}\left(\hat{\gamma}^{1}\hat{\mathbf{G}}_{-\sigma\omega'}^{f(0)}\hat{\gamma}^{1}\right) + \Gamma_{\sigma,-\sigma}^{f,A}\left(\hat{\gamma}^{1}\hat{\mathbf{G}}_{-\sigma\omega'}^{f(0)}\hat{\gamma}^{2}\right) + \Gamma_{\sigma,-\sigma}^{f,R}\left(\hat{\gamma}^{2}\hat{\mathbf{G}}_{-\sigma\omega'}^{f(0)}\hat{\gamma}^{1}\right)\right\}, \tag{26}$$

where $\hat{\gamma}^{1} = \mathbf{1}$ and $\hat{\gamma}^{2} = \hat{\sigma}_1$.

Here, the dual-fermion vertex components are defined by the matrix

$$\hat{\mathbf{\Gamma}}_{\sigma,-\sigma}^{f} = \left[\mathbf{1} - \tilde{U}\hat{\sigma}_1\hat{\mathbf{\Pi}}_{\sigma,-\sigma}^{f(0)}(\omega)\right]^{-1}\hat{\sigma}_1\tilde{U} = \begin{pmatrix} \Gamma_{\sigma,-\sigma}^{f,K} & \Gamma_{\sigma,-\sigma}^{f,R} \\ \Gamma_{\sigma,-\sigma}^{f,A} & 0 \end{pmatrix}, \tag{27}$$

where the polarisation matrix is defined by

$$\hat{\mathbf{\Pi}}^{f(0)}_{\sigma,-\sigma}(\omega) = \begin{pmatrix} 0 & \Pi^{A}_{\sigma,-\sigma}(\omega) \\ \Pi^{R}_{\sigma,-\sigma}(\omega) & \Pi^{K}_{\sigma,-\sigma}(\omega) \end{pmatrix}. \tag{28}$$

The different components of the polarization insertion in the trigonal representation are given by the expressions [38]

$$\begin{aligned}
\Pi^{f,K}_{\sigma,-\sigma}(\omega) &= -\frac{1}{2}\int_{-\infty}^{+\infty}\frac{d\omega'}{2i\pi}\left\{G^{f(0),K}_{\sigma,\omega+\omega'}G^{f(0),K}_{-\sigma\omega'} + G^{f(0),R}_{\sigma,\omega+\omega'}G^{f(0),A}_{-\sigma\omega'} + G^{f(0),A}_{\sigma,\omega+\omega'}G^{f(0),R}_{-\sigma\omega'}\right\}, &(29)\\
\Pi^{f,R}_{\sigma,-\sigma}(\omega) &= -\frac{1}{2}\int_{-\infty}^{+\infty}\frac{d\omega'}{2i\pi}\left\{G^{f(0),R}_{\sigma,\omega+\omega'}G^{f(0),K}_{-\sigma\omega'} + G^{f(0),K}_{\sigma,\omega+\omega'}G^{f(0),A}_{-\sigma\omega'}\right\}, &(30)\\
\Pi^{f,A}_{\sigma,-\sigma}(\omega) &= \left[\Pi^{f,R}_{\sigma,-\sigma}(\omega)\right]^{*}. &(31)
\end{aligned}$$

In particular, at $\omega = 0$, $T = 0$, $V = 0$, and to zero order in $\tilde{U}$, one has

$$\Pi^{f(0),R}_{\sigma,-\sigma}(0) = \Pi^{f(0),A}_{\sigma,-\sigma}(0) = \frac{1}{\pi\tilde{\Delta}} + \frac{1}{\pi\tilde{\Delta}}\frac{\tilde{\epsilon}^{2}_{d,-\sigma}\tan^{-1}(\tilde{\epsilon}_{d\sigma}) - \tilde{\epsilon}^{2}_{d\sigma}\tan^{-1}(\tilde{\epsilon}_{d,-\sigma})}{\tilde{\epsilon}_{d\sigma}\tilde{\epsilon}_{d,-\sigma}\left(\tilde{\epsilon}_{d\sigma} - \tilde{\epsilon}_{d,-\sigma}\right)}. \tag{32}$$

The dual fermion vertex components are thus obtained as

$$\begin{aligned}
\Gamma^{f,R}_{\sigma,-\sigma} &= \frac{\tilde{U}_s}{1 - \tilde{U}_s\Pi^{f,R}_{\sigma-\sigma}(\omega)},\\
\Gamma^{f,A}_{\sigma,-\sigma} &= \frac{\tilde{U}_s}{1 - \tilde{U}_s\Pi^{f,A}_{\sigma,-\sigma}(\omega)},\\
\Gamma^{f,K}_{\sigma,-\sigma} &= -\frac{\tilde{U}_s^{2}\Pi^{f,K}_{\sigma,-\sigma}(\omega)}{\left(1 - \tilde{U}_s\Pi^{f,R}_{\sigma,-\sigma}(\omega)\right)\left(1 - \tilde{U}_s\Pi^{f,A}_{\sigma,-\sigma}(\omega)\right)}. &(33)
\end{aligned}$$

Note that the Keldysh component of the dual fermion vertex is of higher order in the effective quasiparticle interaction $\tilde{U}_s$ than the retarded and advanced components.

In order to obtain analytical approximations to the transport coefficients, defined in Eq. (2), which are exact up to $O(\tilde{u}^3)$, we define an effective dual-fermion interaction from the expression for the dual-fermion vertex at $\omega = 0$, $V = 0$, and $T = 0$ by

$$\tilde{U}^{f}_{\sigma,-\sigma} = \Gamma^{f,A}_{\sigma,-\sigma}\Big|_{\omega=0,V=0,T=0} = \Gamma^{f,R}_{\sigma,-\sigma}\Big|_{\omega=0,V=0,T=0} = \frac{\tilde{U}_s}{1 - \tilde{U}_s\Pi^{f,R}_{\sigma,-\sigma}(0)}. \tag{34}$$

Up to order $O(\tilde{u}^3)$ we have that the self-energy of the non-symmetric system defined by Eq.(25) reduces to the expression

$$\tilde{\Sigma}^{R}_{E_d,\sigma\omega} = \tilde{\Sigma}^{R}_{\sigma\omega} + \Sigma^{f,R}_{\sigma\omega} + O(\tilde{u}^3). \tag{35}$$

5. Transport coefficients

With the results of the previous sections, we are in a position to calculate the experimentally accessible transport coefficients α and γ defined in Eq. (2). From the Meir-Wingreen expression [39] for the charge current through the quantum dot,

$$I = \frac{e}{2\hbar}\sum_{\sigma}\int d\omega\frac{4\Delta_L\Delta_R}{\Delta_L + \Delta_R}A_\sigma(\omega,T,V)\left[f_L(\omega) - f_R(\omega)\right] \tag{36}$$

where $f_{L,R} = f(\omega - \mu_{L,R})$ is the Fermi function for the left/right lead, respectively, and $A_\sigma(\omega, T, V) = -\pi^{-1}\Im G^R_{\sigma\omega}$ is the local spectral function. Expanding the differential conductance $dI/dV \equiv G(T, V, B)$ up to second order in V, T, and B, one obtains the expression [1]

$$\frac{G(T,0,B) - G(T,V,B)}{G(0,0,0)} = c_V \left(\frac{eV}{\tilde{\Delta}}\right)^2 - c_{TV}\left(\frac{eV}{\tilde{\Delta}}\right)^2\left(\frac{k_BT}{\tilde{\Delta}}\right)^2 - c_{VE_d}\left(\frac{eV}{\tilde{\Delta}}\right) + c_{TVE_d}\left(\frac{eV}{\tilde{\Delta}}\right)^2. \tag{37}$$

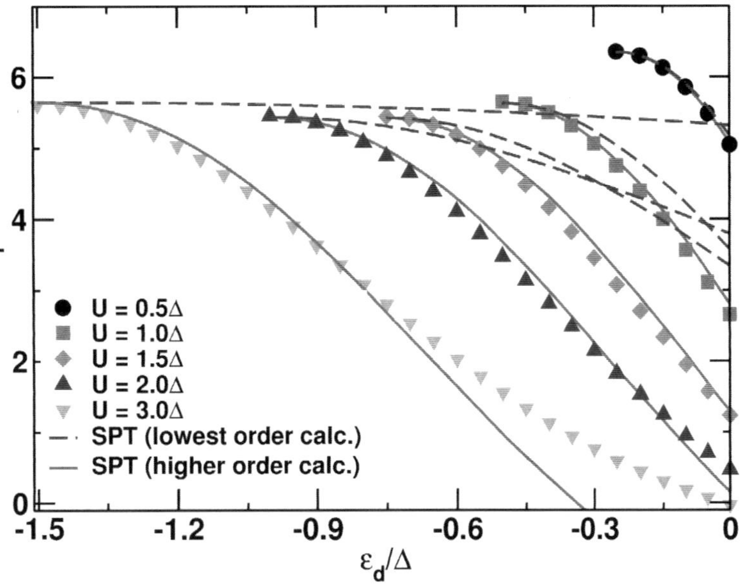

Figure 1. The transport coefficient c_T versus the asymmetry parameter ϵ_d/Δ: Continuous lines are the renormalized rSPT results including higher-order terms. Dashed lines are the renormalized rSPT results at the lowest order. Symbols are the NRG results for the same set of parameters. The particle-hole symmetric reference system is located at $\epsilon_d = -U/2$.

In particular, for the linear-response conductance ($V \to 0$), we have the relations

$$\begin{aligned} \frac{G(T, V=0, B=0)}{G(0,0,0)} &= 1 - c_T\left(\frac{k_BT}{\tilde{\Delta}}\right)^2, \\ \frac{G(T=0, V=0, B)}{G(0,0,0)} &= 1 - c_B\left(\frac{g\mu_BB}{\tilde{\Delta}}\right)^2. \end{aligned} \tag{38}$$

In Fig. 1, we show the renormalized rSPT result for c_T, defined in Eq.(38), obtained by including the higher-order dual-fermion contributions as described in section 4. The linear-response conductance can also be calculated via the NRG, which yields essentially exact results for this quantity [13, 34]. Clearly, the renormalized rSPT results for c_T are in excellent agreement with NRG calculations, even at relatively large values of the asymmetry parameter ϵ_d/Δ and Coulomb interaction U. We also display the results (dashed lines in Fig. 1) obtained from the rSPT calculation at the lowest order [34], where the improvement upon inclusion of higher order contributions is evident. A similar level of agreement with the NRG is obtained for the transport coefficient c_B, as shown in Fig. 2.

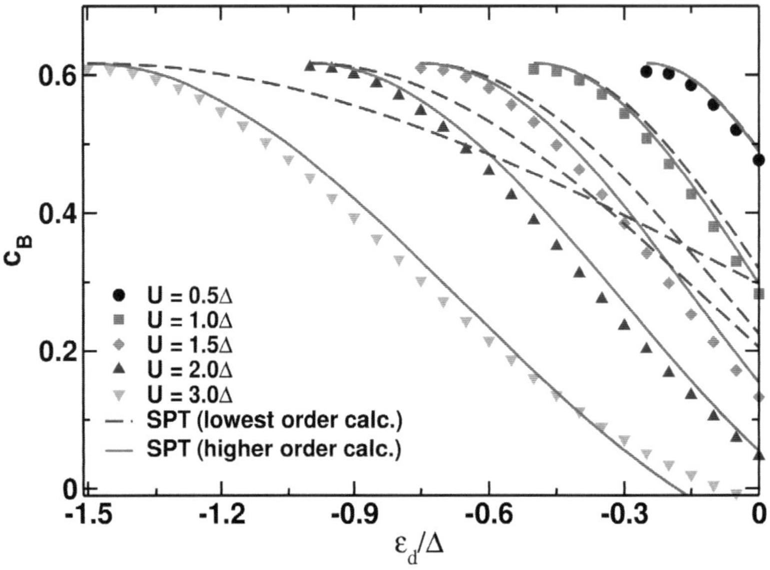

Figure 2. The transport coefficient c_B as a function of the asymmetry parameter ϵ_d/Δ: Continuous lines are the renormalized rSPT results including higher-order terms. Dashed lines are the renormalized rSPT results at the lowest order. Symbols are the NRG results for the same set of parameters.

6. Conclusion

In summary, we have reviewed the renormalized rSPT approach to non-thermal steady states in the Anderson model. By construction, this approach captures the strong-coupling nature of the fixed point. We present an extension to higher orders and compared these higher-order results to NRG calculations. As a result, the renormalized rSPT predictions are in excellent agreement with the (essentially) exact NRG results. The renormalized rSPT is a versatile approach for the out-of-equilibrium properties at sufficiently low energies and temperatures that can also be applied to more complicated models or more general steady states, that ensue by applying finite voltage and temperature differences across the system [40]. We also elucidated the role of Ward identities in calculating the out-of-equilibrium properties and clarified several controversial statements that appeared in the literature.

Acknowledgments

E. Muñoz acknowledges financial support by Fondecyt (Chile) under contract No.1141146. S. Kirchner acknowledges partial support by the National Science Foundation of China, grant No.11474250, and the National Key R&D Program of the MOST of China (No.2016YFA0300202).

Appendix A: Violation of Oguri's Ward identity in certain RPT schemes

In this appendix we briefly clarify a few controversial statements that appeared in the literature on the validity and the range of applicability of Oguri's Ward identity. Particle number conservation and the conservation of spin in a magnetic field lead to relations among derivatives of the proper local selfenergy $\Sigma(\omega)$ of the Anderson Hamiltonian, Eq. (1), with respect to $E_{d\sigma}$, magnetic field B, and ω through Ward identities. These identities were first obtained by Yamada and Yosida [16]. The extensions to finite voltages is due to Oguri [36] who showed that

$$\frac{\partial}{\partial V}\tilde{\mathbf{\Sigma}}(\omega,T,V)\Big|_{V=0} = -\gamma\left(\frac{\partial}{\partial\omega}+\frac{\partial}{\partial\tilde{\epsilon}_d}\right)\tilde{\mathbf{\Sigma}}(\omega,T,V=0), \tag{39}$$

here given in terms of renormalized quantities and where $\tilde{\mathbf{\Sigma}}$ is the selfenergy matrix on the Keldysh contour as in Eq.(21). The parameter $\gamma = (\Delta_L\alpha_L - \Delta_R\alpha_R)/(\Delta_L+\Delta_R)$. To demonstrate that the relation (39) is valid at each order of perturbation theory, we here explicitly demonstrate that the sunset diagram, shown in Fig. 3, fulfills Oguri's Ward identity. A general proof can be obtained following similar arguments [27].

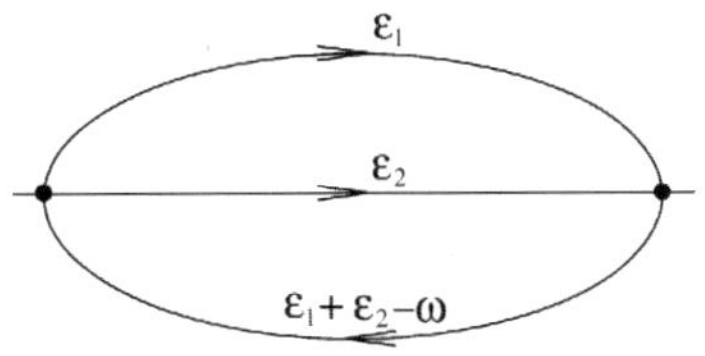

Figure 3. The sunset diagram appears in second order in the renormalized coupling constant and is part of the 'reference system' of the renormalized rSPT approach.

Specifically, the lesser component of $\tilde{\mathbf{\Sigma}}$ is considered which, on the Keldysh contour, is simply given by

$$\tilde{\Sigma}^<(\tau) = \tilde{U}^2\tilde{G}^<(\tau)\tilde{G}^<(\tau)\tilde{G}^>(-\tau) \tag{40}$$

or after Fourier transformation to frequency variables

$$\tilde{\Sigma}^<(\omega,T,V) = -2\pi i\tilde{U}^2\int d\epsilon_1\int d\epsilon_2\tilde{f}(\epsilon_1)\tilde{f}(\epsilon_2)\tilde{f}(\omega-\epsilon_1-\epsilon_2)\tilde{A}(\epsilon_1)\tilde{A}(\epsilon_2)\tilde{A}(\epsilon_1+\epsilon_2-\omega),$$

where

$$\tilde{A}(x) = \frac{1}{\pi}\frac{\tilde{\Delta}}{(x-\tilde{\epsilon}_d)^2+\tilde{\Delta}^2} \tag{41}$$

is the bare local renormalized spectral density which obeys the identity

$$\left(\frac{\partial}{\partial x}+\frac{\partial}{\partial\tilde{\epsilon}_d}\right)\tilde{A}(x) = 0. \tag{42}$$

Note that here, $\tilde{\epsilon}_d = zE_d$ is the renormalized, voltage and temperature independent local resonance level. The effective distribution function $\tilde{f}(\omega,V)$ is given by the expression

$$\tilde{f}(\omega,V) = \sum_{\lambda=L,R}\frac{\Delta_\lambda}{\Delta_L+\Delta_R}f_0(\omega-\alpha_\lambda V), \tag{43}$$

where the chemical potential at lead $\lambda = L,R$ is $\mu_L = \alpha_L V$, $\mu_R = -\alpha_R V$, with $\alpha_L+\alpha_R = 1$. Equation (40) implies

$$\begin{aligned} \lim_{V\to 0}\tilde{f}(\omega,V) &= f_0(\omega), \\ \frac{\partial\tilde{f}(\epsilon,V)}{\partial V}\Big|_{V=0} &= \gamma\left(-\frac{\partial f_0(\epsilon)}{\partial\epsilon}\right), \end{aligned} \tag{44}$$

where f_0 is the usual Fermi function.

Consider the voltage derivative of the self-energy expression in Eq. (40), which can be written

$$\frac{\partial}{\partial V}\tilde{\Sigma}^{<}(\omega,T,V)\Big|_{V=0} = -2\pi i\tilde{U}^2\gamma\left(I_1+I_2+I_3\right), \tag{45}$$

with the help of Eq. (44) and where we have defined

$$\begin{aligned} I_1 &= \int d\epsilon_1 \int d\epsilon_2 \tilde{A}(\epsilon_2)\tilde{A}(\epsilon_1+\epsilon_2-\omega)f_0(\epsilon_2)f_0(\omega-\epsilon_1-\epsilon_2)\left\{\tilde{A}(\epsilon_1)\left(-\frac{\partial f_0(\epsilon_1)}{\partial\epsilon_1}\right)\right\} \\ I_2 &= \int d\epsilon_1 \int d\epsilon_2 \tilde{A}(\epsilon_1)\tilde{A}(\epsilon_1+\epsilon_2-\omega)f_0(\epsilon_1)f_0(\omega-\epsilon_1-\epsilon_2)\left\{\tilde{A}(\epsilon_2)\left(-\frac{\partial f_0(\epsilon_2)}{\partial\epsilon_2}\right)\right\} \\ I_3 &= \int d\epsilon_1 \int d\epsilon_2 \tilde{A}(\epsilon_1)\tilde{A}(\epsilon_2)f_0(\epsilon_1)f_0(\epsilon_2)\left\{\tilde{A}(\epsilon_1+\epsilon_2-\omega)\left(-\frac{\partial}{\partial\epsilon_1}f_0(\omega-\epsilon_1-\epsilon_2)\right)\right\}. \end{aligned} \tag{46}$$

We now focus on the term I_1 of Eq.(45) and apply the identity

$$\tilde{A}(\epsilon_1)\left(-\frac{\partial f_0(\epsilon_1)}{\partial\epsilon_1}\right) = -\tilde{A}(\epsilon_1)\left(\frac{\partial}{\partial\epsilon_1}+\frac{\partial}{\partial\tilde{\epsilon}_d}\right)f_0(\epsilon_1) = -\left(\frac{\partial}{\partial\epsilon_1}+\frac{\partial}{\partial\tilde{\epsilon}_d}\right)\left\{\tilde{A}(\epsilon_1)f_0(\epsilon_1)\right\}. \tag{47}$$

Thus, inserting Eq. (47) into the integral I_1 we have

$$I_1 = -\int d\epsilon_1 \int d\epsilon_2 \tilde{A}(\epsilon_2)\tilde{A}(\epsilon_1+\epsilon_2-\omega)\tilde{f}(\epsilon_2)\tilde{f}(\omega-\epsilon_1-\epsilon_2)\left(\frac{\partial}{\partial\epsilon_1}+\frac{\partial}{\partial\tilde{\epsilon}_d}\right)\left\{\tilde{A}(\epsilon_1)f_0(\epsilon_1)\right\}. \tag{48}$$

Shifting the integration variable $\epsilon_1 \to \epsilon_1+\omega$, leads to

$$\begin{aligned} I_1 &= -\int d\epsilon_1 \int d\epsilon_2 \tilde{A}(\epsilon_2)\tilde{A}(\epsilon_1+\epsilon_2)f_0(\epsilon_2)f_0(-\epsilon_1-\epsilon_2)\left(\frac{\partial}{\partial\omega}+\frac{\partial}{\partial\tilde{\epsilon}_d}\right)\left\{\tilde{A}(\epsilon_1+\omega)f_0(\epsilon_1+\omega)\right\} \\ &= -\left(\frac{\partial}{\partial\omega}+\frac{\partial}{\partial\tilde{\epsilon}_d}\right)\int d\epsilon_1 \int d\epsilon_2 \tilde{A}(\epsilon_2)\tilde{A}(\epsilon_1+\epsilon_2)f_0(\epsilon_2)f_0(-\epsilon_1-\epsilon_2)\tilde{A}(\epsilon_1+\omega)f_0(\epsilon_1+\omega) \\ &= -\left(\frac{\partial}{\partial\omega}+\frac{\partial}{\partial\tilde{\epsilon}_d}\right)\int d\epsilon_1 \int d\epsilon_2 \tilde{A}(\epsilon_2)\tilde{A}(\epsilon_1+\epsilon_2-\omega)f_0(\epsilon_2)f_0(\omega-\epsilon_1-\epsilon_2)\tilde{A}(\epsilon_1)f_0(\epsilon_1), \end{aligned} \tag{49}$$

where, in the last line, we have shifted back the integration variable $\epsilon_1 \to \epsilon_1-\omega$.

The same type of manipulations can be applied to integrals I_2 and I_3. Adding the three terms together, we therefore obtain

$$\frac{\partial}{\partial V}\tilde{\Sigma}^{<}(\omega,T,V)\Big|_{V=0} = -\gamma\left(\frac{\partial}{\partial\omega}+\frac{\partial}{\partial\tilde{\epsilon}_d}\right)\tilde{\Sigma}^{<}(\omega,T,V=0), \tag{50}$$

which proves the validity of Oguri's Ward identity explicitly for the second order diagram of Fig. 3. A general perturbative proof can be constructed along similar lines [36]. Nonetheless, it has been argued that the identity is only valid at $V=0$ and that the limits $\omega\to 0$ and $V\to 0$ do not commute at $T=0$ but do so at non-zero temperature [41].

While there are no convincing indications that the strong-coupling fixed point at $T=0$, $\omega=0$, $V=0$ is singular, it should be clear that an important element in the explicit proof provided above, is the fact that the renormalized resonance level $\tilde{\epsilon}_d$ is independent of voltage and temperature. If this were not the case, there would be additional terms proportional to first-order-in-voltage derivatives of the type $\partial\tilde{\epsilon}_d/\partial V|_{V=0}$. As a result, it would no longer be possible to collect all terms in such a way as to satisfy the Ward identity. This is exactly what happens in the non-equilibrium self-consistent perturbative scheme presented in [42]. There, the bare local renormalized spectral density of Eq. (41) is taken to be

$$\tilde{A}(x) = \frac{1}{\pi} \frac{\tilde{\Delta}}{(x - \tilde{\Delta}\cot(\pi\langle n_{d\sigma}\rangle))^2 + \tilde{\Delta}^2}, \tag{51}$$

where $\langle n_{d\sigma}\rangle$ is the average local occupation number of spin projection σ and is obtained from

$$\langle n_{d\sigma}\rangle = -i \int d\omega\, \tilde{G}^<_\sigma(\omega, T, V). \tag{52}$$

Notice that $\tilde{G}^<$ is related to $\tilde{\Sigma}^<$ via the identity $\tilde{G}^< = \tilde{G}^A\tilde{\Sigma}^<\tilde{G}^R$, where the hybridization term $i\tilde{\Delta}$ is included in the self-energy components [43]. Thus, in such a perturbative scheme, the Ward identity is violated as already pointed out in reference [43]. Moreover, this violation occurs even if only the Hartree value for $\langle n_{d\sigma}\rangle$ is used in Eq.(51).

7. References

[1] Muñoz E, Bolech C J and Kirchner S 2013 *Phys. Rev. Lett.* **110** 016601
[2] Goldhaber-Gordon D, Göres J, Kastner M A, Shtrikman H, Mahalu D and Meirav U 1998 *Phys. Rev. Lett.* **81** 5225
[3] Cronenwett S M, Oosterkamp T H and Kouwenhoven L P 1998 *Science* **24** 540–544
[4] Potok R M, Rau I G, Shtrikman H, Oreg Y and Goldhaber-Gordon D 2007 *Nature* **446** 167
[5] Grobis M, Rau I G, Potok R M, Shtrikman H and Goldhaber-Gordon D 2008 *Phys. Rev. Lett.* **100** 246601
[6] Scott G D, Keane Z K, Ciszek J W, Tour J M and Natelson D 2009 *Phys. Rev. B* **79** 165413
[7] Hewson A C 1993 *The Kondo Problem to Heavy Fermions* (Cambridge: Cambridge University Press)
[8] Glazman L I and Raikh M E 1988 *JETP Lett.* **47** 452
[9] Ng T K and Lee P A 1988 *Phys. Rev. Lett.* **61** 1768
[10] Kaminski A, Nazarov Y V and Glazman L I 2000 *Phys. Rev. B* **62** 8154
[11] Doyon B and Andrei N 2006 *Phys. Rev. B* **73** 245326
[12] Schiller A and Hershfield S 1995 *Phys. Rev. B* **51** 12896
[13] Costi T A, Hewson A C and Zlatić V 1994 *J. Phys. C* **6** 2519
[14] Kirchner S, Kroha J and Wölfle P 2004 *Phys. Rev. B* **70** 165102
[15] Yosida K and Yamada K 1970 *Prog. Theor. Phys. Suppl.* **46** 244
[16] Yamada K 1975 *Prog. Theor. Phys.* **54** 316
[17] Yamada K 1979 *Prog. Theo. Phys.* **62** 354
[18] Andrei N, Furuya K and Lowenstein J H 1983 *Rev. Mod. Phys.* **55** 331
[19] Tsvelick A and Wiegmann P 1983 *Adv. Phys.* **32** 453
[20] Zlatić V and Horvatić B 1983 *Phys. Rev. B* **28** 6904
[21] Rosch A, Paaske J, Kroha J, and Wölfle P 2003 *Phys. Rev. Lett.* **90** 076804
[22] Rosch A, Paaske J, Kroha J, and Wölfle P 2005 *J. Phys. Soc. Jpn.* **74** 118
[23] Jakobs S G, Meden V, and Schoeller H 2007 *Phys. Rev. Lett.* **99** 150603
[24] Hewson A C 1993 *Phys. Rev. Lett.* **70** 4007
[25] Hewson A C 2001 *J. Phys.: Condens. Matter* **13** 10011
[26] Hewson A C 2011 *J. Phys.: Condens. Matter* **23** 045601
[27] Oguri A 2005 *J. Phys. Soc. Jpn* **74** 110
[28] Hewson A C, Bauer J and Oguri A 2005 *J. Phys.:Condens. Matter* **17** 5413
[29] Hafermann H, Jung C, Brenner S, Katnelson M I, Rubtsov A N and Lichtenstein A I 2009 *EPL* **85** 27007
[30] Rubtsov A N, Katsnelson M I and Lichtenstein A I 2008 *Phys. Rev. B* **77** 033101
[31] Jung C, Lieder A, Brener S, Hafermann H, Baxevanis B, Chudnovskiy A, Rubtsov A N, Katsnelson M I and Lichtenstein A I 2011 *Ann. Phys.* **1-16**
[32] Muñoz E, Bolech C J and Kirchner S 2013 Supplemental material to Ref. [1]
[33] Baym G 1962 *Phys. Rev.* **127** 1391
[34] Merker L, Kirchner S, Muñoz E and Costi T A 2013 *Phys. Rev. B* **87** 165132
[35] Scott G D, Natelson D, Kirchner S and Muñoz E 2013 *Phys. Rev. B* **87** 241104(R)
[36] Oguri A 2001 *Phys. Rev. B* **64** 153305
[37] Hewson A C 2006 *J. Phys.: Condens. Matter* **18** 1815

[38] Kamenev A 2011 *Field Theory of Non-Equilibrium Systems* (Cambridge University Press) chapter 9, sections 5–8

[39] Meir Y and Wingreen N 1992 *Phys. Rev. Lett.* **68** 2512–2515

[40] Kirchner S, Zamani F and Muñoz E 2012 *New Materials for Thermoelectric Applications: Theory and Experiment* (Springer) chap 10 Nonlinear thermoelectric response of quantum dots: renormalized dual fermions out of equilibrium ISBN: 978-9400749863 (2012)

[41] Aligia A A 2013 Reply to the reply to 'comment on universal out-of-equilibrium transport in Kondo-correlated quantum dots' arXiv:1310.8324

[42] Aligia A A 2012 *J. Phys.:Condens. Matter* **24** 015306

[43] Muñoz E, Bolech C J and Kirchner S 2013 *Phys. Rev. Lett.* **111** 089702

Current inversion in a chiral *d*-wave superconductor due to surface roughness

Shu-Ichiro Suzuki[1] **and Yasuhiro Asano**[1,2,3]

[1] Department of Applied Physics, Hokkaido University, Kita 13 Nishi 8, Sapporo 060-8628, Japan

[2] Center of Topological Science and Technology, Hokkaido University, Kita 13 Nishi 8, Sapporo 060-8628, Japan

[3] Moscow Institute of Physics and Technology, 141700 Dolgoprudny, Russia

E-mail: shu.suzuki@eng.hokudai.ac.jp

Abstract. We report the effects of surface roughness on the edge current in a chiral *d*-wave superconductors. By solving the quasiclassical Eilenberger equation and the Maxwell equation simultaneously, we obtained self-consistent solutions of the pair potential and the vector potential. We calculate numerically the spatial distribution of the chiral edge current in a small superconductor with and without surface roughness. The edge current in a *d*-wave superconductor is robust against surface roughness. Its direction, however, is inverted by surface roughness.

1. Introduction

The experimental detection of a spontaneous edge current is essential to demonstrate chiral superconductivity. The angular momentum of Cooper pairs give rise to the edge current in chiral superconductors [1, 2]. A leading candidate for a chiral superconductor is Strontium ruthenate Sr_2RuO_4 [3–5] in which chiral *p*-wave superconductivity is believe to realized. In addition to chiral *p*-wave superconductors, the possibilities of chiral *d*-wave superconductivity have been discussed in several materials [6–16]. Several theories have concluded that the amount of edge current becomes smaller in a chiral *d*-wave superconductor compared to the chiral *p*-wave case even in the clean limit [17, 18].

To detect the spontaneous edge current, we can not avoid the problem of surface roughness. Experimentally it is difficult to make a superconducting sample with a specular surface. The surface roughness of a sample would greatly affect the chiral edge current in a chiral *d*-wave superconductor. In the case of *non-chiral* *d*-wave superconductors, the surface Andreev bound states [21,22] are fragile against surface roughness [19,20]. We can infer from this conclusion that the edge states in a chiral *d*-wave superconductor are also fragile against the surface roughness. Thus we study the effects of surface roughness on the chiral edge current in chiral *d*-wave superconductors.

In this paper, we theoretically study the spontaneous edge currents and the spontaneous field in a small chiral-*d*-wave superconducting disk based on the quasiclassical Eilenberger formalism. By solving the Eilenberger and Maxwell equations self-consistently and simultaneously, we obtain the spatial profiles of the chiral edge currents and the temperature dependence of a spontaneous

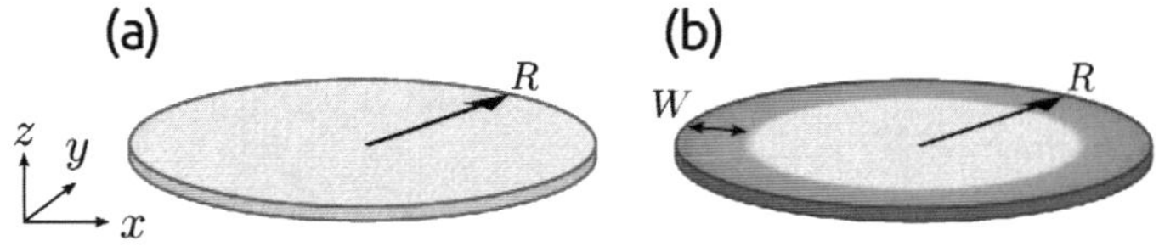

Figure 1. Schematics of two-dimensional superconducting disks. The disk with a specular surface and that with a rough surface are shown in (a) and (b), respectively. The radius of the disk is denoted by R. The width of the disordered region is denoted by W in (b).

magnetization. The surface roughness is considered in terms of the impurity self-energy near a surface. The edge current in a d-wave superconductor is robust against surface roughness. However, the edge current flows in the opposite direction compared with that in the clean limit.

2. Formulation

We consider a small two-dimensional disk of a chiral superconductor as shown in Fig. 1. We assume there is no chiral-domain wall in a disk by choosing the radius of the disk R to be comparable to the coherence length. We apply the quasiclassical Green-function theory [23] to investigate the spontaneous chiral current in a chiral superconductor. The Eilenberger equation for spin-singlet superconductors in equilibrium is represented as

$$iv_F \boldsymbol{k} \cdot \boldsymbol{\nabla}_{\boldsymbol{r}} \hat{g} + \left[\hat{H} + \hat{\Sigma} , \hat{g} \right]_- = 0, \tag{1}$$

where v_F and $\boldsymbol{k}$ are the Fermi velocity and the unit wave vector on the Fermi surface, respectively. We assume the Fermi surface is isotropic. Throughout this paper, we use the set of units $\hbar = k_B = c = 1$, where $2\pi\hbar$ is the Planck constant, k_B is the Boltzmann constant, and c is the speed of light. The matrices $\hat{g}$ and $\hat{H}$ are defined as follows,

$$\hat{g}(\boldsymbol{r}, \boldsymbol{k}, i\omega_n) = \begin{bmatrix} g(\boldsymbol{r}, \boldsymbol{k}, i\omega_n) & f(\boldsymbol{r}, \boldsymbol{k}, i\omega_n) \\ -\underset{\sim}{f}(\boldsymbol{r}, \boldsymbol{k}, i\omega_n) & -\underset{\sim}{g}(\boldsymbol{r}, \boldsymbol{k}, i\omega_n) \end{bmatrix}, \tag{2}$$

$$\hat{H}(\boldsymbol{r}, \boldsymbol{k}, i\omega_n) = \begin{bmatrix} \xi(\boldsymbol{r}, \boldsymbol{k}, i\omega_n) & \Delta(\boldsymbol{r}, \boldsymbol{k}) \\ \underset{\sim}{\Delta}(\boldsymbol{r}, \boldsymbol{k}) & \underset{\sim}{\xi}(\boldsymbol{r}, \boldsymbol{k}, i\omega_n) \end{bmatrix}, \tag{3}$$

$$\xi(\boldsymbol{r}, \boldsymbol{k}, i\omega_n) = [i\omega_n + ev_F \boldsymbol{k} \cdot \boldsymbol{A}(\boldsymbol{r})], \tag{4}$$

where $\omega_n = (2n+1)\pi T$ is the Matsubara frequencies with n being an integer, T is the temperature, Δ represents the pair potential, and $\boldsymbol{A}$ is the vector potential. All of the functions including $\boldsymbol{k}$ satisfy the relation $\underset{\sim}{K}(\boldsymbol{r}, \boldsymbol{k}, i\omega_n) = K^*(\boldsymbol{r}, -\boldsymbol{k}, i\omega_n)$. The symbol $\hat{\cdot}$ represents 2×2 matrix structure.

We take into account the effects of surface roughness through the impurity self-energy, which is defined by

$$\hat{\Sigma}(\boldsymbol{r}, i\omega_n) = \Theta(r - R + w) \frac{i}{2\tau_0} \int \frac{d\theta}{2\pi} \hat{g}(\boldsymbol{r}, \theta, i\omega_n) \tag{5}$$

where $r = |\boldsymbol{r}|$, $k_x = \cos\theta$, $k_y = \sin\theta$, Θ is the step function, and τ_0 is the mean free time due to the impurity scatterings. The self-energy has finite values only near the surface as shown in Fig. 1(b), where W denotes the width of the disordered region.

We consider the spin-singlet chiral d-wave superconductor. The pair potential for a chiral d-wave superconductor is described by

$$\Delta(\boldsymbol{r},\theta) = \Delta_1(\boldsymbol{r})\cos(2\theta) + i\nu\Delta_2(\boldsymbol{r})\sin(2\theta), \tag{6}$$

where Δ_1 and Δ_2 are the local amplitude of two independent components. The doubly degenerate chiral superconducting states are indicated by $\pm\nu$. In this study, we consider superconducting states with a positive ν. In the simulations, Δ_1 and Δ_2 are self-consistently determined by the gap equation,

$$\begin{bmatrix} \Delta_1(\boldsymbol{r}) \\ \Delta_2(\boldsymbol{r}) \end{bmatrix} = N_0 g_0 \pi T \sum_{\omega_n} \int \frac{d\theta'}{2\pi} f(\boldsymbol{r},\theta',i\omega_n) \begin{bmatrix} V_1(\theta') \\ V_2(\theta') \end{bmatrix} \tag{7}$$

where N_0 is the density of states per spin at the Fermi level. The coupling constant g_0 is determined by

$$(N_0 g_0)^{-1} = \ln\left(\frac{T}{T_c}\right) + \sum_{n=0}^{n_c} \frac{1}{n+1/2}, \tag{8}$$

where $n_c = (\omega_c/2\pi T)$ with ω_c being the cutoff energy. The attractive interactions are given by $V_1(\theta) = 2\cos(2\theta)$ and $V_2(\theta) = 2\sin(2\theta)$. The electric current $\boldsymbol{j}(\boldsymbol{r})$ is calculated from the Green function

$$\boldsymbol{j}(\boldsymbol{r}) = 2i\pi|e|v_F N_0 T \sum_{\omega_n} \int \frac{d\theta}{2\pi} \boldsymbol{k}\, g(\boldsymbol{r},\theta,i\omega_n) \tag{9}$$

The vector potential should be determined by solving the Maxwell equations $\nabla \times \boldsymbol{A}(\boldsymbol{r}) = \boldsymbol{H}(\boldsymbol{r})$ and $\nabla \times \boldsymbol{H}(\boldsymbol{r}) = 4\pi\boldsymbol{j}(\boldsymbol{r})$. In a finite size superconductor, we define the amplitude of a spontaneous magnetization M in terms of the spontaneous magnetic field $\boldsymbol{H}(\boldsymbol{r})$ as

$$M = \frac{1}{\pi R^2} \int_{r<R} d\boldsymbol{r} H(\boldsymbol{r}). \tag{10}$$

We iterate the Eilenberger equation and the Maxwell equation to obtain the self-consistent solutions of $\Delta_1(\boldsymbol{r})$, $\Delta_2(\boldsymbol{r})$, $\boldsymbol{A}(\boldsymbol{r})$, and $\hat{\Sigma}(\boldsymbol{r}, i\omega_n)$. To solve the Eilenberger equation, we apply the Riccati parametrization [24–26]. Moreover, we apply the technique discussed in Ref. [27] to analize disk-shaped superconductors. We start all of the simulations with the initial condition $\Delta_1(\boldsymbol{r}) = \Delta_2(\boldsymbol{r}) = |\bar{\Delta}(T)|$ and $\boldsymbol{A}(\boldsymbol{r}) = 0$, where $|\bar{\Delta}(T)|$ is the amplitude of the pair potential in a homogeneous superconductor at a temperature T.

3. Results

3.1. Chiral edge current

We first discuss the effects of surface roughness on the chiral edge current. The spatial profiles of the edge current are shown in Fig. 2. The edge current under a specular (rough) surface is indicated by the solid (broken) line, where we show the results only at $y = 0$ because they are circular symmetric in a disk. The current densities are measured in units of $j_0 = \hbar c^2/4\pi|e|\xi_0^3$. We fix several parameters: the temperature $T = 0.2T_c$, the radius of a disk $R = 10\xi_0$, the penetration depth $\lambda_L = 5\xi_0$, the cutoff energy $\omega_c = 6\pi T_c$ with T_c and $\xi_0 = \hbar v_F/2\pi T_c$ being the critical temperature and the coherence length, respectively. In the clean limit, the current density is negative (clockwise) for $|x| > 8\xi_0$ and is positive (counterclockwise) for $|x| < 8\xi_0$. This characteristic current profiles are related to the topological number of a chiral d-wave

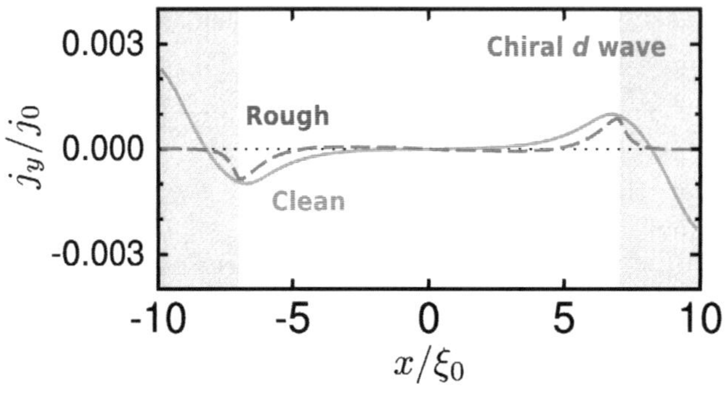

Figure 2. Current densities in a chiral d-wave superconductor. The solid and broken lines indicate the results in a disk with and without surface roughness, respectively. The current densities are measured in units of $j_0 = \hbar c^2/4\pi|e|\xi_0^3$. We fix several parameters: the radius of a disk $R = 10\xi_0$, the penetration depth $\lambda_L = 5\xi_0$, and the cutoff energy $\omega_c = 6\pi T_c$ with T_c and $\xi_0 = \hbar v_F/2\pi T_c$ being the critical temperature and the coherence length, respectively.

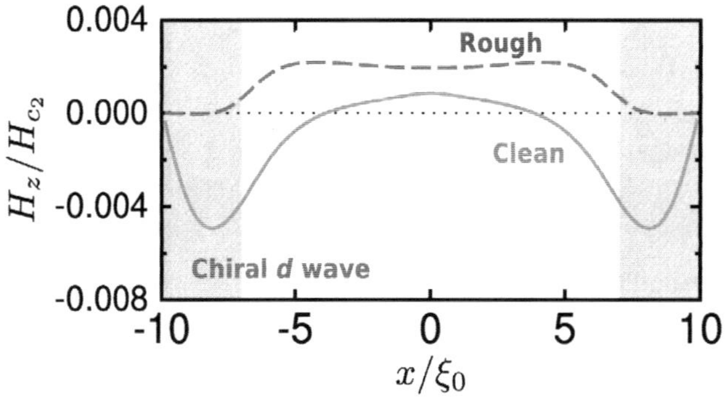

Figure 3. Spontaneous fields in a chiral d-wave superconductor. The solid and broken lines indicate the results in a disk with and without surface roughness, respectively. The magnetic fields are normalized to the second critical magnetic field $H_{c_2} = \hbar c/|e|\xi_0^2$. The parameters are set to the values used in Fig. 2.

superconductor. Because the topological number for a chiral d-wave superconductor is two, two edge states should appear at the surface of a disk. Figure 2 shows that there are edge states with different decay lengths in a disk, and they flows in the opposite directions each other.

When there is surface roughness, on the other hand, the current flows only in the counterclockwise direction as shown in Fig. 2. The outer edge current vanishes due to the surface roughness. The integrated current flows in the opposite direction compared to that in the clean limit. We confirm that the inner chiral current can survive even under much stronger roughness such as $\xi/\ell = 30$. Therefore we conclude that the chiral current in a chiral d-wave superconductor is robust against surface roughness. Its direction, however, flips due to the surface roughness.

3.2. Spontaneous field

We show the spatial profiles of spontaneous fields in a chiral d-wave disk. The result with a specular (rough) surface is indicated by the solid (broken) line, The magnetic field is measured in units of the second critical field $H_{c_2} = \hbar c/|e|\xi_0^2$. The parameters are set to the same values used in Fig. 2. In a chiral d-wave disk with a specular surface, the spontaneous magnetic field is mainly localized in the region $r > 5\xi_0$. Since the edge currents in a chiral d-wave disk cancel intrinsically each other. The amplitude of the spontaneous magnetic field is smaller than that of a chiral p [1]. The magnetic field generated by the outer chiral current is intrinsically screened by the inner chiral current in addition to the Meissner current.

When there is surface roughness, the spontaneous magnetic field is positive everywhere in the disk. In a disk with a rough surface, there is a current flowing in the counterclockwise direction. Thus the sign of a spontaneous field is opposite to that in the clean limit.

4. Conclusion

We have studied the effects of surface roughness on the spontaneous edge current in a small chiral superconductor with chiral d-wave pairing symmetry. On the basis of the quasiclassical Eilenberger formalism, we numerically calculated the chiral current and the spontaneous magnetization. The edge current of a chiral d-wave superconductor is robust against the roughness. The direction of the total edge current, however, is opposite to that obtained in the clean limit.

Acknowledgments

The authors are grateful to A. A. Golubov, Y. Tanaka, S. Kashiwaya, Ya. V. Fominov, M. Yu. Kupriyanov, T. Habe, and S. Ikegaya for useful discussions. This work was supported by Topological Materials Science (No. 15H05852) and KAKENHI (Nos. 26287069 and 15H03525) from the Ministry of Education, Culture, Sports, Science and Technology (MEXT) of Japan, and by the Ministry of Education and Science of the Russian Federation (Grant No. 14Y.26.31.0007). S.-I. S. is supported in part by Grant-in-Aid for JSPS Fellows (Grant No. 15J00797) by Japan Society for the Promotion of Science (JSPS).

[1] M. Matsumoto and M. Sigrist. J. Phys. Soc. Jpn. **68**, 994 (1999).
[2] A. Furusaki, M. Matsumoto, and M. Sigrist. Phys. Rev. B **64**, 054514 (2001).
[3] Y. Maeno, H. Hashimoto, K. Yoshida, S. Nishizaki, T. Fujita, J.G. Bednorz, and F. Lichtenberg, Nature **372**, 532 (1994).
[4] T.M. Rice and M. Sigrist, J. Phys.: Condens. Matter **7**, L643 (1995).
[5] A. P. Mackenzie and Y. Maeno, Rev. Mod. Phys. **75**, 657 (2003).
[6] K. Takeda, H. Sakurai, E. Takayama-Muromachi, F. Izumi, R. A. Dilanian, and T. Sakai, Nature **422** 53 (2003).
[7] G. Baskaran, Phys. Rev. Lett. **91**, 097003 (2003).
[8] M. Ogata, J. Phys. Soc. Jpn. **72**, 1839 (2003).
[9] A. Tanaka and X. Hu, Phys. Rev. Lett. **91**, 257006 (2003).
[10] R. Nandkishore, L. S. Levitov, and A. V. Chubukov, Nature Physics **8**, 158 (2012).
[11] M. L. Kiesel, C. Platt, W. Hanke, D. A. Abanin, and R. Thomale, Phys. Rev. B **86**, 020507(R) (2012).
[12] M. L. Kiesel, C. Platt, W. Hanke, and R. Thomale, Phys. Rev. Lett. **111**, 097001 (2013).
[13] M. H. Fischer, T. Neupert, C. Platt, A. P. Schnyder, W. Hanke, J. Goryo, R. Thomale, and M. Sigrist, Phys. Rev. B **89**, 020509(R) (2013).
[14] Y. Kasahara, H. Shishido, T. Shibauchi, Y. Haga, T. D. Matsuda, Y Onuki, and Y. Matsuda, New Journal of Physics, **11**, 055061 (2009).
[15] H. Kusunose, J. Phys. Soc. Jpn. **81**, 023704 (2012).
[16] M. Tsuchiizu, Y. Yamakawa, S. Onari, Y. Ohno, and H. Kontani, Phys. Rev. B **91**, 155103 (2015).
[17] Y. Tada, W. Nie, and M. Oshikawa, Phys. Rev. Lett. **114**, 195301 (2015).
[18] W. Huang, E. Taylor, and C. Kallin, Phys. Rev. B **90**, 224519 (2014).

IOP Conf. Series: Journal of Physics: Conf. Series **807** (2017) 102001 doi:10.1088/1742-6596/807/10/102001

[19] K. Yamada, Y. Nagato, S. Higashitani, and K. Nagai, J. Phys. Soc. Jpn. **65**, 1540 (1996).
[20] S.-I. Suzuki and Y. Asano, Phys. Rev. B **91**, 214510 (2015).
[21] L. J. Buchholtz and G. Zwicknagl, Phys. Rev. B **23**, 5788 (1981).
[22] J. Hara and K. Nagai, Progress of Theoretical Physics **74**, 1237 (1986).
[23] G. Eilenberger, Z. Phys. **214**, 195 (1968).
[24] N. Schopohl and K. Maki, Phys. Rev. B **52**, 490 (1995).
[25] N. Schopohl, arXiv:cond-mat/9804064.
[26] M. Eschrig, Phys. Rev. B **80**, 134511 (2009).
[27] Y. Nagai, K. Tanaka, and N. Hayashi, Phys. Rev. B **86**, 094526 (2012).

Fractional Electric Charge and Massive Quasiparticles on the Domain Wall between Topological Insulators and Spin Ice Compounds

I Kanazawa, E Imai, T Sasaki, R Maeda

Department of Physics, Tokyo Gakugei University, Koganeishi, Tokyo 184-8501, Japan

E-mail: kanazawa@u-gakugei.ac.jp

Abstract. We have proposed the mass-creation mechanism on massless Dirac fermions on the domain wall between topological insulators and spin ice compounds through the interaction between the massless Dirac fermion and excited magnetic monopoles

1. Introduction

The quantum Hall state gives the rst example of topological states of matter which have topological quantum numbers di erent from ordinary states of matters [1, 2]. In the presence of many-body interactions and disorder, the Berry curvature and the Chern number can be de ned over the space of twisted boundary condition [3]. The quantum Hall e ect of the (2+1)-dimensional time reversal-breaking insulator has been generalized to time reversal-invariant insulators in various dimensions. Time reversal- invariant topological insulators have been classi ed in (3+1) dimensions [4]. Recently the three-dimensional topological insulator $Bi_{1\ x}Sb_x$ in a certain range of composition [5] has been predicted. Really massless Dirac fermions on the surface of three-dimensional topological insulator have been observed by angle-resolved photoemission spectroscopy [6]. Spin ices are frustrated magnets. The spin ice family [$Dy_2Ti_2O_7$ and $Ho_2Ti_2O_7$] are predicted to support sharply de ned magnetic monopole excitations [7, 8]. In addition, spin ices have residual entropy at low temperature, which is well-approximated by the Pauling entropy for water ice [9]. One(I.K) of the present authors has reported the importance of the hole-induced domain wall and the hedgehog-like magnetic soliton in magnetoresistance in diluted magnetic semiconductors [10, 11, 12],and doped manganites [13, 14]. In addition, one(I.K) of the present authors has proposed exotic quasiparticle with fractional charges in semiconductor-dot from collectively induced-charge e ect on a domain wall shell [15, 16]. Recently the present authors [17, 18] have proposed that there might be emergent quasiparticles with fractional electronic charge such dyons on the domain wall between topological insulators and spin ice compounds through the Witten e ect. In this study, we shall propose the mass-creation mechanism of Dirac fermions on the domain wall between topological insulators and spin ice components through the interaction between Dirac fermions and excited magnetic monopoles.

2. Exotic particle and mass-creation on the domain wall between the topological insulator and spin ice

The time reversal-invariant (TRI) topological insulator in (3+1) dimension can be obtained as descendants from the fundamental TRI insulator in (4+1) dimension through a dimensional reduction procedure. The e ective topological eld theory and Z_2 topological classi cation for the TRI insulators in (3+1) dimension are naturally obtained from this procedure [19]. The e ective action of gauge eld A^μ is obtained by the following path integral.

$$e^{iS_{eff}} = \int D[c]D[c^+]exp\left[i\int dt[\sum_m c^+_{m\alpha}(i\partial_t)c_{m\alpha} - H(A)]\right]$$

which determines the response of the fermionic system through the equastion,

$$j_\mu(x) = \frac{S_{eff}}{A_\mu(x)}$$

In the case of (4+1)-dimensional insulator, a topological term is in general present in the e ective action, which is the second Cherm-Simons term. To study the response properties of the (3+1) dimension system, the e ective action $S_{3D}(A,\theta)$ with an adiabatic parameter θ can be de ned as

$$e^{iS_{3D}(A,\theta)} = \int D\psi D\overline{\psi}\exp\left[i\int dt\sum_{\vec{x}}\overline{\psi}_{\vec{x}}(i\partial_\tau - A_{\vec{x}0})\psi_{\vec{x}} - \mathcal{H}[A,\theta]\right].$$

$\mathcal{H}(A,\theta)$ is the (3+1) dimensional Hamiltonian of Dirac model coupled to an external $U(1)$ gauge eld A [10]. ψ is the wavefunction of the Dirac electron. Where $\vec{x}$ stands for the three-dimension coordinates. A Taylor expansion of S_{3D} can be carried out around the gauge eld con guration $A_s(\vec{x},t)$, whose $s = 1,2,3$ stands for the x,y,z directions, $\theta(\vec{x},t) \equiv \theta_0$, which contains a nonlinear-response term derived from the (4+1) dimensional Chern-Simon action,

$$S_{3D}(A,\theta) = \frac{G_3(\theta_0)}{4\pi}\int d^3x\,dt\,\varepsilon^{\mu\nu\sigma\tau}\ \theta\partial_\mu A_\nu\partial_\sigma A_\tau\ .$$

The eld $\theta(\vec{x},t) = \theta(\vec{x},t) - \theta_0$ plays the role of A_4 in the (4+1) dimension, and the coefficient $G_3(\theta_0)$ is determined by Goldstone and Wilczek-type Feynman diagram [11] in Fig. 1. Consequently, $G_3(\theta_0)$ can be calculated as follows.

$$G_3(\theta_0) = -\frac{\pi}{6}\int\frac{d^3k\,d\omega}{(2\pi)^4}Tr\varepsilon^{\mu\nu\sigma\tau}\left[\left(G\frac{\partial G^{\ 1}}{\partial q^\mu}\right)\left(G\frac{\partial G^{\ 1}}{\partial q^\nu}\right)\left(G\frac{\partial G^{\ 1}}{\partial q^\sigma}\right)\left(G\frac{\partial G^{\ 1}}{\partial q^\tau}\right)\left(G\frac{\partial G^{\ 1}}{\partial \theta_0}\right)\right]$$

where $q^\mu = (\omega, k_x, k_y, k_z)$ and $G(q_\mu) = [\omega + i\ \ - h(k_i)]^{\ 1}$ is the single-particle Green's function. In addition, $G_3(\theta_0)$ is determined from the Berry phase curvature as

$$G_3(\theta_0) = \frac{1}{8\pi^2}\int d^3k\,\varepsilon^{ijh}Tr[f_{\theta i}f_{jh}]$$

in which the Berry phase gause eld is de ned in the four-dimensional space $(k_x, k_y, k_z, \theta_0)$, ie.,

$$a_i^{\alpha\beta} = -i\left\langle\vec{k},\theta_0;\alpha|\left(\frac{\partial}{\partial k_i}\right)|\vec{k},\theta_0;\beta\right\rangle$$

and

$$a_\theta^{\alpha\beta} = -i\left\langle\vec{k},\theta_0;\alpha|\left(\frac{\partial}{\partial\theta_0}\right)|\vec{k},\theta_0;\beta\right\rangle.$$

IOP Conf. Series: Journal of Physics: Conf. Series **807** (2017) 102002 doi:10.1088/1742-6596/807/10/102002

A generalized polarization $P_3(\theta_0)$ can also be de ned in (3+1) dimensions so that $G_3(\theta_0) = \frac{\partial P_3(\theta_0)}{\partial \theta_0} = \int d^3k \partial_\theta K^\theta$, with

$$P_3(\theta_0) = \int d^3k\, K_\theta = \frac{1}{16\pi^2} \int d^3k\, \varepsilon^{\theta ijk} Tr\left\{\left(f_{ij} - \frac{1}{3}[a_i, a_j]\right) \cdot a_k\right\}.$$

The e ective action for the (3+1)-dimensional system is nally written as

$$S_{3D}(A, \theta) = \frac{1}{16\pi^2} \int d^3x\, dt\, \theta_0(\vec{x}, t) \varepsilon^{\mu\nu\sigma\tau} \partial_\mu A_\nu \partial_\sigma A_\tau$$

In the case of the (3+1) dimensional topological insulator, $\theta_0(\vec{x}, t)$ correspond to π. Here we shall consider the domain wall between the topological insulator and spin ice as shown in Fig. 1. We must consider the (3+1) dimensional Dirac fermion with a domain wall con guration of the $\theta_0(z)$ eld given by

$$\theta_0(z) = \frac{\pi}{2}\left[1 - \tanh\left(\frac{z}{4\xi}\right)\right]$$

which has the asymptotic behavior $\theta_0(z \to -\infty) = \pi$, $\theta_0(z \to \infty) = 0$. The domain wall width is ξ.

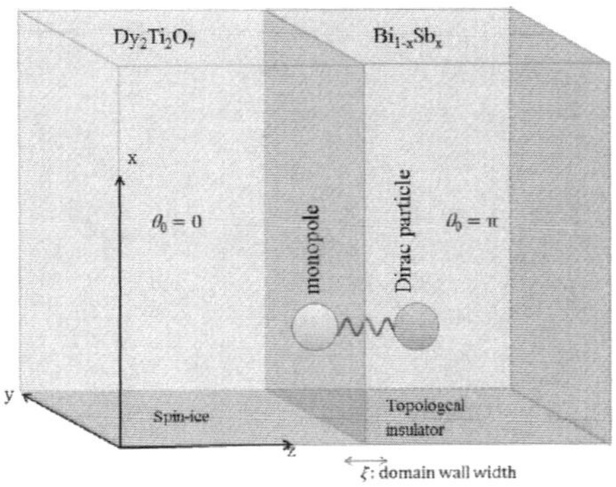

Figure 1. The bilayer of spin ice $Dy_2Ti_2O_7$ and topological insulator $Bi_{1\ x}Sb_x$

It is proposed that magnetic charges can exist in certain materials(spin ices) in the form of emergent excitations that manifest like point charges of magtnetic monopoles. Spin ice is governed by a model dipolar Hamiltonian as follows,

$$\begin{aligned} H &= J_{ex}^{nm} \sum_{<ij>} S_i \cdot S_j \\ &+ \frac{\mu_0}{4\pi} \sum_{i<j} \left[\frac{S_i \cdot S_j}{r_{ij}^3} - \frac{3(S_i \cdot r_{ij})(S_j \cdot r_{ij})}{r_{ij}^5}\right] \end{aligned} \tag{1}$$

Where J_{ex}^{nm} is the exchange interaction truncated at the nearest-neighbor level, the spins S_i point parallel to the local [111] axis, and μ_0 is the vacuum permeability. The rare-earth spins S_j have typical a dipole moment of approximately $\mu \sim 10\ \mu_B$ (μ_B is Bohr magneton). The magnetic charge, g_M, of excited magnetic monopole in the spin-ice takes the valves $\pm 2\mu/b$, $\mu \sim 10\ \mu_B$ being the dipole moment of a spin and b the distance between the center of adjacement tetrahedra. Now we shall introduce the e ective gauge eld A_{icc}(SU(2)), the scalar eld φ_{ice}^a (Higgs triplet),

IOP Conf. Series: Journal of Physics: Conf. Series **807** (2017) 102002 doi:10.1088/1742-6596/807/10/102002

in order to discuss in more detail the e ect of excited magnetic monopoles in the spin-ice. That is,

$$A^{a,cl}_{i,ice} = \frac{1}{gr}\varepsilon^{aij}n^j(1 - F_{ice}(r)) \quad (2)$$

$$A^{cl}_{0,ice} = 0 \quad (3)$$

$$\varphi^{cl}_{ice} = n^a\tau^a v(1 - H_{ice}(r)). \quad (4)$$

where $g = \frac{1}{g_M}$, $r = \sqrt{x^2}$, $n = x/r$, $F_{ice}(r)$ and $H_{ice}(r)$ obey the following boundary conditions, $F_{ice}(0) = H_{ice}(0) = 1, \quad F_{ice}(\infty) = H_{ice}(\infty) = 0$. We shall immediately write down the Dirac equation.

$$(i\gamma^\mu D_\mu - h\frac{\tau^a}{2}\varphi^a_{ice})\psi = 0 \quad (5)$$

ψ is the wavefunction of the Dirac eld on the domain wall between topological insulator and the spin ice. h is a real coupling constant between the excited magnetic monopole and massless Dirac fermion on the domain wall. It is noted that we have set explicit mass parameter equal to zero. The covariant derivative in eq.(5) is

$$D_\mu = \partial_\mu - ig\frac{\tau^a}{2}A_\mu \quad (6)$$

In the covariant derivative (6), we retain the electromagnetic eld A^3_μ and we set $\varphi^a_{ice} = v\ ^{a3}$. Then equation (5) will have the form

$$[i\gamma^\mu(\partial_\mu - ig\frac{\tau^a}{2}A^3_\mu) - h\frac{\tau^3}{2}v]\psi = 0$$

The equation for the upper and lower components of SU(2)-doublet,

$$\psi = \begin{pmatrix} \chi_1 \\ \chi_2 \end{pmatrix}$$

decouple and have the form

$$[i\gamma^\mu(\partial_\mu - i\frac{g}{2}A^3_\mu) - h\frac{v}{2}]\chi_1 = 0$$

$$[i\gamma^\mu(\partial_\mu + i\frac{g}{2}A^3_\mu) + h\frac{v}{2}]\chi_2 = 0$$

That is, the fermions χ_1 and χ_2 have the same mass $\frac{hv}{2}$. Taking into account the monopole density in the spin ice [21], we might be able to measure the mass of fermions on the domain wall between the spin ice in $Dy_2Ti_2O_7$ and the topological insulator $Bi_{1\ x}Sb_x$ at ~1K by means of angle-resolved photoemission spectroscopy.

3. Conclusion

The mass-creation mechanism of massless Dirac fermions on the domain wall between topological insulators and spin ice compounds has been proposed. The interaction between excited magnetic monopoles and massless Dirac fermions is of signi cance in the mechanism.

References

[1] Klitzing K V, Dorda G and Pepper M 1980 *Phys. Rev. Lett.* **45** 494
[2] Tsui D C, Stormer H L and Gossard A C1982 *Phys. Rev. Lett.* **48** 1559
[3] Niu Q, Thouless D J and Wie Y S 1985 *Phys. Rev. B.* **31** 3372
[4] Fu L, Kane C L and Mele E J, 2007 *Phys. Rev. Lett.* **98** 106803
[5] Fu L and Kane C L 2007 *Phys. Rev. B.* **76** 045302
[6] Hsieh D et al. 2008 *Nature* **452** 970
[7] Bramniev S T and Gingras M J P 2001 *Science* **294** 1495
[8] Castelnovo C, Moessner R and Sondhi S L 2008 *Nature* **451** 42
[9] Ramirez A P et al. 1999 *Nature* **399** 333
[10] Kanazawa I 2005 *Physica E* **29** 647
[11] Kanazawa I 2006 *Phys. Lett. A.* **355** 460
[12] Kanazawa I 2007 *Physica E* **40** 277
[13] Kanazawa I 2010 *Phys. Stat. Sol. B* **247** 644
[14] Kanazawa I 2013 *J. Low Temp. Phys.* **172** 177
[15] Kanazawa I 2012 *J. Phys. Conf. Ser.* **400** 042028
[16] Kanazawa I 2015 *J. Phys. Conf. Ser.* **592** 012135
[17] Sasaki T, Imai E and Kanazawa I 2014 *J. Phys. Conf. Ser.* **568** 052029
[18] Kanazawa I, Sasaki T, 2015 *Physics Procedia* **75** 967
[19] Qi X L, Hughes T L and Zhang S C 2008 *Phys. Rev. B.* **78** 195424
[20] GoldStone J and Wilczek F 1981 *Phys. Rev. Lett.* **47** 986
[21] Castelnovo C, Moessner R and Sondhi S L 2011 *Phys. Rev. B.* **84** 144435

IOP Conf. Series: Journal of Physics: Conf. Series **807** (2017) 102003 doi:10.1088/1742-6596/807/10/102003

Multi-critical behaviour in a self-dual Josephson junction array –fixed dimension renormalization versus large-N technique

S Sakhi

Physics department, American University of Sharjah, United Arab emirates, POBOX 26666

ssakhi@aus.edu

Abstract. Self-dual Josephson junction arrays are modelled by a theory of N components complex fields coupled to gauge fields. The multi-critical behaviour is analysed through a one loop renormalization group investigation at fixed dimension, which reveals the existence of charged infrared-stable fixed point solutions for large N. Further analysis of the model in the framework of the 1/N expansion confirms that the decoupled fixed point that was stable in the massive regime within the approximate one-loop renormalization group is destabilized. This is ascribed to the special interaction mediated by the mixed Chern-Simons term.

1. Introduction

Two-dimensional Josephson junction arrays (JJA) continue to draw interest due to their rich underlying physics [1]. In these systems, Cooper pairs and vortices have conserved currents, which in 2+1 dimension can be embodied as the curls of fictitious gauge fields. The resultant gauge field formulation [2] has Maxwell terms that express kinetic terms for charges and vortices and a mixed Chern-Simons term that describes the Lorentz force exerted by the vortices on the charges and the Magnus force exerted by the charges on the vortices. Together with the emerging gauge fields, complex scalar fields are added to account for quantum disorder due to electric and magnetic excitations in these systems [3]. The ensuing low energy effective model is a Ginzburg-Landau theory with two charged disorder fields coupled to Maxwell-mixed-Chern-Simons gauge fields. These competing scalar fields along with the fluctuating gauge fields lead to multi-critical behavior, which we report on here.

2. The model

The Euclidean Lagrangian of interest [4] is

$$L = |(\partial_\mu - ia_\mu)\Psi|^2 + |(\partial_\mu - ib_\mu)\Phi|^2 + r_1|\Psi|^2 + r_2|\Phi|^2 + \frac{u}{6}(|\Phi|^4 + |\Psi|^4) + \frac{w}{3}|\Psi|^2|\Phi|^2 + \frac{1}{4e_b{}^2} f_{\mu\nu}^2 + \frac{1}{4e_a^2} g_{\mu\nu}^2 + i\kappa b_\mu \varepsilon^{\mu\nu\lambda}\partial_\nu a_\lambda \qquad (1)$$

The gauge fields a_μ, b_μ are associated with the currents of Cooper pairs and vortices and implicate two different gauge couplings. These are defined by the parameters of the JJA system [3]: $e_a = \sqrt{8\mathrm{E_C}}$ and $e_b = 2\pi\sqrt{E_J}$ with E_J the Josephson coupling energy measuring the strength for the phase-coupling between two superconductors and $\mathrm{E_C}$ the charging energy required to add extra charges to neutral islands. The mixed Chern-Simons term has a coefficient $\kappa = 1/\pi$. The fields Φ and Ψ represent the disorder of electric and magnetic charges which control the superconducting and insulating phases of the model. The superconducting phase emerges when the charge disorder field condenses while the magnetic disorder field resides in a vacuum; the insulating phase appears as a condensate of magnetic charges and a vacuum for electric disorder fields [3]. In this self-dual model, both types of excitations (electric-like and magnetic-like) are treated on an equal footing, leading to the distinctive feature of the action in Eq. (1) with equal quartic coupling constants of the matter fields. The other feature of the action in Eq. (1) is the presence of a mixed Chern-Simons term, which mediates the interaction between the electric and magnetic excitations. This term describes both the Lorentz force exerted by the vortices on the charges and the Magnus force exerted by the charges on the vortices [3]. This term also incorporates the symmetry between the two species a_μ, b_μ: integration by parts changes the mixed Chern-Simons term into something of the same form with a_μ and b_μ interchanged. To facilitate a more general theoretical description, we introduce a large N generalization of the model by allowing the fields Φ and Ψ to have each N complex components.

3. Renormalization analysis at fixed dimension

3.1. Massless regime

To study the critical properties of the model, I develop a renormalization analysis at fixed dimension. This approach is suitable here because of the Chern-Simons term, which is an inherently three dimensional object. The calculation is done when both r_1 and r_2 are tuned to their critical values (*i.e.* $r_1 = 0$, $r_2 = 0$) by appropriately choosing the renormalization conditions for the vertex functions. Infrared divergences resulting from massless scalar propagators are handled by defining the renormalized coupling constants at finite values of the external momenta, and the renormalization beta functions are obtained by taking derivatives of the couplings with respect to the logarithm of the external momentum.

Up to one-loop order, the renormalized dimensionless coupling constants of the model satisfy the following flow equations [4]:

$$\beta(\hat{u}) = \left(2\eta_\psi - 1\right)\hat{u} + \frac{N+4}{24}\hat{u}^2 + \frac{N}{24}\hat{w}^2 + \hat{e}_a^4 f(\hat{m}) \tag{2}$$

$$\beta(\hat{w}) = \left(\eta_\psi + \eta_\phi - 1\right)\hat{w} + \frac{N+1}{12}\hat{u}\hat{w} + \frac{\hat{w}^2}{12} - \hat{e}_a^2\hat{e}_b^2 g(\hat{m}) \tag{3}$$

$$\beta(\hat{e}_a^2) = -\hat{e}_a^2\left(1 - \frac{N}{16}\hat{e}_a^2\right), \quad \beta(\hat{e}_b^2) = -\hat{e}_b^2\left(1 - \frac{N}{16}\hat{e}_b^2\right) \tag{4}$$

where $\eta_\psi = d\ln(Z_\psi)/d\ln(p)$ and $\eta_\phi = d\ln(Z_\phi)/d\ln(p)$ are the anomalous dimensions of the scalar fields. η_ψ, η_ϕ, $f(\hat{m})$, $g(\hat{m})$ are function of $\hat{m} = \hat{e}_a\hat{e}_b/\pi$ and have well defined limits.

The fixed points of the model are obtained by finding the zeros of the beta functions. Since the beta functions (4) associated with the gauge couplings have nontrivial solutions when $\hat{e}_a^2 = \hat{e}_b^2 = 16/N$, these lead to the fully charged fixed points (FP). One important observation about these charged FP is that the decoupled fixed point at $w = 0$ which was possible in the neutral case is no longer attainable. The result of the flow equations (2) and (3) is summarized in Fig. 1 for two different values of N.
As Fig. 1 illustrates, the structure of the flow diagram in the critical plane depends critically on the number of components N. When N<12, there are two fixed points, one with two repulsive eigenvalues and one with one attractive and one repulsive eigenvalue. When N>12, there are four fixed points one of which has all attractive eigenvalues and should be identified with the critical point of the model.

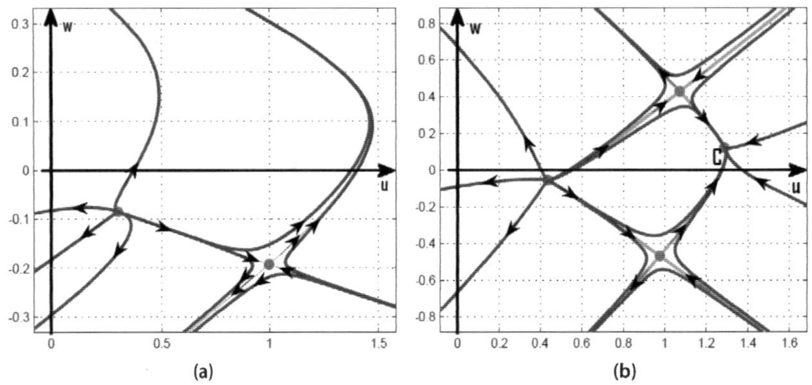

FIG. 1 Flow diagrams in the $w-u$ plane in the massless regime with initial condition $\hat{e}_a^2 = \hat{e}_b^2 = 16/N$. The arrows show the direction of the flow as $p \to 0$. (a) $N < 12$. (b) $N > 12$.

3.2. Massive regime

It is also useful to examine the properties of this field theory from the point of view of perturbative RG in the massive regime. This will be compared with 1/N expansion predictions. Construction of the beta functions of the couplings is done in the standard fashion [5], and the study of the stability of the fixed points reveals that for N < 2 the stable fixed point is the one with the enlarged U(N) symmetry (Fig. 2a). For 2 < N < 4, on the other hand, the mixed fixed point becomes the stable one (Fig 2b). Finally, for N > 4 the decoupled fixed point is stable (Fig. 2c).

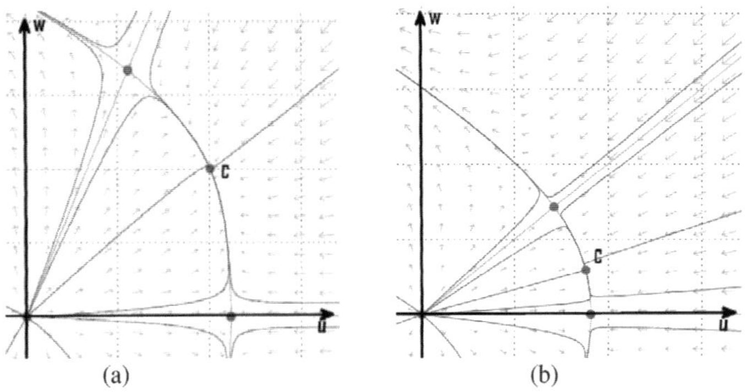

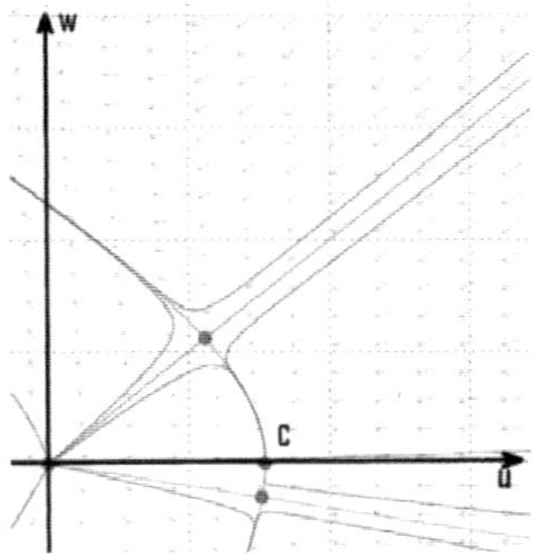

(c)

FIG. 2 One-loop renormalization group flows in the (u,w) plane at fixed dimension d = 3 in the massive regime. (a) N < 2, (b) 2 < N < 4, (c) N > 4.

4. Large N analysis

The fixed-dimension renormalization approach presented in the previous section has the drawback that it lacks a small parameter to control the calculation. To rectify this, I investigate the renormalized zero-momentum four-point couplings of the model by using large-N technique. The advantage of this analysis is that 1/N provides an expansion parameter which allows the summation of infinite classes of Feynman graphs. It turns out that this model is renormalizable in the 1/N expansion, and its critical exponents can be obtained in a systematic way [5].

The construction proceeds in a standard fashion, making the assumption that u and $w$are both of order 1/N. The next-to-leading 1/N corrections to the renormalized couplings are found as a function of the bare couplings and the renormalized mass. These are used to find the 1/N correction to the β-functions and the fixed point values of the couplings.

The renormalization is performed according to the following prescription for the two- and four-point functions of Ψ and Φ:

$$\begin{aligned} \Gamma^{(2)}_{11}(p) &= \Gamma^{(2)}_{22}(p) = Z^{-1}[m_r^2 + p^2] \\ \Gamma^{(4)}_{11}(0) &= Z^{-2}\frac{3\hat{u}}{N}m_r \\ \Gamma^{(4)}_{12}(0) &= Z^{-2}\frac{3\hat{w}}{N}m_r \end{aligned} \qquad (5)$$

The renormalized mass m_r and the two-point function are related to the self-energy Σ(p,m), $\Gamma^{(2)}(p) = p^2 + m^2 - \Sigma(p,m)$, and the dimensionless renormalized couplings $\hat{u}$ and $\hat{w}$ are expressed as

$$\begin{aligned} \hat{u} &= \frac{N}{3m_r}\left[1 + 2\frac{\partial\Sigma(p,m)}{\partial p^2}\right]_{p=0} \Gamma^{(4)}_{1,1}(0,0,0,0) \\ \hat{w} &= \frac{N}{3m_r}\left[1 + 2\frac{\partial\Sigma(p,m)}{\partial p^2}\right]_{p=0} \Gamma^{(4)}_{1,2}(0,0,0,0) \end{aligned} \qquad (6)$$

The large-N limit of the β-functions are found to be

$$\begin{aligned}\beta_u(\hat{u},\hat{w}) &= m_r\frac{d\hat{u}}{dm_r} = -\hat{u} + \frac{1}{16\pi}\left(\hat{u}^2+\hat{w}^2\right)\\ \beta_w(\hat{u},\hat{w}) &= m_r\frac{d\hat{w}}{dm_r} = -\hat{w} + \frac{\hat{u}\hat{w}}{8\pi}\end{aligned} \tag{7}$$

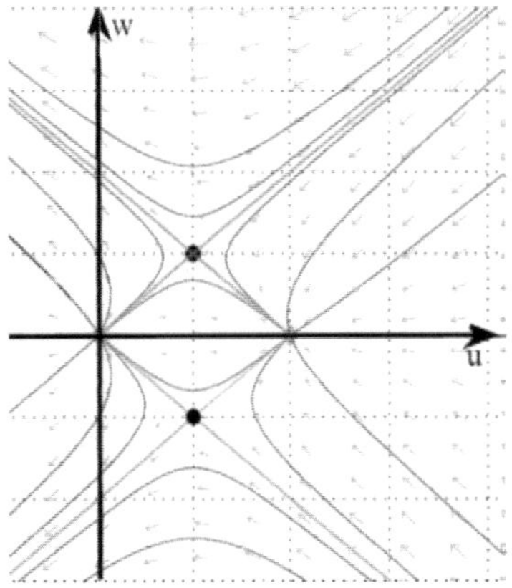

FIG.3 Flow of the interaction coupling constants for N = ∞.

As Fig. 3 shows, there are four fixed points: the Gaussian fixed point $(u,w) = (0,0)$, which is unstable in both directions in the u-w plane; the decoupled fixed point, $(u,w) = (16\pi, 0)$ which is stable; the enlarged U(N) symmetric fixed point $(u,w) = (8\pi, 8\pi)$, which is unstable, and $(u,w) = (8\pi, -8\pi)$, which is also unstable. In the next-to-leading order, the β-functions can be expanded in a systematic way, and a numerical analysis reveals that the fixed point values of the couplings at the next order are shifted to $\hat{u} = 16\pi(1 - 3.55/N)$ and $\hat{w} = 34/N$. This shows that the decoupled fixed point which was stable in the massive regime within the approximate one-loop renormalization group is destabilized in the framework of the 1/N expansion. This is attributed to the interaction mediated by the mixed Chern-Simons term which generates a sequence of Feynman diagrams as shown in Fig. 9 and Fig 10 of reference [5], and which efficiently couples the fields Φ and ψ.

5. Conclusion

In summary, I have formulated a field-theoretic approach that investigates the multi-critical behaviour of the low energy electric and magnetic excitations near the quantum critical point of a Josephson junction array system. In the massless regime, a one loop renormalization group analysis at fixed dimension of the beta functions reveals the existence of a rich phase diagram consisting of two sets of infrared-stable charged fixed points: partially charged solutions with respect to the gauge fields; and fully charged solutions above some critical value of N. Furthermore in the massless regime, the approximate one loop analysis shows that the decoupled fixed point which is stable in the neutral case (no gauge contribution) is no longer attainable in the presence of fluctuating gauge fields. When we examined the model in the massive regime, both the one-loop renormalization group at large N and the leading order of the 1/N expansion lead to a compatible result, which is the existence of a stable decoupled fixed point as in the neutral gauge. At this fixed point, the scalar fields Φ and Ψ are effectively decoupled (w = 0). However, analysis of the next-to-leading contribution of the 1/N expansion in the same regime reveals that the decoupled fixed point (w = 0) that was stable within the approximate one-loop renormalization group in the massive regime (Fig. 2) is destabilized. This is attributed to the interaction mediated by the mixed Chern-Simons term which generates a sequence of

Feynman diagrams which efficiently couples the fields Φ and ψ at the next-to-leading order of the 1/N expansion. Another way to view this result is to note that in the large N generalization of the model there are N components of each of the scalar fields and only one component of the gauge fields. Accordingly the effect of the mixed Chern-Simons term, which favors coupling the scalar fields, is underestimated at the leading order of the 1/N expansion and starts becoming noticeable at the next-to-leading order contribution.

References

[1] Fazio R and Van der Zant H 2001 *Phys. Rep.* **355** 235
[2] Diamantini M C, Sodano P and Trugenberger C A 1996 *Nucl. Phys B* **474** 641
[3] Sakhi S 2006 *Europhys. Lett.* **73** 267; 2006 *Phys. Rev. B* **73** 132505
[4] Sakhi S 2013 *Physica A* **392** 6255
[5] Sakhi S 2014 *Phys. Rev. D* **90** 045028

IOP Conf. Series: Journal of Physics: Conf. Series **807** (2017) 102004 doi:10.1088/1742-6596/807/10/102004

Induced Topological-charges on a Domain Wall of the Semiconductor Dot and the D-brane-like Dot

I Kanazawa

Department of Physics, Tokyo Gakugei University, Koganeishi, Tokyo 184-8501, Japan

E-mail: `kanazawa@u-gakugei.ac.jp`

Abstract. Anomalous excitations in the electron state of the semiconductor-dot have been analyzed from viewpoint of eld-theoretical formula. We have introduced the toy-model of the quantum-dot, which is composed of three degenerated $D4$-branes-like solitons. The relation to low energy hadron in QCD has been suggested.

1. Introduction

Recent fusion between condensed matter physics and elementary particle physics is remarkable. For example, the integral and fractional quantum Hall e ect can be described by the topological Cherm-Simons eld theory [1] in 2 + 1 dimensions. This e ective topological eld theory in eld-theoretical formula captures the topological e ects, including the quantization of the Hall conductance, the fractional charge, and statistics of quasiparticles [2, 3]. Furthermore, the study of anomalous properties of Dirac fermions on surfaces of topological insulators [3] and Majorana fermions on the domain wall between anomalous superconductors and the metals [4] is one of most important subjects in condensed matter physics. During the past decade, self-assembled InAs quantum dots(SAQD) have proven to be very important quantum structures both from a fundamental physics point of view as well as applied elds. SAQD is an ideal tool to study the energy states of quasi-zero- dimensional electron system [5, 6]. Furthermore, due to the relatively strong con nement in lateral direction, SAQD can be regarded as arti cial atoms. With resonant inelastic scattering, one has direct access to the electronic elementary excitations of low-dimensional electron systems, the so-called spin-density and charge-density excitations, Brocke et al. [7] have reported very interesting results in SAQD by resonant inelastic light scattering. They observe excitations, which might be identi ed as transition of electrons from the s-shell to the p-one(s-p transition) and from the p-shell to the d-one(p-d transition) of the quasi-atoms. Taking into account the e ect of many-body interactions, they explain the shift and broadening of the s-p transition of collective excitations as additional excitations at lower energies. But the origin of additional excitations is not con rmed so far. Recently the present author has indicated the importance of the hole-induced domain-wall in magnetoresistance in diluted magnetic semiconductors [8, 9] and doped manganites [10, 11]. In this study, we will discuss an origin of the anomalous excitations in the quantum dot, which might correspond to exotic excitations with fractional charges, extending the previous formula [12, 13, 14] from Callen-Harvey theory [15].

2. A model system

Brocke et al. [7] observed the shift and broadening of the s-p transition of collective excitations in InAs SAQD by resonant inelastic light scattering. An interesting point is that the shift and broadening e ects are enhanced when the Fermi energy level is located near the p-shell energy level. The p-shell energy level will be distorded strongly near the domain wall shell. It is likely that the crossing between the p-shell energy level and the Fermi energy level may occur near the domain wall shell. In this case, we assume that the low-energy excitations are described approximately in terms of massive relativistic fermions (with a mass, m, of the order of the energy gap). Thus we will introduce the e ective Lagrangian L_{eff} as follows,

$$L_{eff} = \overline{\Psi}(i\gamma_i\partial_i - M(r) - i\gamma^5 K(r))\Psi$$

Ψ is the Fermi wavefunction. Where we have introduced the e ect of an antiphase boundary (domain wall), through the position dependence mass term $M(r)$. We will assume that $M(r = +\infty) = +\mathrm{m}$ ($\mathrm{m} \geq 0$). $K(r)$ is the background eld with the domain wall shell. Then, the Dirac Hamiltonian is as follows,

$$H = \begin{pmatrix} M(r) & -\gamma_i\partial_i + K(r) \\ \gamma_i\partial_i + K(r) & -M(r) \end{pmatrix}$$

$K(r) \to +v$ when $r \gg r_0$. r_0 is the radius of the dot. $K(r) \to -v$, when $r_0 \gg r \sim 0$. Using the theoretical formula [15], the zero mode is given by

$$\Psi_0 = \chi_0(r)exp[-\int_0^r K(z)\, dz], \tag{1}$$

with $i\gamma_i\partial_i\chi_0 = 0$ so that the zero modes are solutions of the massless Dirac equation on the domain wall shell. Using the Mellin transformation of the odd spactral density [13], the Fermi number N corresponding to the Hamiltonian H is given by

$$\begin{aligned} N &= -\frac{1}{2}\eta_H = T_{3+1} + Index(D) \\ &+ \frac{1}{2\pi}[arctan[v/m] - arctan[-v/m]] \\ &= T_{3+1} + Index(D) + \frac{1}{\pi}\theta. \end{aligned} \tag{2}$$

$$\theta = arctan[\frac{v}{m}]. \tag{3}$$

η_H is the η - invariant term of the Atiya-Patodi-Singer index theorem [16]. Index (D) is the index of the Dirac operator D. T_{3+1} is the Pontryagin index of the background elds with the domain wall shell. The fractional part of the fermion number in eq.(2) is $\frac{1}{\pi}\theta$. This suggests strongly that excitations with the fractional charges exist in this dot-system. It is likely that these exotic excitations with the fractional charges might create new energy-levels in the dot. In addition, it is expected that the present model will be applicable to the dot of a narrow-gap semiconductor such as PdTe [14]. Now we shall discuss the wavefunction $\psi(r,t)$ of fermions in this system.

$$\psi(r,t) = \frac{\chi(r)}{r}\eta_{jm}exp(-iEt). \qquad (4)$$

Where η_{jm} is modi ed harmonic function[25-28], and E is the energy.

$$\chi(r) = \begin{bmatrix} F(r) \\ G(r) \end{bmatrix} \qquad (5)$$

Where $\chi(r)$ is an ordinary two-component wavefunctions $\chi_1(r)$ and $\chi_2(r)$.

$$\chi_1(r) = \begin{bmatrix} F_1(r) \\ G_1(r) \end{bmatrix} \qquad (6)$$

$$\chi_2(r) = \begin{bmatrix} F_2(r) \\ G_2(r) \end{bmatrix} \qquad (7)$$

$$\chi_1(E,r) = \begin{bmatrix} \frac{ik}{E\ \ 1/r_0} sinkr \\ coskr \end{bmatrix} \qquad (8)$$

$$\chi_2(E,r) = \begin{bmatrix} coskr \\ \frac{ik}{E+1/r_0} sinkr \end{bmatrix} \qquad (9)$$

The boundary condition is of the form as follow,

$$\chi(0) \propto \begin{bmatrix} isin(\frac{1}{2}arctan[v/m] + \frac{\pi}{4}) \\ cos(\frac{1}{2}arctan[v/m] + \frac{\pi}{4}) \end{bmatrix} \qquad (10)$$

Now, we shall consider the quantum-dot from the viewpoint of the soliton such as $D4$-branes [17, 18]. The $D4$-brane(a soliton), given by the one-pole 't Hooft ansatz [19], $(r) = 1 + {}^2/|x-a|^2$ with width , position a=0, and $x = (x_1, x_2, x_3)$ generates a Skyrme eld [20] of the hedgehog form $U(x) = exp[if(r)x \cdot \tau]$ with a pro le function given by

$$f(r) = \pi[1 - (1 + {}^2/r^2)\ {}^{1/2}]. \qquad (11)$$

Here we shall consider the toy-model of a quantum-dot, which is composed of three degenerated $D^{i=1,2,3}4$-branes with the pro le $f(r)$. It is assumed that a width of the quantum-dot is $\sim$. From three degenerated $D^{i=1,2,3}4$-branes [21], it is known that the SU(3) gauge elds are created among di erent D^i4-branes. As a consequence of ground-state degenerary of three degenereted D^i4-branes, stable elementary excitations of fractional charges $\pm(1/3)$e and $\pm(2/3)$e can be created [12, 22, 23, 24] in the toy-model, which is composed of three degenerated D^i4-branes. Where e means the charge value of a positron. Because relativistic eld-theory of meson and hadrons is very complex, one cannot help but notice the analog with the quark model. If it is assumed that "i" of D^i4-branes correspond to the colours, the SU(3) gauge elds are analogous to the gluon elds. In the trapping of electron-particle as lepton in the quantum-dot, the excitations with +(2/3)e and -(1/3)e in three degenerated D^i4-branes are analogous to the up-quark and down-quark, respectively. In the trapping of μ-particle as lepton in the quantum-dot, the excitations with +(2/3)e and -(1/3)e in three degenerated D^i4-branes are analogous to the charm-quark and strange-quark, respectively. In the trapping of τ-particle as lepton in the quantum-dot, the excitations with +(2/3)e and -(1/3)e in three degenerated D^i4-branes are analogous to the top-quark and botom-quark, respectively. Taking into account the $D4$-brane model [18, 29], The action of the gluon gauge elds, which connect between these quarks on D^i4-branes, is

$$\begin{aligned} S &= S_{YM} + S_{CS}, \\ S_{YM} &= -\kappa \int tr[\frac{1}{2}k(z)^{\ 1/3} F_{\mu\nu}^2 + k(z) F_{\mu z}^2]\, d^4x\, dz \\ S_{CS} &= \frac{3}{24\pi^2} \int_{M^4 \times R} \omega_5(A). \end{aligned}$$

Here μ, ν = 0,1,2,3 are four-dimensional Lorentz indices, and z is the coordinatice of the fth dimension. The quantity $A = A_\alpha dx^\alpha = A_\mu dx^\mu + A_z dz (\mu = 0,1,2,3,z)$ is the ve-dimensional gauge elds and $F = \frac{1}{2}F_{\alpha\beta}dx^\alpha \times dx^\beta = dA + iA \times A$ is its eld strength. The constant κ is related to the 't Hooft coupling g and colour number $N_c = 3$ as $\kappa = \frac{3g}{216\pi^3}$. $k(z) = 1 + z^2$ and $\omega_5(A)$ is the Cherm - Simons 5-form de ned as $\omega_5(A) = tr(AF^2 - \frac{1}{2}A^3F - \frac{1}{10}A^5)$. Because the present model of a quantum-dot is too simple and the toy-model, the present model might not correspond to the real states in mesons and hadrons. However, these analogies are very fascinating.

3. Conclusion

We have analyzed anamalous excitations in the electron state in the semiconductor-dot from viewpoint of eld-theoretical formula. We have introduced the toy-model of the quantum-dot, which is composed three degenerated $D^{i=1,2,3}4$-branes-like solitons. In addition, the relation to low energy hadron in QCD has been suggested.

References

[1] Zhang S C, Hansson T H, Kivelson S 1989 *Phy. Rev. Lett* **62** 82
[2] Zhang S C 1992 *Int. J. Mod. Phys.* **B6** 25
[3] Qi X L, Hughes T L, Zhang S C 2008 *Phys. Rev.* **B78** 195424
[4] Alice J 2012 *Rep. Prog. Phys.* **75** 076501
[5] Marzin J A et al. 1994 *Phy. Rev. Lett* **73** 716
[6] Drexler H et al. 1994 *Phy. Rev. Lett* **73** 2257
[7] Broche T et al. 2004 *Physica* **E 22** 478
[8] Kanazawa I 2006 *Phys.Lett* **A 355** 460
[9] Kanazawa I 2007 *Physica* **E 40** 277
[10] Kanazawa I 2010 *Phys. Status Solids* **B 247** 644
[11] Kanazawa I 2013 *J. Low Temp. Phys.* **172** 177
[12] Goldstone J and Wilczek F 1981 *Phys. Rev. Lett.* **47** 986
[13] Niemi A 1985 *Nucl. Phys.* **B 253** 14
[14] Fradkin E et al. 1986 *Phys. Rev. Lett.* **57** 2967
[15] Callen Jr C G and Harvey J A 1985 *Nucl. Phys.* **B 250** 427
[16] Atiyah M, Patodi V, and Singer I 1973 *Bull. London. Math. Soc.* **5** 229
[17] Polchinski J 1995 *Phys. Rev. Lett.* **75** 4724
[18] Sakai T and Sugimoto S 2005 *Prog. Theor. Phys.* **113** 847
[19] 't Hooft G 1976 *Phys. Rev.* **D 14** 3432
[20] Atiyah M F and Manton N S 1986 *Phys. Lett.* **B 222** 438

[21] Polchinski J 1998 String Theory I, II, Cambridge Univ. Press.
[22] Jackiw R and Rebbi C 1976 *Phys. Rev.* **D 13** 3398
[23] Su W P, Schriefer J R, and Heeger A J 1979 *Phys. Rev. Lett.* **42** 1698
[24] Su W P and Schriefer J R 1981 *Phys. Rev. Lett.* **46** 738
[25] Wu T T and Yang C N 1975 *Nucl. Phy.***B 107** 365
[26] Rubakov V A 1982 *Nucl. Phy.***B 203** 311
[27] Callen,Jr C G 1982 *Phy. Rev.* **D 25** 2141
[28] Yamagishi H 1983 *Phy. Rev.* **D 27** 2383
[29] Hashimoto K, Sakai T, Sugimoto S 2008 *Prog. Theor. Phys.* **120** 1093

IOP Conf. Series: Journal of Physics: Conf. Series **807** (2017) 102005 doi:10.1088/1742-6596/807/10/102005

Simulated floating zone method

Ryo Ozawa, Yasuyuki Kato, and Yukitoshi Motome

Department of Applied Physics, University of Tokyo, Tokyo 113-8656, Japan

E-mail: ozawa@aion.t.u-tokyo.ac.jp

Abstract. This paper provides the simulated floating zone (SFZ) method that is an efficient simulation technique to obtain thermal equilibrium states, especially useful when domain formation prevents the system from reaching a spatially-uniform stable state. In the SFZ method, the system is heated up locally, and the heated region is steadily shifted, similar to the floating zone method for growing a single crystal with less lattice defect and impurity in experiments. We demonstrate that the SFZ method suppresses domain formation and accelerates the optimization of the state, taking an example of a magnetic vortex crystal state realized in itinerant magnets. We show that the efficiency is maximized when the local heating temperature is tuned to be comparable to the characteristic energy scale of the ordered state.

1. Introduction

Numerical simulation is one of the most powerful methods for searching thermal equilibrium states of the system. It has elucidated numerous novel properties that one could hardly deal with analytically. One of the standard techniques is the Markov-chain Monte Carlo sampling on the basis of the Metropolis algorithm, e.g., the single-spin flip update used for classical spin systems. A common difficulty in such numerical simulations is that the state is often frozen into some configuration and hardly updated, especially when the system has a peculiar energy landscape with multiple local minima or almost flat energy dependence in the phase space. For instance, in some classical spin systems, the spin configuration is frozen in a multiple-domain state, and hardly escapes from the metastable state by local spin-flip update. For avoiding such freezing problems, a variety of the simulation techniques have been proposed, such as the simulated annealing [1], cluster update [2, 3], and replica exchange methods [4, 5].

In this paper, we propose an efficient technique, which will be useful for preventing the system from being frozen into multiple-domain states. Our method is similar to the floating zone method for growing a single crystal in experiments. Specifically, a part of the system is heated up to 'melt' the frozen structure, and the molten region is shifted smoothly to sweep away local defects or domain boundaries during the simulation. The procedure helps to grow a single-domain state efficiently, similar to the crystal growth in experiments. That is the reason why we call this method the "simulated floating zone (SFZ) method". We demonstrate the efficiency of the SFZ method by applying it to a complicated magnetic ground state in the Kondo lattice model with classical localized spins. We show that the multiple-domain structures are quickly swept away and the system reaches a single-domain state. We find that the efficiency of the SFZ method is optimized by setting the local heating temperature comparable to the characteristic energy scale of the ordered phase.

The rest of the paper is organized as follows. In Sec. 2, we introduce the fundamental idea of the SFZ method. In Sec. 3, we demonstrate the efficiency of the SFZ method by applying it to the Langevin dynamics simulation for the Kondo lattice model. We summarize our results in Sec. 4.

2. Method

The idea of the SFZ method is simple. It follows the floating zone method used in the single crystal growth. In the floating zone method, a narrow region of the crystal is melted by heating, and this molten region is shifted along the crystal in a rod shape. The procedure suppresses the grains and domains efficiently, and results in a purified single crystal. We can introduce the similar technique in numerical simulations for obtaining spatially-uniform equilibrium state. Thus, the SFZ method consists of the following steps: (1) a part of the system is heated up so as to melt the metastable configuration, (2) the molten region is moved smoothly along the system, and (3) the procedures are repeated a sufficient number of times, until a spatially-uniform stable state is reached.

Figure 1 shows a schematic picture of the SFZ method. The red region is the heated region, whose temperature is set at a higher temperature ($T = T_{\rm loc}$) than the other parts of the system ($T = T_{\rm bulk}$). The heated region is shifted along the system in the direction of the arrow. The size and shape of the heated region as well as the speed and direction of its movement are arbitrary: one can tune them so as to reach the stable configuration quickly.

The SFZ method is easy to implement and generally compatible with many simulation techniques. It is also applicable to a wide range of systems, both classical and quantum, in any spatial dimensions.

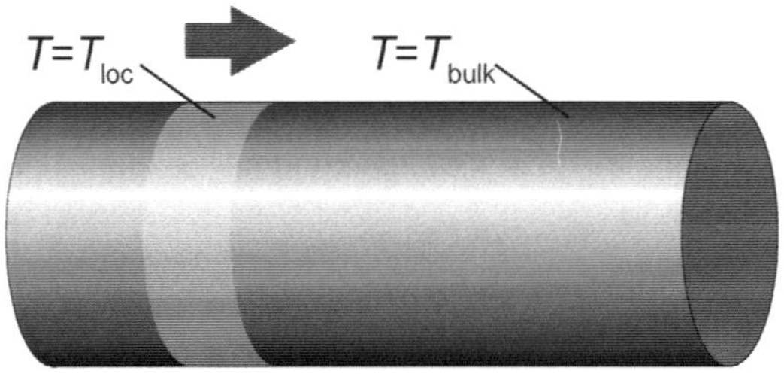

Figure 1. Schematic picture of the SFZ method. The red region is the heated region, which is smoothly shifted in the direction of the arrow.

3. Application

In this section, we demonstrate the efficiency of the SFZ method in the numerical search of the stable ground state, where domain formation makes difficult to reach a spatially-uniform state. As a typical example, we here consider a noncoplanar spin texture emergent from the interplay between charge and spin degrees of freedom in electrons. Specifically, we focus on a magnetically-ordered state composed of a periodic arrangement of spin vortices (meron, i.e., half-skyrmion), which we recently found as a candidate for the ground state of the Kondo lattice model with classical localized spins on a square lattice [6]. The Hamiltonian is given by

$$\mathcal{H} = -\sum_{i,j}\sum_{s} t_{ij}(\hat{c}^{\dagger}_{is}\hat{c}_{js} + \text{h.c.}) - J\sum_{i}\sum_{s,s'} \hat{c}^{\dagger}_{is}\boldsymbol{\sigma}_{ss'}\hat{c}_{is'}\cdot\mathbf{S}_i, \quad (1)$$

where $\hat{c}^{\dagger}_{is}(\hat{c}_{is})$ is a creation (annihilation) operator of an itinerant electron with spin $s = \uparrow, \downarrow$ at site i, $\boldsymbol{\sigma} = (\sigma_x, \sigma_y, \sigma_z)$ is the vector representation of the Pauli matrices, and $\mathbf{S}_i$ is the classical localized spin at site i with $|\mathbf{S}_i| = 1$. The first term in Eq. (1) is the hopping term of itinerant electrons with transfer integral t_{ij}: here, we consider only the nearest- and third-neighbor hoppings, $t_1 = 1$ and $t_3 = -0.5$, respectively, on the square lattice. The second term in Eq. (1) is the onsite Hund's-rule coupling with the coupling constant J. We set the lattice constant $a = 1$, the reduced Planck constant $\hbar = 1$, and the Boltzmann constant $k_{\rm B} = 1$.

The vortex crystal state that we found in Ref. [6] exhibits stripes of the spin scalar chirality in the diagonal direction of the square lattice. The spin scalar chirality is defined for each square plaquette p as

$$\chi_p = \frac{1}{4}\left[\left(\mathbf{S}_{p_1}\times\mathbf{S}_{p_2}\right)\cdot\mathbf{S}_{p_3} + \left(\mathbf{S}_{p_4}\times\mathbf{S}_{p_1}\right)\cdot\mathbf{S}_{p_2} + \left(\mathbf{S}_{p_3}\times\mathbf{S}_{p_4}\right)\cdot\mathbf{S}_{p_1} + \left(\mathbf{S}_{p_2}\times\mathbf{S}_{p_3}\right)\cdot\mathbf{S}_{p_4}\right], \tag{2}$$

where sites $p_1, \cdots, p_4$ are the vertices of the plaquette p in the counterclockwise direction. Thus, the vortex crystal state has trivial twofold degeneracy with respect to the directions of chirality stripes, which are connected by in-plane fourfold rotation, in addition to the global rotation in the spin space. It is this twofold degeneracy that leads to domain formation in the current system, as discussed later in Fig. 2(b).

We employ the SFZ method in the numerical simulation using the Langevin dynamics for the update of the configurations of localized spins [7]. The Langevin dynamics of the localized spins is described by the Landau-Lifshitz-Gilbert equation [8] as

$$\frac{d\mathbf{S}_i}{d\tau} = -\mathbf{S}_i \times \mathbf{H}_i - \gamma \mathbf{S}_i \times (\mathbf{S}_i \times \mathbf{H}_i), \tag{3}$$

where the first (second) term describes a precession (dumping) of spins. $\mathbf{H}_i$ is the internal effective magnetic field acting on $\mathbf{S}_i$, which is given by

$$\mathbf{H}_i = \frac{\partial\Omega(\{\mathbf{S}_j\})}{\partial\mathbf{S}_i} + \mathbf{h}_i(\tau, T). \tag{4}$$

Here, $\Omega(\{\mathbf{S}_i\})$ is the grand potential for a given spin configuration $\{\mathbf{S}_i\}$, which is calculated by using the kernel polynomial method [9]; $\mathbf{h}_i$ describes thermal fluctuations in the Langevin dynamics, which satisfies $\langle h_i^\mu(\tau_1, T) h_j^\nu(\tau_2, T)\rangle_\tau = 2T\delta(\tau_1 - \tau_2)\delta_{ij}\delta_{\mu\nu}$ [$\langle\cdots\rangle_\tau$ is the time average and h_i^μ is the μ component of $\mathbf{h}_i$ ($\mu = x, y, z$)]. In this thermal-fluctuation term, we introduce the SFZ method. The system within a narrow diagonal strip is heated up to a temperature $T = T_{\rm loc}$, and moved along another diagonal direction; see the schematic picture in Fig. 2(a); we take the heated strip in the diagonal direction because the chiral stripes appear along the diagonal (or $\pi/4$ rotated) direction, as shown below.

Figure 2 shows the results of the Langevin dynamics simulation accelerated by the SFZ method. We consider the model in Eq. (1) with $J = 0.05$ on the square lattice with 96^2 sites and periodic boundary conditions. We set the electron filling so as to realize the vortex crystal with ordering vectors $\sim(\pm\pi/4, \pi/4)$ [6]. The Langevin dynamics simulation is performed by taking $\gamma = 1$ and the time interval as $\Delta\tau = 30$. In the kernel polynomial method, we take the Chebyshev order $M = 800$ and the number of random vectors $R = 144$. We use the heated strip with four-site width (we do not impose the periodic boundary conditions on the strip shape), and set the temperature at $T_{\rm loc} = 0.004$, while taking $T_{\rm bulk} = 0$. The heated strip is shifted in a constant speed along the system so that one sweep from the bottom-left to top-right corner takes a time of 4.8×10^4: the Langevin dynamics update is performed for 1.6×10^3 times during the single sweep. The next sweep starts from the same bottom-left corner just after the previous sweep ends.

Figure 2(b) shows a typical multiple-domain state obtained by the Langevin dynamics simulation without the SFZ method staring from a random spin configuration. We here plot the real-space distribution of the scalar chirality χ_p defined in Eq. (2). The system has several domains; each domain has chiral stripes running in the diagonal-upward or downward direction. In this state, the spin configuration is almost frozen: the domain walls are hardly removed by the update in the Langevin dynamics. Once we turn on the SFZ method, however, the multiple domains are smeared and the system reaches a single-domain state. Figures 2(c)-2(h) show the evolution of the real-space distributions of χ_p during the simulation with the use of the SFZ method. As shown in the figures, when we repeat the sweep, one type of the two degenerated domains grows and merges into larger domains. After eight cycles, we successfully obtain the spatially-uniform state, as shown in Fig. 2(h).

We note that the direction of the chiral stripes in the final uniform state is always parallel to the heated strip: when we rotate the strip and its motion by $\pi/2$ in the SFZ simulation, the other domain with

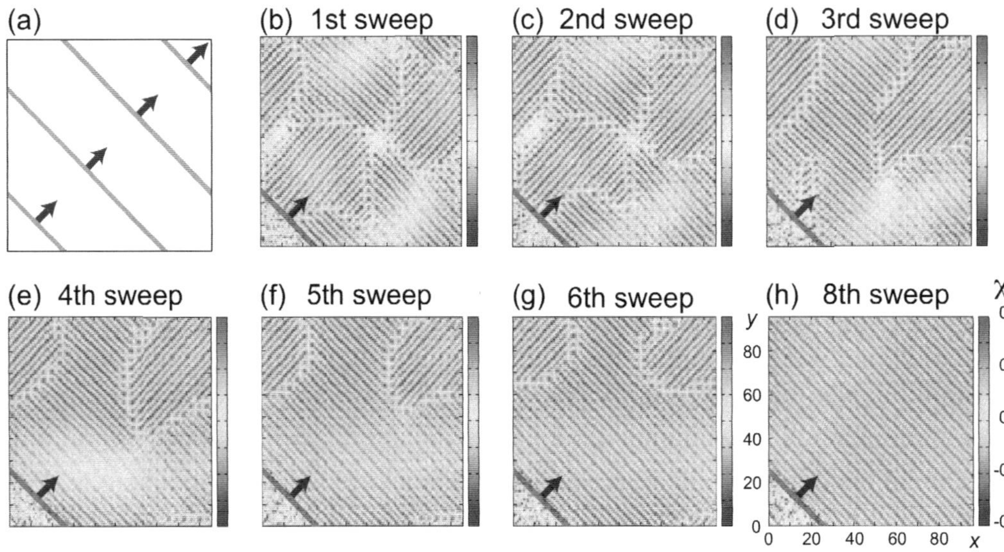

Figure 2. (a) Schematic of the SFZ method applied to the model in Eq. (1) on the square lattice. The red region represents the heated strip. The arrow indicates the direction of the movement of the strip: the strip moves from the bottom-left to top-right corner, and restarts from the bottom-left corner in the next cycle. (b)-(h) Real-space distributions of the spin scalar chirality [Eq. (2)] in the simulation with the use of the SFZ method after the sweeps indicated in each figure. The simulation was performed at $t_1 = 1$, $t_3 = -0.5$, and $J = 0.05$ on the square lattice with 96^2 sites.

$\pi/2$-rotated stripes dominates the system in the final uniform state. Such a selection of energetically-degenerate domains might occur in the SFZ method, depending on the microscopic energetics in the system as well as the details of the floating zone. In the current case, we note that the selected domain has the dominant spin helix parallel to the heated strip [6]. The reason why this domain is chosen is presumably because the disturbance of electron motion along the spin helix leads to a substantial energy loss in the spin state.

Let us discuss the efficiency of the SFZ method while changing the heating temperature $T_{\rm loc}$. In order to measure the efficiency, we monitor the time evolution of the dominant peaks of the structure factors for spin and chirality degrees of freedom. The structure factors are defined by

$$S(\mathbf{q}) = \frac{1}{N}\sum_{i,j}\mathbf{S}_i\cdot\mathbf{S}_j e^{i\mathbf{q}\cdot(\mathbf{r}_i-\mathbf{r}_j)}, \quad \chi(\mathbf{q}) = \frac{1}{N}\sum_{p,q}\chi_p\chi_q e^{i\mathbf{q}\cdot(\mathbf{r}_{p_1}-\mathbf{r}_{q_1})}, \tag{5}$$

where N is the system size and $\mathbf{r}_i$ is the real-space coordinate of site i.

Figure 3 shows the time evolution of the largest peaks of $S(\mathbf{q})$ and $\chi(\mathbf{q})$ in the SFZ simulation for several different $T_{\rm loc}$ by starting from the same initial state shown in Fig. 2(b). Here, we plot the data at every SFZ sweep. When we set $T_{\rm loc} = 0.001$, both peaks grow very slowly and do not change substantially from the initial values after the long simulation time. This is presumably because $T_{\rm loc} = 0.001$ is too low to melt the frozen domain structure. For $T_{\rm loc} \geq 0.002$, both peaks grow during the simulation, and finally saturate at the same values, which are expected for the single-domain state, after the sufficiently long simulation time. When further increasing $T_{\rm loc}$, however, the time necessary for the convergence becomes longer. For both $S(\mathbf{q})$ and $\chi(\mathbf{q})$, $T_{\rm loc} \simeq 0.002$-0.004 optimizes the efficiency of the SFZ simulation in the current case.

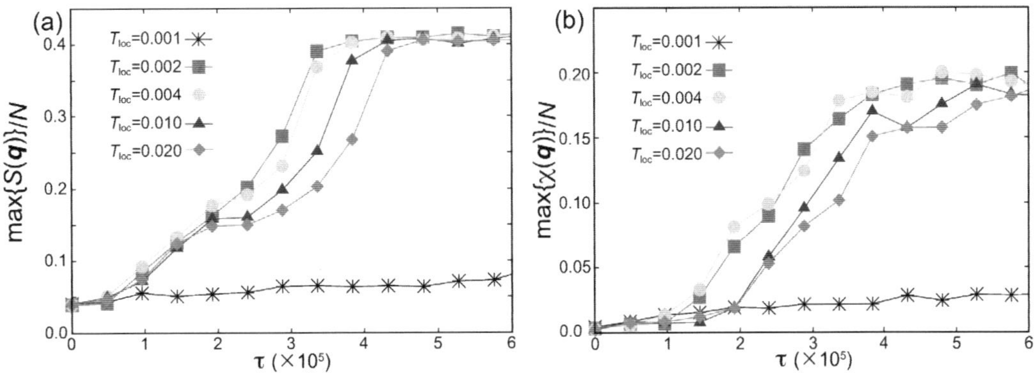

Figure 3. Time evolution of the largest peak (a) in the spin structure factor, $S(\mathbf{q})$, and (b) in the chirality structure factor, $\chi(\mathbf{q})$, divided by the system size for different temperatures of the heated region in the SFZ method, $T_{\rm loc} = 0.001, 0.002, 0.004, 0.010$, and 0.020.

We note that the optimal value of $T_{\rm loc}$ is comparable to the energy scale of the magnetic ordering. In the Kondo lattice model in the small J region, the energy scale is set by the so-called Ruderman-Kittel-Kasuya-Yosida (RKKY) interaction, which is given by the second-order perturbation in terms of J [10, 11, 12]. In the current case, the energy scale is about $J^2 = 0.0025$, which is close to the optimal $T_{\rm loc} \simeq 0.002$-0.004. Thus, the result suggests that the SFZ method works most efficiently when the heating temperature is comparable to the melting temperature of the corresponding ordered phase. This is presumably because the small part of the system included in the heated strip is easily aligned in a spatially-uniform ordered state owing to the growing correlation length near the melting temperature.

4. Summary

In summary, we have introduced the SFZ method as an efficient technique to obtain the thermal equilibrium state in numerical simulations. In this method, a part of the system is locally heated up and shifted smoothly all over the system. The main purpose of using this method is to accelerate the numerical convergence, preventing the state from being frozen in a local minimum or a flat energy landscape in the phase space. The advantage is that this method is simple and applicable to many situations. In this paper, we have applied the SFZ method to the Langevin dynamics simulations for the square Kondo lattice model as an example where domain formation is often problematic for obtaining the ground state in the simulations without the SFZ method. We have demonstrated that the introduction of the SFZ method sweeps away the domain structures and substantially accelerates the convergence to the spatially-uniform state. In addition, we have found that the convergence is most accelerated by setting the local heating temperature comparable to the characteristic energy scale of the ordered state.

5. Acknowledgments

The SFZ methods based on the Langevin dynamics simulations were carried out at the Supercomputer Center, Institute for Solid State Physics, University of Tokyo. R.O. is supported by the Japan Society for the Promotion of Science through a research fellowship for young scientists and the Program for Leading Graduate Schools (ALPS). This research was supported by KAKENHI (No. 24340076, 26800199), the Strategic Programs for Innovative Research (SPIRE), MEXT, and the Computational Materials Science Initiative (CMSI), Japan.

References

[1] Kirkpatrick S 1984 J. Stat. Phys. **34** 975
[2] Wang J S and Swendsen R H 1990 Physica A **167** 565
[3] Wolff U 1989 Phys. Rev. Lett. **62**(4) 361
[4] Swendsen R H and Wang J S 1986 Phys. Rev. Lett. **57**(21) 2607
[5] Hukushima K and Nemoto K 1996 J. Phys. Soc. Jpn. **65** 1604
[6] Ozawa R, Hayami S, Barros K, Chern G W, Motome Y and Batista C D arXiv:1510.06830
[7] Barros K and Kato Y 2013 Phys. Rev. B **88**(23) 235101
[8] Gilbert T L 2004 IEEE Trans. Magn. **40** 3443
[9] Weiße A, Wellein G, Alvermann A and Fehske H 2006 Rev. Mod. Phys. **78**(1) 275
[10] Ruderman M A and Kittel C 1954 Phys. Rev. **96**(1) 99
[11] Kasuya T 1956 Prog. Theor. Phys. **16** 45
[12] Yosida K 1957 Phys. Rev. **106**(5) 893

Thermal Hall conductivity of spin-triplet superconductor with time reversal symmetry breaking

Yoshiki Imai[1], Katsunori Wakabayashi[2], Manfred Sigrist[3]
[1]Department of Physics, Saitama University, Saitama 338-8570, Japan
[2]Department of Nanotechnology for Sustainable Energy, Kwansei Gakuin University, Gakuen 2-1, Sanda 669-1337, Hyogo, Japan
[3]Institut für Theoretische Physik, ETH-Zürich, CH-8093 Zürich, Switzerland

E-mail: imai@phy.saitama-u.ac.jp

Abstract. The thermal Hall conductivity is investigated for the superconductor Sr_2RuO_4 assuming a time reversal symmetry breaking order parameter. The γ band of Sr_2RuO_4 has its Fermi level near the van Hove points, close to the Lifshitz transition. Within a Bogoliubov-de Gennes approach we calculate the thermal Hall conductivity for pairing symmetries including two exemplary cases, the usual chiral p-wave phase and a $f_{x^2-y^2}$-wave phase of the structure $\boldsymbol{d} = \Delta\hat{z}(k_x^2 - k_y^2)(k_x \pm ik_y)$. In the chiral p-wave phase the thermal Hall conductivity consists of a universal temperature-linear term and an exponential correction due to quasiparticle activation with a full excitation gap. Due to line-nodes in the gap we find a correction which is quadratic in temperature for $f_{x^2-y^2}$-wave state. This difference in the corrections allows to analyze the gap structure of the superconducting phase.

1. Introduction

The transition metal oxide Sr_2RuO_4 [1, 2] is often viewed as a good candidate for a topological superconductor. Experimental studies indicate that the superconducting state has magnetic properties indicating broken time-reversal symmetry [3, 4]. The symmetry analysis has led to the identification of the pairing channel as chiral p-wave with a gap function written in $\boldsymbol{d}$-vector notation as $\boldsymbol{d} = \Delta\hat{z}(k_x \pm ik_y)$ with an orbital angular momentum $L_z = \pm 1$ along the z axis [5]. This is the two-dimensional analog of the Anderson-Brinkman-Morel (ABM) state, and has generally a nodeless excitation gap. Recently $f_{x^2-y^2}$-wave was proposed an alternative candidate [6], whose gap function has the basic form $\boldsymbol{d} = \Delta\hat{z}(k_x^2 - k_y^2)(k_x \pm ik_y)$, having vertical line nodes in the excitation gap.

It is well known that the electronic low-energy physics of Sr_2RuO_4 is governed by the α, β and γ bands composed of the Ru $4d$ t_{2g} orbitals. The topological properties originate from the two-dimensional electron-like γ-band and are characterized by a Chern number [7]. The Fermi energy of the γ-band lies very close to the van Hove point ($\boldsymbol{k} = (\pi, 0)$), such that the topology of the Fermi surface may be changed by some form of tuning through a Lifshitz transition, which is then accompanied by the switching of the Chern number [7, 8].

It turns out that the edge currents induced by the time reversal symmetry breaking phase does not change strongly at the Lifshitz transition counter naive expectation [8]. In contrast, the thermal Hall effect being directly connected to the Chern number in topological superconductors is certainly more suitable to probe the change of topological features [9]. We have discussed that the temperature dependence of the thermal Hall conductivity in the chiral p-wave phase is composed of a temperature linear term and an exponential term. The former is proportional to the Chern number and the latter gives the information on the magnitude of the quasiparticle excitation gap and the topological properties [10].

In our present report we propose to examine the thermal Hall conductivity in the superconducting phases with time-reversal symmetry breaking in order to distinguish the pairing symmetry around the Lifshitz transition.

2. Model

To demonstrate the connection between the topological property and the thermal Hall effect, we employ the single band tight-binding model on the two-dimensional square lattice with a structure like the γ band, described by the following Bogoliubov de-Gennes (BdG) Hamiltonian

$$H_{\mathrm{BdG}} = \sum_{\boldsymbol{k}} \begin{pmatrix} c^{\dagger}_{\boldsymbol{k}\uparrow}, c_{-\boldsymbol{k}\downarrow} \end{pmatrix} \begin{pmatrix} \varepsilon_{\boldsymbol{k}} & \Delta_{\boldsymbol{k}} \\ \Delta^{*}_{\boldsymbol{k}} & -\varepsilon_{\boldsymbol{k}} \end{pmatrix} \begin{pmatrix} c_{\boldsymbol{k}\uparrow} \\ c^{\dagger}_{-\boldsymbol{k}\downarrow} \end{pmatrix}, \qquad (1)$$

where $c^{\dagger}_{\boldsymbol{k}\sigma}$ ($c_{\boldsymbol{k}\sigma}$) is the creation (annihilation) operator of an electron with wave vector $\boldsymbol{k}[=(k_x, k_y)]$ and spin $\sigma(=\uparrow,\downarrow)$. $\varepsilon_{\boldsymbol{k}}$ and $\Delta_{\boldsymbol{k}}$ denote the energy dispersion of the γ-band in the normal phase and the gap function, respectively, given by,

$$\varepsilon_{\boldsymbol{k}} = -2t(\cos k_x + \cos k_y) - 4t' \cos k_x \cos k_y - \mu, \qquad (2)$$

$$\Delta_{\boldsymbol{k}} = \begin{cases} \Delta(\sin k_x \pm i \sin k_y), & (\text{chiral } p\text{-wave}) \\ \Delta(\cos^2 k_x - \cos^2 k_y)(\sin k_x \pm i \sin k_y), & (f_{x^2-y^2}\text{-wave}) \end{cases} \qquad (3)$$

where t (t') are the hopping amplitudes between nearest-neighbor (next-nearest-neighbor) sites. μ is the chemical potential and Δ is the amplitude of the gap function.

Figure 1 displays the Fermi surface for several choices of the chemical potential. The result for $\mu = 1.3t$ reproduces the electron-like γ-band of bulk Sr_2RuO_4 well. With increasing μ, the Fermi surface structure switches from the electron-like to hole-like shape when passing through the van Hove points at $\mu = 1.4t(\equiv \mu_c)$. Here we implement the change of the Fermi surface topology through the tuning of the chemical potential instead of the direct consideration of physical and/or chemical tunings in Sr_2RuO_4, for simplicity.

The quasiparticle energies of the BdG Hamiltonian are given by $E_{\boldsymbol{k}\pm} = \pm\sqrt{\varepsilon^2_{\boldsymbol{k}} + \Delta^2_{\boldsymbol{k}}}$. Thus while the chiral p-wave phase has fully gapped excitation spectrum, $E_{\boldsymbol{k}}$ vanishes at

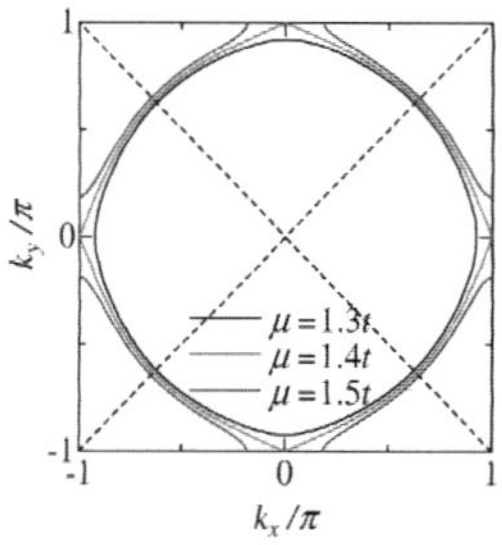

Figure 1. (Color online) Fermi surface for several choices of μ with $t' = 0.35t$ in the normal phase ($\Delta = 0$). The dashed lines represent line nodes with $|k_x| = |k_y|$ in the $f_{x^2-y^2}$-wave phase.

$\boldsymbol{k} = \boldsymbol{k}_0 \equiv (\pm k_0, \pm k_0)$ and $(\pm k_0, \mp k_0)$ for the $f_{x^2-y^2}$-wave phase due to line nodes, where k_0 is given as follows $k_0 = \arccos\left\{\left(-t + \sqrt{t^2 - t'\mu}\right)/(2t')\right\}$. With increasing μ, k_0 changes continuously and monotonically even while passing through the Lifshitz transition.

3. Results

In this study we use the following model parameters: the next-nearest-neighbor hopping matrix element and the gap function are $t' = 0.35t$ and $\Delta = 0.1t$, respectively, given in units of t. For our further calculation we choose the chiral phases with Chern number $+1$, noting that the case of -1 would yield simply analogous results.

Figure 2 displays the quasiparticle energy dispersions and the density of states (DOS) for the superconducting states. For the chiral p-wave ($f_{x^2-y^2}$-wave) phase, DOS reveals a full gap (a pseudogap) in the low-energy sector, whose magnitude is parametrized as ε_0 in both pairing phases. Around the node at $\boldsymbol{k} = \boldsymbol{k}_0$, in the $f_{x^2-y^2}$-wave phase $E_{\boldsymbol{k}}$ disperse linearly in the momentum approximated by

$$E_{\boldsymbol{k}\pm} \approx \pm\sqrt{a + b\sin 2\theta}\, q, \tag{4}$$

with $a = 2\sin^2 k_0 \left\{2(t + 2t'\cos k_0)^2 + \Delta^2 \sin^2 k_0\right\}$ and $b = 2\sin^2 k_0 \left\{2(t + 2t'\cos k_0)^2 - \Delta^2 \sin^2 k_0\right\}$. The angle θ denotes the orientation of the momentum relative to $\boldsymbol{k}_0$ and $q = |\boldsymbol{k} - \boldsymbol{k}_0|$. This linear dispersion leads to a DOS linear in energy within the range $|\varepsilon| \leq \varepsilon_0$.

The topological properties are characterized by the Chern number N_C, which is defined as

$$N_C = \frac{4\pi}{N} \sum_{n=\pm} \sum_{\substack{\boldsymbol{k} \\ (E_{\boldsymbol{k}n}<0)}} \mathrm{Im} \left\langle \frac{\partial u_{\boldsymbol{k}n}}{\partial k_x} \middle| \frac{\partial u_{\boldsymbol{k}n}}{\partial k_y} \right\rangle, \tag{5}$$

where N stands for the number of sites and $\boldsymbol{u}_{\boldsymbol{k}n}$ is the periodic part of the Bloch wave function of the BdG Hamiltonian.

Let us now discuss the temperature dependence of the thermal Hall conductivity. The formula in the superconducting phase with broken time-reversal symmetry is given by

$$\kappa_{xy} = -\frac{1}{4\pi T} \int d\varepsilon\, \varepsilon^2 \Lambda(\varepsilon) f'(\varepsilon), \tag{6}$$

$$\Lambda(\varepsilon) = \frac{4\pi}{N} \sum_{\boldsymbol{k}n=\pm} \mathrm{Im} \left\langle \frac{\partial u_{\boldsymbol{k}n}}{\partial k_x} \middle| \frac{\partial u_{\boldsymbol{k}n}}{\partial k_y} \right\rangle \theta(\varepsilon - E_{\boldsymbol{k}n}), \tag{7}$$

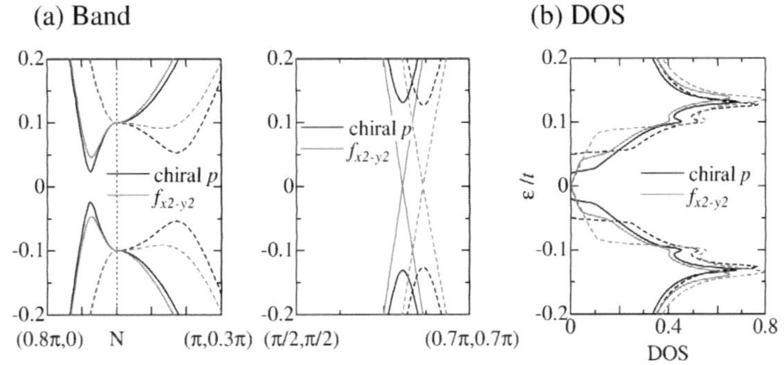

Figure 2. (Color online) (a) Quasiparticle energy dispersion and (b) DOS in the low-energy region. The solid (dashed) lines denote the $\mu = 1.3t$ ($\mu = 1.5t$) result.

where $f'(\varepsilon)$ and $\theta(\varepsilon)$ stand for the derivative of the Fermi distribution function and the Heaviside step function, respectively [9]. Note that $\Lambda(\varepsilon = 0)$ corresponds to the Chern number for both superconducting phases. In the low-temperature limit, the thermal Hall conductivity is linear in temperature to lowest order and proportional to the Chern number [9, 11, 12],

$$\kappa_{xy} \approx \kappa_{xy}^{L} = \frac{\pi N_C}{12} T. \tag{8}$$

We first consider the structure of $\Lambda(\varepsilon)$, which is important for the correction to the linear temperature dependence of κ_{xy}. Figure 3 displays $\Lambda(\varepsilon)$ for $\mu = 1.3t$ and $\mu = 1.5t$. $\Lambda(\varepsilon)$ has a basically rectangular-like shape and switches from convex downward to convex upward around $\mu = \mu_c$ for both pairing states. With increasing $|\varepsilon|$, $\Lambda(\varepsilon)$ shows an abrupt change at $|\varepsilon| = \varepsilon_0$. For the larger $|\varepsilon|$ ($> \varepsilon_0$) region, $\Lambda(\varepsilon)$ decreases towards zero gradually, whose actual structure depends on details of the model, and is not associated with the unique topological aspects and the Lifshitz transition directly.

Since the chiral p-wave phase has a fully gapped excitation spectrum, $\Lambda(\varepsilon)$ becomes constant within $|\varepsilon| \leq \varepsilon_0$, at value of the Chern number N_C.

On the other hand, in the $f_{x^2-y^2}$-wave phase, $\Lambda(\varepsilon)$ deviates from N_C except at $\varepsilon = 0$ due to gapless excitations around $\boldsymbol{k} = \boldsymbol{k}_0$. The derivative of $\Lambda(\varepsilon)$ in the low-energy region is written as

$$\Lambda'(\varepsilon) = \frac{\alpha}{\Delta}\left(2\theta(\varepsilon) - 1\right). \quad \left(\alpha \equiv \frac{\cos k_0}{2\sqrt{2}\sin^3 k_0}\right) \tag{9}$$

This formula is analytically obtained by the expansion with respect to wave vector around $\boldsymbol{k} = \boldsymbol{k}_0$, and indicates that $\Lambda(\varepsilon)$ becomes linear in $|\varepsilon|$ in the low-energy region. Figure 3 shows that the result from Eq. (9) agrees with the $\Lambda(\varepsilon)$ obtained numerically from Eq. (7). Note that the sign of the coefficient α becomes negative due to $\cos k_0 < 0$ and $\sin k_0 > 0$ within the used parameter range and does not change around the Lifshitz transition.

In order to investigate the low-temperature behavior of κ_{xy} analytically, we calculate Eq. (6) by means of $\tilde{\Lambda}(\varepsilon)$ instead of the exact $\Lambda(\varepsilon)$, which is given by

$$\tilde{\Lambda}(\varepsilon) = \begin{cases} N_C + \frac{\alpha}{\Delta}|\varepsilon| & (|\varepsilon| \leq \varepsilon_0) \\ 0 & (|\varepsilon| > \varepsilon_0) \end{cases}, \tag{10}$$

$$\alpha = \begin{cases} 0 & (\text{chiral } p\text{-wave}) \\ \frac{\cos k_0}{2\sqrt{2}\sin^3 k_0} & (f_{x^2-y^2}\text{-wave}) \end{cases}. \tag{11}$$

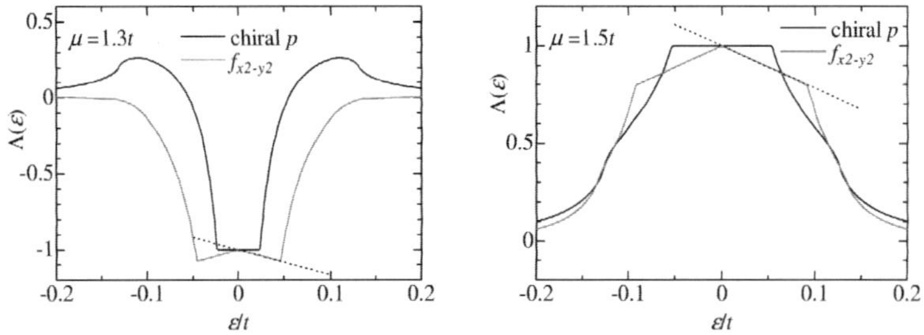

Figure 3. (Color online) $\Lambda(\varepsilon)$ for $\mu = 1.3t$ and $\mu = 1.5t$. The dotted lines stand for the results from Eq. (9) for the $f_{x^2-y^2}$-wave state for $\varepsilon > 0$.

Then the approximate thermal Hall conductivity $\tilde{\kappa}_{xy}$ is obtained as follows,

$$\kappa_{xy} \approx \tilde{\kappa}_{xy} = \kappa_{xy}^L + \frac{9\zeta(3)}{4\pi}\frac{\alpha}{\Delta}T^2 - \{\gamma(T) + \gamma'(T)\}\, e^{-\frac{\varepsilon_0}{T}}, \quad (12)$$

$$\gamma(T) \equiv \frac{N_C T}{\pi}\left\{2\left(\frac{\varepsilon_0}{2T}\right)^2 + 2\left(\frac{\varepsilon_0}{2T}\right) + 1\right\}, \quad (13)$$

$$\gamma'(T) \equiv \frac{2\alpha T^2}{\pi\Delta}\left\{2\left(\frac{\varepsilon_0}{2T}\right)^3 + 3\left(\frac{\varepsilon_0}{2T}\right)^2 + 3\left(\frac{\varepsilon_0}{2T}\right) + \frac{3}{2}\right\}, \quad (14)$$

where $\zeta(s)$ is Riemann ζ function and $\zeta(3) \sim 1.202$. Note that $|\gamma'(T)/\gamma(T)| \sim |\alpha\varepsilon_0/(N_C\Delta)| \sim |\alpha/N_C| \sim 0.15 - 0.20$ in the low-temperature region, so that the $\gamma(T)$ term is dominant by the exponential factor in Eq. (12).

This result indicates that the thermal Hall conductivity is expected to sustain the temperature linear behavior at lowest temperature, while having correction terms quadratic in T and exponentially activated. The sign of the exponential term $-\gamma(T)e^{-\frac{\varepsilon_0}{T}}$ in Eq. (12) switches from positive for $\mu < \mu_c$ to negative for $\mu > \mu_c$. In contrast the quadratic term remains negative regardless of the chemical potential in our example. Thus while $\tilde{\kappa}_{xy} - \kappa_{xy}^L$ changes the sign around $\mu = \mu_c$ in the chiral p-wave phase, $\tilde{\kappa}_{xy} - \kappa_{xy}^L$ has no drastic change due to the contribution of the quadratic term in the $f_{x^2-y^2}$-wave phase in the low-temperature region.

The difference $\kappa_{xy} - \kappa_{xy}^L$ as a function of temperature is depicted in Fig. 4 where κ_{xy} is obtained numerically from Eq. (6). $\kappa_{xy} - \kappa_{xy}^L$ is well described by the exponential function in the chiral p-wave phase [10]. In both amplitudes of the chemical potential ($\mu = 1.3t$ and $\mu = 1.5t$) the low-temperature behavior fits well to the T-quadratic term expected for the $f_{x^2-y^2}$-wave phase, validating our low-temperature approximation.

For the electron-like Fermi surface with $\mu < \mu_c$, $\kappa_{xy} - \kappa_{xy}^L$ in the chiral p-wave phase and the $f_{x^2-y^2}$-wave phase have opposite sign. This result indicates that the low-temperature behavior in κ_{xy} may give the information concerning the pairing symmetry in bulk Sr_2RuO_4.

4. Conclusions

Motivated by the transition metal compound Sr_2RuO_4 we have investigated the thermal Hall conductivity in the spin-triplet superconducting phases with broken time-reversal symmetry. By means of the two-dimensional tight-binding model describing the γ band, we obtain the temperature dependence of the thermal Hall conductivity in the chiral p-wave phase and the $f_{x^2-y^2}$-wave phase. The latter state has gapless quasiparticle excitations due to line nodes yielding a T^2 correction to the thermal Hall conductivity on top of the universal T-linear term.

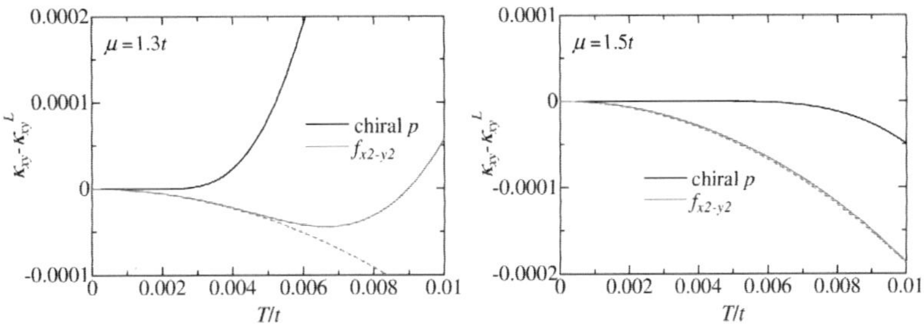

Figure 4. (Color online) $\kappa_{xy} - \kappa_{xy}^L$ for $\mu = 1.3t$ and $\mu = 1.5t$. The dotted lines stand for the quadratic term in Eq. (12) in the $f_{x^2-y^2}$-wave phase.

We emphasize here that for the electron-like Fermi surface (as realized in Sr_2RuO_4) the sign of the deviation from universal low-temperature thermal Hall conductivity for the $f_{x^2-y^2}$-wave state is different from that for the chiral p-wave state. This qualitative result may, in principle, enable us to experimentally test the pairing state for nodal gaps in Sr_2RuO_4.

References

[1] Y. Maeno, H. Hashimoto, K. Yoshida, S. Nishizaki, T. Fujita, J. G. Bednorz, and F. Lichtenberg. *Nature*, 372:532, 1994.
[2] A. P. Mackenzie and Y. Maeno. *Rev. Mod. Phys.*, 75:657, 2003.
[3] G. M. Luke, Y. Fudamoto, K. M. Kojima, M. I. Larkin, J. Merrin, B. Nachumi, Y. J. Uemura, Y. Maeno, Z. Q. Mao, Y. Mori, H. Nakamura, and M. Sigrist. *Nature (London)*, 394:558, 1998.
[4] J. Xia, Y. Maeno, P. T. Beyersdorf, M. M. Fejer, and A. Kapitulnik. *Phys. Rev. Lett.*, 97:167002, 2006.
[5] T. M. Rice and M. Sigrist. *J. Phys. Condens. Matter*, 7:643, 1995.
[6] Y. Hasegawa, K. Machida, and M. Ozaki. *J. Phys. Soc. Jpn*, 69:336, 2000.
[7] Y. Imai, K. Wakabayashi, and M. Sigrist. *Phys. Rev. B*, 88:144503, 2013.
[8] Y. Imai, K. Wakabayashi, and M. Sigrist. *J. Phys. Soc. Jpn*, 83:124712, 2014.
[9] H. Sumiyoshi and S. Fujimoto. *J. Phys. Soc. Jpn.*, 82:023602, 2013.
[10] Y. Imai, K. Wakabayashi, and M. Sigrist. *Phys. Rev. B*, 93:024510, 2016.
[11] N. Read and D. Green. *Phys. Rev. B*, 61:10267, 2000.
[12] T. Scaffidi and S. Simon. *Phys. Rev. Lett.*, 115:087003, 2015.

IOP Conf. Series: Journal of Physics: Conf. Series **807** (2017) 112001 doi:10.1088/1742-6596/807/11/112001

Coherence Factors and Quantum Interferences in Excitonic Condensation of Ta_2NiSe_5

Koudai Sugimoto[1], **Tatsuya Kaneko**[2] **and Yukinori Ohta**[2]

[1]Center for Frontier Science, Chiba University, Chiba 263-8522, Japan
[2]Department of Physics, Chiba University, Chiba 263-8522, Japan

E-mail: sugimoto@chiba-u.jp

Abstract. In order to elucidate whether Ta_2NiSe_5 is in an excitonic condensation state or not, we study macroscopic quantum interferences in ultrasonic attenuation rate and nuclear magnetic resonance relaxation rate. Using the three-chain model describing the excitonic condensation of Ta_2NiSe_5, we demonstrate analytically that the ultrasonic attenuation rate shows a characteristic peak just below the transition temperature of the excitonic condensation, while the nuclear magnetic resonance relaxation rate shows a rapid drop. In particular, we find that the constructive interference originates from the hybridization between the conduction and valence bands induced by an external field.

1. Introduction

It is known that, in narrow-gap semiconductors or semimetals, pairs of electrons and holes (excitons) are spontaneously formed and go into a condensed state with macroscopic phase coherence. This state is referred to as an excitonic condensation [1], and it entails the flattering of the band edges in the case of semiconductors, and the opening of the band gap in the case of semimetals. It was predicted half a century ago, but actual materials of this phase are not known. However, thanks to the recent development of angle-resolved photoemission spectroscopy experiments that enable us to observe band dispersions directly, the realizations of excitonic condensation in some materials have been suggested [2, 3, 4, 5, 6]. Especially, Ta_2NiSe_5, which is known as a direct-gap semiconductor [7, 8], is regarded as one of the most promising candidates of excitonic condensation [2, 3, 9].

A measurement of transformation of the band dispersion is, however, merely an indirect evidence for excitonic condensation. In this respect, there have been some proposals that the ultrasonic attenuation rate shows a coherence peak in a simple electron-gas model [10, 11] although the connection to an actual material lacks. In order to elucidate whether Ta_2NiSe_5 is actually in the excitonic phase or not, we have recently proposed that measurements of the ultrasonic attenuation rate and nuclear magnetic resonance (NMR) relaxation rate can lead to a direct evidence for excitonic condensation in Ta_2NiSe_5 [12]. In our proposal, we predicted that the ultrasonic attenuation rate shows a characteristic peak just below the transition temperature, while the NMR relaxation rate shows a rapid drop just below it. These behaviors originate from the quantum interferences due to the superposition of the wave functions of electrons and holes in the excitonic condensation state.

The purpose of the present paper is then to provide an analysis of the coherence factors just below the transition temperature from an analytical (rather than numerical) point of view, which has not been given in Ref. [12]. The origins of the peak in the ultrasonic attenuation rate and the drop in the NMR relaxation rate are thus elucidated. In Sect. 2, we will introduce the three-chain model describing Ta_2NiSe_5 and apply the BCS-like mean-field approximation to obtain the Bogoliubov transformation for excitonic condensation. In Sect. 3, we will calculate the coherence factors appearing in the ultrasonic attenuation rate and NMR relaxation rate and discuss the origins of the peak and drop analytically. In Sect. 4, we will summarize our results.

2. Three-chain model and mean-field approximation

In order to describe the electronic properties in the excitonic condensation state of Ta_2NiSe_5, we make use of the three-chain model that we have previously introduced [9, 12]. There are doubly degenerate conduction bands (CBs) of two Ta chains (labeled $\alpha = 1, 2$) and a valence band (VB) of one Ni chain in our model. The Hamiltonian is written as $H = H_0 + H_{\mathrm{e-e}} + H_{\mathrm{lat}}$, where H_0 is a kinetic term, $H_{\mathrm{e-e}}$ is an electron-electron interaction term, and H_{lat} is a lattice-distortion term relating to the structural phase transition in Ta_2NiSe_5. The creation (annihilation) operator of a CB electron at site j with spin σ on the chain α is defined as $c^{\dagger}_{j,\alpha,\sigma}$ ($c_{j,\alpha,\sigma}$), and the creation (annihilation) operator of a VB electron at site j with spin σ is defined as $f^{\dagger}_{j,\sigma}$ ($f_{j,\sigma}$). Also, the number operators of the CB and VB electrons are defined as $\hat{n}^{\mathrm{c}}_{j,\alpha,\sigma} = c^{\dagger}_{j,\alpha,\sigma} c_{j,\alpha,\sigma}$ and $\hat{n}^{\mathrm{f}}_{j,\sigma} = f^{\dagger}_{j,\sigma} f_{j,\sigma}$, respectively. The kinetic term is written as

$$H_0 = t_{\mathrm{c}} \sum_{j,\alpha,\sigma} \sum_{\delta=\pm 1} c^{\dagger}_{j+\delta,\alpha,\sigma} c_{j,\alpha,\sigma} + \varepsilon_{\mathrm{c}} \sum_{j,\alpha,\sigma} \hat{n}^{\mathrm{c}}_{j,\alpha,\sigma} + t_{\mathrm{f}} \sum_{j,\sigma} \sum_{\delta=\pm 1} f^{\dagger}_{j+\delta,\sigma} f_{j,\sigma} + \varepsilon_{\mathrm{f}} \sum_{j,\sigma} \hat{n}^{\mathrm{f}}_{j,\sigma}, \tag{1}$$

where $t_{\mathrm{c(f)}}$ and $\varepsilon_{\mathrm{c(f)}}$ are the hopping integrals and the on-site energies of the CB (VB) electrons, respectively. The electron-electron interaction term is expressed as $H_{\mathrm{e-e}} = H_{U_{\mathrm{c}}} + H_{U_{\mathrm{f}}} + H_V$ with

$$H_{U_{\mathrm{c}}} = U_{\mathrm{c}} \sum_{j,\alpha} \hat{n}^{\mathrm{c}}_{j,\alpha,\uparrow} \hat{n}^{\mathrm{c}}_{j,\alpha,\downarrow}, \quad H_{U_{\mathrm{f}}} = U_{\mathrm{f}} \sum_{j} \hat{n}^{\mathrm{f}}_{j,\uparrow} \hat{n}^{\mathrm{f}}_{j,\downarrow}, \tag{2}$$

and

$$H_V = V \sum_{j,\sigma,\sigma',\alpha} \hat{n}^{\mathrm{c}}_{j,\alpha,\sigma} \hat{n}^{\mathrm{f}}_{j,\sigma'} + V \sum_{j,\sigma,\sigma'} \left(\hat{n}^{\mathrm{c}}_{j+1,1,\sigma} + \hat{n}^{\mathrm{c}}_{j-1,2,\sigma} \right) \hat{n}^{\mathrm{f}}_{j,\sigma'}, \tag{3}$$

where $H_{U_{\mathrm{c(f)}}}$ is the on-site Coulomb interaction $U_{\mathrm{c(f)}}$ of the CB (VB) electrons, and H_V is the intersite interaction V between the CB and the VB electrons. The lattice-distortion term consists of a term of harmonic oscillators for Ta atoms and a hybridization term between CB and VB electrons caused by the lattice distortion, and is written as

$$H_{\mathrm{lat}} = \sum_{j,\alpha} \left\{ \frac{K}{2} X^2_{j,\alpha} - \gamma_{\alpha} X_{j,\alpha} \sum_{\sigma} \left(c^{\dagger}_{j,\alpha,\sigma} f_{j,\sigma} + f^{\dagger}_{j,\sigma} c_{j,\alpha,\sigma} \right) \right\}, \tag{4}$$

where $X_{j,\alpha}$ is the displacement of a Ta atom at site j in the α-th chain measured from its equilibrium position, and K is a spring constant of the harmonic oscillators. γ_{α} is the strength of hybridization between the CB and VB electrons caused by the lattice distortion. We assume $\gamma_1 = \gamma$ and $\gamma_2 = -\gamma$.

The mixed order parameters of excitonic condenstation and lattice distortion are defined as

$$\tilde{\Delta}_1(k) = V \left\langle c^{\dagger}_{j,1,\sigma} f_{j,\sigma} \right\rangle + V \left\langle c^{\dagger}_{j+1,1,\sigma} f_{j,\sigma} \right\rangle \mathrm{e}^{\mathrm{i}ka} + \gamma_1 X_{j,1}, \tag{5}$$

$$\tilde{\Delta}_2(k) = V \left\langle c^{\dagger}_{j,2,\sigma} f_{j,\sigma} \right\rangle + V \left\langle c^{\dagger}_{j-1,2,\sigma} f_{j,\sigma} \right\rangle \mathrm{e}^{-\mathrm{i}ka} + \gamma_2 X_{j,2}, \tag{6}$$

where a is the lattice constant, which is set to be unity hereafter. We assume that the system is homogeneous, and thus the order parameters are independent of the lattice site. In the mean-field approximation, the Hamiltonian in k space without a constant term reads

$$H^{\mathrm{MF}} = \sum_{k,\alpha,\sigma} \tilde{\varepsilon}_{\mathrm{c}}(k) c^{\dagger}_{k,\alpha,\sigma} c_{k,\alpha,\sigma} + \sum_{k,\sigma} \tilde{\varepsilon}_{\mathrm{f}}(k) f^{\dagger}_{k,\sigma} f_{k,\sigma} - \sum_{k,\sigma,\alpha} \left\{ \tilde{\Delta}_{\alpha}(k) f^{\dagger}_{k,\sigma} c_{k,\alpha,\sigma} + \tilde{\Delta}^{*}_{\alpha}(k) c^{\dagger}_{k,\alpha,\sigma} f_{k,\sigma} \right\}, \quad (7)$$

where $\tilde{\varepsilon}_{\mathrm{c(f)}}(k)$ is the same as that in the Supplemental Material of our previous work [12]. Diagonalizing the mean-field Hamiltonian given in Eq. (7), we obtain

$$H^{\mathrm{MF}} = \sum_{k,\sigma} \sum_{\epsilon=c,\pm} E_{k,\epsilon} \gamma^{\dagger}_{k,\epsilon,\sigma} \gamma_{k,\epsilon,\sigma}, \quad (8)$$

where

$$E_{k,c} = \tilde{\varepsilon}_{\mathrm{c}}(k), \quad E_{k,\pm} = \eta_k \pm E_k \quad (9)$$

with

$$\eta_k = \frac{\tilde{\varepsilon}_{\mathrm{c}}(k) + \tilde{\varepsilon}_{\mathrm{f}}(k)}{2}, \quad \xi_k = \frac{\tilde{\varepsilon}_{\mathrm{c}}(k) - \tilde{\varepsilon}_{\mathrm{f}}(k)}{2}, \quad E_k = \sqrt{\xi_k^2 + \left|\tilde{\Delta}_1(k)\right|^2 + \left|\tilde{\Delta}_2(k)\right|^2}. \quad (10)$$

$\gamma_{k,\epsilon,\sigma}$ ($\gamma^{\dagger}_{k,\epsilon,\sigma}$) is the annihilation (creation) operator of the quasiparticle. They satisfy the relation $c_{k,\mu,\sigma} = \sum_{\epsilon} \psi_{k,\sigma;\mu,\epsilon} \gamma_{k,\epsilon,\sigma}$, where $\psi_{k,\sigma;\mu,\epsilon}$ is the Bogoliubov transformation coefficient and $f_{k,\sigma} = c_{k,3,\sigma}$. The coefficients $\psi_{k,\sigma;\mu,\epsilon}$ are determined as

$$\begin{pmatrix} c_{k,1,\sigma} \\ c_{k,2,\sigma} \\ f_{k,\sigma} \end{pmatrix} = \begin{pmatrix} \frac{\tilde{\Delta}^{*}_{1}(k)}{|\tilde{\Delta}_1(k)|} \frac{|\tilde{\Delta}_2(k)|}{\sqrt{|\tilde{\Delta}_1(k)|^2 + |\tilde{\Delta}_2(k)|^2}} & -\frac{\tilde{\Delta}^{*}_{1}(k)}{2E_k v_k} & \frac{\tilde{\Delta}^{*}_{1}(k)}{2E_k u_k} \\ -\frac{\tilde{\Delta}^{*}_{2}(k)}{|\tilde{\Delta}_2(k)|} \frac{|\tilde{\Delta}_1(k)|}{\sqrt{|\tilde{\Delta}_1(k)|^2 + |\tilde{\Delta}_2(k)|^2}} & -\frac{\tilde{\Delta}^{*}_{2}(k)}{2E_k v_k} & \frac{\tilde{\Delta}^{*}_{2}(k)}{2E_k u_k} \\ 0 & v_k & u_k \end{pmatrix} \begin{pmatrix} \gamma_{k,c,\sigma} \\ \gamma_{k,+,\sigma} \\ \gamma_{k,-,\sigma} \end{pmatrix} \quad (11)$$

with

$$u_k = \sqrt{\frac{1}{2}\left(1 + \frac{\xi_k}{E_k}\right)}, \quad v_k = \sqrt{\frac{1}{2}\left(1 - \frac{\xi_k}{E_k}\right)}. \quad (12)$$

The detailed studies of this model are given in our previous papers [9, 12].

Figure 1 shows the band dispersions of this model. In the excitonic phase, one of the doubly degenerate CBs ($\epsilon = +$) hybridizes with the VB ($\epsilon = -$), leading to the flattering of these bands and the lifting of the degeneracy in the CBs. The other CB ($\epsilon = \mathrm{c}$) sustains its dispersion even in the excitonic phase.

In this paper, we are interested in a behavior just below the transition temperature, where the order parameters $\left|\tilde{\Delta}_{\alpha}(k)\right|$ are sufficiently small compared to the typical energy scales such as ξ_k. To the second order of $\left|\tilde{\Delta}_{\alpha}(k)\right|$, the Bogoliubov trasnformation in Eq. (11) becomes

$$\begin{pmatrix} c_{k,1,\sigma} \\ c_{k,2,\sigma} \\ f_{k,\sigma} \end{pmatrix} \simeq \begin{pmatrix} \frac{\tilde{\Delta}^{*}_{1}(k)}{|\tilde{\Delta}_1(k)|} \frac{|\tilde{\Delta}_2(k)|}{|\tilde{\Delta}(k)|} & -\frac{\tilde{\Delta}^{*}_{1}(k)}{|\tilde{\Delta}(k)|}\left(1 - \frac{|\tilde{\Delta}(k)|^2}{2\xi_k^2}\right) & \frac{\tilde{\Delta}^{*}_{1}(k)}{2|\xi_k|} \\ -\frac{\tilde{\Delta}^{*}_{2}(k)}{|\tilde{\Delta}_2(k)|} \frac{|\tilde{\Delta}_1(k)|}{|\tilde{\Delta}(k)|} & -\frac{\tilde{\Delta}^{*}_{2}(k)}{|\tilde{\Delta}(k)|}\left(1 - \frac{|\tilde{\Delta}(k)|^2}{2\xi_k^2}\right) & \frac{\tilde{\Delta}^{*}_{2}(k)}{2|\xi_k|} \\ 0 & \frac{|\tilde{\Delta}(k)|}{2|\xi_k|} & 1 - \frac{|\tilde{\Delta}(k)|^2}{8\xi_k^2} \end{pmatrix} \begin{pmatrix} \gamma_{k,c,\sigma} \\ \gamma_{k,+,\sigma} \\ \gamma_{k,-,\sigma} \end{pmatrix}, \quad (13)$$

where $\left|\tilde{\Delta}(k)\right|^2 = \left|\tilde{\Delta}_1(k)\right|^2 + \left|\tilde{\Delta}_2(k)\right|^2$.

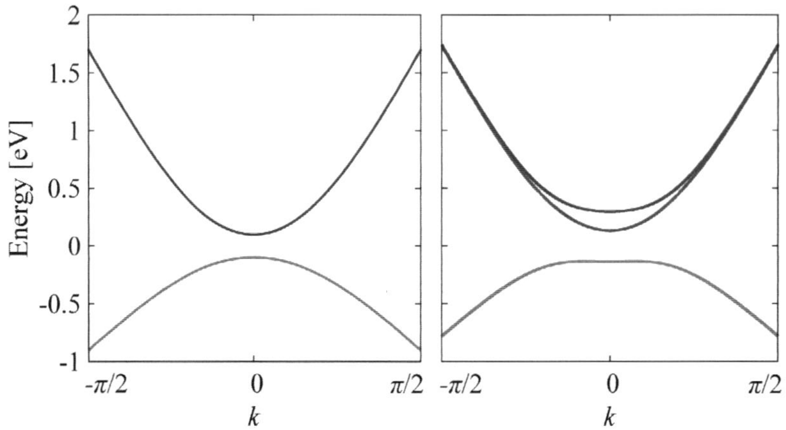

Figure 1. (Left) The band dispersions in the normal phase. Hopping integrals of CB and VB are $t_{\mathrm{c}} = -0.8$ eV and $t_{\mathrm{f}} = 0.4$ eV, respectively. The energy gap is set to be 0.2 eV. Blue line indicates the doubly degenerate CBs and red line indicates the VB. (Right) The band dispersions in the excitonic phase. We set $V = 0.8$ eV, $U_{\mathrm{c}} = U_{\mathrm{f}} = 4V$, and $\lambda = 0.01$ eV, where V is larger than that of our previous work ($V = 0.6$ eV) [12] in order to emphasize the flattening of the bands. Green, blue, and red lines indicate the bands c, +, and −, respectively.

3. Coherence factors as the response to external fields

3.1. *Ultrasonic attenuation rate*

We define the operator of the phonon of Ta atoms on α-th chain as $A_{q,\alpha} = a_{q,\alpha} + a^{\dagger}_{-q,\alpha}$. The electron-phonon interaction Hamiltonian by the ultrasonic wave is then written as

$$H' = \sum_{k,q,\sigma,\alpha} \left\{ M^{\mathrm{cc}}_{-q} A_{-q,\alpha} c^{\dagger}_{k_-,\alpha,\sigma} c_{k_+,\alpha,\sigma} - M^{\mathrm{cf}}_{-q} A_{-q,\alpha} \left(c^{\dagger}_{k_-,\alpha,\sigma} f_{k_+,\sigma} + f^{\dagger}_{k_-,\alpha,\sigma} c_{k_+,\sigma} \right) \right\}, \tag{14}$$

where $k_{\pm} = k \pm \frac{q}{2}$, M^{cc}_{-q} term is the coupling between the phonon and electron-charge density, and M^{cf}_{-q} term is the hybridization of CB and VB electrons due to the phonon oscillation. For simplicity, we neglect the phonon of Ni atoms. In the shear-mode ultrasonic wave, the phonon does not couple to the charge density, and the M^{cf}_{-q} term may be the dominant contribution to the electron scattering. The ultrasonic attenuation rate $\alpha_{q=0,\alpha}$ then reads [12]

$$\begin{aligned}\alpha_{q=0,\alpha} =& 2\pi\omega_{\mathrm{ph}} \sum_{k,\sigma} \sum_{\epsilon_1,\epsilon_2} \frac{\beta}{4\cosh^2\left(\beta\left(E_{k,\epsilon_1} - \mu\right)/2\right)} \delta\left(E_{k,\epsilon_1} - E_{k,\epsilon_2}\right) \\ &\times \sum_{\mu,\nu=1}^{3} \left\{ \frac{1}{4} \left(M^{\mathrm{cc}}_0\right)^2 \delta_{\alpha,\mu}\delta_{\alpha,\nu} + \left(M^{\mathrm{cf}}_0\right)^2 \delta_{\mu,3}\delta_{\nu,3} - M^{\mathrm{cc}}_0 M^{\mathrm{cf}}_0 \delta_{\alpha,\mu}\delta_{\nu,3} \right\} \\ &\times 2\mathrm{Re}\left\{ \Psi_{\alpha,\mu;\epsilon_1}(k,\sigma)\Psi_{\alpha,\nu;\epsilon_2}(k,\sigma) + \Psi_{\alpha,\alpha;\epsilon_1}(k,\sigma)\Psi_{\mu,\nu;\epsilon_2}(k,\sigma) \right\},\end{aligned} \tag{15}$$

where $\Psi_{\nu,\kappa;\epsilon}(k,\sigma) = \psi_{k,\sigma;\nu,\epsilon}\psi^{*}_{k,\sigma;\kappa,\epsilon}$ and ω_{ph} is the phonon frequency. The interband contribution to the ultrasonic attenuation rate is prohibited since the energy of the phonon is too small to scatter an electron from one band to another, and thus we can assume $\epsilon_1 = \epsilon_2$ in the summation.

There are three contributions to the ultrasonic attenuation rate; the charge-density term (proportional to $(M^{\mathrm{cc}}_0)^2$), the CB-VB hybridization term (proportional to $\left(M^{\mathrm{cf}}_0\right)^2$), and the cross term (proportional to $M_0^{\mathrm{cc}} M_0^{\mathrm{cf}}$). We examine the coherence factors up to the second order of the order parameters. The coherence factors are written as products of four wave functions given in Eq. (13). We focus on the attenuation of phonons in the Ta $\alpha = 1$ chain below.

3.1.1. Charge-density term The orbitals are restricted to $\alpha = \mu = \nu = 1$ in this case. For $\epsilon = \mathrm{c}$ and $\epsilon = -$, we have

$$\Psi_{1,1;\mathrm{c}}(k,\sigma)\Psi_{1,1;\mathrm{c}}(k,\sigma) = \frac{\left|\tilde{\Delta}_2(k)\right|^4}{\left|\tilde{\Delta}(k)\right|^4}, \tag{16}$$

$$\Psi_{1,1;-}(k,\sigma)\Psi_{1,1;-}(k,\sigma) = O\left(\frac{\left|\tilde{\Delta}_1(k)\right|^4}{|\xi_k|^4}\right), \tag{17}$$

respectively, and they are almost irrelevant to the quantum interference. However, for $\epsilon = +$, we have

$$\Psi_{1,1;+}(k,\sigma)\Psi_{1,1;+}(k,\sigma) \simeq \frac{\left|\tilde{\Delta}_1(k)\right|^4}{\left|\tilde{\Delta}(k)\right|^4}\left(1 - \frac{2\left|\tilde{\Delta}(k)\right|^2}{\xi_k^2}\right), \tag{18}$$

where the second term leads to the destructive interference. As a result, the coupling between the phonons and charge density leads to the drop in the attenuation rate.

3.1.2. CB-VB hybridization term The orbitals are restricted to $\alpha = 1$ and $\mu = \nu = 3$ in this case. For $\epsilon = \mathrm{c}$, we have

$$\Psi_{3,1;\mathrm{c}}(k,\sigma)\Psi_{3,1;\mathrm{c}}(k,\sigma) = \Psi_{1,1;\mathrm{c}}(k,\sigma)\Psi_{3,3;\mathrm{c}}(k,\sigma) = 0, \tag{19}$$

which is completely irrelevant to the attenuation rate. On the other hand, for $\epsilon = +$ and $\epsilon = -$, we have

$$\Psi_{3,1;+}(k,\sigma)\Psi_{3,1;+}(k,\sigma) \simeq \frac{\tilde{\Delta}_1(k)^2}{4|\xi_k|^2}, \qquad \Psi_{1,1;+}(k,\sigma)\Psi_{3,3;+}(k,\sigma) \simeq \frac{\left|\tilde{\Delta}_1(k)\right|^2}{4|\xi_k|^2}, \tag{20}$$

$$\Psi_{3,1;-}(k,\sigma)\Psi_{3,1;-}(k,\sigma) \simeq \frac{\tilde{\Delta}_1(k)^2}{4|\xi_k|^2}, \qquad \Psi_{1,1;-}(k,\sigma)\Psi_{3,3;-}(k,\sigma) \simeq \frac{\left|\tilde{\Delta}_1(k)\right|^2}{4|\xi_k|^2}, \tag{21}$$

all of which lead to the constructive interferences. This fact indicates that the peak appearing in ultrasonic attenuation rate just below the transition temperature originates from the CB-VB hybridization by the phonons.

3.1.3. Cross term The orbitals are restricted to $\alpha = \mu = 1$ and $\nu = 3$. As in the case of the charge-density term, we have

$$-\Psi_{1,1;\mathrm{c}}(k,\sigma)\Psi_{1,3;\mathrm{c}}(k,\sigma) = 0 \tag{22}$$

$$-\Psi_{1,1;-}(k,\sigma)\Psi_{1,3;-}(k,\sigma) = O\left(\frac{\tilde{\Delta}_1(k)^3}{|\xi_k|^3}\right) \tag{23}$$

for $\epsilon = \mathrm{c}$ and $\epsilon = -$, respectively, which are also irrelevant to the quantum interference. For $\epsilon = +$, however, we have

$$-\Psi_{1,1;+}(k,\sigma)\Psi_{1,3;+}(k,\sigma) \simeq -\frac{\left|\tilde{\Delta}_1(k)\right|^2 \tilde{\Delta}_1^*(k)}{2\left|\xi_k\right|\left|\tilde{\Delta}(k)\right|^2}, \tag{24}$$

which leads to the destructive interference.

3.2. NMR relaxation rate

The NMR relaxation rate $1/T_1$ for μ-th chain atom is obtained from the spin-transverse correlation function. It is written as [12]

$$\frac{1}{T_1} \propto \pi \sum_{k,q} \sum_{\epsilon_1,\epsilon_2} \Psi_{\mu,\mu;\epsilon_1}(k_-,\uparrow)\Psi_{\mu,\mu;\epsilon_2}(k_+,\downarrow)\frac{1}{4\cosh^2\left(\beta\left(E_{k_-,\epsilon_1}-\mu\right)/2\right)}\delta\left(E_{k_-,\epsilon_1}-E_{k_+,\epsilon_2}\right). \tag{25}$$

There appears a summation over the wave number q, which is different from the case of the ultrasonic attenuation rate where the wave vector of the phonon is fixed. The interband scattering can therefore occur. Note that, as shown in Fig. 1, the $\epsilon = \mathrm{c}$ band and $\epsilon = +$ band are above the Fermi level, while $\epsilon = -$ band is below the Fermi level. Thus, only the scattering between c band and + band is allowed.

3.2.1. For Ta atom NMR We consider the Ta $\mu = 1$ chain here. In the case of the intraband scattering $\epsilon_1 = \epsilon_2$, we have

$$\Psi_{1,1;\mathrm{c}}(k_-,\uparrow)\Psi_{1,1;\mathrm{c}}(k_+,\downarrow) = \frac{\left|\tilde{\Delta}_2(k_-)\right|^2\left|\tilde{\Delta}_2(k_+)\right|^2}{\left|\tilde{\Delta}(k_-)\right|^2\left|\tilde{\Delta}(k_+)\right|^2} \tag{26}$$

$$\Psi_{1,1;+}(k_-,\uparrow)\Psi_{1,1;+}(k_+,\downarrow) \simeq \frac{\left|\tilde{\Delta}_1(k_-)\right|^2\left|\tilde{\Delta}_1(k_+)\right|^2}{\left|\tilde{\Delta}(k_-)\right|^2\left|\tilde{\Delta}(k_+)\right|^2}\left(1-\frac{\left|\tilde{\Delta}(k_+)\right|^2}{\xi_{k_+}^2}-\frac{\left|\tilde{\Delta}(k_-)\right|^2}{\xi_{k_-}^2}\right) \tag{27}$$

$$\Psi_{1,1;-}(k_-,\uparrow)\Psi_{1,1;-}(k_+,\downarrow) = O\left(\frac{\left|\tilde{\Delta}_1(k_\pm)\right|^4}{\xi_{k_\pm}^4}\right). \tag{28}$$

This implies that the the intraband scattering of + band leads to the destructive interference.

In the case of the interband scattering $\epsilon_1 = \mathrm{c}$ and $\epsilon_2 = +$, we have

$$\Psi_{1,1;\mathrm{c}}(k_-,\uparrow)\Psi_{1,1;+}(k_+,\downarrow) \simeq \frac{\left|\tilde{\Delta}_2(k_-)\right|^2\left|\tilde{\Delta}_1(k_+)\right|^2}{\left|\tilde{\Delta}(k_-)\right|^2\left|\tilde{\Delta}(k_+)\right|^2}\left(1-\frac{\left|\tilde{\Delta}(k_+)\right|^2}{\xi_{k_+}^2}\right), \tag{29}$$

which also leads to the destructive one.

3.2.2. For Ni atom NMR $\mu = 3$ indicates the Ni chain. In this case, there is no scattering in the band relating to $\epsilon = c$ since the coherence factors in this case are necessarily zero.

In the case of $\epsilon_1 = \epsilon_2 = +$, we have

$$\Psi_{3,3;+}(k_-, \uparrow)\Psi_{3,3;+}(k_+, \downarrow) = O\left(\frac{\left|\tilde{\Delta}(k_\pm)\right|^4}{\xi_{k_\pm}^4}\right), \tag{30}$$

which is negligible. On the other hand, in the case of $\epsilon_1 = \epsilon_2 = -$, we have

$$\Psi_{3,3;-}(k_-, \uparrow)\Psi_{3,3;-}(k_+, \downarrow) \simeq 1 - \frac{\left|\tilde{\Delta}(k_-)\right|^2}{4\xi_{k_-}^2} - \frac{\left|\tilde{\Delta}(k_+)\right|^2}{4\xi_{k_-}^2}, \tag{31}$$

which leads to the destructive interference.

4. Summary

Introducing the three-chain model describing the excitonic condensation in Ta_2NiSe_5, we have obtained the Bogoliubov transformation for the excitonic condensation in the mean-field approximation. From the transformation coefficients, we have calculated the coherence factors in the ultrasonic attenuation rate and the NMR relaxation rate and have expanded them to the second order of the order parameters. The results indicate that the CB-VB hybridization term in the ultrasonic attenuation rate leads to the constructive interference, giving rise to the coherence peak just below the transition temperature. On the other hand, the charge-density term in the ultrasonic attenuation rate leads to the destructive interference, resulting in the drop of the rate below the transition temperature. In the NMR relaxation rate, only the destructive interference terms appear, and therefore the rate inevitably decreases just below the transition temperature. These results imply that the coherence peak can be observed only if the external field hybridizes the CB electrons and VB electrons in excitonic condensation.

Acknowledgments

This work was supported in part by Grants-in-Aid for Scientific Research (No. 26400349 and No. 15H06093) from JSPS of Japan. T. K. acknowledges support from the JSPS Research Fellowship for Young Scientists.

References

[1] Jérome D, Rice T M and Kohn W 1967 *Phys. Rev.* **158** 462
[2] Wakisaka Y *et al.* 2009 *Phys. Rev. Lett.* **103** 026402
[3] Seki K *et al.* 2014 *Phys. Rev.* B **90**, 155116
[4] Pillo T, Hayoz J, Berger H, Lévy F, Schlapbach L and Aebi P, 2000 *Phys. Rev.* B **61**, 16213
[5] Kidd T E, Miller T, Chou M Y and Chiang T C 2002 *Phys. Rev. Lett.* 88, 226402
[6] Cercellier H *et al.* 2007 *Phys. Rev. Lett.* **99** 146403
[7] Sunshine S A and Ibers J A 1985 *Inorg. Chem.* **24**, 3611
[8] Di Salvo F J, Chen C H, Fleming R M, Waszczak J V, Dunn R G, Sunshine S A and Ibers J A 1986 *J. Less Common Met.* **116** 51
[9] Kaneko T, Toriyama T, Konishi T and Ohta Y 2013 *Phys. Rev.* B **87** 035121
[10] Maki K and Nakanishi K 1971 *J. Low Temp. Phys.* **5** 55
[11] Amritkar R E and Kumar N 1978 *Solid State Commun.* **26** 627
[12] Sugimoto K, Kaneko T and Ohta Y 2016 *Phys. Rev.* B **93** 041105(R)

Sine–Gordon Model in the Homogeneous Higher Grading

Alexander Zuevsky

Institute of Mathematics, Czech Academy of Sciences, Prague

E-mail: zuevsky@yahoo.com

Abstract. A construction of equations and solutions for the sine–Gordon model in the homogeneous grading as an example of higher grading affine Toda models are considered.

1. Introduction and preliminaries

The problem of construction of exactly solvable models and their solutions is a very important problem in the theory of integrable models in general and in application to real dynamical systems in Physics. One of the ways to proceed is to us the the Lie-algebraic method to construct non-linear exactly solvable models in classical regions. This method is very well known and elaborated [10]. Applying the zero-curvature conditions on elements of connection containing Lie algebra generators in appropriate grading subspaces, we obtain systems of equations of motion associated to a specific Lie algebra.

The main motivation to use the theory of Lie algebras in exactly solvable models is its effectiveness. We are able not only to re-construct the equations of motion but also to find exact solutions starting from internal algebraic symmetries based on deep algebraic symmetries of systems under consideration.

In [4] the higher grading generalization to the conformal affine Toda models was considered. Elements of the higher (then number one) grading subspaces are taking into account while connection elements are constructed. The main example was the principal grading case.

In this paper we conside an alternative, new case, which is corresponds to the homogeneous grading of the Lie algebra. We derive the systems of equations generalizing the case of the sine-Gordon equation and provide quantum group solutions.

1.1. *Affine Kac–Moody Lie algebras*

In this subsection we recall some facts about affine Kac–Moody algebras [7], [4]. Consider an untwisted affine Kac-Moody algebra $\widehat{\mathcal{G}}$ endowed with an integral grading $\widehat{\mathcal{G}} = \bigoplus_{n\in\mathbb{Z}} \widehat{\mathcal{G}}_n$, and denote $\widehat{\mathcal{G}}_\pm = \bigoplus_{n>0} \widehat{\mathcal{G}}_{\pm n}$. By an affine Lie algebra we mean a loop algebra corresponding to a finite dimensional simple Lie algebra $\mathcal{G}$ of rank r, extended by the center C and the derivation D. According to [7], integral gradings of $\widehat{\mathcal{G}}$ are labelled by a set of co-prime integers $\mathbf{s} = (s_0, s_1, \dots s_r)$, and the grading operators are given by

$$Q_{\mathbf{s}} \equiv H_{\mathbf{s}} + N_{\mathbf{s}}\, D - \frac{1}{2N_{\mathbf{s}}} Tr H_{\mathbf{s}}^2\, C\,. \tag{1}$$

Here $H_{\mathbf{s}} \equiv \sum_{a=1}^{r} s_a l_a^v \cdot H^0$, $N_{\mathbf{s}} \equiv \sum_{i=0}^{r} s_i m_i^{\psi}$, $\psi = \sum_{a=1}^{r} m_a^{\psi} \alpha_a$, $m_0^{\psi} = 1$. H^0 is an element of Cartan subalgebra of $\mathcal{G}$; α_a, $a = 1, 2, \ldots r$, are its simple roots; ψ is its maximal root; m_a^{ψ} the integers in expansion $\psi = \sum_{a=1}^{r} m_a^{\psi} \alpha_a$; and l_a^v are the fundamental co-weights satisfying the relation $\alpha_a \cdot l_b^v = \delta_{ab}$. The principal grading operator Q_{ppal} is given by 1 where $N_{\mathbf{s}} = h$ is Coxeter number. Therefore $\widehat{\mathcal{G}}_0 = \{H_a^0, a = 1, 2, \ldots r; C; Q_{\mathrm{ppal}}\}$, $\widehat{\mathcal{G}}_m = \{E^0_{\alpha^{(m)}}, E^1_{-\alpha^{(h-m)}}\}$, $\widehat{\mathcal{G}}_{-m} = \{E^0_{-\alpha^{(m)}}, E^{-1}_{\alpha^{(h-m)}}\}$ where $0 < m < h$, and $\alpha^{(m)}$ are positive roots of height m, and C is the center. The element B is parameterized as $B = e^{\varphi \cdot \tilde{H}^0} e^{\nu C} e^{\eta Q_{\mathrm{ppal}}} = e^{\varphi \cdot H^0} e^{\tilde{\nu} C} e^{\eta Q_{\mathrm{ppal}}}$, where $\tilde{H}^0$ was defined in [4] as $\tilde{H}_a^0 = H_a^0 - \frac{1}{N_{\mathbf{s}}} Tr\left(H_{\mathbf{s}} H_a^0\right) C = H_a^0 - \frac{2}{\alpha_a^2} \frac{s_a}{N_{\mathbf{s}}} C$, and $\tilde{\nu} = \nu - \frac{2}{h} \widehat{\delta} \cdot \varphi$, with $\widehat{\delta} = \sum_{a=1}^{r} \frac{l_a}{\alpha_a^2}$, and l_a being the fundamental weights of $\mathcal{G}$. Let us denote by H^n, $E^n_{\pm}$, D, C the Chevalley basis generators of $\widehat{sl}_2$. The commutation relations are

$$[H^m, H^n] = 2\, m\, C\, \delta_{m+n,0}, \quad [H^m, E^n_{\pm}] = \pm 2\, E^{m+n}_{\pm},$$
$$[E^m_{+}, E^n_{-}] = H^{m+n} + m\, C\, \delta_{m+n,0}, \quad [D, T^m] = m\, T^m, \quad T^m \equiv H^m, E^m_{\pm},$$

where C is the center. The grading operator for the principal grading ($\mathbf{s} = (1,1)$) is $Q \equiv \frac{1}{2} H^0 + 2D$. Then the eigensubspaces are $\widehat{\mathcal{G}}_0 = \{H^0, C, Q\}$, $\widehat{\mathcal{G}}_{2n+1} = \{E^n_{+}, E^{n+1}_{-}\}$, $n \in \mathbb{Z}$, $\widehat{\mathcal{G}}_{2n} = \{H^n\}, n \in \{\mathbb{Z} - 0\}$.

1.2. Quantized universal enveloping algebra $U'q(\widehat{sl_2})$

In the spirit of [2], [5], the quantised enveloping algebra $U_q(sl_2)$ is an associative algebra generated by X^+, X^-, H with q-deformed commutation relations $X^+X^- - X^-X^+ = \left(q^H - q^{-H}\right)\left(q - q^{-1}\right)^{-1}$, $HX^{\pm} - X^{\pm}H = \pm 2X^{\pm}$. It possesses a Hopf algebra structure with the deformed adjoint action $(ad_{X^{\pm}})_q a = X^{\pm} a q^{H/2} - q^{\mp 1} q^{H/2} a X^{\pm}$, $(ad_H)_q a = Ha - aH$, for all $a \in U_q(sl_2)$. Let us recall the second Drinfeld realization of the quantized universal enveloping algebra $U_q'(\widehat{sl_2})$, (i.e., $U_q(\widehat{sl_2})$ without grading operator) [2], [6], which is a natural quantum analogue of the algebra $\widehat{sl_2}$ in the loop realizations. $U_q'(\widehat{sl_2})$ is an associative algebra generated by $\{x_k^{\pm}, k \in \mathbb{Z}; a_n, n \in \{\mathbb{Z} - 0\}; \gamma^{\pm \frac{1}{2}}, K\}$, where $\gamma^{\pm\frac{1}{2}}$ belong to the center of the algebra, satisfying the commutation relations $[K, a_k] = 0$, $[a_k, a_l] = \delta_{k,-l} \frac{[2k]}{k} \frac{\gamma^k - \gamma^{-k}}{q - q^{-1}}$, $K x_k^{\pm} K^{-1} = q^{\pm 2} x_k^{\pm}$, $[a_n, x_k^{\pm}] = \pm \frac{[2n]}{n} \gamma^{\mp \frac{|n|}{2}} x^{\pm}_{n+k}$, $[x_k^+, x_n^-] = \frac{\gamma^{(k-n)/2} \psi_{k+n} - \gamma^{(n-k)/2} \phi_{k+n}}{q - q^{-1}}$, $x^{\pm}_{k+l} x^{\pm}_l - q^{\pm 2} x^{\pm}_l x^{\pm}_k = q^{\pm 2} x^{\pm}_k x^{\pm}_{l+1} - x^{\pm}_{l+1} x^{\pm}_k$. The generators ϕ_k and ψ_{-k}, $k \in \mathbb{Z}_+$ are related to a_k and a_{-k} by means of the expressions $\sum_{k=0}^{\infty} \psi_m z^{-m} = K exp\left((q - q^{-1}) \sum_{k=1}^{\infty} a_k z^{-k}\right)$, $\sum_{k=0}^{\infty} \phi_{-m} z^m = K^{-1} exp\left(-(q - q^{-1}) \sum_{k=1}^{\infty} a_{-k} z^k\right)$, i.e., $\psi_m = 0$, $m < 0$; $\phi_m = 0$, $m > 0$. Here $[k] = \frac{q^k - q^{-k}}{q - q^{-1}}$.

It is easy to define the grading operators corresponding to the principal and homogeneous grading of $U_q'(\widehat{sl_2})$ by analogy with the grading of $U_q'(\mathcal{G})$ where $\mathcal{G}$ is a simple Lie algebra. The principal grading can be realized with the help of the operator $D_p x = \frac{1}{2} q K^{-1} \left(\frac{d}{dq}(K x K^{-1})\right) K + 2\lambda \frac{d}{d\lambda} x$, where $x \in U_q(\widehat{sl_2})$ and λ is an affinization parameter. The power of λ is denoted by the subscript of $U_q'(\widehat{sl_2})$ generators. Then the grading subspaces are ${}_q\widehat{\mathcal{G}}_0 = \{K, \gamma\}$, ${}_q\widehat{\mathcal{G}}_{2n+1} = \{x_n^+, x_{n+1}^-, n \in \mathbb{Z}\}$, ${}_q\widehat{\mathcal{G}}_{2n} = \{a_n, n \in \{\mathbb{Z} - 0\}\}$. The grading operator for the homogeneous grading is $D_h x = 2\lambda \frac{d}{d\lambda} x$, so that the grading subspaces are ${}_q\widehat{\mathcal{G}}_0 = \{K, \gamma, x_0^+, x_0^-\}$, ${}_q\widehat{\mathcal{G}}_n = \{x_n^+, x_n^-, a_n, n \in \{\mathbb{Z} - 0\}\}$.

The level one irreducible integrable highest weight representation of $U_q'(\widehat{sl_2})$ can be constructed in the following way [6]. Let $P = \mathbb{Z}\frac{\alpha}{2}$, $Q = \mathbb{Z}\alpha$ be the weight/root lattice of sl_2. Consider the group algebras $F[P]$, $F[Q]$ of P and Q. The multiplicative basis of $F[P]$

is formed by $e^{\frac{\alpha}{2}n}$, $n \in \mathbb{Z}$. The $F[Q]$-module is split into $F[P] = F[P]_0 \oplus F[P]_1$ where $F[P]_n = F[Q]e^{\frac{\alpha}{2}n}$. The sl_2-module structure on the space $W = F[a_{-1}, a_{-2}, ...] \otimes F[P]$ is given by the action of the $a_k, k \in \{\mathbb{Z} - 0\}$ and e^{α}, $\partial_\alpha = a_0$ generators in accordance with the rules $a_k(f \otimes e^{\beta}) = (a_k f \otimes e^{\beta})$, $k < 0$, $a_k(f \otimes e^{\beta}) = ([a_k, f] \otimes e^{\beta})$, $k > 0$, $e^{\alpha}(f \otimes e^{\beta}) = (f \otimes e^{\alpha+\beta})$, $\partial_\alpha(f \otimes e^{\beta}) = (\alpha, \beta)(f \otimes e^{\beta})$, $K = 1 \otimes q^{\partial_\alpha}$, $\gamma = q \otimes id$. Then W is a $U'_q(\widehat{sl_2})$-module. Its submodules are isomorphic to irreducible highest weight modules $V(\Lambda_n)$ with the highest vectors $|\Lambda_n\rangle = |1 \otimes e^{\frac{\alpha n}{2}}\rangle$, $n = 0, 1$.

2. Higher grading affine Toda system

In this and the next sections we recall [4] the affine Toda system consrtuction. Consider a two dimensional manifold $\mathcal{M}$ with local coordinates $z_\pm$. Up to a gauge transformation, $(1,0)$-component lying in (see subsection 1.1) $\bigoplus_{n=0}^{l} \widehat{\mathcal{G}}_{+n}$ and $(0,1)$-component in $\bigoplus_{n=0}^{l} \widehat{\mathcal{G}}_{-n}$ of a flat connection $\mathcal{A}$ in the trivial holomorphic principal fibre bundle $\mathcal{M} \times \widehat{G} \longrightarrow \mathcal{M}$ ($l > 0$ is fixed integer) satisfy the zero curvature condition

$$[\partial_+ + A_+, \partial_- + A_-] = 0. \tag{2}$$

The components $A_\pm$ are the following (we keep notations of [4]) for $k = \pm$:

$$A_k = -\left(\partial_-^{\delta_{k,-}}\right) B \left(F^+\right)^{\delta_{k,+}} B^{-1} + \delta_{k,-} F^-. \tag{3}$$

Here B is a mapping $\mathcal{M} \longrightarrow \widehat{G}_0$ ($\widehat{G}_0$ is a group with the Lie algebra $\widehat{\mathcal{G}}_0$) and $F^\pm$ $(1 \le m \le l-1)$ are mappings to $\bigoplus_{n=1}^{l} \widehat{\mathcal{G}}_{\pm n}$

$$F^\pm = E_{\pm l} + \sum_{m=1}^{l-1} F_m^\pm,$$

where $E_{\pm l}$ are some fixed elements of $\widehat{\mathcal{G}}_{\pm l}$ and $F_m^\pm \in \widehat{\mathcal{G}}_{\pm m}$, $(1 \le m \le l-1)$. Substituting 3 into 2 one arrives at the equations of motion

$$\partial_+\partial_- B\, B^{-1} = [E_{-l}\,, B\, E_l\, B^{-1}] + \sum_{n=1}^{l-1} [F_n^-\,, B\, F_n^+\, B^{-1}], \tag{4}$$

$$\partial_\pm F_m^\mp = \mp[E_{\mp l}\,, B^{\pm 1}\, F_{l-m}^\pm\, B^{\mp 1}] \mp \sum_{n=1}^{l-m-1} [F_{n+m}^\mp\,, B^{\pm 1}\, F_n^\pm\, B^{\mp 1}]\,. \tag{5}$$

Since $Q_{\mathbf{s}}$, $C \in \widehat{\mathcal{G}}_0$ then B can be parameterized as $B = b\, e^{\eta Q_{\mathbf{s}}}\, e^{\nu C}$ where b is a mapping to G_0, the subgroup of $\widehat{G}_0$ generated by all elements of $\widehat{\mathcal{G}}_0$ other than $Q_{\mathbf{s}}$ and C. Substituting B into the equations of motion (4–5) one has

$$\partial_+\partial_- b b^{-1} + \partial_+\partial_-\, \nu\, C = e^{l\eta}[E_{-l}\,, b\, E_l\, b^{-1}] + \sum_{n=1}^{l-1} e^{n\eta}\, [F_n^-\,, b\, F_n^+ b^{-1}]\,, \tag{6}$$

$$\partial_- F_m^+ = e^{(l-m)\eta}\, [E_l\,, b^{-1}\, F_{l-m}^-\, b] + \sum_{n=1}^{l-m-1} e^{n\eta}[F_{m+n}^+\,, b^{-1}\, F_n^-\, b]\,, \tag{7}$$

$$\partial_+ F_m^- = -e^{(l-m)\eta}\, [E_{-l}\,, b\, F_{l-m}^+\, b^{-1}] - \sum_{n=1}^{l-m-1} e^{n\eta}[F_{m+n}^-\,, b\, F_n^+\, b^{-1}]\,, \tag{8}$$

$$\partial_+\partial_-\, \eta\, Q_{\mathbf{s}} = 0\,. \tag{9}$$

2.1. The case $l = 1$

Now consider the case $l = 1$. Let us parameterize the element B in the homogeneous grading of $\widehat{\mathcal{G}}$, [4]. From the equations (6–9) for an infinite dimensional Lie algebra $\widehat{\mathcal{G}}$ in the principal grading we obtain the affine Toda field theory systems of equations

$$\partial_+\partial_-\phi + \frac{4\mu}{\beta}\sum_{i=1}^{r}\left(m_i\frac{\alpha_i}{\alpha_i^2}\exp(\beta\alpha_i\cdot\phi) - \frac{\psi}{2}\exp(-\beta\cdot\phi)\right) = 0. \tag{10}$$

Let define

$$J(g,|\xi\rangle) = \frac{\langle\xi_i|g|\xi_i\rangle}{(\langle\xi_0|g|\xi_0\rangle)^{m_j}},$$

for representation vectors $|\xi_i\rangle$ (see below) and a group-like element g. The formal general solution to the above equation was introduced in [12]:

$$e^{-\beta\lambda_i\cdot\phi} = e^{-\beta\lambda_i\cdot\phi_0}J((\gamma_0^+)^{-1}\mu_+^{-1}(z_+)\mu_-(z_-)(\gamma_0^-),|\lambda_i\rangle^{(1)}) = e^{-\beta\lambda_i\cdot\phi_0}J(B,|\lambda_i\rangle^{(1)}) \tag{11}$$

The general solutions to the matter fields $F_i^{\pm}$ may be written in the following form. For $m = 1$ in (6–9) one has [4] for $g_\mu = \mu_+^{-1}\mu_-$,

$$\langle i|F_1^+|i;i\rangle = f_i^+ = e^{\sum_{l=0}^{r}k_{il}(\phi^- - \phi)_l}e^{\nu_0}\partial_+\left(\langle i|g_\mu|i;i\rangle J^{-1}(g_\mu,|i\rangle)\right).$$

Here $|i;i\rangle$ denotes an element of the Verma module which is result of the action of the lowering generator on the highest state vector. The fact that (11) is indeed a solution to (10) may be checked by using the representation theory of $\widehat{\mathcal{G}}$. A map $g : \mathcal{M} \longrightarrow G$ appearing in the gradient form of the flat connection $A_\pm = g^{-1}\partial_\pm g$, may be factorized (according to the Lie algebra decomposition $\mathcal{G} = \mathcal{G}_- \oplus \mathcal{G}_0 \oplus \mathcal{G}_+$) by the modified Gauss decomposition $g = \mu_-\nu_+\gamma_{0-}$ or $g = \mu_+\nu_-\gamma_{0+}$ with maps $\gamma_{0\pm} : \mathcal{M} \longrightarrow G_0$, $\mu_\pm, \nu_\pm : \mathcal{M} \longrightarrow G_\pm$. The grading condition provides the holomorphic property of $\mu_\pm$, i.e., they satisfy the initial value problem

$$\partial_\pm\mu_\pm(z_\pm) = \mu_\pm(z_\pm)\tilde{\mathcal{E}}_\pm(z_\pm), \tag{12}$$

$$\tilde{\mathcal{E}}_\pm(z_\pm) = \sum_{m=1}^{M}\tilde{\mathcal{E}}_m^{\pm}(\Phi^\pm), \qquad \tilde{\mathcal{E}}_m^{\pm}(\Phi^\pm) = \sum_{\alpha\in\Delta_m^+}\Phi_\alpha^{\pm m}(z_\pm)X_{\pm\alpha}, \tag{13}$$

with arbitrary functions $\Phi_\alpha^{\pm m}(z_\pm)$ determining the general solution to the system. Note that the summations in (13) are performed over the set of positive roots Δ_m^+ of $\mathcal{G} = \sum_{m\in\mathbb{Z}}\mathcal{G}_m$ in the subspace $\mathcal{G}_m$.

3. Soliton solution for the sine–Gordon in homogeneous grading

Another way to construct soliton solutions [13] to the sine–Gordon equation is to consider the formal general solution (10) in the homogeneous grading and to use vertex operators [7] which are related to the homogeneous Heisenberg subalgebra of $\widehat{sl_2}$. Take the general solution (11) to the affine Toda system (10). In the homogeneous grading the mappings $\gamma_\pm$ can be parameterized as $\gamma_\pm = e^{d\phi_d}e^{c\phi_c}e^{\phi_0^\pm x_0^\pm}$, where d is the grading operator, c is the center of $\widehat{sl_2}$ and $x_k^\pm$ are generators of the subspaces $\widehat{\mathcal{G}}_k$ corresponding to the homogeneous grading. The mappings $\mu_\pm$ satisfy (12) where $\kappa_\pm(z^\pm) = a_{\pm1} + \phi^\pm x_1^\pm$. In order to obtain a soliton solution we put $\phi^\pm = 0$, $\phi_0^\pm = 0$. Then the general solution reduces to

$$e^{-\beta\phi(z^+,z^-)} = J(g_{a,\mu},|\Lambda_1\rangle) \tag{14}$$

where $g_{a,\mu} = e^{a_{+1}z^+}\mu(0)e^{a_{-1}z^-}$ The following group element $\mu(0)$ in (14) $\mu(0) = e^{-\frac{\alpha}{2}N}\prod_{n=1}^{N}\left[exp\left((-1)^{\partial_\alpha+1}iQ_n\Phi(\zeta_n)\right)e^{\frac{\alpha}{2}}\zeta_n^{\frac{1}{2}\partial_\alpha}\right]$, generates an N–soliton solution. Here the action of the operators $\frac{1}{2}\partial_\alpha$ and $e^{\frac{\alpha}{2}}$ on the highest vectors $|\Lambda_n\rangle = |1 \otimes e^{\frac{\alpha}{2}n}\rangle$, $n = 0, 1$ is the same as in the case of $U'_q(\widehat{sl_2})$ [7] when $q = 1$. The operator $\Phi(\zeta)$ is given by $\Phi(\zeta) = exp\left(\sum_{k=1}^{\infty}\frac{a_{-n}}{n}\zeta^n\right)exp\left(-\sum_{k=1}^{\infty}\frac{a_{+n}}{n}\zeta^{-n}\right)$, and diagonalises the action of $a_{\pm k}$, $k \in \mathbb{N}$, i.e., $[a_{\pm k}, \Phi(\zeta)] = \zeta^{\pm k}\Phi(\zeta)$. The product of two vertex operators can be normal ordered as $\Phi(\zeta_1)\Phi(\zeta_2) = X(x) : \Phi(\zeta_1)\Phi(\zeta_2) :$, where $X(x) = exp(-\sum_{n=1}^{\infty}x^{2n}/n) = \exp\left(\log(1-x^2)\right)$. When $x = 1$, $X(x)$ vanishes which results in $\Phi(\zeta)\cdot\Phi(\zeta) = 0$. Therefore the exponential of $\Phi(\zeta)$ operator terminates after the first order.

In the limit $q \longrightarrow 1$ soliton–soliton, antisoliton–antisoliton and soliton–-antisoliton scattering reduce to the classical case, i.e., $F^a(\zeta_1)F^b(\zeta_2) = \frac{1}{x}\frac{X(x)}{X(x^{-1})}F^b(\zeta_2)F^a(\zeta_1)$, where $x^2 = \zeta_2/\zeta_1$, a, b denote soliton (antisoliton), and the factor $1/x$ comes from the commutation of $e^{\frac{\alpha}{2}}\zeta_1^{\frac{1}{2}\partial_\alpha}$ and $e^{\frac{\alpha}{2}}\zeta_2^{\frac{1}{2}\partial_\alpha}$ operators. Therefore the vertex operator generating a classical soliton solution is $F(\zeta) = Q\ \Phi(\zeta)\ e^{\frac{\alpha}{2}}\ \zeta_2^{\frac{1}{2}\partial_\alpha}$. Taking into account the properties of the operator $F(\zeta)$ we rewrite the solution (14) for $g_\zeta = \left(1 + (-1)^{\partial_\alpha+1}iQ\Phi(\zeta)\right)e^{\frac{\alpha}{2}}\zeta^{\frac{1}{2}\partial_\alpha}$

$$e^{-\beta\phi(z^+,z^-)} = J\left(g_\zeta, |\Lambda_1\rangle\right) = \left(1 + iQe^{\zeta z^+ - \zeta^{-1}z^-}\right)\left(1 - iQe^{\zeta z^+ - \zeta^{-1}z^-}\right)\zeta.$$

The antisoliton solution can be associated with the vertex operator $\bar{F}(\zeta) = -Q\ \Phi(\zeta)\ e^{\frac{\alpha}{2}}\ \zeta^{\frac{1}{2}\partial_\alpha}$.

4. Homogeneous higher grading generalization of the affine Toda model

The Lie-algebraic way to construct non-linear exactly solvable models in classical regions is very well known and elaborated [10]. Applying the zero-curvature conditions on elements of connection containing Lie algebra generators in appropriate grading subspaces, we obtain systems of equations of motion associated to a specific Lie algebra. In [4] (of which we keep notations) the higher grading generalization to the conformal affine Toda models was considered. Elements of the higher (then number one) grading subspaces are taking into account while connection elements are constructed. The main example of [4] is the principal grading case. In this paper we consider an alternative, the homogeneous grading case. We derive the systems of equations generalizing the case of the sine–Gordon equation and provide quantum group solutions.

We start with the equations (6–9) of [4] (see subsection 2). Consider the case $l = 1$. In the principal grading we obtain from (6) the sine–Gordon equation. Recall that In the homogeneous grading of $\widetilde{\mathcal{G}}$ the grading subspaces are $\widetilde{\mathcal{G}}_n = \{H^n, E^n_\pm\}$. We take

$$E_1 = E^1_+ + E^1_-, \quad E_{-1} = E^{-1}_+ + E^1_-. \tag{15}$$

Consider a particular case when we parameterize the group element b as

$$b = e^{\phi H^0}. \tag{16}$$

Then, substituting (15) and (16) into (6–8) we get the following system of equations

$$\partial_\pm\phi = e^\eta\left(e^{-2\phi} - e^{2\phi}\right), \quad \partial_\pm\nu = -e^\eta\left(e^{2\phi} + e^{-2\phi}\right), \quad \partial_\pm\eta = 0,$$

i.e., in the first equation is again the sine–Gordon equation. The solution to the field ϕ is then the standard classical solution (11), [12] (see subsection 2).

Now consider the case $l = 2$. The equations corresponding to the principal grading can be found in [4]. Here again we take $b = e^{\phi H^0}$ though it this is not the most general choice of the group element parameterization since it does not contain dependence on the $E^0_\pm$ elements from $\widehat{\mathcal{G}}$, i.e., we have send corresponding fields near those generators to zero. Let us also put $F_1^+ = \kappa^+ H^1 + f_+^+ E_+^1 + f_-^+ E_-^1$, $F_1^- = \kappa^- H^1 + f_+^- E_+^{-1} + f_-^- E_-^{-1}$. Then the system (6–8) gives the following system of equations:

$$\begin{gathered}\partial_\pm \phi = e^\eta \left(e^{-2\phi} - e^{2\phi}\right) + e^\eta \left(f_-^+ f_+^- e^{-2\phi} - f_-^- f_+^+ e^{2\phi}\right),\\ \partial_\pm \nu = 2e^\eta \left(e^{2\phi} - e^{-2\phi}\right) + e^\eta \left(-2\kappa^+\kappa^- + f_-^- f_+^+ e^{2\phi} - f_-^+ f_+^- e^{-2\phi}\right),\ \partial_\pm \eta = 0\end{gathered} \tag{17}$$

$$\begin{gathered}\partial_+ \kappa^- = -e^\eta \left(f_-^+ e^{-2\phi} - f_+^+ e^{2\phi}\right),\quad \partial_- \kappa^+ = e^\eta \left(f_-^- e^{-2\phi} - f_+^- e^{2\phi}\right),\\ \partial_+ f_+^- = 2e^\eta \kappa^+,\ \partial_- f_+^+ = -2e^\eta \kappa^-,\ \partial_+ f_-^- = -2e^\eta \kappa^+,\ \partial_- f_-^+ = 2e^\eta \kappa^-,\end{gathered} \tag{18}$$

$$e^{2\phi} f_+^+ \kappa^- = \kappa^+ f_+^-,\quad e^{-2\phi} f_-^+ \kappa^- = \kappa^+ f_-^-. \tag{19}$$

The formal general solution to (17–19) age given in [4]:

$$e^{-\phi} = e^{\phi_0^+ - \phi_0} J(g_\mu|\Lambda_1\rangle), \tag{20}$$

for the ϕ field and for the $F_1^\pm$ elements $\langle i|F_i^+|i;i\rangle = e^{k_{i1}(\phi_{0i}^- - \phi_{0i})}\partial_+ \left(\langle i|\gamma_0^- b \left(\gamma_0^+\right)^{-1} \mu_+^{-1}\mu_-|i;i\rangle\right)$. In the homogeneous grading case, taking into account the parameterization of b element, we get

$$\langle 1|F_i^+|1;1\rangle = e^{2(\phi_0^- - \phi_0)}\partial_+ \left(\langle 1|e^{\phi_0^- H^0} e^{\phi H^0} e^{-\phi_0^+ H^0} g_\mu|1;1\rangle\right).$$

Here we have made use of the properties of the i-th fundamental representation (corresponding to the homogeneous grading) of the Lie algebra $\widehat{sl}_2$. Thus, using (20) we get

$$\langle 1|F_i^+|1;1\rangle = e^{2(\phi_0^- - \phi_0)}\partial_+ \left(\langle 1|g_\mu|1;1\rangle \cdot J^{-1}(g_\mu, |\Lambda_1\rangle)\right).$$

5. Dirac equations

Let's switch notations similarly to Dirac field components, i.e., $\psi_R = f_+^+$, $\psi_L = f_+^-$, $\widetilde{\psi}_R = f_-^+$, $\widetilde{\psi}_L = f_-^-$. Now use the extra conditions (19), substituting them into (17). Then we see that the second summands in the first two formulae in (17) vanish, i.e., the final equations are

$$\partial_+\partial_-\phi = e^\eta \left(e^{-2\phi} - e^{2\phi}\right),\quad \partial_+\partial_-\nu = 2e^\eta \left(e^{2\phi} - e^{-2\phi}\right) - 2e^\eta \kappa^+\kappa^-,\quad \partial_+\partial_-\eta = 0, \tag{21}$$

i.e., equations (21) do not differ much from the corresponding equations with $l = 1$.

Now suppose that $\eta = \eta_0 = const$. Then substitute the last four equations of (18) on $f = \psi$ fields into the first two on $\kappa^\pm$ fields. Then we get

$$\partial_+\partial_-\psi_R = 2e^{2\eta_0}\left(\widetilde{\psi}_R e^{-2\phi} - \psi_R e^{2\phi}\right),\quad \partial_+\partial_-\widetilde{\psi}_R = -2e^{2\eta_0}\left(\widetilde{\psi}_R e^{-2\phi} - \psi_R e^{2\phi}\right), \tag{22}$$

$$\partial_+\partial_-\psi_L = 2e^{2\eta_0}\left(\widetilde{\psi}_L e^{-2\phi} - \psi_L e^{2\phi}\right),\quad \partial_+\partial_-\widetilde{\psi}_L = -2e^{2\eta_0}\left(\widetilde{\psi}_L e^{-2\phi} - \psi_L e^{2\phi}\right), \tag{23}$$

that can be rewritten as

$$\partial_+\partial_-\omega_R = 0,\ \partial_+\partial_-\tau_R = 2e^{2\eta_0}\omega_R \left(e^{-2\phi} - e^{2\phi}\right),\ \partial_+\partial_-\omega_L = 0,\ \partial_+\partial_-\tau_L = 2e^{2\eta_0}\omega_L \left(e^{-2\phi} - e^{2\phi}\right), \tag{24}$$

where $\omega_{R,L} = \psi_{R,L} + \widetilde{\psi}_{R,L}$, $\tau_{R,L} = \psi_{R,L} - \widetilde{\psi}_{R,L}$. The upshot is that using such a parametrization $b = e^{\phi H^0}$ we arrive at three systems of sine–Gordon like systems when η is a constant.

5.1. $l = 2$. *The general case*

Let's consider such $b \in G_0$ that involves all generators of the $\mathcal{G}_0$ in the homogenous gradation. Take for instance $b = e^{\phi_+ E_+^0} e^{\phi H^0} e^{\phi_- E_-^0}$, Then for $l = 2$ the equations are

$$\partial_+ \left(\partial_- b b^{-1}\right) + \partial_+ \partial_- \nu C = e^{2\eta} \left[E_{-2}, b E_2 b^{-1}\right] + \left[F_1^-, b F_1^+ b^{-1}\right], \tag{25}$$

$$\partial_- F_1^+ = e^{\eta} \left[E_2, b^{-1} F_1^- b\right], \quad \partial_+ F_1^- = -e^{\eta} \left[E_{-2}, b F_1^+ b^{-1}\right], \quad \partial_+ \partial_- \eta = 0, \tag{26}$$

where $F_1^+ = \kappa^+ H^1 + f_+^+ E_+^1 + f_-^+ E_-^1$, $F_1^- = \kappa^- H^1 + f_+^- E_+^{-1} + f_-^- E_-^{-1}$. Let us take $\kappa^+ = \kappa_- = 0$. Then we have

$$\begin{aligned}
&\partial_+ \left(\partial_- \phi + \phi_+ \partial_- \phi_-\right) = e^{2\eta} \left[e^{-2\phi}(1 - \phi_-^2) - e^{2\phi} - 2\phi_+ \phi_-)\right] \\
&\qquad\qquad + e^{\eta} \left[e^{-2\phi}(f_-^+ - f_+^+ \phi_-^2)(f_+^- + f_-^- \phi_+^2) - 2 f_+^+ f_-^- \phi_- \phi_+ - f_+^+ f_-^- e^{2\phi}\right], \\
&\partial_+ \partial_- \nu = 2 e^{2\eta} \left[-e^{-2\phi}(1 - \phi_-^2) - e^{2\phi} - 2\phi_+ \phi_-)\right] \\
&\qquad\qquad + e^{\eta} \left[e^{-2\phi}(f_-^+ - f_+^+ \phi_-^2)(f_+^- + f_-^- \phi_+^2) - 2 f_+^+ f_-^- \phi_- \phi_+ - f_+^+ f_-^- e^{2\phi}\right], \\
&\partial_+ (\partial_- \phi_+ - 2(\partial_- \phi)\phi_+ - \phi_+^2 (\partial_- \phi_-) e^{-2\phi}) = e^{2\eta} \left[2\phi_+ e^{-2\phi}(1 - \phi_-^2) - 2\phi_-\right] \\
&\qquad\qquad + e^{\eta} \left[-2 f_+^- (e^{-2\phi}(f_-^+ - f_+^+ \phi_-^2)\phi_+ - \phi_- f_+^+\right], \\
&\partial_+ (\partial_- \phi_- e^{-2\phi}) = e^{2\eta} \left[2\phi_+ e^{-2\phi}(1 - \phi_-^2) - \phi_-\right] + e^{\eta} \left[2 f_+^+ (e^{-2\phi}(f_-^+ - f_+^+ \phi_-^2)\phi_+ - \phi_- f_+^+\right], \\
&\partial_+ \partial_- \eta = 0,
\end{aligned} \tag{27}$$

$$\begin{aligned}
&\partial_+ f_+^- = e^{\eta} \left[-2(-f_+^+ \phi_- + e^{-2\phi}(f_+^+ \phi_-^2 + f_-^+ \phi_+\right], \ \partial_+ f_-^- = e^{\eta} \left[2(-f_+^+ \phi_- + e^{-2\phi}(f_+^+ \phi_-^2 + f_-^+ \phi_+\right] \\
&e^{-2\phi}(f_+^+ \phi_-^2 + f_-^+)(1 + \phi_+^2) - 2 f_+^+ \phi_- \phi_+ - f_+^+ e^{2\phi} = 0
\end{aligned} \tag{28}$$

$$\begin{aligned}
&\partial_- f_+^+ = e^{\eta} \left[-2(-\phi_- e^{-2\phi}(f_+^- - \phi_+^2 f_-^-) - \phi_+ f_-^-\right], \partial_- f_-^+ = \left[2(-\phi_- e^{-2\phi}(f_+^- - \phi_+^2 f_-^-) - \phi_+ f_-^-\right], \\
&-e^{-2\phi}(f_+^- - \phi_+^2 f_-^-)(1 + \phi_-^2) e^{2\phi} f_-^- + \phi_- \phi_+ f_-^- = 0.
\end{aligned} \tag{29}$$

When $\kappa^+ \neq 0$ and $\kappa^- \neq 0$ the system of equations is more complicated.

6. Solitonic solutions from general solutions

In [12] it was shown how to extract solitonic solutions from the formal general solutions of the affine Toda field equations. Let's take $\gamma_0^\pm = 1$ in (11) to be a constant function. Then the mappings $\mu_\pm$ are $\mu_\pm = \mu_\pm^0 e^{z_\pm \mathcal{E}_\pm}$ with $\mu_\pm^0$ being some fixed mappings independent of $z_\pm$. Next take $\tilde{\mathcal{E}}_\pm$ in 13 as $\mathcal{E}_\pm \equiv E_{\pm l} + \sum_{N=1}^{l-1} c_N^\pm E_{\pm N}$ where $E_\pm$ are elements of a Heisenberg subalgebra of $\widehat{\mathcal{G}}$, namely $[\mathcal{E}_+, \mathcal{E}_-] = \Omega C$. One can consider principal of homogeneous Heisenberg subalgebras for that purpose. In this paper we only deel with the principal case while the homogeneous case will be discussed elsewhere. Thus, we arrive at a special solution to (10)

$$e^{-\beta \lambda_i \cdot \phi} = e^{-\beta \lambda_i \cdot \phi_0} J \langle (g_{\mathcal{E},\mu}, |\lambda_i\rangle^{(1)} \rangle), \tag{30}$$

for $g_{\mathcal{E},\mu} = e^{x_\pm \mathcal{E}_\pm} \mu^0 e^{x_\pm \mathcal{E}_\pm}$. In order to compute these solutions explicitly we have to remove $\mathcal{E}_\pm$-dependence from (30) moving $\mathcal{E}_+$ to the right and $\mathcal{E}_-$ to the left . Then we should find such a $\mu_0 = \prod_{i=1}^N e^{\mathcal{V}_i}$ so that $\mathcal{V}_i$ would be eigenvectors with respect to the adjoint action of $\mathcal{E}_\pm$, i.e., $[\mathcal{E}_\pm, \mathcal{V}_i] = \omega_\pm^{(i)} \mathcal{V}_i$. Then it turns out [12] that resulting expressions provide us with solitonic solutions to the equations under considerations while parameters $\omega_\pm^{(i)}$ characterize solitons.

7. Quantum group soliton solution for sine–Gordon in homogeneous grading

As in [11] one can show that the affine Toda models are co-invariant with respect to the light-cone quantization. Namely, the equation of motion are preserved in form though a standard normal ordering has to be introduced as well as some infinite constant comming from quantum versions of Lax pair to generate equations using Lie algebra elements in quantum case. At the same time infinite constants do not appear in final formal solutions to the light-cone quantized versions of equations. In order to find quantum solutions, one has to replace [3], [8], [9] group elements as well as state vectors formal general solutions by their quantum group counterparts. In this subsection we write examples of quantum group solutions to the quantized affine Toda model in the specific case of the higher grading sine–Gordon equation (the cases $l = 1, 2, 3$). Recall [13], that the homogeneous grading subspaces of $U_q(\widehat{sl}_2)$ are ${}_q\widehat{\mathcal{G}}_0 = \{K, \gamma, x_0^+, x_0^-\}$, ${}_q\widehat{\mathcal{G}}_n = \{x_n^+, x_n^-, a_n, n \in \{\mathbb{Z} - 0\}\}$.

7.1. The case $l = 1$

From the commutation relations for $x_m^\pm$ and a_m (see subsection 1.2) it follows that in this realization of the quantum group $U_q(\widehat{sl}_2)$, the generators $x_m^\pm, a_m \in \mathcal{G}_m$, $x_0^\pm \in \mathcal{G}_0$. The solution

$$e^{-\beta\lambda_j\cdot\phi} = e^{-\beta\lambda_j\cdot\phi_0} J(e^{-a_1 z_+} e^{Q\phi_-} e^{a_{-1}z_-}, |\Lambda_j\rangle),$$

where $|\Lambda_0\rangle = |1 \otimes 1\rangle$, $|\Lambda_1\rangle = |1 \otimes e^{\frac{\alpha}{2}}\rangle$ and the homogeneous grading quantum vertex operator is

$$\phi_- = exp\left(\sum_{k=1}^{\infty} \frac{a_{-k}}{[2k]} q^{\frac{7k}{2}} \zeta^k\right) exp\left(-\sum_{k=1}^{\infty} \frac{a_k}{[2k]} q^{-\frac{5k}{2}} \zeta^{-k}\right) \otimes e^{\frac{\alpha}{2}} (-q^3\zeta)^{\frac{(\partial_\alpha + i)}{2}}.$$

Using the fact that [5] $[a_k, \phi_-] = q^{\frac{7k}{2}} \frac{[k]}{k} \zeta^k$, $\phi_-, k > 0$, $[a_k, \phi_-] = q^{-\frac{5k}{2}} \frac{[k]}{k} \zeta^{-k}$, $\phi_-, k > 0$, we commute $\exp(-a_1 z_+)$ with $\exp(Q\phi_-)$ to the right and $\exp(Q\phi_-)$ with $\exp(a_{-1}z_-)$ to the left. The commutation of $\exp(-a_1 z_+)$ withexp$(a_{-1}z_-)$ gives $\exp(-z_+z_-[2])$). Thus it follows that

$$\langle\Lambda_j| e^{-a_1 z_+} e^{Q\phi_-} e^{a_1 z_-} |\Lambda_j\rangle = \langle\Lambda_j| exp\left(Qe^{-q^{\frac{7}{2}} z_+ \zeta - q^{-\frac{5}{2}} z_- \zeta^{-1}} \phi_-\right) exp\left(-z_+ z_- [2]\right) |\Lambda_j\rangle,$$

(recall that $\exp(-a_1 z_+)$ and $\exp(a_{-1}z_-)$ act on $|\Lambda_j\rangle$ and $\langle\Lambda_j|$ as identities). Then we expand $exp\left(Qe^{-q^{\frac{7}{2}} z_+ \zeta - q^{-\frac{5}{2}} z_- \zeta^{-1}} \phi_-\right)$ as a series and apply the action of powers of operators $exp\left(\sum_{k=1}^{\infty} \frac{a_{-k}}{[2k]} q^{\frac{7k}{2}} \zeta^k\right)$ and $exp\left(-\sum_{k=1}^{\infty} \frac{a_k}{[2k]} q^{-\frac{5k}{2}} \zeta^{-k}\right)$, to the left and to the right. Powers of operators ϕ_- act on the second part of tensor product as follows:

$$(\phi_-)^n |1 \otimes e^{\frac{\alpha}{2}}\rangle = (-q^3\zeta)^{\frac{i+n}{2}} |1 \otimes e^{\frac{\alpha(n+1)}{2}}\rangle,\ (\phi_-)^n |1 \otimes 1\rangle = (-q^3\zeta)^{i + \frac{n-1}{2}} |1 \otimes e^{\frac{\alpha n}{2}}\rangle.$$

Thus we have for $g_Q = e^{-z_+ z_- [2]} exp\left(Qe^{-q^{\frac{7}{2}} z_+ \zeta - q^{-\frac{5}{2}} z_- \zeta^{-1}} (-q^3\zeta)^{\frac{i}{2}}\right) (-q^3\zeta)^{\frac{i}{2}} \otimes exp\left(e^{\frac{\alpha}{2}}\right)$,

$$e^{-\beta\lambda_j\cdot\phi} = e^{-\beta\lambda_j\cdot\phi_0} J(g_Q, |\Lambda_j\rangle). \tag{31}$$

In the limit $q \to 1$ we obtain ordinary soliton solutions.

7.2. The case $l = 2$

As in [13], if we put $\widehat{\phi}^\pm_{x,1} = 0$, then $E_\pm = a_{\pm 2} + a_{\pm 1}$, and one can integrate the equations for ${}_q\mu_\pm$ to obtain ${}_q\mu_\pm(z^\pm) = {}_q\mu_\pm(0) e^{(a_{\pm 2} + a_{\pm 1}) z^\pm}$. Then the quantum soliton solution to the quantized (17) is

$$: e^{-\beta\widehat{\phi}(z^+, z^-)} :=: e^{-\beta\widehat{\phi}_0(z^+, z^-)} : J(g_a, |\Lambda_1\rangle_q),$$

where $g_a = e^{(a_{+1} + a_{+2}) z^+} {}_q\mu(0) e^{(a_{-1} + a_{-2}) z^-}$, and ${}_q\mu(0)$ should be chosen the same as in [13]. Then we have

$$: e^{-\beta\widehat{\phi}(z^+, z^-)} :=: e^{-\beta\widehat{\phi}_0(z^+, z^-)} : J(g_\alpha, |\Lambda_1\rangle_q) =: e^{-\beta\widehat{\phi}_0(z^+, z^-)} : \frac{1 + iW_2 Q}{1 - iW_2 Q} \zeta^{\frac{1}{2}},$$

where $g_\alpha = e^{-\frac{\alpha}{2}} exp\left(i(-1)^{\partial_\alpha+1} Q\, W_2 \cdot {}_q\Phi(\zeta)\right) e^{\frac{\alpha}{2}} \zeta^{\frac{1}{2}\partial_\alpha}$, $W_2 = \exp\left(\sum_{k=1}^{2} \frac{7k}{2}\frac{[k]}{k}\zeta^k z^+ - \sum_{k=1}^{2} \frac{5k}{2}\frac{[k]}{k}\zeta^{-k} z^-\right)$.
Similarly,

$$ {}_q\langle 1|F_1^+|1;1\rangle_q = e^{2(\widehat{\phi_0^-}-\widehat{\phi})}\partial_+\left({}_q\langle 1|g_a|1;1\rangle_q J(g_a, |\Lambda_1\rangle_q)\right). $$

Thus,

$$ {}_q\langle 1|F_1^+|1;1\rangle_q = e^{2(\widehat{\phi_0^-}-\widehat{\phi})}\partial_+\Big({}_q\langle 1|e^{(a_{+1}+a_{+2})z^+} g_\alpha e^{(a_{-1}+a_{-2})z^-}|1;1\rangle_q \\ \times J(g_\alpha, |\Lambda_1\rangle_q)\Big). $$

Finally,

$$ {}_q\langle 1|F_1^+|1;1\rangle_q = e^{2(\widehat{\phi_0^-}-\widehat{\phi})}\partial_+\left((1+iW_2Q[2])\cdot \tfrac{1-iW_2Q}{1+iW_2Q}\zeta^{-\frac{1}{2}}\right). $$

7.3. Case l=3
The states

$$ |\Lambda_0\rangle^{(m)} = |\prod_{k=1}^{m+1} a_{-(m-k)} \otimes 1\rangle,\ |\Lambda_1\rangle^{(m)} = |\prod_{k=1}^{m+1} a_{-(m-k)} \otimes e^{\frac{\alpha}{2}}\rangle,\ |\Lambda_0\rangle^{(1)} = |\Lambda_0\rangle,\ |\Lambda_1\rangle^{(1)} = |\Lambda_1\rangle, $$

are annihilated by the action of $\mathcal{G}_n, n \geq m$. Therefore for F_m^+ we have

$$ {}^{(1)}\langle\Lambda_1|e^{-\sum_{k=1}^{3} a_k z_+} e^{Q\phi_-} e^{\sum_{k=1}^{3} a_{-k} z_-}|\Lambda_1\rangle^{(m)} \\ = {}^{(1)}\langle\Lambda_1|exp\left(Qe^{-q^{\frac{7}{2}}z_+\zeta - q^{\frac{-5}{2}}z_-\zeta^{-1}}\phi_-\right) exp\left(-z_+z_-\sum_{k=1}^{3}\frac{[2k]}{k}[k]\right) e^{-\sum_{k=1}^{3} a_k z_+}|\Lambda_1\rangle^{(m)}. $$

Action by the operators $e^{-\sum_{k=1}^{3} a_k z_+}$ on $|\Lambda_1\rangle^{(m)}$, $m = 1,\, 2,\, 3$ we get for instance,

$$ e^{-\sum_{k=1}^{3} a_k z_+}|\Lambda_1\rangle^{(3)} = |\Lambda_1\rangle^{(3)} - z_+(C_2 + a_{-2})|\Lambda_1\rangle^{(1)} + z_+^2 C_1 C_2|\Lambda_1\rangle^{(1)}, $$

where $C_k = \frac{[2k]}{k}[k]$. Then we expand e^{ϕ_-} again and act on the states. Therefore we get an infinite series over $|\Lambda_1\rangle^{(3)}, |\Lambda_1\rangle^{(2)}, |\Lambda_1\rangle^{(1)}$ which contain $C_k, (k = 1, 2, 3)$, z_+ and tensor $\otimes$-part due to powers of $e^{\frac{\alpha}{2}}(-q^3\zeta)^{\frac{(\partial_\alpha+i)}{2}}$.

8. Conclusions
In this paper we considered the alternative case of the homogeneous grading of the symmetry algebra of the affine Toda systems and, in particular, the sine-Gordon equations. The Lie-algebraic method helps us to understand the use pure algebraic nature of exact integrability of the dynamical system under consideration. We show explicitly that even in the new case of homogeneous grading it is possible to derive the equations of motion and provider explicit solutions both in classical and quantum regions. The knowledge of the quantum group structures underlying the homogeneous higher grading case leads to specific form of the quantum group soliton solutions for the sine-Gordon equation. As possible way to develop, we could mention the search for the explicit solutions associated to all higher grading subspaces of the affine Lie algebra.

Acknowledgments
We would like to thank the Orgnizers of the International Conference of Strongly Correlated Electron Systems, 2016.

Alexander Zuevsky
Institute of Mathematics, Czech Academy of Sciences, Prague, Czech Republic
zuevsky@yahoo.com

References

[1] Delius G W and Hueffmann A 1996 J. Phys.A29
[2] Drinfeld V G 1985 Sov. Math. Dokl. 32
[3] Fedoseev I A and Leznov A N 1984 Phys. Lett. 141B, 1-2 , 100–104
[4] Ferreira L A, Gervais J-L, Guillen, J S and Saveliev M V 1996 Nuclear Phys. B 470 1-2
[5] Jimbo M 1985 Lett. Math. Phys. 10, 63
[6] Jimbo M, Miki K, Miwa T and Nakayashiki A 1992 Phys. Let. A. 168 256–263
[7] Kac V G 1990 *Infinite dimensional Lie algebras.* Third edition (Cambridge university Press, Cambridge)
[8] Leznov A N and Fedoseev I A 1982 Theor. Math. Phys. 53 3, 1175
[9] Leznov A N and Mukhtarov M A 1987 Theor. Math. Phys. 71 1
[10] Leznov A N and Saveliev M V 1992 *Group-Theoretical Methods for Integration of Non–Linear Dynamical Systems.* Progress in Physics Series, v. 15, (Birkhauser–Verlag, Basel)
[11] Mansfield P 1983 Nucl. Phys. B 222
[12] Olive D I Turok N and Underwood J W R 1993 Nucl. Phys. B401
[13] Saveliev M V and Zuevsky A B 2000 Internat. J. Modern Phys. A 15 24

AUTHOR INDEX

AUTHOR INDEX

AUTHOR INDEX